AF601487

ALLE ZEIT WACH
1842

Attention all "Enzyme Handbook" users:

A file with the complete volume indexes Vols. 1 through 6 in delimited ASCII format is available for downloading at no charge from the Springer EARN mailbox. Delimited ASCII format can be imported into most databanks.

The file has been compressed using the popular shareware program "PKZIP" (Trademark of PKware Inc., PKZIP is available from most BBS and shareware distributors).

This file is distributed without any expressed or implied warranty.

To receive this file send an e-mail message to:
SVSERV@DHDSPRI6.BITNET

The message must be:
GET /CHEMISTRY/ENZ_HB.ZIP

SVSERV is an automatic data distribution system. It responds to your message. The following commands are available:

HELP	returns a detailed instruction set for the use of SVSERV,
DIR *(name)*	returns a list of files available in the directory "name",
INDEX *(name)*	same as "DIR"
CD <*name*>	changes to directory "name",
SEND <*filename*>	invokes a message with the file "filename",
GET <*filename*>	same as "SEND".

D. Schomburg · M. Salzmann · D. Stephan (Eds.)
GBF – Gesellschaft für Biotechnologische Forschung

Enzyme Handbook 7

Class 1.5–1.12: Oxidoreductases

Springer-Verlag
Berlin Heidelberg New York
London Paris Tokyo
Hong Kong Barcelona
Budapest

Professor Dr. Dietmar Schomburg
Margit Salzmann
Dr. Dörte Stephan
GBF – Gesellschaft für Biotechnologische Forschung mbH
Mascheroder Weg 1
38124 Braunschweig
FRG

This collection of datasheets was generated from the database „BRENDA"

Library of Congress Cataloging-in-Publication Data. (Revised for volume 7). Enzyme handbook. Vol. 7 edited by D. Schomburg, M. Salzmann, D. Stephan. Loose-leaf. Includes bibliographical references and indexes. Contents: 1. Class 4: Lyases – 2. Class 5: Isomerases. Class 6: Ligases – 4–5. Class 3: Hydrolases – 6. Class 1.5–1.12: Oxidoreductases. 1. Enzymes – Handbooks, manuals, etc. I. Schomburg, D. (Dietmar) II. Salzmann, M. (Margit) I. Stephan, D. (Dörte)
QP601.E5158 1990 660'.634 91-145566

ISBN 978-3-642-48964-8 ISBN 978-3-642-78521-4 (eBook)
DOI 10.1007/978-3-642-78521-4

This work is subject to copyright. All rights are reserved, whether the whole or part of the material is concerned, specifically the rights of translation, reprinting, reuse of illustrations, recitation, broadcasting, reproduction on microfilms or in other ways, and storage in data banks. Duplication of this publication or parts thereof ist only permitted under the provisions of the German Copyright Law of September 9, 1965, in its current version, and a copyright fee must always be paid. Violations fall under the prosecution act of the German Copyright Law.

© Springer-Verlag Berlin Heidelberg 1994

The use of registered names, trademarks, etc. in this publication does not imply, even in the absence of a specific statement, that such names are exempt from the relevant protective laws and regulations and therefore free for general use.

The publisher cannot assume any legal responsibility for given data, especially as far as directions for the use and the handling of chemicals and biological materials are concerned. This information can be obtained from the instructions on safe laboratory practice and from the manufacturers of chemicals and laboratory equipment.

Media conversion, printing and bookbinding: Brühlsche Universitätsdruckerei, Giessen
Production of the plasticfiles: Lux-Plastik oHG, Murnau
51/3130 - 5 4 3 2 1 0 – Printed on acid-free paper

Preface

Recent progress on enzyme immobilisation, enzyme production, coenzyme regeneration and enzyme engineering has opened up fascinating new fields for the potential application of enzymes in a large range of different areas. As more progress in research and application of enzymes has been made the lack of an up-to-date overview of enzyme molecular properties has become more apparent. Therefore, we started the development of an enzyme data information system as part of protein-design activities at GBF. The present book "Enzyme Handbook" represents the printed version of this data bank. In future a computer searchable version will be also available.

The enzymes in this Handbook are arranged according to the Enzyme Commission list of enzymes. Some 3000 "different" enzymes will be covered. Frequently enzymes with very different properties are included under the same EC number. Although we intend to give a representative overview on the characteristics and variability of each enzyme the Handbook is not a compendium. The reader will have to go to the primary literature for more detailed information. Naturally it is not possible to cover all the numerous literature references for each enzyme (for special enzymes up to 40000) if the data representation is to be concise as is intended.

It should be mentioned here that the literature data are extracted from literature and critically evaluated by qualified scientists. On the other hand the original authors' nomenclature for enzyme forms and subunits is retained as is their nomenclature for organisms and strains even if the organism is reclassified in the meantime. The cross references to the protein sequence data bank and to the Brookhaven protein 3D structure data bank are taken directly from their data files without further verification by the authors. In order to keep the tables concise redundant information is avoided as far as possible (e.g. if K_m values are measured in the presence of an obvious cosubstrate, only the name of the cosubstrate is given in parentheses as a commentary without reference to its specific role).

The authors are grateful to the following biologists and chemists for invaluable help in the compilation of data: Cornelia Munaretto, Dr. Ida Schomburg, Dr. Astrid Beermann. In addition we would like to thank Mrs. C. Munaretto and Dr. I. Schomburg for the correction of the final manuscript.

Braunschweig, Autumn 1993

Margit Salzmann
Dörte Stephan
Dietmar Schomburg

BRENDA – Compilation of Enzyme Data

To collect basic characteristics of enzymes – is that not a kind of archaic activity in the times of molecular biology and computer-aided data banks providing sequences of nucleic acids and proteins with little more delay than a few days as well as their three-dimensional structures? What should be the purpose of compiling turnover numbers, Michaelis constants, substrate specificities, sources, synonyms etc. of enzymes from sometimes remote publications? The answer sounds as simple as surprising: The aim of the compilation of data is to make use of the overwhelming abundance of structural knoweldge we owe to the new techniques of molecular biology.

Admittedly, it was not primarily enzymology which caused the explosion of knowledge in biology during the last decade. This was due to the advance of molecular biology which enabled us to isolate genes, to amplify them *ad libidum* and to elucidate their primary structure within days only. Also, the optimization and automatization of techniques for the analysis of macromolecules has provided detailed insights into a large variety of complex biomolecules nobody would have anticipated in the early seventies. Due to powerful computers it has now become feasible to propose fairly realistic models of macromolecules based solely on primary structures and homology considerations.

Nevertheless – or therefore – it appears as mandatory as rewarding to know the brave world of enzymology in which one had and often still has to come along without any detailed structural knowledge. We should not ignore that nature has not generated the multiplicity of structures, because it simply felt obliged to the principle of diversification or because it wanted to test our computing capacity to handle sequence data. It had to create new structures to cope with the steadily changing demands of a variable environment. Thus, amino acid sequences, folding of peptide chains and conformational details are only the technical tools of nature to catalyse specific biological functions. In consequence, *it is the functional profile of an enzyme which enables a biologist or physician to analyze a metabolic pathway and its disturbance; it is the substrate specificity of an enzyme which tells an analytical biochemist how to design an assay; it is the stability, specificity and efficiency of an enzyme which determines its usefulness in the biotechnical transformation of a molecule.* And the sum of all these functional data will have to be considered when the designer of artificial biocatalysts has to choose the optimum prototype to start with.

Unfortunately, it is by no means as simple to design (organize) a meaningful and systematic compilation of functional enzymological data as to enter sequences of amino acids or nucleotides into a data base. Functional data are less well defined, are never devoid of a trace of ambiguity, their selection remains inevitably subjective, and their complexity requires simplification. The present compilation of enzymological data, therefore, can and will not be a substitute for original publications but rather offer a key to the literature. But I do think that the Enzyme Handbook is indeed an excellent key to open or reopen the mysterious world of enzyme to all those who there have to find the solutions of their problems: to biologists, physicians, structural biochemists, biochemical analysts, biotechnologists and also to the molecular biologists.

Braunschweig, Spring 1993

Leopold Flohé
GBF, Scientific Director

List of Abbreviations

A	adenosine
Ac	acetyl
ADP	adenosine 5'-diphosphate
Ala	alanine
All	allose
Alt	altrose
AMP	adenosine 5'-monophosphate
Ara	arabinose
Arg	arginine
Asn	asparagine
Asp	aspartic acid
ATP	adenosine 5'-triphosphate
Bicine	N,N'-bis(2-hydroxyethyl) glycine
C	cytidine
cal	calorie
CDP	cytidine 5'-diphosphate
CDTA	trans-1,2-diaminocyclohexane-N,N,N,N-tetra-acetic acid
CHAPS	3-[(3-cholamidopropyl)-dimethylammonio]-1-propanesulfonate
CHAPSO	3-[(3-cholamidopropyl)-dimethylammonio]-2-hydroxy-1-propane-sulfonate
CMP	cytidine 5'-monophosphate
CoA	coenzyme A
CTP	cytidine 5'-triphosphate
Cys	cysteine
d	deoxy-
D- and L-	prefixes indicating configuration
DFP	diisopropyl fluorophosphate
DNA	deoxyribonucleic acid
DPN	diphosphopyridinium nucleotide (now NAD)
DTNB	5,5'-dithiobis (2-nitrobenzoate)
DTT	dithiothreitol (i.e. Cleland's reagent)
e	electron
EC	number of enzyme in Enzyme Commission's system
E. coli	Escherichia coli
EDTA	ethylene diaminetetraacetate
EGTA	ethylene glycol bis (β-aminoethyl ether) tetraacetate
EPR	electron paramagnetic resonance
ER	endoplasmic reticulum
Et	ethyl
EXAFS	extended X-ray absorption fine structure
FAD	flavin-adenine dinucleotide
FMN	flavin mononucleotide (riboflavin 5'-monophosphate)
Fru	fructose
Fuc	fucose
G	guanosine
Gal	galactose
GDP	guanosine 5'-diphosphate
Glc	glucose
GlcN	glucosamine
GlcNAc	N-acetylglucosamine
Gln	glutamine
Glu	glutamic acid
Gly	glycine
GMP	guanosine 5'-monophosphate

GSH glutathione
GSSG oxidized glutathione
GTP guanosine 5'-triphosphate
Gul gulose
h hour
H_4 tetrahydro
HEPES 4-(2-hydroxyethyl)-1-piperazineethane sulfonic acid
His histidine
HPLC high performance liquid chromatography
Hyl hydroxylysine
Hyp hydroxyproline
IAA iodoacetamide
Ig immunoglobulin
Ile isoleucine
Ido idose
IDP inosine 5'-diphosphate
IMP inosine 5'-monophosphate
ir irreversible
ITP inosine 5'-triphosphate
K_m Michaelis constant
L- see D-
Leu leucine
Lys lysine
Lyx lyxose
M mol/l
m- meta-
Man mannose
MES 2-(N-morpholino)ethane sulfonate
Met methionine
min minute
MOPS 3-(N-morpholino) propane sulfonate
Mur muramic acid
MW molecular weight
NAD nicotinamide-adenine dinucleotide
NADH reduced NAD
NADP NAD phosphate
NADPH reduced NADP
NAD(P)H indicates either NADH or NADPH
NDP nucleoside 5'-diphosphate
NEM N-ethylmaleimide
Neu Neuraminic acid
NMN nicotinamide mononucleotide
NMP nucleoside 5'-monophosphate
NTP nucleoside 5'-triphosphate
o- ortho-
Orn ornithine
p- para-
PCMB p-chloro-mercuribenzoate
PEP phosphoenolpyruvate
pH $-\log_{10}[H^+]$
Ph phenyl
Phe phenylalanine
PIXE proton-induced X-ray emission
PMSF phenylmethane-sulfonylfluoride
Pro proline
Q_{10} factor for the change in reaction rate for a 10° temperature increase
r reversible
Rha rhamnose
Rib ribose
RNA ribonucleic acid
mRNA messenger RNA
rRNA ribosomal RNA
tRNA transfer RNA
Sar N-methylglycine (sarcosine)
SDS-PAGE sodium dodecyl sulphate polyacrylamide gel electrophoresis
Ser serine
SFK-525A 2-diethylaminoethyl-2,2-diphenylvalerate
sp. species

T	ribosylthymine
$t_{1/2}$	time for half-completion of reaction
Tal	talose
TDP	ribosylthymine 5'-diphosphate
TEA	triethanolamine
THF	tetrahydrofolate
Thr	threonine
TMP	ribosylthymine 5'-monophosphate
Tos-	tosyl- (p-toluenesulfonyl-)
TPN	triphosphopyridinium nucleotide (now NADP)
Tris	tris(hydroxymethyl)-aminomethane
Trp	tryptophan
TTP	ribosylthymine 5'-triphosphate
Tyr	tyrosine
U	uridine
U/mg	μmol/(mg·min)
UDP	uridine 5'-diphosphate
UMP	uridine 5'-monophosphate
UTP	uridine 5'-triphosphate
Val	valine
Xaa	symbol for an amino acid of unknown constitution in peptide formula
XAS	X-ray absorption spectroscopy
XTP	xanthosine 5'-triphosphate
Xyl	xylose

Index

(Alphabetical order of Enzyme names)

EC-No.	Name
1.10.1.1	trans-Acenaphthene-1,2-diol dehydrogenase
1.7.3.2	Acetylindoxyl oxidase
1.8.99.2	Adenylylsulfate reductase
1.5.1.17	Alanopine dehydrogenase
1.5.1.26	beta-Alanopine dehydrogenase
1.10.3.4	o-Aminophenol oxidase
1.6.99.8	Aquacobalamin reductase
1.6.99.11	Aquacobalamin reductase (NADPH)
1.10.3.3	L-Ascorbate oxidase
1.11.1.11	L-Ascorbate peroxidase
1.10.2.1	L-Ascorbate-cytochrome-b_5 reductase
1.6.4.7	Asparagusate reductase (NADH)
1.6.6.7	Azobenzene reductase
1.6.4.9	Bis-gamma-glutamylcystine reductase (NADPH)
1.5.1.24	N^5–(Carboxyethyl)ornithine synthase
1.11.1.6	Catalase
1.10.3.1	Catechol oxidase
1.11.1.10	Chloride peroxidase
1.6.4.10	CoA-disulfide reductase (NADH)
1.6.4.6	CoA-glutathione reductase (NADPH)
1.6.99.9	Cob(II)alamin reductase
1.12.99.1	Coenzyme F_{420} hydrogenase
1.12.99.2	Coenzyme-M-7-mercapto-heptan-oylthreonine-phosphate-heterodisulfide hydrogenase
1.6.99.12	Cyanocobalamin reductase (NADPH, cyanide-eliminating)
1.6.4.1	Cystine reductase (NADH)
1.6.2.2	Cytochrome-b_5 reductase
1.9.3.1	Cytochrome-c oxidase
1.11.1.5	Cytochrome-c peroxidase
1.12.2.1	Cytochrome-c_3 hydrogenase
1.11.1.14	Diarylpropane peroxidase
1.5.1.3	Dihydrofolate reductase
1.8.1.4	Dihydrolipoamide dehydrogenase
1.6.99.7	Dihydropteridine reductase
1.6.6.12	4-(Dimethylamino)phenyl-azobenzene reductase
1.5.99.2	Dimethylglycine dehydrogenase
1.5.3.10	Dimethylglycine oxidase
1.5.5.1	Electron-transferring-flavoprotein dehydrogenase
1.8.4.7	Enzyme-thiol transhydrogenase (oxidized-glutathione)
1.5.1.18	Ephedrine dehydrogenase
1.11.1.3	Fatty-acid peroxidase
1.7.7.2	Ferredoxin-nitrate reductase
1.7.7.1	Ferredoxin-nitrite reductase
1.6.99.13	Ferric-chelate reductase
1.5.1.6	Formyltetrahydrofolate dehydrogenase
1.8.5.1	Glutathione dehydrogenase (ascorbate)
1.8.3.3	Glutathione oxidase
1.11.1.9	Glutathione peroxidase
1.6.4.2	Glutathione reductase (NADPH)
1.8.4.3	Glutathione-CoA-glutathione transhydrogenase
1.8.4.4	Glutathione-cystine transhydrogenase
1.8.4.1	Glutathione-homocystine transhydrogenase
1.6.6.8	GMP reductase
1.12.1.2	Hydrogen dehydrogenase
1.8.99.3	Hydrogensulfite reductase
1.10.3.5	3-Hydroxyanthranilate oxidase
1.7.3.4	Hydroxylamine oxidase
1.7.99.1	Hydroxylamine reductase
1.6.6.11	Hydroxylamine reductase (NADH)
1.5.3.5	(S)-6-Hydroxynicotine oxidase
1.5.3.6	(R)-6-Hydroxynicotine oxidase
1.6.6.13	N-Hydroxy-2-acetamido-fluorene reductase
1.6.6.6	Hyponitrite reductase

EC-No.	Name
1.8.1.3	Hypotaurine dehydrogenase
1.11.1.8	Iodide peroxidase
1.9.99.1	Iron-cytochrome-c reductase
1.10.3.2	Laccase
1.6.2.6	Leghemoglobin reductase
1.5.1.16	D-Lysopine dehydrogenase
1.11.1.13	Manganese peroxidase
1.8.3.4	Methanethiol oxidase
1.8.4.5	Methionine-S-oxide reductase
1.5.1.5	Methylenetetrahydrofolate dehydrogenase ($NADP^+$)
1.5.1.15	Methylenetetrahydrofolate dehydrogenase (NAD^+)
1.7.99.5	5,10-Methylenetetrahydrofolate reductase ($FADH_2$)
1.5.1.20	Methylenetetrahydrofolate reductase (NADPH)
1.5.99.9	Methylenetetrahydromethan-opterin dehydrogenase
1.5.99.5	Methylglutamate dehydrogenase
1.5.3.4	N^6–Methyl-lysine oxidase
1.5.3.2	N-Methyl-L-amino-acid oxidase
1.6.5.4	Monodehydroascorbate reductase (NADH)
1.6.99.3	NADH dehydrogenase
1.6.99.5	NADH dehydrogenase (quinone)
1.6.5.3	NADH dehydrogenase (ubiquinone)
1.11.1.1	NADH peroxidase
1.6.99.1	NADPH dehydrogenase
1.6.8.2	NADPH dehydrogenase (flavin)
1.6.99.6	NADPH dehydrogenase (quinone)
1.11.1.2	NADPH peroxidase
1.6.2.5	NADPH-cytochrome c_2 reductase
1.6.2.4	NADPH-ferrihemoprotein reductase
1.6.8.1	NAD(P)H dehydrogenase (FMN)
1.6.99.2	NAD(P)H dehydrogenase (quinone)
1.6.1.2	$NAD(P)^+$ transhydrogenase (AB-specific)
1.6.1.1	$NAD(P)^+$ transhydrogenase (B-specific)
1.5.1.13	Nicotinate dehydrogenase
1.5.99.4	Nicotine dehydrogenase
1.7.99.4	Nitrate reductase
1.9.6.1	Nitrate reductase (cytochrome)
1.6.6.1	Nitrate reductase (NADH)
1.6.6.3	Nitrate reductase (NADPH)
1.6.6.2	Nitrate reductase (NAD(P)H)
1.7.99.7	Nitric-oxide reductase
1.7.99.3	Nitrite reductase
1.7.2.1	Nitrite reductase (cytochrome)
1.6.6.4	Nitrite reductase (NAD(P)H)
1.7.3.1	Nitroethane oxidase
1.7.3.5	3-aci-Nitropropanoate oxidase
1.6.6.10	Nitroquinoline-N-oxide reductase
1.7.99.6	Nitrous-oxide reductase
1.5.1.19	D-Nopaline dehydrogenase
1.5.1.11	D-Octopine dehydrogenase
1.11.1.7	Peroxidase
1.11.1.12	Phospholipid-hydroperoxide glutathione peroxidase
1.5.99.3	L-Pipecolate dehydrogenase
1.5.3.7	L-Pipecolate oxidase
1.5.1.21	$DELTA^1$-Piperideine-2-carboxylate reductase
1.10.99.1	Plastoquinol-plastocyanin reductase
1.5.3.11	Polyamine oxidase
1.5.99.8	Proline dehydrogenase
1.8.4.2	Protein-disulfide reductase (glutathione)
1.6.4.4	Protein-disulfide reductase (NAD(P)H)
1.8.4.6	Protein-methionine-S-oxide reductase
1.9.3.2	Pseudomonas cytochrome oxidase
1.5.4.1	Pyrimidodiazepine synthase
1.5.1.1	Pyrroline-2-carboxylate reductase
1.5.1.12	1-Pyrroline-5-carboxylate dehydrogenase
1.5.1.2	Pyrroline-5-carboxylate reductase
1.5.3.9	Reticuline oxidase
1.10.3.6	Rifamycin-B oxidase
1.5.1.10	Saccharopine dehydrogenase ($NADP^+$, L-glutamate forming)
1.5.1.8	Saccharopine dehydrogenase ($NADP^+$, L-lysine forming)
1.5.1.9	Saccharopine dehydrogenase (NAD^+, L-glutamate-forming)

EC-No.	Name	EC-No.	Name
1.5.1.7	Saccharopine dehydrogenase (NAD^+, L-lysine forming)	1.5.1.23	Tauropine dehydrogenase
1.5.99.1	Sarcosine dehydrogenase	1.5.3.8	(S)-Tetrahydroprotoberberine oxidase
1.5.3.1	Sarcosine oxidase	1.8.3.2	Thiol oxidase
1.5.99.6	Spermidine dehydrogenase	1.5.1.25	Thiomorpholine-carboxylate dehydrogenase
1.5.1.22	Strombine dehydrogenase	1.6.4.5	Thioredoxin reductase (NADPH)
1.8.2.1	Sulfite dehydrogenase	1.8.2.2	Thiosulfate dehydrogenase
1.8.3.1	Sulfite oxidase	1.5.99.7	Trimethylamine dehydrogenase
1.8.99.1	Sulfite reductase	1.6.6.9	Trimethylamine-N-oxide reductase
1.8.7.1	Sulfite reductase (ferredoxin)	1.6.4.8	Trypanothione reductase
1.8.1.2	Sulfite reductase (NADPH)	1.10.2.2	Ubiquinol-cytochrome-c reductase
1.10.3.8	Sulochrin oxidase ((-)-bisdechlorogeodin-forming)	1.7.3.3	Urate oxidase
1.10.3.7	Sulochrin oxidase ((+)-bisdechlorogeodin-forming)		

1 NOMENCLATURE

EC number
1.5.1.1

Systematic name
L-Proline: $NAD(P)^+$ 2-oxidoreductase

Recommended name
Pyrroline-2-carboxylate reductase

Synonymes
Reductase, pyrroline-2-carboxylate
DELTA[1]-pyrroline-2-carboxylate reductase

CAS Reg. No.
9029-16-7

2 REACTION AND SPECIFICITY

Catalysed reaction
1-Pyrroline-2-carboxylate + NAD(P)H →
→ L-proline + $NAD(P)^+$

Reaction type
Redox reaction

Natural substrates
DELTA[1]-pyrroline-2-carboxylate + NAD(P)H

Substrate spectrum
1 DELTA[1]-pyrroline-2-carboxylate + NAD(P)H (ir [2]) [1–3]
2 DELTA[1]-piperidine-2-carboxylate + NAD(P)H (ir [2]) [1–3]

Product spectrum
1 L-Proline + $NAD(P)^+$ [1, 2]
2 L-Pipecolic acid + $NAD(P)^+$ [1, 2]

Inhibitor(s)
N-Formyl-L-methionine [1]; More (not: p-hydroxymercuribenzoate) [1]

Cofactor(s)/prostethic group(s)
NADH [1, 2]; NADPH [2]

Metal compounds/salts

Enzyme Handbook © Springer-Verlag Berlin Heidelberg 1994
Duplication, reproduction and storage in data banks are only allowed with the prior permission of the publishers

Turnover number (min^{-1})

Specific activity (U/mg)

K_m-value (mM)

0.86 (DELTA[1]-pyrroline-2-carboxylate) [1]; 0.43 (NADH (+ DELTA[1]-pyrroline-2-carboxylate)) [1]; 0.21 (DELTA[1]-piperidine-2-carboxylate) [1]; 0.02 (NADH (+ DELTA[1]-piperidine-2-carboxylate)) [1]; 0.082 (DELTA[1]-pyrroline-2-carboxylate (+ NADH)) [2]; 0.12 (DELTA[1]-pyrroline-2-carboxylate (+ NADPH)) [2]; 0.06 (DELTA[1]-piperidine-2-carboxylate (+ NADH)) [2]; 0.043 (NADH (+ DELTA[1]-pyrroline-2-carboxylate)) [2]; 0.074 (NADH (+ DELTA[1]-piperidine-2-carboxylate)) [2]; 0.038 (NADPH (+ DELTA[1]-pyrroline-2-carboxylate)) [2]

pH-optimum

5.2 (DELTA[1]-pyrroline-2-carboxylate) [1];
5.4 (DELTA[1]-piperidine-2-carboxylate) [1]; 6 [2]

pH-range

Temperature optimum (°C)

Temperature range (°C)

---

3 ENZYME STRUCTURE

Molecular weight

Subunits

Glycoprotein/Lipoprotein

–

---

4 ISOLATION/PREPARATION

Source organism

Mouse [1]; Dog (beagle) [1]; Monkey [1]; Rat [2, 3]; Pisum sativum [2]; Phaseolus radiatus [2]; Aerobacter aerogenes [2]

Source tissue

Brain (prosencephalic region [1]) [1, 2]; Liver [2]; Spleen [2]; Testis [2]; Heart [2]; Skeletal muscle [2]; Germinated seeds [2]

Localisation in source

Purification

Crystallization

–

Cloned

–

Renaturated

–

5 STABILITY

pH

Temperature (°C)

Oxidation

Organic solvent

General stability information

Storage

6 CROSSREFERENCES TO STRUCTURE DATABANKS

PIR/MIPS code

Brookhaven code

7 LITERATURE REFERENCES

[1] Garweg, G., Von Rehren, D., Hintze, U.: J. Neurochem.,35,616–621 (1980)
[2] Meister, A., Radhakrishnan, A.N., Buckley, S.D.: J. Biol. Chem.,229,789–800 (1957)
[3] Meister, A.: Methods Enzymol. (Colowick, S.P., Kaplan, N.O., Eds.) 5,878–882 (1962)

Enzyme Handbook © Springer-Verlag Berlin Heidelberg 1994
Duplication, reproduction and storage in data banks are only allowed with the prior permission of the publishers

1 NOMENCLATURE

EC number
1.5.1.2

Systematic name
L-Proline: NAD(P)$^+$ 5-oxidoreductase

Recommended name
Pyrroline-5-carboxylate reductase

Synonymes
Reductase, pyrroline-5-carboxylate
Proline oxidase
L-Proline oxidase
1-Pyrroline-5-carboxylate reductase
NADPH-L-DELTA'-pyrroline carboxylic acid reductase
L-Proline-NAD(P)$^+$ 5-oxidoreductase [19, 20]

CAS Reg. No.
9029-17-8

---

2 REACTION AND SPECIFICITY

Catalysed reaction
1-Pyrroline-5-carboxylate + NAD(P)H →
→ L-proline + NAD(P)$^+$

Reaction type
Redox reaction

Natural substrates
Pyrroline-5-carboxylate + NAD(P)H (discussion of role in vivo [1–3], transfer of reducing equivalents from host plants to symbiotic partner [5], transfer of reducing equivalents into mitochondria, linkage to pentose phosphate pathway [10], linkage to hexose monophosphate pathway [16]) [1–3, 5, 10, 16]

Enzyme Handbook © Springer-Verlag Berlin Heidelberg 1994
Duplication, reproduction and storage in data banks are only allowed with the prior permission of the publishers

Substrate spectrum

1 1-Pyrroline-5-carboxylate + NAD(P)H (ir [3, 17, 19, 21, 26, 27], ir at pH 7.4, r at pH 10.3 [25], specific for L-isomer [17, 18]) [3, 8, 14, 17–19, 21, 25–27]
2 L-Proline methyl ester + NAD^+ (16% of activity with proline) [8]
3 L-Proline t-butyl ester + NAD^+ (10% of activity with proline) [8]
4 L-Proline benzyl ester + NAD^+ (24% of activity with proline) [8]
5 1-Pyrroline-3-hydroxy-5-carboxylate + NAD(P)H [27]
6 More (not: D-proline, L-hydroxyproline, L-azatidine-2-carboxylate, no other amino acids) [8]

Product spectrum

1 L-Proline + $NAD(P)^+$
2 1-Pyrroline-5-carboxy methyl ester + NADH
3 1-Pyrroline-5-carboxy t-butylester + NADH
4 L-Pyrroline-5-carboxy benzyl ester + NADH
5 L-Hydroxyproline + $NAD(P)^+$ [27]
6 ?

Inhibitor(s)

NaCl [3]; Proline [3, 7, 17, 20–23, 25, 26]; ATP [7, 12, 20, 22, 25, 26]; ADP [7, 20, 22, 26]; $NADP^+$ [7, 20, 21]; p-Chloromercuribenzoate (not: Cucurbita moschata [22], Cucurbita maxima [25]) [7, 17, 20, 21]; Hg^{2+} [7, 17, 22, 25, 26]; Cu^{2+} (not: Cucurbita moschata [22]) [2, 26]; Cd^{2+} [7, 26]; p-Hydroxymercuribenzoate [14]; N-Ethylmaleimide [14]; 5,5'-Dithiobis-(2-nitrobenzoate) [14]; 3-Acetylpyrimidine analogue of NAD^+ [17]; Thio-NAD^+ [17]; Thio-$NADP^+$ [17]; Ag^+ [17, 26]; NH_2OH [20, 22, 25]; Thiazolidine-4-carboxylic acid [20, 26]; Phosphate [20]; GTP [20]; CTP [20]; NADPH (above 0.13 mM) [21]; NADH (above 0.16 mM) [21]; Cysteine [22]; Thiol compounds [27]; Zn^{2+} (not: Cucurbita moschata [22]) [26]; NAD^+ [26]; Mn^{2+} [26]

Cofactor(s)/prostethic group(s)

NADPH (preferentially [6, 7, 21], no reaction [18]) [6–8, 12, 19–21, 25–27]; NADH (specific for [3], preferentially [17, 19, 20, 25, 26]) [3, 8, 12, 17–21, 25–27]

Metal compounds/salts

KCl (increase of activity) [2, 12]; $MgCl_2$ (increase of activity) [2]; NH_4Cl (increase of activity) [2]; Ammonium acetate (increase of activity) [2]; Phosphate (10 mM, reaction enhancement) [12]

Turnover number (min^{-1})

76000 (pyrroline-5-carboxylate) [8]

Specific activity (U/mg)
507 [8]; 1528 [11]; 70.1 [18]; More (assay method [24]) [12, 13, 17, 20–24, 27]

K_m-value (mM)
0.09–0.16 (NADPH [2], NADH [3, 20], pyrroline-5-carboxylate [6, 12, 17, 21]) [2, 3, 6, 12, 17, 20, 21]; 0.17–0.2 (NADH [2, 12], NADPH [17], pyrroline-5-carboxylate [3, 7, 17, 20, 21, 25]) [2, 3, 7, 12, 17, 20, 21, 25]; 1.62–1.64 (NADH, depending on buffer) [6]; 0.006–0.0071 (NADPH) [7, 12]; 0.37–0.62 (NADH [7, 19], pyrroline-5-carboxylate [19]) [7, 19]; 0.05–0.051 (DL-pyrroline-5-carboxylate (+ NADPH) [7], NADPH [20]) [7, 20]; 0.02–0.022 (L-pyrroline-5-carboxylate (+ NADH) [7], NADPH (depending on buffer [6]) [6, 26]) [6, 7, 26]; 0.62 (DL-pyrroline-5-carboxylate (+ NADH)) [7]; 0.21–0.3 (L-pyrroline-5-carboxylate (+ NADH) [7], NADH [21]) [7, 21]; 0.048–0.069 (NADH [14, 17, 25], NADPH [14, 21, 25], DL-pyrroline-5-carboxylate (+ NADPH) [7, 26]) [7, 14, 17, 25, 26]; 31 (L-proline, pH 8.0) [18]; 1.25 (L-proline, pH 10.2) [18]; 1.0–1.2 (NAD^+) [18]; 0.33 (L-pyrroline-5-carboxylate, pH 8.0) [18]; More [26, 28]

pH-optimum
6.0 (cofactor NADH) [28]; 6–6.5 [14]; 6.2–6.5 [17]; 6.2–7. 5 (pyrroline-5-carboxylate reductase activity) [18]; 6.5 (pyrroline-5-carboxylate reductase activity [22]) [2, 22]; 6.5–7.0 [7]; 6.7–7.4 [26]; 6.8–7.5 (pyrroline-5-carboxylate reductase activity) [8]; 7.0 (cofactor NADPH) [28]; 7.0–7.5 [12, 20, 25]; 7.0–8.0 [3, 27]; 9.8–10.4 (proline dehydrogenase activity) [22]; 10.2 (proline dehydrogenase activity) [18]

pH-range
6.0–8.5 (more than 50% of maximal activity at pH 6.0 and 8.5) [3]; 4–8 (pyrroline-5-carboxylate reductase activity) [22]; 8.5 (no proline dehydrogenase activity below) [8]; 9.2–10.5 (proline dehydrogenase activity) [22]

Temperature optimum (°C)

Temperature range (°C)

Enzyme Handbook © Springer-Verlag Berlin Heidelberg 1994
Duplication, reproduction and storage in data banks are only allowed with the prior permission of the publishers

3 ENZYME STRUCTURE

Molecular weight

420000 (Hordeum vulgare, native gel electrophoresis) [9]
320000 (E. coli, gel filtration) [21]
280000 (E. coli, gel chromatography) [11]
240000 (rat, gel filtration) [7]
125000 (Saccharomyces cerevisiae, gel chromatography) [14]
100000 (Cucurbita moschata, gel filtration) [22]
94000–95000 (Pseudomonas aeruginosa, sucrose gradient sedimentation, gel filtration) [17]
More (quarternary structure changes drastically with buffer environment) [3, 8]

Subunits

Polymer (12–16 subunits, x × 30000, Hordeum vulgare, SDS-PAGE) [8]
Decamer (10 × 28112, E. coli, calculation from nucleotide sequence, SDS-PAGE) [11]
Octamer (8 × 30000, rat, SDS-PAGE) [7]

Glycoprotein/Lipoprotein

–

---

4 ISOLATION/PREPARATION

Source organism

Glycine max [1, 5, 20]; Vigna unguiculata [1]; Cajanus cajan [1]; Arachis hypogaea [1]; Lupinus luteus [1]; Phaseolus vulgaris [1]; Pisum sativum [1, 29]; Chlorella autotrophica [3]; Chlorella saccharophila [3]; Aspergillus nidulans [4]; Rat [6, 7, 19, 23, 29]; Hordeum vulgare [8]; Human [10, 16, 23]; E. coli [12, 21]; Bovine (calf [26]) [12, 26, 27]; Saccharomyces cerevisiae [13, 14]; Aldrichina grahami (blowfly) [15]; Pseudomonas aeruginosa [17]; Clostridium sporogenes [18]; Cucurbita moschata [22, 25]; Chinese hamster [23]; Cucurbita maxima [25]; Neurospora crassa [28]; Aeromonas aerogenes [29]; Phaseolus radiatus [29]; Tetrahymena pyriformis [30]; Nicotiana tabacum [31]; More (in all protists, animals, plants known to synthesize proline) [32]

Source tissue

Root nodules [1, 5]; Leaves (from etiolated shoots [2]) [2, 20]; Cell [3, 11, 21]; Liver [6, 26, 29]; Lens [8]; Seedlings [8, 9]; Erythrocytes [10, 16]; Retina (distribution in bovine cornea and lens) [12]; Larvae (fat body, muscle) [15]; Cotyledons [22, 25]; Kidney [29]; Ovary cells [23]; Lung [23]; Fibroblasts [23]; Hepatoma cells [23]; More (distribution in foetal and tumor tissues) [19]

Localisation in source

Chloroplasts [2]; Bacteroids [5]; Mitochondria (matrix space) [15]; Soluble part of cell [19, 25, 28]

Purification

Chlorella sp. [3]; Rat [7]; E. coli [11, 21]; Bovine (bull) [12]; Saccharomyces cerevisiae [13, 14]; Pseudomonas aeruginosa (partial) [17]; Glycine max [20]; Cucurbita moschata (copurification with proline oxidase, both activities located on the same protein) [22]; Neurospora crassa (partial) [28]

Crystallization

–

Cloned

[4, 11]

Renaturated

–

5 STABILITY

pH

6.0 (irreversible inactivation below) [26]; 7.2 (best value for stability) [3]

Temperature (°C)

0 (30 min, 16%–99% loss of activity, highly dependent on buffer [6], complete inactivation [19]) [6, 19]; 22 (96 h stable) [7]; 4–40 [21]; 60 (inactivation above [21], 10 min, complete inactivation [18]) [18, 21]; 65 (15 min stable) [22]

Oxidation

Organic solvent

General stability information

Dithiothreitol stabilizes during purification [14]; Thiol reagents stabilize [3]; Sorbitol stabilizes [3]; Gycine betaine stabilizes [3]; Mercaptoethanol stabilizes [7]; EDTA stabilizes [7]; Glycerol stabilizes [7]; High ionic strength stabilizes [18]; Loss of activity after freezing and thawing [18]

Storage

–80°C, 5% glycerol, 2 mM dithiothreitol, 1 month [8]; –70 °C, 50 mM triethanolamine buffer pH 7.4, 9% w/v sucrose, 1 mM EDTA, 2 mM $MgCl_2$, 30 mM mercaptoethanol [6]; –20°C, 1 mg protein/ml, 50% glycerol, 0.1 mM dithiothreitol, unstable [13]; –20°C, inactivation in few days [12]; –18 °C, 0.15–0.2 M potassium phosphate buffer pH 7.4, 2 months [18]; –15°C, glutathione, EDTA [27]; 4°C or –10°C, 50% loss of activity in 1–4 weeks [20]; 3°C, 0.1 M sodium phosphate buffer pH 7.6, 10% $(NH_4)_2SO_4$, 7 days [22]

Enzyme Handbook © Springer-Verlag Berlin Heidelberg 1994
Duplication, reproduction and storage in data banks are only allowed with the prior permission of the publishers

6 CROSSREFERENCES TO STRUCTURE DATABANKS

PIR/MIPS code

RDECC (Escherichia coli); JQ0418 (precursor Pseudomonas aeruginosa PAO1)

Brookhaven code

7 LITERATURE REFERENCES

[1] Kohl, D.H., Lin, J.-J., Shearer, G., Schubert, K.R.: Plant Physiol.,94,1258–1264 (1990)
[2] Rayapati, P.J., Stewart, C.R., Hack, E.: Plant Physiol.,91,581–586 (1989)
[3] Laliberte, G., Hellebust, J.A.: Plant Physiol.,91,917–923 (1989)
[4] Hull, E.P., Green, P.M., Arst, H.N., Scazzocchio, C.: Mol. Microbiol.,3,553–559 (1989)
[5] Kohl, D.H., Schubert, K.R., Carter, M.B., Hagedorn, C. H., Shearer, G.: Proc. Natl. Acad. Sci. USA,85,2036–2040 (1988)
[6] Hagedorn, C.H.: Biochim. Biophys. Acta,884,11–17 (1986)
[7] Shiono, T., Kador, P.F., Kinoshita, J.J.: Biochim. Biophys. Acta,881,72–78 (1986)
[8] Krueger, R., Jäger, H.-J., Hintz, M., Pahlich, E.: Plant Physiol.,80,142–144 (1986)
[9] Elthon, T.E., Stewart, C.R.: Plant Physiol.,74,213–218 (1984)
[10] Hagedorn, C.H., Phang, J.M.: Arch. Biochem. Biophys.,225,95–101 (1983)
[11] Deutch, A.H., Smith, C.J., Rushlow, K.E., Kretschmer, P.J.: Nucleic Acids Res.,10,7701–7714 (1982)
[12] Matsuzawa, T.: Biochim. Biophys. Acta,717,215–219 (1982)
[13] Matsuzawa, T., Ishiguro, I.: Biochim. Biophys. Acta,616,381–383 (1980)
[14] Matsuzawa, T., Ishiguro, I.: Biochim. Biophys. Acta,613,318–323 (1980)
[15] Tsuyama, S., Higashimo, T., Miura, K.: Experientia,36,1037–1038 (1980)
[16] Yeh, G.C., Phang, J.M.: Biochem. Biophys. Res. Commun.,94,450–457 (1980)
[17] Krishna, R.V., Beilstein, P., Leisinger, T.: Biochem. J.,181,223–230 (1979)
[18] Costilow, R.N., Cooper, D.: J. Bacteriol.,134,139–146 (1978)
[19] Herzfeld, A., Mezl, V.A., Knox, W.E.: Biochem. J.,166,95–103 (1977)
[20] Miler, P.A., Stewart, C.R.: Phytochemistry,15,1855–1857 (1976)
[21] Rossi, J.R., Vender, J., Berg, C.M., Coleman, W.H.: J. Bacteriol.,129,108–114 (1977)
[22] Rena, A.B., Splittstoesser, W.E.: Phytochemistry,14,657–661 (1975)
[23] Valle, D., Downing, S.J., Phang, J.M.: Biochem. Biophys. Res. Commun.,54,1418–1424 (1973)
[24] Phang, J.M., Downing, S.J., Valle, D.: Anal. Biochem.,55,266–271 (1973)
[25] Splittstoesser, S.A., Splittstoesser, E.E.: Phytochemistry,12,1565–1568 (1973)
[26] Strecker, H.J.: Methods Enzymol. (Colowick, S.P., Ed.) 17B,258–261 (1971)
[27] Adams, E., Goldstone, A.: J. Biol. Chem.,235,3499–3503 (1960)
[28] Yura, T., Vogel, H.J.: J. Biol. Chem.,234,335–338 (1959)
[29] Meister, A., Radhakrishnan, A.N., Buckley, S.D.: J. Biol. Chem.,229,789–800 (1957)
[30] Hill, D.L., Chambers, P.: Biochim. Biophys. Acta,148,435–447 (1967)
[31] Noguchi, M., Koiwai, A., Tamaki, E.: Agric. Biol. Chem.,30,452 (1966)
[32] Adams, E., Frank, L.: Annu. Rev. Biochem.,49,1005–1061 (1980)

1 NOMENCLATURE

EC number
1.5.1.3

Systematic name
5,6,7,8-Tetrahydrofolate:$NADP^+$ oxidoreductase

Recommended name
Dihydrofolate reductase

Synonymes
Tetrahydrofolate dehydrogenase
DHFR [1]
Pteridine reductase:dihydrofolate reductase [24]
Dihydrofolate reductase:thymidylate synthase [25]
Thymidylate synthetase-dihydrofolate reductase [48]
Dehydrogenase, tetrahydrofolate
Folic acid reductase
Folic reductase
Dihydrofolic acid reductase
Dihydrofolic reductase
Reductase, dihydrofolate
EC 1.5.1.4 (now included with EC 1.5.1.3)
7,8-Dihydrofolate reductase
NADPH-dihydrofolate reductase

CAS Reg. No.
9002-03-3

2 REACTION AND SPECIFICITY

Catalysed reaction
5,6,7,8-Tetrahydrofolate + $NADP^+$ →
→ 7,8-dihydrofolate + NADPH (mechanism [1, 2, 6], rapid equilibrium random mechanism [1], the enzyme from animals and some micro-organisms also slowly reduces folate to 5,6,7,8-tetrahydrofolate)

Reaction type
Redox reaction

Enzyme Handbook © Springer-Verlag Berlin Heidelberg 1994
Duplication, reproduction and storage in data banks are only allowed with the prior permission of the publishers

Natural substrates

7,8-Dihydrofolate + NADPH (maintainance of adequate levels of fully reduced folate in metabolism of proliferating cells [1, 18], key enzyme in biosynthesis of purines, pyrimidines and several amino acids [43]) [1, 18, 43]

Substrate spectrum

1 7,8-Dihydrofolate + NADPH (equilibrium strongly favors tetrahydrofolate production [18], reverse reaction at only one tenth of forward reaction [26], irreversible [31], equilibrium constant: 320000000000 [42]) [18, 26, 31, 42]
2 Folate + NADPH (at a slow rate, reaction not catalyzed by all dihydrofolate reductases [18], not [22, 23]) [18, 29, 32, 36, 68]
3 8-Methylpterin + NADPH [19]
4 More (equilibrium strongly favors tetrahydrofolate production [18], NADPH + dihydrofolate: steady-state random mechanism [20], 3-acetylpyridine adenine dinucleotide + dihydrofolate: rapid equilibrium random mechanism [20]) [18, 20]

Product spectrum

1 5,6,7,8-Tetrahydrofolate + $NADP^+$
2 5,6,7,8-Tetrahydrofolate + $NADP^+$
3 8-Methyl-7,8-dihydropterin + $NADP^+$ (subsequently reduced more slowly and incompletely to 8-methyl-5,6,7,8-tetrahydropterin) [19]
4 ?

Inhibitor(s)

Aminopterin [31, 34, 39, 50, 54, 66]; Folic acid [31]; N^{10}–Methylpteroylglutamic acid [31]; Urea (3 M, 15% inhibition [32]) [32, 38, 43]; Spermine (12 mM) [32]; Spermidine (12 mM) [32]; Ethylenediamine (12 mM) [32]; Diaminobutane (12 mM) [32]; Diaminopentane (12 mM) [32]; 2, 4-Diaminopyrimidine [34]; Chaotropic salts (e.g. SCN^-, ClO_4^-, Cl^-, F^-) [34]; p-Hydroxymercuribenzoate [39, 50]; 5, 5-Dithiobis(2-nitrobenzoic acid) [39, 50]; 10-Formylaminopterin [39, 42, 50, 66]; Dichloromethotrexate [54]; 4-(N-[(2,4-Diamino-6-pteridyl)methyl]-N-methylamino)benzoate [54]; Triampterene [54]; Pteroate [54]; Mercuric chloride [60]; Triazinate [66]; Citrate (in acid pH range: inhibitor of reduction of dihydrofolate but not folate) [68]; Methotrexate [1, 2, 4, 18, 22–24, 26, 34, 37–39, 42, 43, 47, 49, 50, 54, 57–60, 66, 69]; Trimethoprim [1, 2, 23, 24, 26, 37, 38, 42, 43, 47, 58]; 2,4-Diamino-5-p-chlorophenyl-6-ethylpyridine (pyrimethamine) [1]; 2,4-Diamino-6-butylpyrido[2,3-d]pyrimidine [1]; 2,4-Diamino-5-methyl-butylpyrido[2,3-d]pyrimidine [1]; 2,4-Diamino-5-phenyl-s-triazines [2]; 2,4-Diamino-5-adamantylpyrimidine [2]; 2,4-Diamino-5-benzylpyrimidines [2]; N-Bromosuccinimide [2]; Phenylglyoxal [2]; Diethyldicarbonate [2]; Dansyl chloride [2]; 2,4-Pentanedione

[2]; 2,4,6-Trinitrobenzenesulfonic acid [2]; 2,4-Diaminoquinazoline (derivatives) [38]; Organic mercurials (animal enzyme: activated, bacterial enzyme: unaffected or inhibited) [2, 21]; 4,6-Diamino-1,2-dihydro-2,2-dimethyl-1-(phenylalkylphenyl)-s-triazines (bearing a terminal sulfonyl fluoride) [2]; 2,4-Diamino-5,6-dihydro-6, 6-dimethyl-5(4'-methoxyphenyl)-s-triazine [4]; Na^+ [22, 29, 32]; K^+ [22, 26, 29, 32, 38, 43]; Pyrimethamine [23, 24, 26, 34, 38, 41, 54]; Folate [24, 34, 38, 42, 49, 54]; L-Erythro-biopterin [24]; $NADP^+$ [24, 42, 68]; Mg^{2+} [32]; Ca^{2+} [32]; Ba^{2+} [32]; p-Chloromercuribenzoate [24, 26, 31, 60]; N-Ethylmaleimide [24, 26, 39, 50]; Chaotrophes (e.g. guanidine-HCl [24, 26, 66], formamide [24, 26], urea [24, 26]) [24, 26, 66]; Li^+ (activation of mutant type enzyme (0.2 mM), decrease of wild type enzyme (above 0.1 mM) [29]) [29, 32]; Ru^+ (activation of mutant type enzyme (0.2 mM), decrease of wild type enzyme (above 0.1 mM) [29]) [29, 32]; Cs^+ (activation of mutant type enzyme (0.2 mM), decrease of wild type enzyme (above 0.1 mM) [29]) [29, 32]; NH_4^+ (activation of mutant type enzyme (0.2 mM), decrease of wild type enzyme (above 0.1 mM)) [29]; p-Substituted mercuribenzoate (mutant enzyme: no effect, wild type: inhibition) [29]; Phenylmercuric acetate (mutant enzyme: no effect, wild type enzyme: inhibition) [29]; More (overview [4, 5], stereochemistry of inhibitor binding [1], overview: design of inhibitors from X-ray crystal structures [12]) [1, 4, 5, 12]

Cofactor(s)/prostethic group(s)

NADPH (NADPH oxidized 5.1-times more rapidly than NADH [22]); NADH (NADPH oxidized 5.1-times more rapidly than NADH [22], not [23, 26, 31], chicken liver: at neutral pH specific for NADPH and dihydrofolate, at acid pH: NADH + dihydrofolate and folate + NADPH [68]) [22, 68]; Urea (Lactobacillus casei: slight activation [21], 1 M activates [52], activates [54, 58, 66, 67]) [21, 52, 54, 58, 66, 67]; p-Chloromercuribenzoate (in presence of NADPH activates) [67]; Thiourea (activates) [67]

Metal compounds/salts

Organic mercurials (animal enzyme: activated, bacterial enzyme: unaffected or inhibited [2, 21], Lactobacillus casei: no activation [21], soybean: no activation [38], cultured mammalian cells: activation [58]) [2, 21, 38, 58]; Na^+ (activation of mutant type enzyme (0.2 mM), decrease of wild type enzyme (above 0.1 mM) [29], activates [36]) [29, 36]; K^+ (activation of mutant type enzyme activity (0.2 mM), decrease of wild type enzyme (above 0.1 mM) [29], stimulation [41, 52, 54, 58, 66, 67]) [29, 41, 52, 54, 58, 66, 67]; Li^+ (activation of mutant type enzyme (0.2 mM), decrease of wild type enzyme (above 0.1 mM)) [29]; Ru^+ (activation of mutant type enzyme (0.2 mM), decrease of wild type enzyme (above 0.1 mM)) [29]; Cs^+ (activation of mutant type enzyme (0.2 mM), decrease of wild type enzyme (above 0.1 mM)) [29]; NH_4^+ (activation of mutant type enzyme (0.2 mM), decrease of wild type enzyme (above 0.1 mM)) [29]

Enzyme Handbook © Springer-Verlag Berlin Heidelberg 1994
Duplication, reproduction and storage in data banks are only allowed with the prior permission of the publishers

Turnover number (min^{-1})

1.2 (folate) [36]; 240 (dihydrofolate + NADPH) [36]; 7.5 (folate, mutant type) [29]; 220 (Lactobacillus casei, pH 6. 5) [2]; 230 (methotrexate resistant cells: L1210 (R), pH 7.5) [2]; 300 (chicken liver, pH 6.5) [2]; 358 (human WIL2 (M4)cells, pH 7.5) [2]; 570 (sarcoma 180 (AT/3000) cells, pH 7.5, bovine liver, pH 6.5) [2]; 600 (E. coli MB1428 cells, pH 7.2) [2, 22]; 810 (pig liver, pH 7.6) [2]; 900 (Streptococcus faecium, methotrexate resistant cells [2], mutant type [29]) [2, 29]; 1360 (human KB cells, pH 7.5) [2]; 1800 (trimethoprim resistant E. coli (RT 500), form I, pH 7.0) [2]; 6000 (methotrexate resistant Streptococcus faecium, isoenzyme I [2]) [2, 30]; 8000 (methotrexate resistant Streptococcus faecium, wild type enzyme) [29]; 14400–16100 (Neisseria gonorrhoeae, 2 enzyme forms) [46]

Specific activity (U/mg)

14 (chicken liver) [2]; 26 (bovine liver [2]) [2, 69]; 11 (methotrexate resistant cells, L1210 (R)) [2]; 27 (methotrexate resistant cells, sarcoma 180 (AT/3000)) [2]; 15 (human WIL2 (M4) cells) [2]; 60 (human KB cells) [2]; 300 (Streptococcus faecium, methotrexate resistant cells [2, 30], isoenzyme I [2]) [2, 30]; 45 (Streptococcus faecium, methotrexate resistant cells, isoenzyme II) [2]; 12 (Lactobacillus casei) [2]; 36 (E. coli (MB 1428), methotrexate resistant [2, 22], pig liver [2]) [2, 22]; 55 (trimethoprim resistant E. coli) [2]; 0.058 (mouse leukemia cells) [10]; 17.8 [15]; 8.5 (chicken) [16]; 22.3 (calf) [16]; 132 [7]; 6.8 [26]; 19.1 [32]; 56.5 [38]; 156.3 [39]; 103 [60]; 28 [66]; More (coupled radiometric assay [13], enzyme assay [18, 53], overview [18]) [13, 18, 27, 35, 36, 40, 43, 44, 47, 48, 50, 51, 53, 54, 59, 62, 63, 67]

K_m-value (mM)

0.030 (8-methylpterin, pH 5.0, chicken, bovine) [19]; 0.200 (8-methylpterin, pH 7.4, chicken, bovine) [19]; 0. 130 (8-methylpterin, pH 5.0, human) [19]; 0.300 (above, 8-methylpterin, pH 7.4, human) [19]; 0.00645 (NADPH) [22]; 0.320 (NADH) [22]; 0.00044 (dihydrofolate) [22]; 0.0048 (dihydrofolate) [24]; 0.0011 (dihydrofolate) [26]; 0.0027 (NADPH) [26]; More (overview [2, 18], K_m of enzymes with decreased binding of folate and antagonists and of drug-sensitive enzymes [2]) [2, 18, 23, 30–39, 41–43, 46–48, 51, 54, 57, 59, 63–68]

pH-optimum

3.8 (chicken, folate + NADPH) [68]; 4.0 (dihydrofolate + NADPH, chicken, 2 optima: 4 and 7.4 [2, 68], 3 optima: 4.0, 7.25–7.5 and 8.25–8.5 [67], human WIL 2 cell line, 2 optima: 4 and 7.3–8.3 [2], methotrexate resistant cells L1210 (R), sarcoma 180 (AT/300), 2 optima: 4 and 7.5 [2]) [2, 67, 68]; 4.2 (Salmonella typhimurium, R-plasmid encoded enzyme, 2 optima, 4.2 and 7.8, in absence of KCl [52]) [37, 52]; 4.5 (E. coli, trimethoprim-resistant, dihydrofolate, form I [33], 2 optima: 4.5 and 7.0 [32, 33], folic acid, 1 optimum at pH 4.5 [32]) [32, 33]; 4.5–5.0 [51]; 4.5–5.5 (2 optima: 4.5 and 4.5–5.5 and 7.6) [63]; 4.7 (folate, mutant type enzyme) [29]; 4.8 (in presence of KCl [52]) [52, 65]; 4.8–5.0 (bovine liver [2, 60], buffers of low ionic strength [60]) [2, 60]; 5 (pig liver, 2 optima: 5 and 7.6 [2], hamster [59]) [2, 59]; 5.2–5.5 (chicken liver, dihydrofolate + NADH) [68]; 5.8 (dihydrofolate, Streptococcus faecium, isoenzyme II, methotrexate resistant cells [2], wild type, mutant type [29]) [2, 29]; 5.8–6.0 (8-methylpterin + NADPH) [19]; 6–7 [23]; 6.2 (bovine, presence of KCl) [60]; 6.3 (Candida albicans) [43]; 6.4 (below, E. coli strain 3747) [34]; 6.5 (Streptococcus faecium [2, 30], isoenzyme I, methotrexate resistant cells [2], Lactobacillus casei [2], E. coli, methotrexate-resistant cells [22]) [2, 22, 30]; 6.7 [31]; 6.8 (E. coli MB 1328 [2], Crithidia fasciculata [24], Saccharomyces cerevisiae [42]) [2, 24, 42]; 7.0 (trimethoprim resistant E. coli, 2 optima: 4.5 and 7.0 [32]) [26, 32, 47, 66]; 7.25–7.5 (3 optima: 4.0, 7.25–7.5 and 8.25–8.5) [67]; 7.2–8.2 (human KB cells, 2 optima: below 5 and 7.2–8.2) [2]; 7.3 [36]; 7. 3–8.3 (human WIL2 (M4) cells, 2 optima: 4 and 7.3–8.3) [2]; 7.4 (dihydrofolate + NADPH, chicken liver, 2 optima: 4 and 7.4 [2, 68], soybean [38]) [2, 38, 68]; 7.5 (methotrexate resistant cells, L1210(R), sarcoma 180 (AT/30000), 2 optima: 4 and 7.5 [2], Scenedesmus obliquus (thymidylate synthase and dihydrofolate reductase cannot be dissociated from each other) [49]) [2, 49]; 7.6 (2 optima: 4.5–5.5 and 7.6 [63], pig liver [2], E. coli (strain 1428) [34]) [2, 34, 63]; 7.8 (Salmonella typhimurium, R-plasmid encoded enzyme, 2 optima: 4.2 and 7.8) [37]; 8.0 (E. coli, strain 3746) [34]; 8.25–8.5 (3 optima: 4.0, 7.25–7.5 and 8.25–8.5) [67]; More (trimethoprim-resistant E. coli, form 2: no detectable pH-optimum, continuous increase of activity as pH decreases from pH 9 to 4) [33]

pH-range

4.2–5.6 (half-maximal activities at pH 4.2 and 5.6) [65]; 4.2–8.8 (half-maximal activities at pH 4.2 and 8.8, acetate buffer, Tris buffer) [22]; 5.3–8.0 (half-maximal activities at pH 5.3 and 8.0) [31]; 5.5–7.8 (half-maximal activities at pH 5.5 and 7.8, Crithidia oncopelti) [23]; 5.9–8.5 (half-maximal activities at pH 5.9 and 8.5) [36]; 6–9 (activity negligible below pH 6 and above pH 9) [49]; More (pH-activity profile in absence and presence of KCl [52]) [29, 52]

Temperature optimum (°C)

25–30 [49]

Temperature range (°C)

Enzyme Handbook © Springer-Verlag Berlin Heidelberg 1994
Duplication, reproduction and storage in data banks are only allowed with the prior permission of the publishers

3 ENZYME STRUCTURE

Molecular weight

14900 (Lactobacillus casei, gel filtration, SDS-PAGE, ultracentrifugation, isoenzyme I) [21]
15000 (Mycobacterium phlei, SDS-PAGE) [47]
16900 (Salmonella typhimurium, polyacrylamide gel electrophoresis under denaturating and nondenaturating conditions) [37]
17900 (Lactobacillus casei, gel filtration) [36]
17300 (E. coli, methotrexate-resistant mutant, equilibrium sedimentation) [22]
17470–18300 (E. coli, trimethoprim-resistant, amino acid composition, sedimentation velocity, SDS-PAGE) [32]
17500 (E. coli, trimethoprim-resistant) [34]
18500 (E. coli, trimethoprim-resistant, form 1 and 2, SDS-PAGE) [33]
18000 (L1210 lymphoma cells, gel filtration [17], Neisseria gonorrhoeae, SDS-PAGE [46]) [17, 46]
18250 (E. coli (MB 1428), amino acid sequence) [2]
18300 (Lactobacillus casei, amino acid sequence) [2]
19600 (Streptococcus faecium, isoenzyme II, amino acid sequence) [2]
20000 (Diplococcus pneumoniae, molecular sieve chromatography, wild type strain [28], Streptococcus faecium, velocity sedimentation, mutant type [29]) [28, 29]
20000–22000 (human placenta, gel filtration) [54]
26000 (hamster, sucrose density gradient centrifugation) [59]
21450 (bovine, pig liver, amino acid sequence) [2]
21460 (methotrexate resistant L1210(R) cells, amino acid sequence) [2]
21500 (bovine, amino acid composition) [60]
21651 (chicken liver, amino acid sequence) [2, 56]
22000 (soybean, SDS-PAGE [38], human, SDS-PAGE [65, 67]) [38, 65, 67]
22500 (E. coli B, SDS-PAGE [35], pig, gel filtration [63], rat Walker 256 carcinoma cells, meniscus depletion method [66], bovine, equilibrium sedimentation [69]) [35, 63, 66, 69]
23000 (bovine, SDS-PAGE in presence of 2-mercaptoethanol [61], chicken, sedimentation equilibrium studies, sucrose gradient experiments, aminopterin titration [68]) [61, 68]
23000–24000 (Pseudomonas cepacia, SDS-PAGE) [44]
24000 (Chinese hamster ovary cells, SDS-PAGE) [51]
25000 (Candida albicans, SDS-PAGE) [43]
28000 (Streptococcus faecium, velocity sedimentation, wild type) [29]
29000 (phage T4, SDS-PAGE) [35]
36000 (E. coli, R-plasmid enzyme, gel filtration) [40]
39000 (Saccharomyces cerevisiae, gel filtration) [42]
45000 (wild carrot, enzyme complex with dihydrofolate reductase activity,

thymidylate synthase activity and three other polypeptides of unknown function, dihydrofolate reductase subunit, SDS-PAGE) [50]
100000 (Scenedesmus obliquus, thymidylate synthase and dihydrofolate reductase cannot be dissociated from each other, sucrose density gradient centrifugation) [49]
107000 (Crithidia fasciculata, bifunctional enzyme: thymidylate synthase-dihydrofolate reductase [2, 25], gel filtration [25]) [2, 25]
110000 (Crithidia fasciculata, gel filtration [24, 26], Leishmania tropica, methotrexate resistant, bifunctional, glycerol density gradient centrifugation [48]) [24, 26, 48]
140000 (soybean, gel filtration) [39]
More (increase in MW due to association of enzyme molecules in mutant strains: MW 60000, MW 600000 [28], overview [18]) [18, 28]

Subunits

Monomer (1 × 16900, Salmonella typhimurium, R-plasmid enzyme, polyacrylamide gel electrophoresis under denaturating and nondenaturating conditions [37], 1 × 18250, E. coli (MB 1428), amino acid sequence [2], 1 × 18300, Lactobacillus casei, amino acid sequence [2], 1 × 19600, Streptococcus faecium, isoenzyme II, amino acid sequence [2], 1 × 21450, bovine liver, pig liver, amino acid sequence [2], 1 × 21460, methotrexate resistant L1210 (R) cells, amino acid sequence [2], 1 × 21650, chicken liver, amino acid sequence [2], 1 × 22500, E. coli B, SDS-PAGE [35], 1 × 29000, phage T4, SDS-PAGE [35], 1 × 39000, Saccharomyces cerevisiae, SDS-PAGE [42], 1 × 17000–22000, Drosophila melanogaster [45], 1 × 21300, bovine, SDS-PAGE [69]) [2, 35, 37, 42, 45, 69]
Dimer (2 × 58000, Crithidia fasciculata, SDS-PAGE [26], 2 × 56700, Crithidia fasciculata, bifunctional enzyme: thymidylate synthase-dihydrofolate reductase, SDS-PAGE [2, 25], 2 × 18000, R-plasmid-coded reductase [2], 2 × 23000, T4 phage-coded reductase [2], 2 × 56400, Leishmania tropica, methotrexate-resistant, bifunctional thymidylate synthase-dihydrofolate reductase, SDS-PAGE [48], 2 × 50000, Scenedesmus obliquus, thymidylate synthase and dihydrofolate reductase cannot be dissociated from each other, SDS-PAGE [49]) [2, 25, 26, 48, 49]
Tetramer (4 × 8500–9000, E. coli, R-plasmid-coded type II reductase) [2, 40]
Polymer (soybean, SDS-PAGE) [39]
More (wild carrot: enzyme complex with dihydrofolate reductase activity, thymidylate synthase activity and three other polypeptides of unknown function, 286000 MW complex, 45000 MW dihydrofolate reductase subunit and 4 other polypeptide chains: 95000, 70000, 50000, 26000) [50]

Glycoprotein/Lipoprotein

–

Enzyme Handbook © Springer-Verlag Berlin Heidelberg 1994
Duplication, reproduction and storage in data banks are only allowed with the prior permission of the publishers

4 ISOLATION/PREPARATION

Source organism

Chicken [2, 16, 18, 55, 56, 68]; Bovine (2 forms [61], calf [16, 18]) [2, 16, 18, 60, 61, 69]; Pig [2, 18, 63]; Human [2, 54, 65, 67]; Lactobacillus casei (2 forms: I, II [21]) [2, 11, 15, 21, 36]; Crithidia fasciculata (bifunctional enzyme: thymidylate synthase-dihydrofolate reductase [2], bifunctional enzyme: pteridine reductase-dihydrofolate reductase [24], dihydrofolate reductase-thymidylate synthase (E.C. 2.1.1.45) [25]) [2, 23–26]; Mouse (multiple forms [57], altered enzyme encoded by amplified genes in cultured fibroblasts [64]) [10, 17, 57, 62, 64]; E. coli (E. coli B [35], various strains [2], methotrexate-resistant mutant [22], trimethoprim-resistant strain [32–34], 2 isoenzymes: form 1, form 2 [33], strains: MB3746, MB 3747 [34]) [2, 6, 11, 22, 32–35, 40]; Crithidia oncopelti [23]; Diplococcus pneumoniae (methotrexate-resistant form [27], wild type and mutant forms [28]) [27, 28]; Streptococcus faecium (2 forms: mutant type and wild type enzyme in a single organism [29], methotrexate-resistant strain [29, 30]) [2, 29, 30]; Streptococcus faecalis R [31]; Phage T4 [35]; Salmonella typhimurium (phage type 179, R-plasmid enzyme) [37]; Soybean [38, 39]; Plasmodium berghei (pyrimethamine-sensitive and pyrimethamine resistant strain) [41]; Saccharomyces cerevisiae [42]; Candida albicans [43]; Pseudomonas cepacia (trimethoprim-susceptible and trimethoprim-resistant strains) [44]; Drosophila melanogaster [45]; Neisseria gonorrhoeae (2 forms) [46]; Mycobacterium phlei [47]; Leishmania tropica (methotrexate resistant bifunctional enzyme: thymidylate synthase-dihydrofolate reductase) [48]; Scenedesmus obliquus (green algae, thymidylate synthase and dihydrofolate reductase cannot be dissociated from each other) [49]; Wild carrot (enzyme complex with dihydrofolate reductase activity, thymidylate synthase activity and three other polypeptides of unknown function) [50]; Chinese hamster (methotrexate-sensitive and methotrexate-resistant cells) [51]; Hamster (baby hamster kidney cells [18, 52], 2 forms: I, II [52], wild type and methotrexate-resistant [59]) [18, 52, 59]; Mammalia (cultured mammalian cells, overview) [58]; Rat (Walker 256 carcinosarcoma cells) [66]; More (overview [2], almost every type of living cell, exception: archaebacteria and a few parasitic protozoa [1]) [1, 2]

Source tissue

Liver [2, 16, 18, 56, 60, 61, 63, 68, 69]; Leukemia L1210 mouse cells [10, 57]; L1210 Lymphoma cells [17, 62]; Thymus [18]; Baby hamster kidney cells [18, 52, 59]; Sarcoma cells [18]; Seedlings [38, 39]; Ovary cells [51]; Placenta [54]; Cultured mammalian cells (overview) [58]; Cultured fibroblasts (3T6 mouse embryo fibroblasts) [64]; Cultured cells (HeLa cells/BU-25, normal and methotrexate-resistant variant, methotrexate variant of VA_2-B cell line [65], Walker 256 carcinosarcoma cells [66], WIL2 lymphoblastoid cells [67]) [65–67]

Localisation in source

Soluble [45]

Purification

Mouse [10, 17]; Lactobacillus casei [15, 21, 36]; Chicken [16, 55, 68]; E. coli (trimethoprim-resistant [32–34], strain MB 3746, MB 3747 [34], methotrexate-resistant mutant: large-scale [22], E. coli B [35], R-plasmid enzyme [40]) [22, 32–35, 40]; Crithidia fasciculata (bifunctional enzyme: pteridine reductase-dihydrofolate reductase [24], dihydrofolate reductase-thymidylate synthase (E.C. 2.1.1.45) [25]) [24–26]; Diplococcus pneumoniae (methotrexate-resistant strain) [27]; Streptococcus faecium (2 forms: mutant-type and wild type enzyme in a single organism [29], methotrexate-resistant strain [29, 30]) [29, 30]; Streptococcus faecalis R (partial) [31]; Phage T4 [35]; Salmonella typhimurium (phage type 179, R-plasmid enzyme) [37]; Soybean [38, 39]; Saccharomyces cerevisiae [42]; Candida albicans [43]; Drosophila melanogaster [45]; Neisseria gonorrhoeae [46]; Scenedesmus obliquus (green algae, thymidylate synthase and dihydrofolate reductase cannot be dissociated from each other) [49]; Wild carrot (enzyme complex with dihydrofolate reductase activity, thymidylate synthase activity and three other polypeptides of unknown function) [50]; Chinese hamster (methotrexate-sensitive and methotrexate resistant cells) [51]; Hamster (baby hamster kidney cells, 2 forms: I, II [52], wild type and methotrexate-resistant cells [59]) [52, 59]; Human [54, 65, 67]; Mammalia (cultured mammalian cells, overview) [58]; Bovine (2 forms [61], calf [16]) [16, 60, 61, 69]; Pig [63]; Rat (Walker 256 carcinosarcoma cells) [66]; More (single-step purification on pteridine matrices [9], affinity chromatography [10, 14–18] with: dihydrofolate-Sepharose [14], pteroyllysine-agarose [15], methotrexate-agarose [16, 17], overview [18], via methotrexate-amino-ethyl starch [55]) [9, 10, 14–18, 55]

Crystallization

(crystal structures) [1, 11]

Cloned

(review [3], into multicopy vector pBR322 [71], Lactobacillus casei gene in E. coli using the multicopy vector pBR322 [70]) [3, 70, 71]

Renaturated

(refolding of enzyme reversibly unfolded in 7 M urea, effect of several peptide fragments, derived from limited proteolytic cleavage of dihydrofolate reductase on the attainment of the folded state [7], kinetic of refolding upon dilution of unfolded enzyme in 4.5 M urea to 1.29 M urea [8]) [7, 8]; More (denaturation with 4 M urea, no renaturation by removing urea by dialysis [26], enzyme denatured with 8 M urea cannot be renatured by removal of urea [39]) [26, 39]

Enzyme Handbook © Springer-Verlag Berlin Heidelberg 1994
Duplication, reproduction and storage in data banks are only allowed with the prior permission of the publishers

5 STABILITY

pH

6–8 (maximal stability) [68]

Temperature (°C)

4 (above, for extended length of time, unstable) [61]; 45 (3.5 h, no loss of activity in presence of bovine serum albumin [47], inactivation constant: 0.14 per min (form I), 0.13 per min (form II) [52]) [47, 52]; 53 (decay rate constant: 0.12 per min (wild type), 0.06–0.08 per min (mutant enzymes)) [51]; 55 (5 min, 70% loss of activity, 15 min, more than 80% loss of activity [39], half-life: 8 min [64]) [39, 64]; 60 (half-life: 2 min) [66]; More (dihydrofolate protects against heat inactivation [39], heat stable [44], 1 mM NADPH protects against inactivation [64]) [39, 44, 64]

Oxidation

Organic solvent

General stability information

Ammonium sulfate: unstable to ammonium sulfate precipitation [61], 55% saturation, 0°C, for at least 1 year [24]; Unstable to dialysis [61]; Unstable to lyophilization [61]; Unstable to concentration by membrane filtration under nitrogen [61]; Bovine serum albumin: 1 mg/ml, stabilizes during storage of purified enzyme preparation at 5°C [25], stabilizes purified enzyme [47], protects against inactivation [52]; Unusually resistant to trypsin [52]; 2-Mercaptoethanol, 4 mM, stabilizes purified preparation, inclusion in purification steps: loss of activity [26]; Protease inhibitors phenylmethylsulfonyl fluoride and leupeptin are essential for stabilization during purification [38]; Glycerol: 25%, essential for stabilization during purification [38], 30%, stabilization of purified preparation [26]; Freezing and thawing: significant loss of activity even in presence of glycerol [38], stable to repeated freezing and thawing [32]; NADPH: protects against ethoxyformic anhydride modification [34], protects against heat inactivation [68]; Dihydrofolate: protects against ethoxyformic anhydride modification [34], protects against heat inactivation [39], protects against inactivation [61]; Folate protects against ethoxyformic anhydride modification [34]; $NADP^+$ protects against heat inactivation [68]; N^5–Formyltetrahydrofolate protects against heat inactivation [68]; Frozen solutions not stable for long periods [36]

Storage

4°C, 0.1 M imidazole, pH 7.2, 25% glycerol, stable [38]; –10°C [32]; 4°C, above 1 mg/ml protein concentration, 1 month [43]; In refrigerator for several weeks [32]; –20°C, freeze-dried, several months [36]; 4°C, a few days [36]; 4°C, pH 7, 20% glycerol, 1 mM EDTA, 1 mM dithiothreitol, 2–3 weeks, at least 90% of activity retained [37]; –20 °C, 40% glycerol, pH 7.4, 5 mM dithiothreitol, 1 mM EDTA, 1 month [48]; 4°C, 40–50% loss of activity after 1 month [48]; 5°C, enzyme after affinity chromatography, half-life: 2 days [49]; –80°C, 30% glycerol [49]; –20°C, 50% glycerol, 0.1 M potassium phosphate, pH 7.0, 6 months, less than 10% loss of activity [54]; –60°C, 0.02 mM NADPH, 3 months, frequent freezing and thawing, stable [68]; 4°C, several weeks [69]; –80°C, 1 year [69]

6 CROSSREFERENCES TO STRUCTURE DATABANKS

PIR/MIPS code

RDHUD (Human); RDMSD (Mouse); RDHY75 (cell line DC-3F/A75 Chinese hamster); RDBOD (Bovine); RDPGD (Pig); RDBE11 (Saimirine herpesvirus 1 strain 11); RDBEHS (Saimirine herpesvirus 1 strain 488); RDCHD (Chicken); RDBYD (Yeast Saccharomyces cerevisiae); RDECD (type I Escherichia coli); RDEBDT (type III Salmonella typhimurium plasmid pAZ1); RDBSD (Bacillus subtilis); RDSODF (Enterococcus faecium); RDNHD (Neisseria gonorrhoeae); RDLBD (Lactobacillus casei); RDECD7 (type I Escherichia coli transposon Tn7); RDBPT4 (Phage T4); RDECD6 (type II Escherichia coli plasmid R67); RDECD8 (type II Escherichia coli plasmid R388); RDECD5 (type II Escherichia coli plasmid R751); RDLNTS (/thymidylate synthase EC 2.1.1.45 Leishmania major Contains: dihydrofolate reductase EC 1.5.1.3; Thymidylate synthase EC 2.1.1.45); RDZQK1 (/thymidylate synthase EC 2.1.1.45 Plasmodium falciparum strain K1); S03651 (type I Escherichia coli plasmid R438); A22241 (type III major chain Salmonella typhimurium fragment); B22241 (type III minor chain Salmonella typhimurium fragment); S02008 (Staphylococcus aureus fragment); S04164 (Staphylococcus aureus plasmid pSK1 transposon Tn4003); A24036 (Lactobacillus casei); A12987 (Lactobacillus casei fragment); S08660 (/thymidylate synthase EC 2.1.1.45 Leishmania amazonensis Contains: dihydrofolate reductase EC 1.5.1.3; Thymidylate synthase EC 2.1.1.45); A24311 (/thymidylate synthase EC 2.1.1.45 Leishmania major); C31262 (Plasmodium falciparum clone 3D7); F31262 (Plasmodium falciparum strain Csl-2); E31262 (Plasmodium falciparum strain K 1); G31262 (Plasmodium falciparum strain Palo-alto); D31262 (Plasmodium falciparum strain V 1); B31262 (/thymidylate synthase EC 2.1.1.45 Plasmodium falciparum clone 7G8); A31263 (/thymidylate synthase EC 2.1.1.45 Plasmodium falciparum clone 3D7 fragment); B31263 (/thymidylate synthase EC 2.1.1.45 Plasmodium falciparum clone HB3 frag-

Enzyme Handbook © Springer-Verlag Berlin Heidelberg 1994
Duplication, reproduction and storage in data banks are only allowed with the prior permission of the publishers

ment); A31262 (/thymidylate synthase EC 2.1.1.45 Plasmodium falciparum clone HB3); A22551 (Human); B39799 (Citrobacter freundii); A39799 (Enterobacter aerogenes); A33004 (type IV Escherichia coli plasmid pUK1123 fragment); S11702 (Escherichia coli plasmid pLMO150); S11706 (Escherichia coli plasmid pLMO229); A37174 (type IIIb Shigella sonnei fragment); A33005 (type IIIb Shigella sonnei fragment); S10715 (Staphylococcus aureus); A34429 (Haloferax volcanii); A36177 (Pneumocystis carinii); A32203 (Imperfect fungus Candida albicans fragment); S15756 (/thymidylate synthase EC 2.1.1.45 Leishmania amazonensis); A33484 (/thymidylate synthase EC 2.1.1.45 Plasmodium chabaudi); A39975 (/thymidylate synthase EC 2.1.1.45 Plasmodium falciparum); S17984 (Forest day mosquito); A33105 (Fruit fly Drosophila melanogaster); S13096 (Mouse mitochondrion SGC1); A21119 (variant Mouse fragment)

Brookhaven code

0DRF (HUMAN (HOMO SAPIENS) RECOMBINANT FORM FROM); 1DRF (HUMAN (HOMO SAPIENS) RECOMBINANT FORM EXPRESSED IN); 5DFR (ESCHERICHIA COLI) /TMP-RESISTANT STRAIN /SK383; 8DFR (CHICKEN (GALLUS GALLUS) LIVER)

7 LITERATURE REFERENCES

[1] Kraut, J., Matthews, D.A. in "Biol. Macromol. Assem.",3,1–71 (1987) (Review)
[2] Freisheim, J.H., Matthews, D.A. in "Folate Antagonists Ther. Agents" (Sirotnak, F.M., Ed.) Vol.1,69–131 (1984) (Review)
[3] Hamlin, J.L., Ma, C.: Biochim. Biophys. Acta,1087,107–125 (1990) (Review)
[4] Roth, B., Bliss, E., Beddell, C.R.: Top. Mol. Struct. Biol.,3,363–393 (1983) (Review)
[5] Baker, B.R.: Acc. Chem. Res.,2,129–136 (1969)
[6] Morrison, J.F., Stone, S.R.: Biochemistry,27,5499–5506 (1988)
[7] Hall, J.G., Frieden, C.: Proc. Natl. Acad. Sci. USA,86,3060–3064 (1989)
[8] Frieden, C.: Proc. Natl. Acad. Sci. USA,87,4413–4416 (1990)
[9] Dann, J.G., Harding, N.G.L., Newbold, P.C.H., Whiteley, J.M.: Biochem. J.,127,28P-29P (1972)
[10] Chello, P.L., Cashmore, A.R., Jacobs, S.A., Bertino, J.R.: Biochim. Biophys. Acta,268,30–34 (1972)
[11] Bolin, J.T., Filman, D.J., Matthews, D.A., Hamlin, R. C., Kraut, J.: J. Biol. Chem.,257,13650–13662 (1982)
[12] Roth, B.: Fed. Proc.,45,2765–2772 (1986)
[13] Reyes, P., Rathod, P.K.: Methods Enzymol.,122,360–367 (1986) (Review)
[14] Then, R.L.: Anal. Biochem.,100,122–128 (1979)
[15] Pastore, E.J., Plante, L.T., Kisliuk, R.L.: Methods Enzymol.,34,281–288 (1974) (Review)
[16] Kaufman, B.T.: Methods Enzymol.,34,272–281 (1974) (Review)
[17] Gauldie, J., Hillicoat, B.L.: Biochim. Biophys. Acta,268,35–40 (1972)
[18] Huennekens, F.M., Vitols, K.S., Whiteley, J.M., Neef, V.G.: Methods Cancer Res.,13,199–225 (1976) (Review)

[19] Thibault, V., Koen, M.J., Gready, J.E.: Biochemistry,28,6042–6049 (1989)
[20] Stone, S.R., Morrison, J.F.: Biochemistry,27,5493–5499 (1988)
[21] Gundersen, L. E., Dunlap, R.B., Harding, N.G.L., Freisheim, J.H., Otting, F., Huennekens, F.M.: Biochemistry,11,1018–1023 (1972)
[22] Poe, M., Greenfield, N.J., Hirshfield, J.M., Williams, M.N., Hoogsteen, K.: Biochemistry,11,1023–1030 (1972)
[23] Gutteridge, W.E., McCormack, J.J., Jaffee, J.J.: Biochim. Biophys. Acta,178,453–458 (1969)
[24] Oe, H., Kohashi, M., Iwai, K.: Agric. Biol. Chem.,47,251–258 (1983)
[25] Ferone, R., Roland, S.: Proc. Natl. Acad. Sci. USA,77,5802–5806 (1980)
[26] Iwai, K., Kohashi, M., Oe, H.: Agric. Biol. Chem.,45,113–120 (1981)
[27] Sirotnak, F.M., Salser, J.S.: Arch. Biochem. Biophys.,145,268–275 (1971)
[28] Sirotnak, F.M., Williams, W.A.: Arch. Biochem. Biophys.,136,580–582 (1970)
[29] Nixon, P.F., Blakley, R.J.: J. Biol. Chem.,243,4722–4731 (1968)
[30] D'Souza, L., Warwick, P.E., Freisheim, J.H.: Biochemistry,11,1528–1534 (1972)
[31] Blakley, R.L., McDougall, B.M.: J. Biol. Chem.,236,1163–1167 (1961)
[32] Baccanari, D., Phillips, A., Smith, S., Sinski, D., Burchall, J.: Biochemistry,14,5267–5273 (1975)
[33] Baccanari, D.P., Averett, D., Briggs, C., Burchall, J.: Biochemistry,16,3566–3572 (1977)
[34] Poe, M., Breeze, A.S., Wu, J.K., Short, C.R., Hoogsteen, K.: J. Biol. Chem.,254,1799–1805 (1979)
[35] Erickson, J.S., Mathews, C.K.: Biochemistry,12,372–380 (1973)
[36] Dann, J.G., Ostler, G., Bjur, R.A., King, R.W., Scudder, P., Turner, P.C., Roberts, G.C.K., Burgen, A.S.V., Harding, N.G.L.: Biochem. J.,157,559–571 (1976)
[37] Joyner, S.S., Fling, M.E., Stone, D., Baccanari, D.P. : J. Biol. Chem.,259,5851–5856 (1984)
[38] Ratnam, S., Delcamp, T.J., Hynes, J.B., Freisheim, J. H.: Arch. Biochem. Biophys.,255,279–289 (1987)
[39] Reddy, V.A., Rao, N.A.: Arch. Biochem. Biophys.,174,675–683 (1976)
[40] Smith, S.L., Stone, D., Novak, P., Baccanari, D.P., Burchall, J.J.: J. Biol. Chem.,254,6222–6225 (1979)
[41] Ferone, R.: J. Biol. Chem.,245,850–854 (1970)
[42] Nagelschmidt, M., Jaenicke, L.: Hoppe-Seyler's Z. Physiol. Chem.,353,773–781 (1972)
[43] Baccanari, D.P., Tansik, R.L., Joyner, S.S., Fling, M. E., Smith, P.L.: J. Biol. Chem.,264,1100–1107 (1989)
[44] Burns, J.L., Lien, D.M., Hedin, L.A.: Antimicrob. Agents Chemother.,33,1247–1251 (1989)
[45] Rancourt, S.L., Walker, V.K.: Biochim. Biophys. Acta,1039,261–268 (1990)
[46] Baccanari, D.P., Tansik, R.L., Paterson, S.J., Stone, D.: J. Biol. Chem.,259,12291–12298 (1984)
[47] Al-Rubeai, M., Dale, J.W.: Biochem. J.,235,301–303 (1986)
[48] Meek, T.D., Garvey, E.P., Santi, D.V.: Biochemistry,24,678–686 (1985)
[49] Bachmann, B., Follmann, H.: Arch. Biochem. Biophys.,256,244–252 (1987)
[50] Toth, I., Lazar, G., Goodman, H.M.: EMBO J.,6,1853–1858 (1987)
[51] Gupta, R.S., Flintoff, W.F., Siminovitch, L.: Can. J. Biochem.,55,445–452 (1977)
[52] Hänggi, U.J., Littlefield, J.W.: J. Biol. Chem.,249,1390–1397 (1974)

Enzyme Handbook © Springer-Verlag Berlin Heidelberg 1994
Duplication, reproduction and storage in data banks are only allowed with the prior permission of the publishers

[53] Stanley, B.G., Neal, G.E., Williams, D.C.: Methods Enzymol.,18B,775–779, Ed. Colowich, S.P. (1971) (Review)
[54] Jarabak, J., Bachur, N.R.: Arch. Biochem. Biophys.,142,417–425 (1971)
[55] Mell, G.P., Whiteley, J.M., Huennekens, F.M.: J. Biol. Chem.,243,6074–6075 (1968)
[56] Kumar, A.A., Blankenship, D.T., Kaufman, B.T., Freisheim, J.H.: Biochemistry,19,667–678 (1980)
[57] Fan, C.C., Vitols, K.S., Huennekens, F.M.: Adv. Enzyme Regul.,18,41–52 (1980) (Review)
[58] Gauldie, J., Marshall, L., Hillcoat, B.L.: Biochem. J.,133,349–356 (1973)
[59] Nakamura, H., Littlefield, J.W.: J. Biol. Chem.,247,179–187 (1972)
[60] Peterson, D.L., Gleisner, J.M., Blakley, R.L.: Biochemistry,14,5261–5267 (1975)
[61] Baumann, H., Wilson, K.L.: Eur. J. Biochem.,60,9–15 (1975)
[62] Gupta, S.V., Greenfield, N.J., Poe, M., Makulu, D.R., Williams, M.N., Moroson, B.A., Bertino, J.R.: Biochemistry,16,3073–3079 (1977)
[63] Smith, S.L., Patrick,, P., Stone, D., Phillips, A.W., Burchall, J.J.: J. Biol. Chem.,254,11475–11484 (1979)
[64] Haber, D.A., Beverley, S.M., Kiely, M.L., Schimke, R.T.: J. Biol. Chem.,256,9501–9510 (1981)
[65] Morandi, C., Attardi, G.: J. Biol. Chem.,256,10169–10175 (1981)
[66] Johnson, S.J., Gupta, S.V., Stevenson, K.J., Freisheim, J.H.: Can. J. Biochem.,60,1132–1142 (1982)
[67] Delcamp, T.J., Susten, S.S., Blankenship, D.T., Freisheim, J.H.: Biochemistry,22,633–639 (1983)
[68] Kaufman, B.T., Gardiner, R.C.: J. Biol. Chem.,241,1319–1328 (1966)
[69] Kaufman, B.T., Kemerer, V.F.: Arch. Biochem. Biophys.,172,289–300 (1976)
[70] Davies, R.W., Gronenborn, A.M.: Gene,17,229–233 (1982)
[71] Rood, J.I., Laird, A.J., Williams, J.W.: Gene,8,255–265 (1980)

1 NOMENCLATURE

EC number
1.5.1.5

Systematic name
5,10-Methylenetetrahydrofolate:NADP$^+$ oxidoreductase

Recommended name
Methylenetetrahydrofolate dehydrogenase (NADP$^+$)

Synonymes
N^5, N^{10}–Methylenetetrahydrofolate dehydrogenase [2]
5,10-Methylenetetrahydrofolate:NADP oxidoreductase [5]
Dehydrogenase, methylenetetrahydrofolate
5,10-Methylenetetrahydrofolate dehydrogenase

CAS Reg. No.
9029-14-5

2 REACTION AND SPECIFICITY

Catalysed reaction
5,10-Methylenetetrahydrofolate + NADP$^+$ →
→ 5,10-methenyltetrahydrofolate + NADPH (ordered bi-bi kinetic mechanism [15])

Reaction type
Redox reaction

Natural substrates
5,10-Methylenetetrahydrofolate + NADP$^+$ (enzyme maintains a physiological balance of tetrahydrofolate coenzymes as required and regulated by metabolic needs of the cell) [1]

Substrate spectrum
1 5,10-Methylenetetrahydrofolate + NADP$^+$ (r [11], equilibrium favors formation of N^5, N^{10}–methenyltetrahydrofolate [11])
2 5,10-Methylenetetrahydropteroyltriglutamate + NADP$^+$ [5, 11]
3 5,10-Methylenetetrahydropteroate + NADP$^+$ [5]
4 Tetrahydropteroyl-L-aspartate + NADP$^+$ [11]
5 More (not: tetrahydropteroyl-D-glutamate, tetrahydropteroic acid) [11]

Enzyme Handbook © Springer-Verlag Berlin Heidelberg 1994
Duplication, reproduction and storage in data banks are only allowed with the prior permission of the publishers

Product spectrum

1 5,10-Methenyltetrahydrofolate + NADPH (r [11])
2 5,10-Methenyltetrahydropteroyltriglutamate + NADPH
3 5,10-Methenyltetrahydropteroate + NADPH
4 ? + NADPH
5 ?

Inhibitor(s)

Mg^{2+} (slight inhibition [1]) [1, 8, 11]; Iodosobenzoate (1 mM) [1]; p-Chloromercuribenzoate (3 mM [1]) [1, 8, 21]; N-Ethylmaleimide [1]; Guanine [1]; N^{10}–Formyltetrahydrofolate [2, 3]; GTP [2, 7]; ATP (inhibits enzyme in crude extract [5]) [2, 5, 7]; IMP [2, 7]; NAD^+ [4, 6, 13]; Formaldehyde (activation below 0.3 mM, inhibition above [8], above 0.4 mM [11]) [8, 11]; ITP [7]; AMP [7]; GMP [7]; Ca^{2+} [8, 11]; Borate [8]; NADPH [13]; Folic acid [13, 15]; Pteroylglutamates (containing 1 to 7 glutamyl residues, overview) [16]; Iodoacetate [21]; alpha,alpha-Dipyridyl (slight) [21]; o-Phenanthroline (slight) [21]; More (overview: regulation in folate pathway) [3]

Cofactor(s)/prostethic group(s)

$NADP^+$ (specific for) [1, 4–6, 8, 11, 14]; NAD^+ (can replace $NADP^+$, 23% of the activity with $NADP^+$) [21]; NADPH [11]; Formaldehyde (activation below 0.3 mM, maximal activity at 0.3 mM, inhibition above [8], no reaction upon omission of either formaldehyde or tetrahydrofolate [11, 21]) [8, 11, 21]

Metal compounds/salts

K^+ (stimulation) [8, 11]; Na^+ (stimulation) [8, 11]; NH_4^+ (stimulation) [8, 11]; Rb^+ (stimulation) [8]; Li^+ (stimulation) [8, 11]; More (no effect of monovalent cations) [19]

Turnover number (min^{-1})

Specific activity (U/mg)

31.2 [2]; 310 [4]; 362 [6]; 91 [14]; 6.53 [18]; 8.1 [22]; More [1, 5, 7–9, 11, 12, 20]

K_m-value (mM)

0.21 ($NADP^+$) [1]; 0.034 ($NADP^+$) [5]; 0.026 (5,10-methylenetetrahydrofolate) [1]; 0.035 (5,10-methylenetetrahydrofolate (37°C [6])) [4–6]; 0.09 ($NADP^+$ [4], $NADP^+$, 55°C, pH 7.0) [4, 6]; 0.19 ($NADP^+$, 20°C and 37°C) [6]; 0.03 (methylenetetrahydrofolate, 55°C) [6]; More [6–8, 11–13, 16, 18–21]

pH-optimum

6.5 [11]; 6.7 [21]; 6.8–8 [5]; 7.4 [8]

pH-range

5–9.8 (half-maximal activity at pH 5 and 9.8) [5]; 6.7–10 (5.7: about 30% of activity maximum, 10: 35% of activity maximum) [8]; 5.8–8 (half-maximal activity at pH 5.8 and 8) [11]

Temperature optimum (°C)

39.5 [11]; 64 (above) [6]

Temperature range (°C)

3 ENZYME STRUCTURE

Molecular weight

55000 (Clostridium thermoaceticum [4, 6], methylenetetrahydrofolate dehydrogenase-methenyltetrahydrofolate cyclohydrolase [4], sedimentation equilibrium ultracentrifugation) [4, 6]
70000 (Clostridium cylindrosporum, sucrose density gradient centrifugation) [5]
150000 (pig, gel filtration, trifunctional enzyme: methylenetetrahydrofolate dehydrogenase-methenyltetrahydrofolate cyclohydrolase-formyltetrahydrofolate synthetase) [13, 20]
201000 (Saccharomyces cerevisiae, gel filtration, trifunctional enzyme contains formyltetrahydrofolate synthetase, methylenetetrahydrofolate cyclohydrolase and methylenetetrahydrofolate dehydrogenase activities) [9, 20]
226000 (sheep, molecular exclusion chromatography) [19, 20]
More (in vitro transcription-translation of human cDNA yields a 101000 MW protein [10], structural properties of the enzyme and its 2 domains [17]) [10, 17]

Subunits

Monomer (1 × 100000, pig, SDS-PAGE, trifunctional enzyme: methylenetetrahydrofolate dehydrogenase-methenyltetrahydrofolate cyclohydrolase-formyltetrahydrofolate synthetase) [13]
Dimer (2 × 30000, Clostridium thermoaceticum, SDS-PAGE [4, 6], 2 × 104000, Saccharomyces cerevisiae, trifunctional enzyme contains formyltetrahydrofolate synthetase, methylenetetrahydrofolate cyclohydrolase and methylenetetrahydrofolate dehydrogenase activities, SDS-electrophoresis [9], 2 × 100000, sheep, SDS-PAGE of proteins cross-linked with dimethylsuberimidate [20]) [4, 6, 9, 20]
More (in vitro transcription-translation of human cDNA yields a 101000 MW protein) [10]

Glycoprotein/Lipoprotein

–

Enzyme Handbook © Springer-Verlag Berlin Heidelberg 1994
Duplication, reproduction and storage in data banks are only allowed with the prior permission of the publishers

4 ISOLATION/PREPARATION

Source organism

Streptococcus faecium [1]; E. coli (complex of N^5,N^{10}–methylenetetrahydrofolate dehydrogenase and N^5,N^{10}–methenyltetrahydrofolate cyclohydrolase) [2]; Clostridium thermoaceticum (methylenetetrahydrofolate dehydrogenase-methenyltetrahydrofolate cyclohydrolase [4]) [4, 6]; Clostridium cylindrosporum [5]; Salmonella typhimurium [7]; Saccharomyces cerevisiae (trifunctional enzyme contains formyltetrahydrofolate synthetase, methylenetetrahydrofolate cyclohydrolase and methylenetetrahydrofolate dehydrogenase activities [9, 20]) [8, 9, 20]; Human (normal and transformed cell lines [22]) [10, 22]; Calf [11]; Pig (trifunctional enzyme: 1. methylenetetrahydrofolate dehydrogenase, 2. methenyltetrahydrofolate cyclohydrolase, 3. formyltetrahydrofolate synthetase) [12–16, 20]; Rabbit (trifunctional enzyme: 1. methylenetetrahydrofolate dehydrogenase, 2. methenyltetrahydrofolate cyclohydrolase, 3. formyltetrahydrofolate synthetase) [17]; Mouse (trifunctional enzyme: 1. methylenetetrahydrofolate dehydrogenase, 2. methenyltetrahydrofolate cyclohydrolase, 3. formyltetrahydrofolate synthetase [18], normal and transformed cell lines [22]) [18, 22]; Sheep (trifunctional enzyme: methylenetetrahydrofolate dehydrogenase-methenyltetrahydrofolate cyclohydrolase-formyltetrahydrofolate synthetase) [19, 20]; Pea [21]; Rat [23]; More (vertebrates, overview) [24]

Source tissue

Thymus [11]; Liver [12–20, 24]; Seedlings [21]; Cotyledons [21]; Mammalian cell lines (from mouse and human, normal and transformed, overview) [22]

Localisation in source

Soluble [24]; Mitochondria [24]

Purification

Streptococcus faecium (partial) [1]; E. coli (complex of N^5,N^{10}–methylenetetrahydrofolate dehydrogenase and N^5,N^{10}–methenyltetrahydrofolate cyclohydrolase) [2]; Clostridium thermoaceticum (methylenetetrahydrofolate dehydrogenase-methenyltetrahydrofolate cyclohydrolase [4]) [4, 6]; Clostridium cylindrosporum [5]; Salmonella typhimurium [7]; Saccharomyces cerevisiae (trifunctional enzyme contains formyltetrahydrofolate synthetase, methylenetetrahydrofolate cyclohydrolase and methylenetetrahydrofolate dehydrogenase activities [9, 20]) [8, 9, 20]; Calf [11]; Pig (trifunctional enzyme: methylenetetrahydrofolate dehydrogenase-methenyltetrahydrofolate cyclohydrolase-formyltetrahydrofolate synthetase) [12–14]; Rabbit (trifunctional enzyme: methylenetetrahydrofolate dehydrogenase-methenyltetrahydrofolate cyclohydrolase-formyltetrahydrofolate synthetase) [17]; Mouse (trifunctional enzyme: methylenetetrahydrofolate dehydrogenase-methenyltetrahydrofolate cyclohydrolase-formyltetrahydrofolate synthetase) [18]; Sheep (trifunctional enzyme: methylenetetrahydrofolate dehydrogenase-methenyltetrahydrofolate cyclohydrolase-formyltetrahydrofolate synthetase) [19, 20]; Pea (partial) [21]; Human (trifunctional enzyme: methylenetetrahydrofolate dehydrogenase-methenyltetrahydrofolate cyclohydrolase-formyltetrahydrofolate synthetase) [22]

Crystallization

–

Cloned

(cDNA encoding human trifunctional enzyme: 5,10-methylene tetrahydrofolate dehydrogenase-5,10-methenyltetrahydrofolate cyclohydrolase-10-formyltetrahydrofolate synthase) [10]

Renaturated

–

5 STABILITY

pH

5.2 (10 min, 40% loss of activity) [8]; 7.4 (maximal stability) [8]; 8 (10 min, 30% loss of activity) [8]

Temperature (°C)

37 (30 min, stable) [6]; 40 (half-life: 49.3 min) [11]; 46 (half-life: 6.7 min) [11]; 50 (30 min, stable) [6]; 60 (3 min, completely destroyed) [8]; 61 (20 min, 20% loss of activity) [6]; 68 (10 min, 50% loss of activity) [6]

Oxidation

Organic solvent

Enzyme Handbook © Springer-Verlag Berlin Heidelberg 1994
Duplication, reproduction and storage in data banks are only allowed with the prior permission of the publishers

General stability information

Thiol compounds protect against inactivation [11]; Glycerol protects against inactivation [11, 12]; EDTA protects against inactivation [11]; NADP$^+$: enzyme extremely labile, 1 mM NADP$^+$ required for protection during purification [1], stabilizes throughout freezing and protects against inactivation [12, 14]; Freezing: inactivates [1, 4], a single freezing and thawing inactivates [1], upon daily freezing and thawing, 30% loss of initial activity in 2 weeks [5], enzyme very unstable even in frozen state, complete loss of activity at –15°C overnight [7], frozen, stable for 12 days [7], repeated freezing and thawing decreases activity [7]; Highly unstable in ammonium sulphate, enzyme can only be preserved as suspension in ammonium sulphate at 50% saturation [7]; Bovine serum albumin: 1% stabilizes [7], protects against inactivation [11]; Lyophilization: 5–15% loss of activity [5], irreversible loss of activity [21]

Storage

–20°C, 0.1% bovine plasma albumin [1]; 4°C, stable for at least 1 month [4]; –20°C, for at least 2 months [5]; 4°C, various protein concentrations, for at least 2 months [6]; –20°C, dilute protein concentration (0.05 mg/ml) for at least 2 months [6]; 0–4°C, 3–4 days, 20–30% loss of activity [8]; 5°C, 60 h, 40% loss of activity [21]

6 CROSSREFERENCES TO STRUCTURE DATABANKS

PIR/MIPS code

A31903 (/methenyltetrahydrofolate cyclohydrolase EC 3.5.4.9 /formate-tetrahydrofolate ligase EC 6.3.4.3 Human); JS0662 (/methenyltetrahydrofolate cyclohydrolase EC 3.5.4.9 precursor Escherichia coli)

Brookhaven code

7 LITERATURE REFERENCES

[1] Albrecht, A.M., Pearce, F.K., Hutchinson, D.J.: J. Bacteriol.,95,1779–1789 (1968)
[2] Dev, I.K., Harvey, R.J.: J. Biol. Chem.,253,4245–4253 (1978)
[3] Harvey, R.J., Dev, I.K.: Adv. Enzyme Regul.,13,99–124 (1975) (Review)
[4] Ljungdahl, L.G., O'Brien, W.E., Moore, M.R., Liu, M.-T.: Methods Enzymol.,66,599–609 (1980) (Review)
[5] Uyeda, K., Rabinowitz, J.C.: J. Biol. Chem.,242,4378–4385 (1967)
[6] O'Brien, W.E., Brewer, J.M., Ljungdahl, L.G.: J. Biol. Chem.,248,403–408 (1973)
[7] Dalal, F.R., Gots, J.S.: J. Biol. Chem.,242,3636–3640 (1967)
[8] Ramasastri, B.V., Blakley, R.L.: J. Biol. Chem.,237,1982–1988 (1962)
[9] Paukert, J.L., Williams, G.R., Rabinowitz, J.C.: Biochem. Biophys. Res. Commun.,77,147–154 (1977)

[10] Hum, D.W., Bell, A.W., Rozen, R., MacKenzie, R.E.: J. Biol. Chem.,263,15946–15950 (1988)
[11] Yeh, Y.-C., Greenberg, D.M.: Biochim. Biophys. Acta,105,279–291 (1965)
[12] MacKenzie, R.E., Tan, L.U.L.: Methods Enzymol.,66,609–615 (1980) (Review)
[13] Tan, L.U.L., Drury, E.J., MacKenzie, R.E.: J. Biol. Chem.,252,1117–1122 (1977)
[14] MacKenzie, R.E.: Biochem. Biophys. Res. Commun.,53,1088–1095 (1973)
[15] Cohen, L., MacKenzie, R.E.: Biochim. Biophys. Acta,522,311–317 (1978)
[16] Ross, J., Green, J., Baugh, C.M., MacKenzie, R.E., Matthews, R.G.: Biochemistry,23,1796–1801 (1984)
[17] Villar, E., Schuster, B., Peterson, D., Schirch, V.: J. Biol. Chem.,260,2245–2252 (1985)
[18] Gardam, M.A., Mejia, N.R., MacKenzie, R.E.: Biochem. Cell Biol.,66,66–70 (1988)
[19] Paukert, J.L., D'Ari Straus, L., Rabinowitz, J.C.: J. Biol. Chem.,251,5104–5111 (1976)
[20] Paukert, J.L., Rabinowitz, J.C.: Methods Enzymol.,66,616–626 (1980) (Review)
[21] Cossins, E.A., Wong, K.F., Roos, A.J.: Phytochemistry,9,1463–1471 (1970)
[22] MacKenzie, R.E., Mejia, N., Yang, X.-M.: Adv. Enzyme Regul.,27,31–39 (1988) (Review)
[23] Perry, J., Deacon, R., Lumb, M., Chanarin, I.: Biochem. Biophys. Res. Commun.,97,1329–1333 (1980)
[24] Yoshida, T., Kikuchi, G.: J. Biochem.,72,1503–1516 (1972)

Enzyme Handbook © Springer-Verlag Berlin Heidelberg 1994
Duplication, reproduction and storage in data banks are only allowed with the prior permission of the publishers

1 NOMENCLATURE

EC number
1.5.1.6

Systematic name
10-Formyltetrahydrofolate:$NADP^+$ oxidoreductase

Recommended name
Formyltetrahydrofolate dehydrogenase

Synonymes
10-Formyl tetrahydrofolate:NADP oxidoreductase [1]
10-Formyl-H_2PtGlu:NADP oxidoreductase [1]
10-Formyl-H_4folate dehydrogenase [6]
N^{10}–Formyltetrahydrofolate dehydrogenase [7]
Dehydrogenase, formyltetrahydrofolate
10-Formyltetrahydrofolate dehydrogenase
Proteins, folate-binding cytosol I
FBP-CI proteins
Folate-binding proteins, cytosol I

CAS Reg. No.
37256-25-0

2 REACTION AND SPECIFICITY

Catalysed reaction
10-Formyltetrahydrofolate + $NADP^+$ + H_2O →
→ tetrahydrofolate + CO_2 + NADPH

Reaction type
Redox reaction

Natural substrates

Enzyme Handbook © Springer-Verlag Berlin Heidelberg 1994
Duplication, reproduction and storage in data banks are only allowed with the prior permission of the publishers

Substrate spectrum

1 10-Formyltetrahydrofolate + $NADP^+$ + H_2O (ir, equilibrium constant: 160000000 [1])

2 More (enzyme is specific for (-)-10-formyltetrahydrofolate and $NADP^+$ [1], not: (+)-10-formyltetrahydrofolate [1] and 5-formyltetrahydrofolate [1, 4], in absence of $NADP^+$ the enzyme catalyzes hydrolytic cleavage of 10-formyltetrahydrofolate to formate and tetrahydrofolate (15–30% the rate of oxidative reaction) [1], bifunctional enzyme: 10-formyltetrahydrofolate dehydrogenase separated from 10-formyltetrahydrofolate hydrolase activity [3]) [1–4]

Product spectrum

1 Tetrahydrofolate + CO_2 + NADPH

2 ?

Inhibitor(s)

Tetrahydrofolate [1, 3, 7]; p-Chloromercuribenzoate (reversed by 2-mercaptoethanol) [1]; Iodoacetamide (reversed by 2-mercaptoethanol) [1]; Hydroxylamine [1]; Tetrahydropteroylhexa-gamma-glutamate [5]

Cofactor(s)/prostethic group(s)

$NADP^+$ (specific for) [1, 4]; More (no effect: NADPH [4], NAD^+: less than 5% of activity with $NADP^+$ activity) [4]

Metal compounds/salts

Turnover number (min^{-1})

Specific activity (U/mg)

980 [1]; 305 [4]; 800 [6]; More [3, 7]

K_m-value (mM)

0.0035 ($NADP^+$) [1]; 0.0082 ((-)-10-formyltetrahydrofolate) [1]; 0.001 ($NADP^+$) [2]; 0.017 (10-formyltetrahydrofolate) [2]; 0.0004–0.0006 ($NADP^+$) [3]; 0.0004–0.00075 ((6R, S)-10-tetrahydrofolate) [3]; More [5–7]

pH-optimum

7.5–8.5 [4]; 7.8 [1]

pH-range

6.5–8.5 (more than 60% of maximal activity at pH 6.5 and 8.5) [1]

Temperature optimum (°C)

25 (assay at) [3]; 30 (assay at) [1]

Temperature range (°C)

3 ENZYME STRUCTURE

Molecular weight

320000 (pig, gel filtration) [1]
413000 (rat, gel filtration, sedimentation analysis) [7]

Subunits

? (x × 92500, pig, SDS-PAGE [3], x × 94000–96000, human, rat, SDS-PAGE [6]) [3, 6]
Tetramer (4 × 108000, rat, SDS-PAGE) [7]

Glycoprotein/Lipoprotein

–

4 ISOLATION/PREPARATION

Source organism

Pig [1, 3–5]; Chicken [1]; Rat [2, 5–7]; Human [6]; More (not found in microorganisms) [1]

Source tissue

Liver [1–5, 7]; Kidney [1]; Heart [1]; Spleen [1]

Localisation in source

Purification

Pig (partial [4], bifunctional enzyme: 10-formyltetrahydrofolate dehydrogenase-hydrolase [3]) [1, 3–5]; Rat [2, 5–7]; Human [6]

Crystallization

–

Cloned

–

Renaturated

–

5 STABILITY

pH

Temperature (°C)

Oxidation

Organic solvent

Enzyme Handbook © Springer-Verlag Berlin Heidelberg 1994
Duplication, reproduction and storage in data banks are only allowed with the prior permission of the publishers

General stability information

Storage

–20°C, dialyzed, stable for several weeks [3]; –70°C, 20% glycerol, $(NH_4)_2SO_4$ containing buffer [7]

6 CROSSREFERENCES TO STRUCTURE DATABANKS

PIR/MIPS code

A23709 (Rat)

Brookhaven code

7 LITERATURE REFERENCES

[1] Kutzbach, C., Stokstad, E.L.R.: Methods Enzymol.,18B,793–798 (1971) (Review)
[2] Case, G.L., Kaisaki, P.J., Steele, R.D.: J. Biol. Chem.,263,10204–10207 (1988)
[3] Rios-Orlandi, E.M., Zarkadas, C.G., MacKenzie, R.E.: Biochim. Biophys. Acta,871,24–35 (1986)
[4] Kutzbach, C., Stokstad, E.L.R.: Biochem. Biophys. Res. Commun.,30,111–117 (1968)
[5] Min, H., Shane, B., Stokstad, E.L.R.: Biochim. Biophys. Acta,967,348–353 (1988)
[6] Johlin, F.C., Swain, E., Smith, C., Tephly, T.R.: Mol. Pharmacol.,35,745–750 (1989)
[7] Scrutton, M.C., Beis, I.: Biochem. J.,177,833–846 (1979)

1 NOMENCLATURE

EC number
1.5.1.7

Systematic name
N^6–(L-1,3-Dicarboxypropyl)-L-lysine:NAD$^+$ oxidoreductase (L-lysine-forming)

Recommended name
Saccharopine dehydrogenase (NAD$^+$, L-lysine forming)

Synonymes
Lysine-2-oxoglutarate reductase
Dehydrogenase, saccharopine (nicotinamide adenine dinucleotide, lysine forming)
epsilon-N-(L-glutaryl-2)-L-lysine:NAD oxidoreductase (L-lysine forming) [3]
N^6–(Glutar-2-yl)-L-lysine:NAD oxidoreductase (L-lysine-forming) [14]

CAS Reg. No.
9073-96-5

2 REACTION AND SPECIFICITY

Catalysed reaction
N^6–(L-1,3-Dicarboxypropyl)-L-lysine + NAD$^+$ + H_2O →
→ L-lysine + 2-oxoglutarate + NADH (mechanism [5, 6, 13, 18])

Reaction type
Redox reaction
Reductive condensation [13]

Natural substrates
N^6–(L-1,3-Dicarboxypropyl)-L-lysine + NAD$^+$ + H_2O (involved in lysine biosynthesis [1, 3, 4], last step of alpha-aminoadipate pathway for lysine biosynthesis [1, 2, 6, 12]) [1–4, 6, 12]

Substrate spectrum
1 N^6–(L-1,3-Dicarboxypropyl)-L-lysine + NAD$^+$ + H_2O (i.e. saccharopine, r [3, 4, 18], high specificity with respect to coenzyme and substrate [3, 19])
2 More (A-stereospecific in hydrogen transfer in the synthesis of saccharopine from alpha-ketoglutarate + L-lysine) [10]

Enzyme Handbook © Springer-Verlag Berlin Heidelberg 1994
Duplication, reproduction and storage in data banks are only allowed with the prior permission of the publishers

Product spectrum

1 L-Lysine + 2-oxoglutarate + NADH (r [3, 4, 18], pyruvate can substitute for oxoglutarate in direction of reductive condensation [13])
2 ?

Inhibitor(s)

Mn^{2+} (lysine synthesis) [4]; Cu^{2+} (lysine synthesis) [4]; Zn^{2+} (lysine synthesis) [4]; Amino acids (e.g. L-ornithine, L-leucine, L-norleucine, L-isoleucine [4, 17], hydrophobic amino acids with 5 or 6 carbon atoms [4, 17] and hydroxylic amino acids [4] are potent inhibitors, not: dicarboxylic amino acids, aspartate, glutamate, 2-aminoadipate [17]) [4, 17]; 2,3-Butanedione (protection by L-leucine, NADH and alpha-ketoglutarate) [7]; p-Chloromercuribenzoate [11, 14]; Mersalyl [11]; 5,5'-Dithiobis(2-nitrobenzoate) [11]; o-Iodosobenzoate [11]; N-Butylmaleimide [11]; N-Ethylmaleimide [11]; Iodoacetamide [11]; Iodoacetate (slight [11], not [3], NAD^+ and NADH protect [14]) [11, 14]; alpha-Ketoglutarate (substrate inhibition) [15, 18]; Lysine (substrate inhibition) [15, 18]; Adenosine (slight) [1]; AMP [1]; ADP [1]; ATP [1]; NADH (substrate inhibition [18], not [15]) [18]; Sn^{2+} (lysine synthesis) [4]; Hg^{2+} (lysine synthesis) [4]; Pyridoxal 5'-phosphate (reversible inactivation) [6, 8]; Diethyldicarbonate [6, 9]; p-Hydroxymercuribenzoate (NAD^+ protects [3], inhibition counteracted by 2-mercaptoethanol or glutathione [19]) [3, 19]; More (no substrate inhibition by NADH, NAD^+ and saccharopine) [15]

Cofactor(s)/prostethic group(s)

NAD^+; NADH; More (not: $NADP^+$ [16], very low reaction rate with NADPH (5% of NADH reaction [19]) [16, 19]) [16, 19]

Metal compounds/salts

Turnover number (min^{-1})

1127 (NADH) [3]

Specific activity (U/mg)

106 [14]; 24.6 [18]; More [3, 19]

K_m-value (mM)

12 (L-lysine) [3, 19]; 1.6 (L-lysine) [4]; 1.34 (L-lysine) [12]; 0.44 (oxoglutarate) [3, 19]; 0.66 (oxoglutarate) [4]; 0.23 (NADH [4], oxoglutarate [12]) [4, 12]; 0.46 (NADH) [3, 19]; 0.120 (NADH) [12]; 0. 10 (NAD^+) [18]; 1.67 (saccharopine) [18]; More [1]

pH-optimum

6.8 (L-lysine + 2-oxoglutarate + NADH) [12]; 7 (L-lysine + 2-oxoglutarate + NADH) [3, 4, 19]; 7.1 (saccharopine + NAD^+ + H_2O) [1]; 10 (or higher, saccharopine + NAD^+ + H_2O) [3, 19]

pH-range

5.2–9.0 (half-maximal activities at pH 5.2 and 9.0, L-lysine + 2-oxoglutarate + NADH) [3]; 5.6–7.8 (half maximal activities at pH 5.6 and 7.8) [12]

Temperature optimum (°C)

25 (assay at) [3, 19]; 40 (increase of activity from 20°C to 40°C) [12]

Temperature range (°C)

20–40 (increase of activity from 20°C to 40°C) [12]

3 ENZYME STRUCTURE

Molecular weight

38000–40000 (Saccharomyces cerevisiae, gel filtration, sedimentation equilibrium centrifugation, SDS-PAGE) [14]

45000 (Candida maltosa, gel filtration [4], Pichia guilliermondii, gel filtration [1]) [1, 4]

49000 (Saccharomyces cerevisiae, sucrose gradient centrifugation) [3, 19]

Subunits

Monomer (1 × 39000, Saccharomyces cerevisiae, SDS-PAGE) [14]

Glycoprotein/Lipoprotein

–

4 ISOLATION/PREPARATION

Source organism

Pichia guilliermondii (yeast) [1]; Schizosaccharomyces pombe (fission yeast) [2]; Phycomyces blakesleeanus [12]; Saccharomyces cerevisiae (baker's yeast) [3, 5–9, 13–15, 18, 19]; Candida maltosa [4]

Source tissue

Localisation in source

Purification

Phycomyces blakesleeanus (partial) [12]; Saccharomyces cerevisiae [14, 18, 19]

Crystallization

–

Cloned

–

Renaturated

–

Enzyme Handbook © Springer-Verlag Berlin Heidelberg 1994
Duplication, reproduction and storage in data banks are only allowed with the prior permission of the publishers

5 STABILITY

pH

5–8 (unstable outside the range) [3]

Temperature (°C)

Oxidation

Organic solvent

General stability information

High salt concentration stabilizes [3, 19] above 0.1 M KCl [19]; 2-Mercaptoethanol, 1 mM, stabilizes [3, 19]; Bovine serum albumin stabilizes [14]; Unstable at enzyme concentration below 0.1 mg/ml [14]

Storage

–20°C, concentration above 0.1 mg protein/ml, pH 6.8, 1 mM EDTA, several months [14]

6 CROSSREFERENCES TO STRUCTURE DATABANKS

PIR/MIPS code

Brookhaven code

7 LITERATURE REFERENCES

[1] Schmidt, H., Bode, R., Birnbaum, D.: Antonie Leeuwenhoek,56,337–347 (1989)
[2] Ye, Z.-H., Bhattacharjee, J.K.: J. Bacteriol.,170,5958–5970 (1988)
[3] Saunders, P.P., Broquist, H.P.: J. Biol. Chem.,241,3435–3440 (1966)
[4] Schmidt, H., Bode, R., Lindner, M., Birnbaum, D.: J. Basic Microbiol.,25,675–681 (1985)
[5] Fujioka, M.: Arch. Biochem. Biophys.,230,553–559 (1984)
[6] Fujioka, M.: Biochem. Soc. Trans.,9,281–282 (1981)
[7] Fujioka, M., Takata, Y.: Biochemistry,20,468–472 (1981)
[8] Ogawa, H., Fujioka, M.: J. Biol. Chem.,255,7420–7425 (1980)
[9] Fujioka, M., Takata, Y., Ogawa, H., Okamoto, M.: J. Biol. Chem.,255,937–942 (1980)
[10] Fujioka, M., Takata, Y.: Biochim. Biophys. Acta,570,210–212 (1979)
[11] Ogawa, H., Okamoto, M., Fujioka, M.: J. Biol. Chem.,254,7030–7035 (1979)
[12] Hanke, R., Hilgenberg, W.: Z. Pflanzenphysiol.,92,23–31 (1979)
[13] Sugimoto, K., Fujioka, M.: Eur. J. Biochem.,90,301–307 (1978)
[14] Ogawa, H., Fujioka, M.: J. Biol. Chem.,253,3666–3670 (1978)
[15] Fujioka, M.: J. Biol. Chem.,250,8986–8989 (1975)
[16] Fujioka, M., Nakatani, Y.: J. Biol. Chem.,249,6886–6891 (1974)
[17] Fujioka, M., Nakatani, Y.: Eur. J. Biochem.,25,301–307 (1972)
[18] Fujioka, M., Nakatani, Y.: Eur. J. Biochem.,16,180–186 (1970)
[19] Broquist, H.P.: Methods Enzymol.,17B,124–129 (1971)

1 NOMENCLATURE

EC number
1.5.1.8

Systematic name
N^6–(L-1,3-Dicarboxypropyl)-L-lysine:NADP$^+$ oxidoreductase (L-lysine forming)

Recommended name
Saccharopine dehydrogenase (NADP$^+$, L-lysine forming)

Synonymes
Lysine-2-oxoglutarate reductase
Lysine-ketoglutarate reductase [1]
L-Lysine-alpha-ketoglutarate reductase [4]
Lysine:alpha-ketoglutarate:TPNH oxidoreductase (epsilon-N-[gultaryl-2]-L-lysine forming) [7]
Dehydrogenase, saccharopine (nicotinamide adenine dinucleotide phosphate, lysine-forming)

CAS Reg. No.
9031-19-0

---

2 REACTION AND SPECIFICITY

Catalysed reaction
L-Lysine + 2-oxoglutarate + NADPH →
→ N^6–(L-1,3-dicarboxypropyl)-L-lysine + NADP$^+$ + H_2O (mechanism [5])

Reaction type
Redox reaction

Natural substrates
L-Lysine + 2-oxoglutarate + NADPH (enzyme of lysine catabolism [1, 3, 8], enzyme associated with lysine breakdown in the endosperm during seed development [3]) [1, 3, 8]

Substrate spectrum
1 L-Lysine + 2-oxoglutarate + NADPH (r [1, 4], reverse reaction not detected [6], lysine formation is only 3–5% of the saccharopine formation [4], very strict substrate specificity for L-lysine, alpha-ketoglutarate and NADPH [4], reacts effectively with delta-hydroxylysine [6])

Enzyme Handbook © Springer-Verlag Berlin Heidelberg 1994
Duplication, reproduction and storage in data banks are only allowed with the prior permission of the publishers

Product spectrum

1 N^6–(L-1,3-Dicarboxypropyl)-L-lysine + $NADP^+$ + H_2O (r [1, 4], lysine formation is only 3–5% of the saccharopine formation [4]) [1, 4]

Inhibitor(s)

L-Glutamic acid [6]; L-Homocitrulline [6]; DL-Pipecolic acid [6]; Cadaverine [6]; L-Lysylglycine [6]; Carbamylphosphate [6]; Hydroxylamine [6]; $CaCl_2$ [6]; $MgSO_4$ (slight) [6]; $MnCl_2$ [6]; L-Ornithine [4, 6]; Leucine [4, 6]; Tryptophan [4]; Saccharopine (product inhibition [6]) [5, 6]; Detergents (e.g. aerosol 22, Ultrawet 60L, deoxycholic acid) [5]; L-Lysine-p-nitroanilide [5]; S-2-Aminoethyl-L-cysteine [5]; $(NH_4)_2SO_4$ (not [5]) [7]

Cofactor(s)/prostethic group(s)

NADPH [1–10]; $NADP^+$ [1, 4]; More (not: NADH) [2, 7]

Metal compounds/salts

EDTA (slight activation, effect not consistant) [6]

Turnover number (min^{-1})

Specific activity (U/mg)

More (HPLC lysine-ketoglutarate reductase assay [1]) [1, 5, 8]; 17.1 [9]; 28.3 [4]; 40.1 [6]

K_m-value (mM)

1.5 (L-lysine) [6]; 1 (alpha-ketoglutarate) [6]; 0.08 (NADPH) [6]; 5.2 (L-lysine) [2]; 1.8 (alpha-ketoglutarate) [2]; 2.2 (L-lysine) [4]; 1.4 (alpha-ketoglutarate) [4]

pH-optimum

6.7–7.0 [7]; 7 [2]; 7.8 [6]

pH-range

7.2–8.8 (about 50% of activity maximum at pH 7.2 and 8.8) [6]; 6.2–8.6 (6.2: about 50% of activity maximum, 8.6: about 60% of activity maximum) [7]

Temperature optimum (°C)

30 [2]

Temperature range (°C)

3 ENZYME STRUCTURE

Molecular weight

230000 (rat, gel filtration, enzyme clearly separated from saccharopine dehydrogenase (E.C. 1.5.1.9)) [4]
467000–468000 (bovine, baboon, gel filtration, a single protein catalyzes both lysine-ketoglutarate reductase (E.C. 1.5.1.8) and saccharopine dehydrogenase activity (E.C. 1.5.1.9) [9], reductase and dehydrogenase domains are separately folded and functionally independent of each other [10], bifunctional enzyme (aminoadipic semialdehyde synthase)) [9, 10]
480000 (human, gel filtration, no separation of L-lysine-alpha-ketoglutarate reductase (E.C. 1.5.1.8) and saccharopine dehydrogenase (E.C. 1.5.1.9)) [5]

Subunits

Tetramer (4 × 52000, rat, SDS-PAGE [4], enzyme clearly separated from saccharopine dehydrogenase E.C. 1.5.1.9, 4 × 115000, bovine, baboon, SDS-PAGE [9]) [4, 9]

Glycoprotein/Lipoprotein

–

4 ISOLATION/PREPARATION

Source organism

Maize (Zea mays L.) [2, 3]; Rat [4, 8]; Human [5–8]; Baboon [9]; Pig [8]; Dog [8]; Cat [8]; Bovine [8–10]; Sheep [8]

Source tissue

Placenta [5]; Heart [6, 7]; Kidney (low activity [6]) [6, 7]; Skin (low activity) [6, 7]; Endosperm [2, 3]; Liver [4, 6–10]

Localisation in source

Mitochondria [4, 8, 9]

Purification

Rat (enzyme clearly separated from saccharopine dehydrogenase E.C. 1.5.1.9) [4]; Human (partial [6, 7], no separation of L-lysine-alpha-ketoglutarate reductase (E.C. 1.5.1.8) and saccharopine dehydrogenase (E.C. 1.5.1.9) [5]) [5–7]

Crystallization

–

Cloned

–

Renaturated

–

Enzyme Handbook © Springer-Verlag Berlin Heidelberg 1994
Duplication, reproduction and storage in data banks are only allowed with the prior permission of the publishers

5 STABILITY

pH

4.4 (10 min, 50% loss of activity) [6]; 4.6–4.9 (10 min) [6]; 7.5 (4°C, highest stability) [5]

Temperature (°C)

40 (10 min, unstable above) [6]

Oxidation

Organic solvent

General stability information

EDTA stabilizes during purification [5]

Storage

–20°C, 2 months [2]; –20°C, 1 month, 50% loss of activity [6]; 4°C, pH 7.5, 5 mM beta-mercaptoethanol, 0.1 mM EDTA [5]; –25°C, months [7]

6 CROSSREFERENCES TO STRUCTURE DATABANKS

PIR/MIPS code

Brookhaven code

7 LITERATURE REFERENCES

[1] Davis, A.T.: J. Chromatogr.,497,263–267 (1989)
[2] Arruda, P., Sodek, L., Da Silva, W.J.: Plant Physiol.,69,988–989 (1982)
[3] Arruda, P., Da Silva, W.J.: Phytochemistry,22,2687–2689 (1983)
[4] Noda, C., Ichihara, A.: Biochim. Biophys. Acta,525,307–313 (1978)
[5] Fjellstedt, T.A., Robinson, J.C.: Arch. Biochem. Biophys.,168,536–548 (1975)
[6] Hutzler, J., Dancis, J.: Biochim. Biophys. Acta,377,42–51 (1975)
[7] Hutzler, J., Dancis, J.: Biochim. Biophys. Acta,158,62–69 (1968)
[8] Fellows, F.C.I., Lewis, M.H.R.: Biochem. J.,136,329–334 (1973)
[9] Markovitz, P.J., Chuang, D.T., Cox, R.P.: J. Biol. Chem.,259,11643–11646 (1984)
[10] Markovitz, P.J., Chuang, D.T.: J. Biol. Chem.,262,9353–9358 (1987)

1 NOMENCLATURE

EC number

1.5.1.9

Systematic name

N^6–(L-1,3-Dicarboxypropyl)-L-lysine:NAD$^+$ oxidoreductase (L-glutamate-forming)

Recommended name

Saccharopine dehydrogenase (NAD$^+$, L-glutamate-forming)

Synonymes

Dehydrogenase, saccharopine (nicotinamide adenine dinucleotide, glutamate-forming)
Saccharopin dehydrogenase
NAD$^+$ oxidoreductase (L-2-aminoadipic-delta-semialdehyde and glutamate forming) [5]
Aminoadipic semialdehyde synthase (lysine-ketoglutarate reductase (E.C. 1. 5.1.8) and saccharopine dehydrogenase activity (E.C. 1.5.1.9) occur on a single protein, called aminoadipic semialdehyde synthase) [1, 7]

CAS Reg. No.

37256-26-1

2 REACTION AND SPECIFICITY

Catalysed reaction

N^6–(L-1,3-Dicarboxypropyl)-L-lysine + NAD$^+$ + H_2O →
→ L-glutamate + 2-aminoadipate 6-semialdehyde

Reaction type

Redox reaction

Natural substrates

N^6–(L-1,3-Dicarboxypropyl)-L-lysine + NAD$^+$ + H_2O (lysine degradation) [2–4, 6]

Substrate spectrum

1 N^6–(L-1,3-Dicarboxypropyl)-L-lysine + NAD$^+$ + H_2O (N^6–(L-1,3-dicarboxypropyl)-L-lysine is identical with saccharopine, high specificity [5]) [2–5]

Enzyme Handbook © Springer-Verlag Berlin Heidelberg 1994
Duplication, reproduction and storage in data banks are only allowed with the prior permission of the publishers

Product spectrum

1 L-Glutamate + 2-aminoadipate 6-semialdehyde + NADH

Inhibitor(s)

NH_4Cl (slight) [5]; Hydroxylamine (slight) [5]; Sodium bisulfite (slight) [5]; L-Glutamate [2]; p-Hydroxybenzoate [5]; $HgCl_2$ [5]; $CuSO_4$ [5]; $ZnCl_2$ [5]; More (not: L-lysine-p-nitroanilide) [2]

Cofactor(s)/prostethic group(s)

NAD^+; $NADP^+$ (9% of the rate with NAD^+) [2]

Metal compounds/salts

Turnover number (min^{-1})

Specific activity (U/mg)

More [5]

K_m-value (mM)

1.15 (saccharopine) [2]; 0.5 (saccharopine) [5]; 0.0645 (NAD^+) [2]; 0.4 (NAD^+) [5]

pH-optimum

8.5 (Tris-buffer) [2]; 8.8–9.0 [5]; 8.9 (2-amino-2-methyl-1,3-propanediol buffer) [2]

pH-range

7.9–9.3 (half maximal activities at pH 7.9 and 9.3) [2]; 7.8–9.7 (half maximal activities at pH 7.8 and 9.7) [5]

Temperature optimum (°C)

30 [5]

Temperature range (°C)

20–37 (20°C: 35% of activity maximum, 37°C: 85% of activity maximum) [5]

---

3 ENZYME STRUCTURE

Molecular weight

467000–468000 (bovine, baboon, gel filtration, lysine-ketoglutarate reductase (E.C. 1.5.1.8) and saccharopine dehydrogenase activity (E.C. 1.5.1.9) occur on a single protein, called aminoadipic semialdehyde synthase [7]

480000 (human, gel filtration) [2]

Subunits

Tetramer (4 × 115000, bovine, baboon, SDS-PAGE, lysine-ketoglutarate reductase (E.C. 1.5.1.8) and saccharopine dehydrogenase activity (E.C. 1.5.1.9) occur on a single protein, called aminoadipic semialdehyde synthase) [7]

Glycoprotein/Lipoprotein

–

4 ISOLATION/PREPARATION

Source organism

Human (not separable from L-lysine-alpha-ketoglutarate reductase [2]) [2–5]; Rat [4, 6]; Pig [4]; Dog [4]; Cat [4]; Bovine (lysine-ketoglutarate reductase (E.C. 1.5.1.8) and saccharopine dehydrogenase activity (E.C. 1.5.1.9) occur on a single protein, called aminoadipic semialdehyde synthase [1, 7]) [1, 4, 6, 7]; Sheep [4]; Baboon (lysine-ketoglutarate reductase (E.C. 1.5.1.8) and saccharopine dehydrogenase activity (E.C. 1.5.1.9) occur on a single protein, called aminoadipic semialdehyde synthase) [7]

Source tissue

Placenta [2]; Liver [1, 3–5, 7]; Kidney [3, 6]; More (no activity in: human heart, muscle, skin, brain, spleen, pancreas, adrenal, lung, gut, leukocytes and erythrocytes) [3]

Localisation in source

Mitochondria [4, 7]

Purification

Human [5]; Baboon (lysine-ketoglutarate reductase (E.C. 1.5.1.8) and saccharopine dehydrogenase activity (E.C. 1.5.1.9) occur on a single protein, called aminoadipic semialdehyde synthase) [7]

Crystallization

–

Cloned

–

Renaturated

–

Enzyme Handbook © Springer-Verlag Berlin Heidelberg 1994
Duplication, reproduction and storage in data banks are only allowed with the prior permission of the publishers

5 STABILITY

pH

4.4 (10 min, 60% activity) [3]; 4.6–4.9 (10 min) [3]; 6.6–9.9 (10 min) [5]; 8.5 (unstable at high pH, beta-mercaptoethanol protects) [2]

Temperature (°C)

37 (5 min) [5]; 44 (5 min, 50% activity) [5]

Oxidation

Organic solvent

General stability information

Storage

–20°C, crude enzyme stable for several months, purified enzyme inactivated within 3 weeks [5]; 2°C, 5–10% loss of activity per week, purified enzyme [5]

6 CROSSREFERENCES TO STRUCTURE DATABANKS

PIR/MIPS code

Brookhaven code

7 LITERATURE REFERENCES

[1] Markovitz, P.J., Chuang, D.T.: J. Biol. Chem.,262,9353–9358 (1987)
[2] Fjellstedt, T.A., Robinson, J.C.: Arch. Biochem. Biophys.,171,191–196 (1975)
[3] Hutzler, J., Dancis, J.: Biochim. Biophys. Acta,377,42–51 (1975)
[4] Fellows, F.C.I., Lewis, M.H.R.: Biochem. J.,136,329–334 (1973)
[5] Hutzler, J., Dancis, J.: Biochim. Biophys. Acta,206,205–214 (1970)
[6] Mukhopadhyay, A., Mungre, S.M., Desmukh, D.R.: Experientia,46,874–876 (1990)
[7] Markovitz, P.J., Chuang, D.T., Cox, R.P.: J. Biol. Chem.,259,11643–11646 (1984)

1 NOMENCLATURE

EC number
1.5.1.10

Systematic name
N^6–(L-1,3-Dicarboxypropyl)-L-lysine:NADP$^+$ oxidoreductase (L-glutamate-forming)

Recommended name
Saccharopine dehydrogenase (NADP$^+$, L-glutamate forming)

Synonymes
Dehydrogenase, saccharopine (nicotinamide adenine dinucleotide phosphate, glutamate-forming)
Aminoadipic semialdehyde-glutamic reductase
Aminoadipate semialdehyde-glutamate reductase
Aminoadipic semialdehyde-glutamate reductase [2]
epsilon-N-(L-Glutaryl-2)-L-lysine:NAD$^+$(P) oxidoreductase (L-2-aminoadipate-semialdehyde forming)
Saccharopine reductase [7]

CAS Reg. No.
9033-55-0

---

2 REACTION AND SPECIFICITY

Catalysed reaction
N^6–(L-1,3-Dicarboxypropyl)-L-lysine + NADP$^+$ + H_2O →
→ L-glutamate + 2-aminoadipate 6-semialdehyde + NADPH

Reaction type
Redox reaction

Natural substrates
2-Aminoadipate 6-semialdehyde + L-glutamate + NADPH (lysine biosynthesis) [3–5, 7]

Substrate spectrum
1 2-Aminoadipate 6-semialdehyde + L-glutamate + NADPH (r [2, 3, 5])

Product spectrum
1 N^6–(L-1,3-Dicarboxypropyl)-L-lysine + NADP$^+$ + H_2O (r [2, 3, 5], N^6–(L-1,3-dicarboxypropyl)-L-lysine is identical with saccharopine)

Enzyme Handbook © Springer-Verlag Berlin Heidelberg 1994
Duplication, reproduction and storage in data banks are only allowed with the prior permission of the publishers

Inhibitor(s)

$HgCl_2$ [5]; 2,2-Bipyridine [5]; p-Hydroxymercuribenzoate [2]; 1,10-Phenanthroline [5]; L-Tryptophan [7]; L-Leucine [7]; More (not: carbonyl reagents) [5]

Cofactor(s)/prostethic group(s)

NADPH (far more effective than NADH in saccharopine formation) [2]; NADH (NADPH far more effective than NADH in saccaropine formation) [2]; NAD$^+$ (NAD$^+$ and NADP$^+$ equally effective in 2-aminoadipate 6-semialdehyde formation) [2]; NADP$^+$ (NAD$^+$ and NADP$^+$ equallly effective in 2-aminoadipate 6-semialdehyde formation) [2]

Metal compounds/salts

Turnover number (min^{-1})

Specific activity (U/mg)

269.39 [5]; More [2, 3]

K_m-value (mM)

0.92 (saccharopine) [2, 3]; 0.22 (NADP$^+$) [2, 3]; 2.32 (L-saccharopine) [5]; 0.054 (NAD$^+$) [5]; 1.25 (saccharopine) [7]; 0.17 (NADP$^+$) [7]

pH-optimum

7 (saccharopine formation) [2, 3]; 8.8 (saccharopine degradation) [7]; 9.5 (saccharopine degradation) [5]; 10 (saccharopine degradation) [2, 3]

pH-range

5.5–7.8 (5.5: about 45% of activity maximum, 7.8: about 35% of activity maximum) [3]; 8.3–10.3 (about 50% of activity maximum at pH 8.3 and 10.3) [3]; 9–10 (about 50% of activity maximum at pH 9 and 10) [5]

Temperature optimum (°C)

25 (assay at) [2]

Temperature range (°C)

---

3 ENZYME STRUCTURE

Molecular weight

69000 (Saccharomyces cerevisiae, gel filtration) [5]
73000 (Saccharomyces cerevisiae, density gradient centrifugation) [2]

Subunits

Monomer (1 × 50000, Saccharomyces cerevisiae, SDS-PAGE) [5]

Glycoprotein/Lipoprotein

No carbohydrate [5]

4 ISOLATION/PREPARATION

Source organism

Rat (low activity) [1]; Bovine (low activity) [1]; Saccharomyces cerevisiae (baker's yeast) [2, 3, 5]; Schizosaccharomyces pombe (fission yeast) [4]; Rhodotorula glutinis [6]; Pichia guilliermondii [7]

Source tissue

Kidney [1]; Cell [2]

Localisation in source

Purification

Saccharomyces cerevisiae (baker's yeast) [2, 3, 5]

Crystallization

–

Cloned

–

Renaturated

–

5 STABILITY

pH

Temperature (°C)

34 (denaturation above) [5]

Oxidation

Organic solvent

General stability information

Storage

–70°C, pH 8.0, 10 mM 2-mercaptoethanol, 5 mM EDTA [5]

Enzyme Handbook © Springer-Verlag Berlin Heidelberg 1994
Duplication, reproduction and storage in data banks are only allowed with the prior permission of the publishers

6 CROSSREFERENCES TO STRUCTURE DATABANKS

PIR/MIPS code

Brookhaven code

7 LITERATURE REFERENCES

[1] Mukhopadhyay, A., Mungre, S.M., Desmukh, D.R.: Experientia,46,874–876 (1990)
[2] Broquist, H.P.: Methods Enzymol. (Ed. Colowick, S.P.) 17B,121–124 (1971) (Review)
[3] Jones, E.E., Broquist, H.P.: J. Biol. Chem.,241,3430–3434 (1966)
[4] Ye, Z.-H., Bhattacharjee, J.K.: J. Bacteriol.,170,5968–5970 (1988)
[5] Storts, D.R., Bhattacharjee, J.K.: J. Bacteriol.,169,416–418 (1987)
[6] Kinzel, J.J., Bhattacharjee, J.K.: J. Bacteriol.,138,410–417 (1979)
[7] Schmidt, H., Bode, R., Birnbaum, D.: Antonie Leeuwenhoek,56,337–347 (1989)

1 NOMENCLATURE

EC number
1.5.1.11

Systematic name
N^2–(D-1-Carboxyethyl)-L-arginine:NAD^+ oxidoreductase (L-arginine forming)

Recommended name
D-Octopine dehydrogenase

Synonymes
D-Octopine synthase
Octopine dehydrogenase
Dehydrogenase, octopine
Octopine:NAD oxidoreductase
ODH [28]

CAS Reg. No.
37256-27-2

2 REACTION AND SPECIFICITY

Catalysed reaction
L-Arginine + pyruvate + NADH →
→ N^2–(D-1-carboxyethyl)-L-arginine + NAD^+ + H_2O (octopine synthesis: partially random mechanism with NADH as the obligatory first substrate [2], octopine oxidation: ordered sequential addition of NAD^+ and octopine to the enzyme [2, 18], bi-ter sequential mechanism where NAD^+ binds first to the enzyme followed by D-octopine, the products are released in the order: L-arginine, pyruvate and NADH [20], enzyme removes pro-S-hydrogen of the dehydronicotinamide ring [8], kinetic mechanism [9], stereospecificity of hydrogen transfer [17])

Reaction type
Redox reaction

Enzyme Handbook © Springer-Verlag Berlin Heidelberg 1994
Duplication, reproduction and storage in data banks are only allowed with the prior permission of the publishers

Natural substrates

L-Arginine + pyruvate + NADH (physiological role in glycolytic energy production [7], mantle muscle enzyme: rapid synthesis of octopine under conditions of muscular work, terminal enzyme of anaerobic glycolysis [21], enyzme occupies the role of lactate dehydrogenase during short-time anaerobiosis in normally aerobic working muscles in Pecten and some cephalopods [26], Loligo vulgaris: octopine is produced in the mantle via the reaction of the muscle isoenzyme and is subsequently flushed out into the blood and transported to other tissues such as the optic lobe, for re-oxidation via the optic lobe enzyme [29]) [7, 21, 26, 29]

N^2–(D-1-Carboxyethyl(-L-arginine + NAD^+ (brain enzyme: oxidation of octopine [21]) [21, 29]

Substrate spectrum

1 L-Arginine + pyruvate + NADH (r [1, 20], in physiological range (pH 7.5) the quotient of octopine synthesis versus oxidation is 13 [1])
2 L-Homoarginine + pyruvate + NADH [1, 6, 28]
3 L-Canavaline + pyruvate + NADH [1, 6, 28]
4 L-Histidine + pyruvate + NADH [6, 27]
5 L-Citrulline + pyruvate + NADH [6, 27]
6 L-Methionine + pyruvate + NADH [6, 27]
7 L-Lysine + pyruvate + NADH [6, 27, 28]
8 L-Ornithine + pyruvate + NADH [6, 27]
9 L-Cysteine + pyruvate + NADH [6]
10 L-Serine + pyruvate + NADH [6]
11 L-Glutamine + pyruvate + NADH [6, 27]
12 More (only amino acids which contain guanidinium group react to appreciable extent [28], no activity with L-alanine, glycine, L-threonine [1], keto acids which can replace pyruvate (with L-arginine as cosubstrate): alpha-ketovalerate (low activity) [1], alpha-ketocaproate (low activity) [1], oxaloacetate [7], alpha-ketobutyrate [7], alpha-keto acids with short (0–3 carbon) alkyl side chain [6], specificity for keto acids is limited to pyruvate [28], overview: amino acid, keto acid and imino acid substrate specificity of marine invertebrate enzymes [7]) [1, 6, 7, 28]

Product spectrum

1 N^2–(D-1-Carboxyethyl)-L-arginine (i.e. octopine) + NAD^+ + H_2O (r [1, 20])
2 N^2–(D-1-Carboxyethyl)-L-homoarginine + NAD^+ + H_2O
3 N^2–(D-1-Carboxyethyl)-L-canavaline + NAD^+ + H_2O
4 N^2–(D-1-Carboxyethyl)-L-histidine + NAD^+ + H_2O
5 N^2–(D-1-Carboxyethyl)-L-citrulline + NAD^+ + H_2O
6 N^2–(D-1-Carboxyethyl)-L-methionine + NAD^+ + H_2O
7 N^2–(D-1-Carboxyethyl)-L-lysine + NAD^+ + H_2O
8 N^2–(D-1-Carboxyethyl)-L-ornithine + NAD^+ + H_2O
9 N^2–(D-1-Carboxyethyl)-L-cysteine + NAD^+ + H_2O
10 N^2–(D-1-Carboxyethyl)-L-serine + NAD^+ + H_2O
11 N^2–(D-1-Carboxyethyl)-L-glutamine + NAD^+ + H_2O
12 ?

Inhibitor(s)

Octopine (noncompetitive [4], substrate inhibition [1, 4, 21, 29, 31] above 5 mM [31], product inhibition [20, 21, 28, 29, 31]) [1, 4, 20, 21, 28, 29, 31]; L-Arginine (noncompetitive [4], substrate inhibition [1, 4], product inhibition [20]) [1, 4, 20, 27, 29]; NADH (noncompetitive [4], substrate inhibition [1, 4, 20]) [1, 4, 20]; Oxamate [6]; Methyl octopine [6]; Diethyldicarbonate (hydroxylamine reverses inhibition) [13]; Carbodiimide [16]; NAD^+ (not [1], competitive [6], product inhibition [20, 31]) [1, 6, 20, 31]; Pyruvate (substrate inhibition [21]) [21, 27, 29]; epsilon-Aminocaproate [27]; $NADP^+$ (uncompetitive with respect to pyruvate) [27]; Homoarginine (octopine production) [28]; Glyoxylate (octopine production) [28]; beta-Hydroxypyruvate (octopine production) [28]; D-Strombine (pyruvate reduction) [31]; MS-Alanopine (pyruvate reduction) [12]; More (product and dead-end inhibition patterns [1], inhibitors of octopine formation: substrate analogs with a guanidyl (guanidinobutane) or carboxyl group (valeric acid) or both of them (delta-guanidinovaleric acid) [22], inhibitors of dehydrogenation reaction: compounds with only one of both groups (guanidyl or carboxyl group [22])) [1, 22]

Cofactor(s)/prostethic group(s)

NAD^+ [1–30]; NADH [1–30]; NADPH (in Vinca W1 tumor extracts 2 enzymes are present: one dependent on NAD^+ the second on $NADP^+$ [25], enzyme can use NADH as well as NADPH [27], activity negligible [28]) [25, 27]; More (influence of ligands on coenzyme dissociation constants [15], temperature dependency of coenzyme dissociation constant [19], brain isoenzyme: increased activity with hypoxanthine derivative of NADH [21]) [15, 19, 21]

Metal compounds/salts

No metal cofactor (as e.g. Fe^{2+}, Mn^{2+}, Zn^{2+}) [22]

Enzyme Handbook © Springer-Verlag Berlin Heidelberg 1994
Duplication, reproduction and storage in data banks are only allowed with the prior permission of the publishers

Turnover number (min^{-1})
500 (NAD^+ + octopine) [20]; 2300 (NADH + L-arginine + pyruvate) [20]

Specific activity (U/mg)
750 [5]; 5.0–102.4 (enzyme from various marine invertebrates) [7]; 500 [26]; 25.9 [27]; 780 [31]; 45.56 [28]; More [6, 22, 23]

K_m-value (mM)
More (temperature dependency of K_m (substrate, coenzyme) [19], pH-dependency of K_m [13], Strombidae, overview [31]) [5, 6, 13, 19–23, 26–28, 31]; 1.4–2.8 (arginine (+ NADH + pyruvate (0.010–5.0 mM))) [1]; 0.5 (pyruvate (+ NADH + 5.0 mM arginine)) [1]; 1.5 (pyruvate (+ NADH + 1.0 mM arginine)) [1]; 0.028 (NADH (+ pyruvate + arginine)) [1]; 0.22 (octopine) [1]; 0.014 (NAD^+) [1]; 0.044–0.055 (NADH (+ pyruvate + L-arginine)) [2]; 1.05–1.66 (pyruvate (+ NADH + L-arginine)) [2]; 1.43–2.08 (arginine (+ NADH+ + pyruvate)) [2]; 0.166 (NAD^+) [2]; 0.05 (NADH) [5]; 0.90 (octopine) [2]; 0.21 (NAD^+) [5]; 1.05 (L-arginine) [5]; 1.18 (pyruvate) [5]; 2.0 (D-octopine) [5]

pH-optimum
5.2 (ornithine (or lysine) + pyruvate + NADH) [6]; 6.0 (arginine + pyruvate + NADH) [6]; 6.5 (octopine formation) [23]; 6.6 (histidine + pyruvate + NADH) [6]; 6.7 (octopine formation) [5]; 6.8 (octopine formation) [26]; 7. 0 (Cerebratulus lacteus, octopine production [5], pyruvate reduction [31]) [5, 31]; 8.7 (octopine oxidation) [26]; 9.2 (octopine oxidation) [6]; 9.4 (pyruvate formation) [31]; 9. 5 (octopine oxidation) [5]; 9.7 (octopine formation) [23]

pH-range
6–8 (pH 6 and 8: about 70% of activity maximum, octopine production) [1]; 7.0–9.0 (steady increase from pH 7.0 to 9.0, in direction of octopine synthesis) [1]

Temperature optimum (°C)
23 (assay at) [7, 27]; 25 (assay at) [1]

Temperature range (°C)

---

3 ENZYME STRUCTURE

Molecular weight
37000 (Cardium edule, gel filtration) [26]
38000 (crown gall [6], Pecten maximus, amino acid composition [24]) [6, 24]
39000 (Cerebratulus lacteus, ultracentrifugation [1], crown gall tumor tissue induced by Agrobacterium tumefaciens Conn strain B6 on sunflower (Helianthus annuus), gel filtration [27], Strombus luhuanus, gel filtration [31]) [1, 27, 31]

40000 (Metridium senile, gel filtration) [28]
43000 (Pecten maximus, gel filtration, SDS-PAGE) [23]
45000 (Pecten jacobeus, sedimentation equilibrium, SDS-PAGE, gel filtration) [5]

Subunits

Monomer (1 × 35000, Metridium senile, SDS-PAGE [28], 1 × 38000, crown gall tumor tissue induced by Agrobacterium tumefaciens Conn strain B6 on sunflower (Helianthus annuus), SDS-PAGE [27], 1 × 41000, Cerebratulus lacteus, SDS-PAGE [1], 1 × 46000, Pecten maximus, SDS-PAGE [24]) [1, 24, 27, 28]

Glycoprotein/Lipoprotein

–

4 ISOLATION/PREPARATION

Source organism

Cerebratulus lacteus (marine nemertean, 2 isoenzymes) [1]; Pecten jacobeus [1, 3, 5, 29]; Concholepas concholepas (sea mollusc) [1]; Pecten maximus (great scallop, shell fish, 2 isoenzymes [4], 2 forms (A and B) from striated adductor muscle which differ only in charge and sensibility to various chemical treatments [23]) [4, 8–10, 13, 14, 16, 18–20, 22–24]; Vinca rosea (crown gall tumor tissue induced by Agrobacterium tumefaciens strain W1) [6, 25]; Helianthus annuus (crown gall tumor tissue induced by Agrobacterium tumefaciens Conn strain B6) [27]; Calliactes parasitica (sea annemone) [7]; Mytilus edulis (bivalve) [7, 9]; Cerastoderma edule (bivalve) [7]; Glycymeris glycymeris (bivalve) [7]; Sephia officinalis (cuttlefish, cephalopod, kinetically distinct tissue-specific isozymic forms from mantle muscle (M4) and brain (H4)) [21]; Arctica islandica (bivalve) [7]; Patella aspera [11]; Monodonta lineata [11]; Gibulla umbilicalis [11]; Buccinum undulatum (gastropod) [11, 29]; Nucella lapillus [11]; Cardium edule (cockle, bivalve) [26]; Metridium senile (plumose sea anemone, 2 allozymic forms distributed in a highly population-specific manner: ODH 100 (northern population), ODH 103 (southern population)) [28]; Cardium tuberculatum (bivalve) [29]; Loligo vulgaris (mollusc, optic lobe and mantle tissue enzyme differ in elution behavior, kinetic data and inhibition pattern) [29]; Strombus luhuanus [31]; Strombus gibberulus [31]; Lambis lambis [31]; More (overview: enzyme present in prosobranchs, absent from opisthobranchs and pulmonates [11], Strombidae [31]) [11, 31]

Enzyme Handbook © Springer-Verlag Berlin Heidelberg 1994
Duplication, reproduction and storage in data banks are only allowed with the prior permission of the publishers

Source tissue

Adductor muscle [1, 5, 7, 23, 26, 29]; Muscle [2, 13, 16, 18, 22]; Crown gall tumor tissue (induced by Agrobacterium tumefaciens strain W1 on Vinca rosea [6, 25], induced by Agrobacterium tumefaciens Conn strain B6 on sunflower (Helianthus annuus) [27]) [6, 8, 25, 27]; Mantle muscle [7, 11, 21, 29]; Brain [7, 21]; Foot muscle [7, 11, 29]; Pedal muscle [7]; Side wall muscle [7]; Retractor muscle [11, 31]; Optic lobe [29]; Digestive gland [29]; Gill [29]

Localisation in source

Purification

Cerebratulus lacteus (marine nemertean, 2 isoenzymes) [1]; Pecten jacobeus (2 isoenzymes) [5]; Vinca rosea (crown gall tumor tissue induced by Agrobacterium tumefaciens strain W1) [6]; Calliactes parasitica [7]; Mytilus edulis [7]; Sepia officinalis [7]; Cerastoderma edule [7]; Glycymeris glycymeris [7]; Arctica islandica [7]; Pecten maximus [22, 23]; Cardium edule [26]; Helianthus annuus (crown gall tumor tissue induced by Agrobacterium tumefaciens Conn strain B6) [27]; Strombus luhuanus [31]; Metridium senile [28]; More (rapid purification methods) [12]

Crystallization

[10]

Cloned

More (octopine synthase located in a central portion of the T-DNA from Ti plasmid of Agrobacterium tumefaciens and expresssed after it has been transferred to the plant cells, octopin synthase mRNA isolated from sunflower crown gall callus) [30]

Renaturated

(folding pathway of completely unfolded enzyme [3], refolding kinetics [5]) [3, 5]

5 STABILITY

pH

6.5–8.5 (4°C, relatively stable in the range, high loss of activity within 24 h outside the range) [6]

Temperature (°C)

41 ($t_{1/2}$: 116 min [5], 40 min [23]) [5, 23]

Oxidation

Photooxidation with rose bengal [14]

Organic solvent

General stability information

Glycerol stabilizes [27]; NADPH stabilizes [27]; 2-Mercaptoethanol stabilizes [27]

Storage

4°C, ammonium sulfate, several weeks, little loss of activity [23]; –30°C, phosphoric acid-imidazole buffer, stable for many weeks [27]

6 CROSSREFERENCES TO STRUCTURE DATABANKS

PIR/MIPS code

Brookhaven code

7 LITERATURE REFERENCES

[1] Gäde, G., Carlsson, K.-H.: Mar. Biol.,79,39–45 (1984)
[2] Carvajal, N., Kessi, E.: Biochim. Biophys. Acta,953,14–19 (1988)
[3] Teschner, W., Rudolph, R., Garel, J.-R.: Biochemistry,26,2791–2796 (1987)
[4] Schrimsher, J.L., Taylor, K.B.: Biochemistry,23,1348–1353 (1984)
[5] Zettlmeissl, G., Teschner, W., Rudolph, R., Jaenicke, R., Gäde, G.: Eur. J. Biochem.,143,401–407 (1984)
[6] Birnberg, P.R., Rao, S.S., Lippincott, J.A.: Phytochemistry,22,1345–1355 (1983)
[7] Storey, K.B., Dando, P.R.: Comp. Biochem. Physiol.,73 B,521–528 (1982)
[8] Schrimsher, J.L., Taylor, K.B.: J. Biol. Chem.,257,8953–8956 (1982)
[9] Monneuse-Doublet, M.-O., Olomucki, A.: Biochem. Soc. Trans.,9,300–302 (1981)
[10] Olomucki, A.: Biochem. Soc. Trans.,9,278–279 (1981)
[11] Barrett, J., Körting, W.: Experientia,37,958–959 (1981)
[12] Gäde, G., Head, E.J.H.: Experientia,35,304–305 (1979)
[13] Huc, C., Olomucki, A., Le-Thi-Lan, Pho, D.B., Van Thoai, N.: Eur. J. Biochem.,21,161–169 (1971)
[14] Thome-Beau, F., Olomucki, A.: Eur. J. Biochem.,39,557–562 (1973)
[15] Baici, A., Luisi, P.L., Olomucki, A., Doublet, M.-O., Klincak, J.: Eur. J. Biochem.,46,59–66 (1974)
[16] Huc, C., Olomucki, A., Thome-Beau, F.: FEBS Lett.,60,414–418 (1975)
[17] Biellmann, J.-F., Branlant, G., Olomucki, A.: FEBS Lett.,32,254–256 (1973)
[18] Pho, D.B., Olomucki, A., Huc, C., Van Thoai, N.: Biochim. Biophys. Acta,206,46–53 (1970)
[19] Luisi, P.L., Baici, A., Olomucki, A., Doublet, M.-O.: Eur. J. Biochem.,50,511–516 (1975)

Enzyme Handbook © Springer-Verlag Berlin Heidelberg 1994
Duplication, reproduction and storage in data banks are only allowed with the prior permission of the publishers

[20] Doublet, M.-O., Olomucki, A.: Eur. J. Biochem.,59,175–183 (1975)
[21] Storey, K.B., Storey, J.M.: Eur. J. Biochem.,93,545–552 (1979)
[22] Van Thoai, N., Huc, C., Pho, D.B., Olomucki, A.: Biochim. Biophys. Acta,191,46–57 (1969)
[23] Monneuse-Doublet, M.-O., Lefebure, F., Olomucki, A.: Eur. J. Biochem.,108,261–269 (1980)
[24] Olomucki, A., Huc, C., Lefebure, F., Van Thoai, N.: Eur. J. Biochem.,28,261–268 (1972)
[25] Birnberg, P.R., Lippincott, B.B., Lippincott, J.A.: Phytochemistry,16,647–650 (1977)
[26] Gäde, G.: Biochem. Soc. Trans.,4,433–436 (1976)
[27] Hack, E., Kemp, J.D.: Plant Physiol.,65,949–955 (1980)
[28] Walsh, P.J.: J. Comp. Physiol.,143,213–222 (1981)
[29] Gäde, G.: Comp. Biochem. Physiol.,67B,575–582 (1980)
[30] Murai, N., Kemp, J.D.: Proc. Natl. Acad. Sci. USA,79,86–90 (1982)
[31] Baldwin, J., England, W.R.: Pac. Sci.,36,381–394 (1982)

1 NOMENCLATURE

EC number
1.5.1.12

Systematic name
1-Pyrroline-5-carboxylate: NAD^+ oxidoreductase

Recommended name
1-Pyrroline-5-carboxylate dehydrogenase

Synonymes
Dehydrogenase, 1-pyrroline-5-carboxylate
DELTA^1-pyrroline-5-carboxylate dehydrogenase
1-Pyrroline dehydrogenase
Pyrroline-5-carboxylate dehydrogenase
Pyrroline-5-carboxylic acid dehydrogenase
L-Pyrroline-5-carboxylate-NAD^+ oxidoreductase [9]

CAS Reg. No.
9054-82-4

2 REACTION AND SPECIFICITY

Catalysed reaction
1-Pyrroline-5-carboxylate + NAD^+ + H_2O →
→ L-glutamate + NADH

Reaction type
Redox reaction

Natural substrates

Substrate spectrum
1 1-Pyrroline-5-carboxylate + NAD^+ + H_2O (ir [1, 3, 9, 15, 18], highly specific [3]) [1, 3, 9–11, 15, 17, 18]
2 DELTA^1-pyrroline-3-hydroxy-5-carboxylate + NAD^+ + H_2O [17, 18]
3 More (not: pyrroline-2-carboxylate [11], not: glyceraldehyde, valeraldehyde [9]) [9, 11]

Product spectrum
1 L-Glutamate + NADH [3, 10, 11, 15, 18]
2 gamma-Hydroxyglutamate + NADH [17]
3 ?

Enzyme Handbook © Springer-Verlag Berlin Heidelberg 1994
Duplication, reproduction and storage in data banks are only allowed with the prior permission of the publishers

Inhibitor(s)

AMP [1, 15, 18]; alpha-Aminovalerate [1]; L-Hydroxyproline [1, 9, 15, 17]; p-Hydroxymercuribenzoate [1, 9, 15, 18]; N-Ethylmaleimide (not [18]) [9]; L-5-Oxoproline [9]; Chlorides [10]; Sulfates [10]; L-Ala (not [11]) [12, 14]; D-Ala [12]; L-Asp (not [11]) [12]; L-Val (not [11]) [12, 14]; Gly (not [11]) [12, 14]; L-Ser (not [11]) [12, 14]; L-Ile (not [11]) [12]; L-Trp (not [11]) [12]; L-His (not [11]) [12]; L-Glu (not [11]) [12]; L-Leu (not [11]) [12, 14]; 4,5-Dehydro-L-pipecolic acid (baikiain, competitive) [13]; ADP [15, 18]; ATP [15, 18]; delta-Aminovalerate [15, 18]; Valerolactam [15]; gamma-Aminobutyrate [15]; delta-Valerolactone [15]; epsilon-Aminocaproate [15]; Arsenite [15, 17]; $DELTA^1$-pyrroline-2-carboxylate [17]; Imidazole [17]; F^- [17]; Arsenate [17]; EDTA [17]; Borate [17]; Cu^{2+} [18]; CN^- [18]; NH_2OH [18]; More (not: Ca^{2+}, Mn^{2+}, Mg^{2+}, Ba^{2+}, Pb^{2+}, N-ethylmaleimide, o-phenanthroline [18], not: amino acids [11]) [11, 18]

Cofactor(s)/prostethic group(s)

NAD^+ (preferentially [9, 11, 15, 18]) [1, 3, 6, 9–11, 15, 18]; $NADP^+$ (8.1 times higher activity compared to NAD^+ [3]) [3, 11, 15, 18]

Metal compounds/salts

Turnover number (min^{-1})

Specific activity (U/mg)

27.1 [1]; 34.1 [3]; 0.55 [15]; More (assay method [4]) [4, 7, 17, 18]

K_m-value (mM)

0.42 (L-pyrroline-5-carboxylate with $NADP^+$) [3]; 0.75 (L-pyrroline-5-carboxylate with NAD^+) [3]; 2.3 (NAD^+) [3]; 0.0095 ($NADP^+$) [3]; 0.18 (L-pyrroline-5-carboxylate) [9]; 0.35 (pyrroline-5-carboxylate) [12]; 0.29 (1-pyrroline-5-carboxylate with NAD^+) [15]; 0.062 (1-pyrroline-5-carboxylate with $NADP^+$) [15]; 0.053 (NAD^+) [15]; 0.12 ($NADP^+$) [15]; More [10]

pH-optimum

6.5–7.0 [3]; 6.7–7.7 [9]; 7.4 (1-pyrroline-3-hydroxy-5-carboxylate) [17]; 7.5–8.0 [11]; 8 (independent of buffer) [10]; 8.0–8.5 [1]; 8.5–8.6 [15, 18]

pH-range

5.5–7.5 (cofactor NAD^+) [3]; 5.5–8 (cofactor $NADP^+$) [3]; 6–9.5 [11]; 7.4–9.4 [18]

Temperature optimum (°C)

35 (cofactor NAD^+) [3]; 45 (cofactor $NADP^+$) [3]

Temperature range (°C)

3 ENZYME STRUCTURE

Molecular weight

135000–160000 (Salmonella typhimurium, form II, native gel electrophoresis) [7]
115000 (rat, sucrose-glycerol density gradient sedimentation) [1]
100000 (Bacillus nidulans, gel filtration) [3]
210000–270000 (Salmonella typhimurium, form I, native gel electrophoresis) [7]

Subunits

Dimer (2 × 59000, rat, SDS-PAGE [1], 2 × 50000, Bacillus nidulans, SDS-PAGE [3], 2 × 132000, Salmonella typhimurium, detergent solubilized, disc gel electrophoresis [7]) [1, 3, 7]
Monomer (1 × 132000, Salmonella typhimurium, membrane bound in vivo, disc gel electrophoresis) [7]

Glycoprotein/Lipoprotein

–

4 ISOLATION/PREPARATION

Source organism

Rat [1, 5, 9, 16]; Aspergillus nidulans [2]; Bacillus sphaericus [3]; Bacillus stearothermophilus [3]; Bacillus thermodenitrificans [3]; Bacillus caldotenax [3]; Saccharomyces cerevisiae [4, 8, 19]; Salmonella typhimurium [6, 7, 13]; Hordeum distichum [10]; Pisum sativum [11]; Zea mays [11]; Ricinus communis [11]; Cucurbita maxima [11]; Saccharomyces sp. (glutamate auxotroph) [12]; Bovine [14, 15, 17, 18]

Source tissue

Liver [1, 15–18]; Brain (highest activity in cerebellum, distribution in various brain regions) [5]; Shoots (etiolated) [10, 11]; Seedlings [11]; Endosperm [11]; Cotyledons [11]; Kidney [14]; More (distribution in rat tissues) [9]

Localisation in source

Mitochondria (matrix) [1, 4, 10, 11, 14, 16]; Membrane bound [6]; Cytosol [16]

Purification

Rat [1]; Bacillus sphaericus [3]; Bovine (partial [17]) [15, 17]; Salmonella typhimurium (copurification with proline oxidase) [7]

Crystallization

[3]

Cloned

[2]

Enzyme Handbook © Springer-Verlag Berlin Heidelberg 1994
Duplication, reproduction and storage in data banks are only allowed with the prior permission of the publishers

Renaturated

–

5 STABILITY

pH

Temperature (°C)

35 (15 min, 40% loss of activity in presence of NAD^+, 2% loss of activity in presence of $NADP^+$) [3]; 40 (15 min, 75% loss of activity in presence of NAD^+, 10% loss of activity in presence of $NADP^+$) [3]

Oxidation

Organic solvent

General stability information

$NADP^+$ stabilizes [3]; Glutathione stabilizes [17]

Storage

–15°C, 2 months, 50% loss of activity [15, 18]; 0°C, 70 mM Tris-HCl buffer, pH 8.2, 30% v/v glycerol, several weeks [7]

6 CROSSREFERENCES TO STRUCTURE DATABANKS

PIR/MIPS code

RDBYC (precursor Yeast Saccharomyces cerevisiae)

Brookhaven code

7 LITERATURE REFERENCES

[1] Small, W.C., Jones, M.E.: J. Biol. Chem.,265,18668–18672 (1990)
[2] Hull, E.P., Green, P.M., Arst, H.N., Scazzocchio, C.: Mol. Microbiol.,3,553–559 (1989)
[3] Isobe, K., Matsuzawa, T., Soda, K.: Agric. Biol. Chem.,51,1947–1953 (1987)
[4] Small, C., Jones, M.E.: Anal. Biochem.,161,380–386 (1987)
[5] Thompson, S.G., Wong, P.T.-H., Leong, S.F., McGeer, E. G.: J. Neurochem.,45,1791–1796 (1985)
[6] Menzel, R., Roth, J.: J. Biol. Chem.,256,9762–9766 (1981)
[7] Menzel, R., Roth, J.: J. Biol. Chem.,256,9755–9761 (1981)
[8] Brandriss, M.C., Magasanik, B.: J. Bacteriol.,143,1403–1410 (1980)
[9] Herzfeld, A., Mezl, V.A., Knox, W.E.: Biochem. J.,166,95–103 (1977)

[10] Boggess, S.F., Paleg, L.G., Aspinall, D.: Plant Physiol.,56,259–262 (1975)
[11] Stewart, C.R., Lai, E.Y.: Plant Sci. Lett.,3,173–181 (1974)
[12] Lundgren, D.W., Ogur, M.: Biochim. Biophys. Acta,297,246–257 (1973)
[13] Dendinger, S., Brill, W.J.: J. Bacteriol.,112,1134–1141 (1972)
[14] Lundgren, D., Ogur, M.: Biochem. Biophys. Res. Commun.,49,147–149 (1972)
[15] Strecker, H.J.: Methods Enzymol. (Colowick, S.P., Ed.) 17B,262–265 (1971)
[16] Brunner, G., Neupert, W.: FEBS Lett.,3,283–286 (1969)
[17] Adams, E., Goldstone, A.: J. Biol. Chem.,235,3504–3512 (1960)
[18] Strecker, H.J.: J. Biol. Chem.,235,3218–3223 (1960)
[19] Brandriss, M.C.: Mol. Cell. Biol.,3,1846–1856 (1983)

Enzyme Handbook © Springer-Verlag Berlin Heidelberg 1994
Duplication, reproduction and storage in data banks are only allowed with the prior permission of the publishers

1 NOMENCLATURE

EC number
1.5.1.13

Systematic name
Nicotinate:$NADP^+$ 6-oxidoreductase (hydroxylating)

Recommended name
Nicotinate dehydrogenase

Synonymes
Dehydrogenase, nicotinate
Nicotinate dehydrogenase
Nicotinic acid hydroxylase [1]

CAS Reg. No.
9059-03-4

2 REACTION AND SPECIFICITY

Catalysed reaction
Nicotinate + H_2O + $NADP^+$ →
→ 6-hydroxynicotinate + NADPH

Reaction type
Redox reaction

Natural substrates
Nicotinate + H_2O + NADPH

Substrate spectrum
1 Nicotinate + H_2O + $NADP^+$ (r, equilibrium lies far in direction of 6-hydroxynicotinate fomation [1], electron acceptor: benzyl viologen and 2,3,5-triphenyltetrazolium dyes can replace $NADP^+$, not NAD^+, high specificity with respect to nicotinate) [1]

Product spectrum
1 6-Hydroxynicotinate + NADPH (r) [1]

Inhibitor(s)
KCN [2]; Atabrine [5]; alpha,alpha'-Dipyridyl [5]; o-Phenanthroline [5]

Enzyme Handbook © Springer-Verlag Berlin Heidelberg 1994
Duplication, reproduction and storage in data banks are only allowed with the prior permission of the publishers

Cofactor(s)/prostethic group(s)

NADP$^+$ (benzyl viologen and 2,3,5-triphenyltetrazolium dyes can replace NADP$^+$, not NAD$^+$ [1]) [1–5]; Benzyl viologen (can replace NADP$^+$) [1]; 2,3,5-Triphenyltetrazolium dyes (can replace NADP$^+$) [1]; NADPH [1–5]; Flavin (primary electron acceptor, 1.5 mol of flavin per mol of enzyme) [5]; Bactopterin (pterin, different from molybdopterin in MW, phosphate content and structure) [4]; FAD (1.5 mol of flavin per mol of enzyme [1], 2.0 mol of flavin per mol of enzyme [2]) [1, 2]

Metal compounds/salts

Molybdenum (1.5 mol of molybdenum per mol of enzyme) [2]; Phosphate (activates) [1]; Iron (11 mol of iron per mol of enzyme [1], 8.3 mol of iron per mol of enzyme, contains Fe/S centers [2], approximately 8 molecules of iron per molecule of enzyme [5]) [1, 2, 5]

Turnover number (min^{-1})

Specific activity (U/mg)

0.54 [2]; 0.83 [5]; More [1]

K_m-value (mM)

0.11 (nicotinate) [1]; 1.0 (nicotinate) [2]; 0.028 (NADP$^+$) [1]

pH-optimum

7.5 [2]; 7.5–8.0 (phosphate buffer) [1]; 8.0–8.5 (triethanolamine buffer, triethanolamine phosphate buffer, dimethylglutarate-phosphate buffer) [1]; 8.3–9.0 (Tris buffer, diphosphate buffer) [1]

pH-range

Temperature optimum (°C)

25 (assay at) [5]; 30 (assay at) [3]

Temperature range (°C)

---

3 ENZYME STRUCTURE

Molecular weight

280000–340000 (Bacillus niacini, gel filtration, non denaturing PAGE) [1]
300000 (Clostridium sp., analytical ultracentrifugation) [1]
400000 (Bacillus sp., gel filtration) [5]

Subunits

Oligomer (x × 85000, x × 34000, x × 20000, Bacillus niacini, SDS-PAGE, additional bands with increasing period of storage) [2]

Glycoprotein/Lipoprotein

–

4 ISOLATION/PREPARATION

Source organism
Clostridium sp. (nicotinic acid-fermenting) [1]: Bacillus niacini [2, 4]; Bacillus sp. (DSM 2923 [3]) [3, 5]

Source tissue
Cell [1]

Localisation in source

Purification
Clostridium sp. (nicotinic acid-fermenting) [1]; Bacillus niacini [2]; Bacillus sp. [5]

Crystallization
–

Cloned
–

Renaturated
–

5 STABILITY

pH

Temperature (°C)

Oxidation

Organic solvent

General stability information
Lyophilization completely inactivates [2]

Storage
–20°C, increase of activity in crude extract [2]

Enzyme Handbook © Springer-Verlag Berlin Heidelberg 1994
Duplication, reproduction and storage in data banks are only allowed with the prior permission of the publishers

6 CROSSREFERENCES TO STRUCTURE DATABANKS

PIR/MIPS code

Brookhaven code

7 LITERATURE REFERENCES

[1] Holcenberg, J.S., Stadtman, E.R.: J. Biol. Chem.,244,1194–1203 (1969)
[2] Nagel, M., Andreesen, J.R.: Arch. Microbiol.,154,605–613 (1990)
[3] Nagel, M., Andreesen, J.R.: FEMS Microbiol. Lett.,59,147–152 (1989)
[4] Krüger, B., Meyer, O., Nagel, M., Andreesen, J.R., Meincke, M., Bock, E., Blümle, S., Zumft, W.G.: FEMS Microbiol. Lett.,48,225–227 (1987)
[5] Hirschberg, R., Ensingn, J.C.: J. Bacteriol.,108,751–756 (1971)

1 NOMENCLATURE

EC number
1.5.1.15

Systematic name
5,10-Methylenetetrahydrofolate:NAD^+ oxidoreductase

Recommended name
Methylenetetrahydrofolate dehydrogenase (NAD^+)

Synonymes
Dehydrogenase, methylenetetrahydrofolate (nicotinamide adenine dinucleotide)
NAD^+-dependent 5,10-methylenetetrahydrofolate dehydrogenase [1]
NAD-dependent 5,10-methylenetetrahydrofolate dehydrogenase [2]

CAS Reg. No.
82062-90-6

2 REACTION AND SPECIFICITY

Catalysed reaction
5,10-Methylenetetrahydrofolate + NAD^+ →
→ 5,10-methenyltetrahydrofolate + NADH (forward reaction: initial velocity kinetics are consistent with sequential reaction mechanism [1], forward and reverse reaction: substituted (ping-pong)reaction mechanism [2], equilibrium ordered kinetic mechanism with methylene tetrahydrofolate as the last reactant to add and NADH as the last product released [10])

Reaction type
Redox reaction

Natural substrates
5,10-Methylenetetrahydrofolate + NAD^+ (physiological reaction in most organisms) [2]
5,10-Methenyltetrahydrofolate + NADH (physiological reaction in acetogens) [2]
More (promotes purine synthesis and perhaps contributes to the methionine dependency and rapid growth observed for many established cell lines) [5]

Enzyme Handbook © Springer-Verlag Berlin Heidelberg 1994
Duplication, reproduction and storage in data banks are only allowed with the prior permission of the publishers

Substrate spectrum

1 5,10-Methylenetetrahydrofolate + NAD^+ (r [2])
2 More (Saccharomyces cerevisiae [1], Acetobacterium woodii [2]: monofunctional enzyme without 5,10-methenyl-tetrahydrofolate cyclohydrolase and 10-formyl-tetrahydrofolate synthase activity [1, 2], transformed mammalian cells: bifunctional methylene tetrahydrofolate dehydrogenase-cyclohydrolase activity [4, 9, 10], bifunctional NAD-dependent methylenetetrahydrofolate-dehydrogenase-methenyltetrahydrofolate cyclohydrolase (in addition to the usual NADP-dependent dehydrogenase-cyclohydrolase-synthase) [5], enzyme active only with L, L-5,10-methylenetetrahydrofolate not with D, L-isomer [3]) [1–5, 9, 10]

Product spectrum

1 5,10-Methenyltetrahydrofolate + NADH (r [2])
2 ?

Inhibitor(s)

Zn^{2+} [2]; $NADP^+$ (slight) [3]

Cofactor(s)/prostethic group(s)

NAD^+ (specific for) [1, 3]; NADH [2]; More (no activity with $NADP^+$) [3]

Metal compounds/salts

Mg^{2+} (required [11], required for binding NAD^+ (bifunctional enzyme from transformed mammalian cells) [4, 10], not dependent on Mg^{2+} [1], no effect of Mn^{2+} and Mg^{2+} (3–50 mM) [3]) [4, 10]; Inorganic phosphate (stimulates) [11]; Arsenate (stimulates) [11]; Mn^{2+} (can substitute for Mg^{2+}) [11]; Ca^{2+} (can substitute to lesser extent for Mg^{2+}) [1]; More (no effect of Mn^{2+} and Mg^{2+} (3–50 mM)) [3]

Turnover number (min^{-1})

Specific activity (U/mg)

5.4 [1]; 670 [2]; 38 [11]; More [3]

K_m-value (mM)

2.0 (NADH) [2]; 1.0 (5,10-methenyltetrahydrofolate) [2]; 1.6 (NAD^+) [1]; 4.0 (NAD^+) [2]; 0.79 (NAD^+) [3]; 0.06 ((6R, S)-5,10-methylene-tetrahydrofolate) [1]; 0.26 (5,10-methylenetetrahydrofolate) [2]; 0.066 (L, L-5,10-methylenetetrahydrofolate) [3]; 0.111 (NAD^+) [11]; 0.019 (methylenetetrahydrofolate) [11]; 0.250 (Mg^{2+}) [11]; 0.170 (phosphate) [11]; 0.793 (arsenate) [11]; 0.155 (methylenetetrahydrofolate) [11]

pH-optimum

8.0 (assay at) [1]; 7.0 (assay at) [2]; 7.3 (assay at) [5]

pH-range

Temperature optimum (°C)
45 [3]

Temperature range (°C)

3 ENZYME STRUCTURE

Molecular weight
60000 (Clostridium formicoaceticum, sedimentation equilibrium centrifugation) [3]
64000 (Saccharomyces cerevisiae, gel filtration) [1]
69000 (Acetobacter woodii, gel filtration) [2]

Subunits
Dimer (2 × 33000–38000, Saccharomyces cerevisiae, SDS-PAGE, cross-linking experiments with dimethyl suberimidate [1], 2 × 26300, Acetobacterium woodii, amino acid analysis [2], 2 × 30000, Clostridium formicoaceticum, SDS-PAGE [3], 2 × 34000, Ehrlich ascites tumor cells, SDS-PAGE, cross-linking studies with bis(sulfosuccinimidyl)suberate [11]) [1–3, 11]

Glycoprotein/Lipoprotein
–

4 ISOLATION/PREPARATION

Source organism
Saccharomyces cerevisiae [1]; Acetobacterium woodii [2]; Clostridium formicoaceticum [3]; Mouse (transformed cells, Ehrlich ascites tumor cells [10, 11]) [4, 8–11]; Human (transformed cells) [4]; Rat (transformed cells) [7]

Source tissue
Cell [1]; Transformed mammalian cells (overview [4, 7]) [4, 7, 8]; Ehrlich ascites tumor cells [10, 11]; More (overview [5–7]: embryonic, undifferentiated or transformed cells, mammalian cell lines (tumorigenic and non-tumorigenic) [6], tissues containing differentiating cells (bone marrow, thymus, spleen, embryonic liver [5])) [5–7]

Localisation in source
Cytoplasm [1, 4]; Mitochondria (not [1]) [8]

Purification
Saccharomyces cerevisiae [1]; Acetobacterium woodii [2]; Clostridium formicoaceticum [3]; Mouse (Ehrlich ascites tumor cells) [11]

Enzyme Handbook © Springer-Verlag Berlin Heidelberg 1994
Duplication, reproduction and storage in data banks are only allowed with the prior permission of the publishers

Crystallization

–

Cloned

(bifunctional NAD-dependent methylenetetrahydrofolate dehydrogenase-methenyltetrahydrofolate cyclohydrolase) [9]

Renaturated

–

5 STABILITY

pH

Temperature (°C)

50 (30 min, 25% loss of activity, stable below 50°C) [3]; 61 (30 min, 65% loss of activity) [3]; 68 (80% loss of activity) [3]

Oxidation

Oxygen-labile, reactivation of oxidized enzyme with dithiothreitol [2]

Organic solvent

General stability information

Freezing completely inactivates [2]

Storage

4°C, several weeks [1]; 4°C, pH 7, for at least 1 month [3]; –20°C, 50% glycerol, for at least 2 years [2]

6 CROSSREFERENCES TO STRUCTURE DATABANKS

PIR/MIPS code

DEHUMT (precursor Human); A33267 (/methylenetetrahydrofolate cyclohydrolase EC 3.5.4.9 precursor Mouse)

Brookhaven code

7 LITERATURE REFERENCES

[1] Barlowe, C.K., Appling, D.R.: Biochemistry,29,7089–7094 (1990)
[2] Ragsdale, S.W., Ljungdahl, L.G.: J. Biol. Chem.,259,3499–3503 (1984)
[3] Moore, M.R., O'Brien, W.E., Ljungdahl, L.G.: J. Biol. Chem.,249,5250–5253 (1974)
[4] MacKenzie, R.E., Mejia, N., Yang, X.-M.: Adv. Enzyme Regul.,27,31–39 (1988) (Review)
[5] Mejia, N.R., MacKenzie, R.E.: J. Biol. Chem.,260,14616–14620 (1985)
[6] Scrimgeour, K.G., Huennekens, F.M.: Biochem. Biophys. Res. Commun.,2,230–233 (1960)
[7] Smith, G.K., Banks, S.D., Monaco, T.J., Rigual, R., Duch, D.S., Mullin, R.J., Huber, B.E.: Arch. Biochem. Biophys.,283,367–371 (1990)
[8] Mejia, N.R., MacKenzie, R.E.: Biochem. Biophys. Res. Commun.,155,1–6 (1988)
[9] Belanger, C., MacKenzie, R.E.: J. Biol. Chem.,264,4837–4843 (1989)
[10] Rios-Orlandi, E.M., MacKenzie, R.E.: J. Biol. Chem.,263,4662–4667 (1988)
[11] Mejia, N.R., Rios-Orlandi, E.M., MacKenzie, R.E.: J. Biol. Chem.,261,9509–9513 (1986)

Enzyme Handbook © Springer-Verlag Berlin Heidelberg 1994
Duplication, reproduction and storage in data banks are only allowed with the prior permission of the publishers

1 NOMENCLATURE

EC number
1.5.1.16

Systematic name
N^2–(D-1-Carboxyethyl)-L-lysine:NADP$^+$ oxidoreductase (L-lysine-forming)

Recommended name
D-Lysopine dehydrogenase

Synonymes
D-Lysopine synthase
Dehydrogenase, D-lysopine
Lysopine dehydrogenase
D(+)-Lysopine dehydrogenase

CAS Reg. No.
65187-41-9

2 REACTION AND SPECIFICITY

Catalysed reaction
N^2–(D-1-Carboxyethyl)-L-lysine + NADP$^+$ + H_2O →
→ L-lysine + pyruvate + NADPH

Reaction type
Redox reaction

Natural substrates

Substrate spectrum
1 N^2–(D-1-Carboxyethyl)-L-lysine + NADP$^+$ + H_2O (r [1], N^2–(D-1-carboxyethyl)-L-lysine is identical with lysopine, in the reverse reaction a number of L-amino acids (e.g. L-methionine, L-ornithine, L-arginine, L-histidine, L-glutamine) can act instead of L-lysine [1], biotype 3 strain-induced enzyme mainly produces lysopine and octopine from L-lysine or L-arginine (+ pyruvate + NADPH), Agrobacterium tumefaciens induced enzyme readily converts all precursors (L-methionine, L-arginine, L-histidine, L-lysine) into the corresponding opines [3])
2 N^2–(D-1-Carboxyethyl)-L-ornithine + NADP$^+$ + H_2O [1]

Enzyme Handbook © Springer-Verlag Berlin Heidelberg 1994
Duplication, reproduction and storage in data banks are only allowed with the prior permission of the publishers

Product spectrum
1 L-Lysine + pyruvate + NADPH (r [1])
2 L-Ornithine + pyruvate + NADPH [1]

Inhibitor(s)

Cofactor(s)/prostethic group(s)
NADPH; $NADP^+$

Metal compounds/salts

Turnover number (min^{-1})

Specific activity (U/mg)
1.045 [1]

K_m-value (mM)
0.56 (L-lysine) [1]; 1.3 (L-methionine) [1]; 2.3 (L-ornithine) [1]; 4.4 (L-arginine) [1]; 6.1 (L-glutamine, L-histidine) [1]; 17.1 (L-citrulline) [1]; 0.23 (pyruvate) [1]; 0.003 (NADPH) [1]; 0.16 (D(+)-lysopine) [1]; 0.8 (D(+)-octopine) [1]; 0.010 ($NADP^+$) [1]

pH-optimum
6.3 (L-methionine) [1]; 6.2 (L-arginine) [1]; 6.7 (L-ornithine) [1]; 8.9 (lysopine) [1]; 9.1 (N^2–(D-1-carboxyethyl)-L-ornithine (octopine)) [1]

pH-range

Temperature optimum (°C)
30 (assay at) [1]

Temperature range (°C)

3 ENZYME STRUCTURE

Molecular weight

Subunits

Glycoprotein/Lipoprotein
–

4 ISOLATION/PREPARATION

Source organism

Nicotiana tabacum (leaf tumors initiated by infection with Agrobacterium tumefaciens, strain B6S3) [1]; Kalanchoe daigremontiana (internode segments infected with Agrobacterium tumefaciens strain B6S3 [2], tumors induced by Agrobacterium tumefaciens strains: AB-3, AT-6, Hm-1 [3]) [2, 3]

Source tissue

Crown gall tumor tissue (of Nicotiana tabacum leafs infected with Agrobacterium tumefaciens strain B6S3 [1], of Kalanchoe daigremontiana internode segments infected with Agrobacterium tumefaciens strain B6S3 [2], induced in stem segments of Kalanchoe daigremontiana with Agrobacterium tumefaciens strains AB-3, AT-6, Hm-1 [3]) [1–3]

Localisation in source

Purification

Nicotiana tabacum (leaf tumors initiated by infection with Agrobacterium tumefaciens, partial) [1]; Kalanchoe daigremontiana (internode segments infected with Agrobacterium tumefaciens strain B6S3) [2]

Crystallization

–

Cloned

(a part of lysopine dehydrogenase [4]) [4, 5]

Renaturated

–

5 STABILITY

pH

Temperature (°C)

Oxidation

Organic solvent

General stability information

Enzyme Handbook © Springer-Verlag Berlin Heidelberg 1994
Duplication, reproduction and storage in data banks are only allowed with the prior permission of the publishers

Storage

–20°C, many weeks [1]; 4°C, some days [1]

6 CROSSREFERENCES TO STRUCTURE DATABANKS

PIR/MIPS code

DEAGLT (Agrobacterium tumefaciens plasmid pTiAch5)

Brookhaven code

7 LITERATURE REFERENCES

[1] Otten, L.A.B.M., Vreugdenhil, D., Schilperoort, R.A.: Biochim. Biophys. Acta,485,268–277 (1977)
[2] Otten, L.: Plant Sci. Lett.,25,15–27 (1982)
[3] Otten, L., Szegedi, E.: Plant Sci.,40,81–85 (1985)
[4] Schröder, J., Hillebrand, A., Klipp, W., Pühler, A.: Nucleic Acids Res.,9,5187–5202 (1981)
[5] Schröder, J., Schröder, G., Huisman, H., Schilperoort, R.A., Schell, J.: FEBS Lett.,129,166–168 (1981)

1 NOMENCLATURE

EC number
1.5.1.17

Systematic name
2,2'-Iminodipropanoate:NAD$^+$ oxidoreductase (L-alanine-forming)

Recommended name
Alanopine dehydrogenase

Synonymes
Dehydrogenase, alanopine
ALPDH [3]
Alanopine[meso-N-(1-carboxyethyl)-alanine]dehydrogenase [6]
Meso-N-(1-carboxyethyl)-alanine:NAD$^+$ oxidoreductase [6]
Alanopine: NAD oxidoreductase [7]
ADH: alanopine:NAD oxidoreductase [9]

CAS Reg. No.
71343-07-2

2 REACTION AND SPECIFICITY

Catalysed reaction
L-Alanine + pyruvate + NADH →
→ 2,2'-iminodipropanoate + NAD$^+$ + H_2O

Reaction type
Redox reaction

Natural substrates
L-Alanine + pyruvate + NADH (terminal dehydrogenase of glycolysis, role in maintaining energy production under the stresses of environmental or functional anoxia [2], maintainance of redox balance during anaerobiosis in adductor muscle and heart of oyster [4], muscle isoenzyme: alanopine synthesis as an end product of anaerobic glycolysis [9]) [2, 4, 9]
2,2'-Iminodipropanoate + NAD$^+$ + H_2O (hepatopancreas isoenzyme: alanopine oxidation in vivo) [9]

Enzyme Handbook © Springer-Verlag Berlin Heidelberg 1994
Duplication, reproduction and storage in data banks are only allowed with the prior permission of the publishers

Substrate spectrum

1 L-Alanine + pyruvate + NADH (r [4, 10, 11])
2 Glycine + pyruvate + NADH [2, 7, 10, 11]
3 L-Serine + pyruvate + NADH [2, 7, 10, 11]
4 L-2-Aminobutyrate + pyruvate + NADH [2, 7, 10, 11]
5 L-Cysteine + pyruvate + NADH [2, 10, 11]
6 L-Valine + pyruvate + NADH [2, 10]
7 L-Threonine + pyruvate + NADH [2, 10]
8 More (not: D-alanine, L-arginine, L-glutamate [6], keto acids that can replace pyruvate (100% activity): 2-oxobutyrate (37% [11], 22% [2]) [2, 6, 7, 10, 11], glyoxylate (44% [2], 10% [11], not [6]) [2, 11], oxaloacetate (106% [10], 94% [11]) [6, 10, 11], alpha-ketovalerate (25%) [10], not: 2-oxoglutarate [6], relative reaction rate with meso-alanopine to D-strombine is 100: 5 [6]) [2, 6, 7, 10, 11]

Product spectrum

1 2,2'-Iminodipropanoate + NAD^+ + H_2O (r [4, 10, 11])
2 N^2-(D-1-Carboxyethyl)-glycine + NAD^+ + H_2O
3 N^2-(D-1-Carboxyethyl)-L-serine + NAD^+ + H_2O
4 N^2-(D-1-Carboxyethyl)-aminobutyrate + NAD^+ + H_2O
5 N^2-(D-1-Carboxyethyl)-L-cysteine + NAD^+ + H_2O
6 N^2-(D-1-Carboxyethyl)-L-valine + NAD^+ + H_2O
7 N^2-(D-1-Carboxyethyl)-L-threonine + NAD^+ + H_2O
8 ?

Inhibitor(s)

Octopine (pyruvate reduction) [12]; Glycine (pyruvate reduction) [12]; Alanine (pyruvate reduction) [12]; ATP [7, 11]; ADP [7, 11]; AMP [7]; Succinate (weak [2], no effect at pH 7.5, inhibition at high concentrations at pH 6.5 [11]) [2, 7, 11]; L-Lactate (weak [2], strong inhibition [10]) [2, 10, 11]; D-Lactate (weak [2], strong inhibition [10]) [2, 10, 11]; 2-Oxoglutarate (slight inhibition [11]) [7, 11]; Pyruvate (substrate inhibition [9, 11], product inhibition [9]) [9, 11]; L-Alanine (substrate inhibition [9, 11], product inhibition [9]) [9, 11]; NAD^+ (product inhibition [9], pyruvate reduction [12]) [7, 9, 12]; meso-Alanopine (product inhibition) [9]; Oxamate (slight inhibition) [11]; Fructose-1,6-biphosphate (slight inhibition) [11]; 3-Phosphoglycerate (slight inhibition) [11]; Dihydroxyacetone-phosphate (slight inhibition) [11]; Citrate (slight inhibition) [11]; L-Arginine [11]; L-Glutamate [11]; L-Lysine [11]; L-Leucine [11]; L-Isoleucine [11]; L-Tryptophan [11]; Mg^{2+} [11]; $(NH_4)_2SO_4$ (slight inhibition) [11]; NH_4Cl [11]; KCl [11]; NaCl [11]

Cofactor(s)/prostethic group(s)

NAD^+; NADH; More (not: $NADP^+$, NADPH) [6, 11]

Metal compounds/salts

Turnover number (min^{-1})

Specific activity (U/mg)

700 [7]; 318–387 [9]; 670 (Glycera alba) [1]; 2 (Scolelepis fuliginosa) [1]; 3 [2]; 90 [10]; 240 [11]; 200 [12]

K_m-value (mM)

1.3 (pyruvate, Glycera alba) [1]; 0.5 (pyruvate, Scolelepis fuliginosa) [1]; 7.0 (L-alanine, Glycera alba) [1]; 90.5 (L-alanine, Scolelepis fuliginosa) [1]; 291 (glycine) [2]; 28 (L-alanine) [2]; 247 (L-serine) [2]; 15 (L-2-aminobutyrate) [2]; 23 (L-cysteine) [2]; 20 (L-valine) [2]; 110 (L-threonine) [2]; 0.48 (pyruvate) [2]; 2.5 (2-oxobutyrate) [2]; 11.9 (glyoxylate) [2]; 0.6 (pyruvate, oxaloacetate) [6]; 2 (2-oxobutyrate) [6]; 10 (alanine) [6]; 50 (glycine) [6]; 1 (alanopine) [6]; 0.03 (NAD^+) [6]; 0. 01 (NADH) [7]; 0.045 (pyruvate) [7]; 100 (alanine, glycine) [7]; More (K_m values of isoenzymes at pH 6.5 and 7.5 [9, 11], tissue-specific isoenzymes with different K_m values [8], overview [12]) [8–12]

pH-optimum

6.4 (foot muscle enzyme, low substrate concentration, 35 mM L-alanine + 1.4 mM pyruvate) [9]; 6.5 (low substrate concentration, 100 mM L-alanine + 1mM pyruvate + 0.1 mM NADH) [11]; 7.0 (foot muscle enzyme, high substrate concentration, 50 mM L-alanine + 2 mM pyruvate [10], alanine + pyruvate + NADH [6], pyruvate reduction [12]) [6, 10, 12]; 7.5 (high substrate concentration, 130 mM L-alanine + 1.3 mM pyruvate + 0.1 mM NADH) [11]; 8.1 (hepatopancreas enzyme, low substrate concentration, 15 mM meso-alanopine + 1.2 mM NAD^+) [9]; 8.5 (foot muscie enzyme, low substrate concentration, 15 mM meso-alanopine + 1.2 mM NAD^+ [9], low substrate concentration, 10 mM meso-alanopine [11], pyruvate formation [12]) [9, 11, 12]; 8.6 (gill, low substrate concentration, 15 mM meso-alanopine + 1.2 mM NAD^+) [9]; 9.0 (alanopine [6, 10], foot muscle, gill, hepatopancreas, high substrate concentration, 40 mM meso-alanopine, 2 mM NAD^+ [9]) [6, 9, 10]; 9.2 (high substrate concentration, 50 mM meso-alanopine) [11]

pH-range

Temperature optimum (°C)

20–25 (assay at) [1, 6, 7, 9–11]

Temperature range (°C)

Enzyme Handbook © Springer-Verlag Berlin Heidelberg 1994
Duplication, reproduction and storage in data banks are only allowed with the prior permission of the publishers

3 ENZYME STRUCTURE

Molecular weight

38500 (Mytilus edulis, gel filtration) [6]
41000–43000 (Busycotypus canaliculatum, isoenzymes, gel filtration) [9]
42000 (Strombus luhuanus, gel filtration) [12]
42200 (Littorina littorea, gel filtration) [11]
44000 (Aphrodite aculeata, gel filtration) [10]
47000 (Crassostrea gigas, gel filtration) [7]

Subunits

Monomer (1 × 42400, Littorina littorea, SDS-PAGE [11], 1 × 47000, Crassostrea gigas, gel filtration, SDS-PAGE in absence or presence of 8 M guanidine hydrochloride [7], Polydora sp. [3], Mytilus edulis [6], Aphrodite aculeata [10]) [3, 6, 7, 10, 11]

Glycoprotein/Lipoprotein

–

---

4 ISOLATION/PREPARATION

Source organism

Polydora ciliata (sea worm) [3]; Polydora commensalis (sea worm) [3]; Polydora glycymerica (sea worm) [3]; Pseudopolydora paucibranchiata (sea worm) [3]; Crassostrea gigas (oyster) [4, 7]; Arenicola marina (marine invertebrate, EC 1.5.1.17 or EC 1.5.1.22 ?) [5]; Mytilus edulis (marine invertebrate, EC 1.5.1.17 or EC 1.5.1.22 ?) [5]; Nucula nitida (marine invertebrate, EC 1.5.1.17 or EC 1.5.1.22 ?) [5]; Busycotypus canaliculatum (channeled whelk, tissue specific isoenzymes) [8, 9]; Aphrodite aculeata (sea mouse) [10]; Littorina littorea (common periwinkle) [11]; Glycera alba [1]; Scolelepis fuliginosa (low activity) [1]; Mercenaria mercenaria (bivalve mollusc) [2]; Strombus luhuanus [12]; Strombus gibberulus [12]; Lambis lambis [12]; More (Strombidae, overview) [12]

Source tissue

Hepatopancreas [8, 9]; Kidney [8]; Ventricle [8]; Longitudinal muscle [10]; Nerve [10]; Elytra [10]; Heart [4, 7]; Muscle [5]; Body wall [1]; Proboscis muscle [1, 8]; Gill (activity barely detectable [7]) [2, 8, 9]; Mantle muscle [2, 8]; Adductor muscle [2, 8]; Foot muscle [6, 8, 9, 11]; Retractor muscle [12]

Localisation in source

Cytoplasm [6, 7, 8, 11]

Purification

Aphrodite aculeata (partial) [10]; Littorina littorea [11]; Mercenaria mercenaria (bivalve mollusc, partial) [2]; Mytilus edulis (partial) [6]; Crassostrea gigas [7]; Busycotypus canaliculatum [9]; Strombus luhuanus [12]

Crystallization

–

Cloned

–

Renaturated

–

5 STABILITY

pH

Temperature (°C)

Oxidation

Organic solvent

General stability information

Storage

4°C, 2 weeks [9, 11]; 4°C, 1 week [10]

6 CROSSREFERENCES TO STRUCTURE DATABANKS

PIR/MIPS code

Brookhaven code

7 LITERATURE REFERENCES

[1] Blackstock, J., Burdass, M.C.: Biochem. Soc. Trans.,15,383–384 (1987)
[2] Fields, J.H.A., Storey, K.B.: J. Exp. Mar. Biol. Ecol.,105,175–185 (1987)
[3] Manchenko, G.P.: Anal. Biochem.,145,308–310 (1985)
[4] Fields, J.H.A., Eng, A.K., Ramsden, W.D., Hochachka, P.W., Weinstein, B.: Arch. Biochem. Biophys.,201,110–114 (1980)
[5] Siegmund, B., Grieshaber, M.K.: Hoppe-Seyler's Z. Physiol. Chem.,364,807–812 (1983)
[6] Dando, P.R.: Biochem. Soc. Trans.,9,297–298 (1981)
[7] Fields, J.H.A., Hochachka, P.W.: Eur. J. Biochem.,114,615–621 (1981)
[8] Plaxton, W.C., Storey, K.B.: Can. J. Zool.,60,1568–1572 (1982)
[9] Plaxton, W.C., Storey, K.B.: Comp. Biochem. Physiol.,76B,321–326 (1983)
[10] Storey, K.B.: J. Exp. Zool.,225,369–378 (1983)
[11] Plaxton, W.C., Storey, K.B.: J. Comp. Physiol.,149,57–65 (1982)
[12] Baldwin, J., England, W.R.: Pac. Sci.,36,381–394 (1982)

Enzyme Handbook © Springer-Verlag Berlin Heidelberg 1994
Duplication, reproduction and storage in data banks are only allowed with the prior permission of the publishers

1 NOMENCLATURE

EC number
1.5.1.18

Systematic name
(-)-Ephedrine:NAD^+ 2-oxidoreductase

Recommended name
Ephedrine dehydrogenase

Synonymes

CAS Reg. No.
73508-06-2

---

2 REACTION AND SPECIFICITY

Catalysed reaction
(-)-Ephedrine + NAD^+ →
→ (R)-2-methylimino-1-phenylpropan-1-ol + NADH

Reaction type
Redox reaction

Natural substrates

Substrate spectrum
1 (-)-Ephedrine + NAD^+ ((-)-ephedrine is identical with 2-methylamino-1-phenyl-1-propanal) [1]
2 (+)-Pseudoephedrine + NAD^+ (pseudoephedrine is identical with DL-threo-2-(methylamino)-1-phenylpropan-1-ol) [1]
3 (-)Sympatol + NAD^+ (sympatol is identical with p-hydroxy-alpha-[(methylamino)methyl]benzyl tartrate) [1]
4 Suprifene + NAD^+ (suprifene is identical with p-hydroxyephedrine) [1]
5 (-)-Norephedrine + NAD^+ (norephedrine is identical with threo-2-amino-1-hydroxy-1-phenylpropane) [1]

Product spectrum
1 (R)-2-Methylimino-1-phenylpropan-1-ol + NADH
2 2-Methylimino-1-phenylpropan-1-ol + NADH
3 p-Hydroxy-alpha-[(methylimino)methyl]benzyl tartrate + NADH
4 p-Hydroxy-alpha-[(1-methylimino)ethyl]benzyl alcohol + NADH
5 2-Imino-1-hydroxy-1-phenylpropane + NADH

Enzyme Handbook © Springer-Verlag Berlin Heidelberg 1994
Duplication, reproduction and storage in data banks are only allowed with the prior permission of the publishers

Inhibitor(s)

Cofactor(s)/prostethic group(s)
NAD^+

Metal compounds/salts

Turnover number (min^{-1})

Specific activity (U/mg)

K_m-value (mM)
0.02 ((-)-ephedrine) [1]; 0.11 (NAD^+) [1]

pH-optimum
11 [1]

pH-range
9.5–12.2 (activity more than 60% of maximum at pH 9.5 and 12.2) [1]

Temperature optimum (°C)

Temperature range (°C)

---

3 ENZYME STRUCTURE

Molecular weight

Subunits

Glycoprotein/Lipoprotein
–

---

4 ISOLATION/PREPARATION

Source organism
Arthrobacter globiformis [1]

Source tissue

Localisation in source

Purification

Crystallization
–

Cloned
–

Renatured

–

5 STABILITY

pH

7.5–8.5 [1]

Temperature (°C)

54 (not stable above) [1]

Oxidation

Organic solvent

General stability information

Storage

6 CROSSREFERENCES TO STRUCTURE DATABANKS

PIR/MIPS code

Brookhaven code

7 LITERATURE REFERENCES

[1] Klamann, E., Lingens, F.: Z. Naturforsch.,35c,80–87 (1980)

Enzyme Handbook © Springer-Verlag Berlin Heidelberg 1994
Duplication, reproduction and storage in data banks are only allowed with the prior permission of the publishers

1 NOMENCLATURE

EC number
1.5.1.19

Systematic name
N^2–(D-1,3-Dicarboxypropyl)-L-arginine:$NADP^+$ oxidoreductase (L-arginine-forming)

Recommended name
D-Nopaline dehydrogenase

Synonymes
D-Nopaline synthase
Dehydrogenase, nopaline
Nopaline dehydrogenase
Nopaline synthase [1]
NOS [2]

CAS Reg. No.
64763-57-1

2 REACTION AND SPECIFICITY

Catalysed reaction
L-Arginine + 2-oxoglutarate + NADPH →
→ N^2–(D-1,3-dicarboxypropyl)-L-arginine + $NADP^+$ + H_2O (ter-bi rapid-equilibrium random-order mechanism [1])

Reaction type
Redox reaction

Natural substrates
L-Arginine + 2-oxoglutarate + NADPH [1]
L-Ornithine + 2-oxoglutarate + NADPH [1]

Substrate spectrum
1 L-Arginine + 2-oxoglutarate + NADPH [1]
2 L-Ornithine + 2-oxoglutarate + NADPH [1]

Product spectrum
1 N^2–(D-1,3-Dicarboxypropyl)-L-arginine (i.e. nopaline) + $NADP^+$ + H_2O
2 N^2–(1,3-Dicarboxypropyl)ornithine (i.e. ornaline) + $NADP^+$ + H_2O

Enzyme Handbook © Springer-Verlag Berlin Heidelberg 1994
Duplication, reproduction and storage in data banks are only allowed with the prior permission of the publishers

Inhibitor(s)
NADPH (substrate inhibition above 10 × K_m) [1]; alpha-Ketoglutarate (substrate inhibition above 10 × K_m) [1]

Cofactor(s)/prostethic group(s)
NADPH [1–3]; NADH [1]

Metal compounds/salts

Turnover number (min^{-1})

Specific activity (U/mg)
0.026 [1]

K_m-value (mM)
0.74 (arginine) [1]; 0.9 (alpha-ketoglutarate) [1]; 3.7 (ornithine) [1]; 0.008 (NADPH, nopaline synthesis) [1]; 0.009 (NADPH, ornaline synthesis) [1]; 1 (NADPH) [1]

pH-optimum
6.0 (nopaline synthesis) [1]; 6.5 (ornaline synthesis) [1]

pH-range

Temperature optimum (°C)
25 (assay at) [1]

Temperature range (°C)

---

3 ENZYME STRUCTURE

Molecular weight
158000 (Helianthus annuus, crown gall tissue induced by Agrobacterium tumefaciens strain C58 or T37, gel filtration) [1]

Subunits
Tetramer (4 × 40000, Helianthus annuus, crown gall tissue induced by Agrobacterium tumefaciens strain C58 or T37, SDS-PAGE) [1]

Glycoprotein/Lipoprotein
–

4 ISOLATION/PREPARATION

Source organism

Brassica juncea (inoculated with Agrobacterium tumefaciens, strain A 208 having plasmid pTi37) [3]; Helianthus annuus L. (crown gall tissue induced by Agrobacterium tumefaciens strain C58 or T37) [1]; Agrobacterium tumefaciens (crown gall tissue induced by Agrobacterium tumefaciens strain C58 or T37 [1], expression of Agrobacterium tumor-inducing plasmid pTiT37 gene (nopaline synthase) in E. coli [2], inoculated with Agrobacterium tumefaciens, strain A 208 having plasmid pTi37 [3]) [1–3]

Source tissue

Crown gall tissue (induced by Agrobacterium tumefaciens, strain C58 or T37 in Helianthus annuus L.) [1]; Axenic tumor tissue (of Brassica juncea inoculated with Agrobacterium tumefaciens, strain A 208 having plasmid pTiT37) [3]; More (no activity in habituated tissue or in octopine-type crown gall tissue) [1]

Localisation in source

Purification

Helianthus annuus L. (crown gall tissue induced by Agrobacterium tumefaciens strain C58 or T37) [1]

Crystallization

–

Cloned

(expression of Agrobacterium tumor-inducing plasmid pTiT37 gene (nopaline synthase) in E. coli) [2]

Renaturated

–

---

5 STABILITY

pH

Temperature (°C)

Oxidation

Organic solvent

General stability information

Storage

Enzyme Handbook © Springer-Verlag Berlin Heidelberg 1994
Duplication, reproduction and storage in data banks are only allowed with the prior permission of the publishers

6 CROSSREFERENCES TO STRUCTURE DATABANKS

PIR/MIPS code

DEAGNT (Agrobacterium tumefaciens plasmid pTiT37)

Brookhaven code

7 LITERATURE REFERENCES

[1] Kemp, J.D., Sutton, D.W., Hack, E.: Biochemistry,18,3755–3760 (1979)
[2] Gafni, Y., Chilton, M.-D.: Gene,39,141–146 (1985)
[3] Mathews, V.H., Bhatia, C.R., Mitra, R., Krishna, T.G., Rao, P.S.: Plant Sci.,39,49–54 (1985)

1 NOMENCLATURE

EC number
1.5.1.20

Systematic name
5-Methyltetrahydrofolate:NADP$^+$ oxidoreductase

Recommended name
Methylenetetrahydrofolate reductase (NADPH)

Synonymes
EC 1.1.1.171 (formerly)
Reductase, methylenetetrahydrofolate (reduced nicotinamide adenine dinucleotide phosphate)
5,10-Methylenetetrahydrofolate reductase (NADPH)
5,10-Methylenetetrahydrofolic acid reductase
5,10-CH_2-H_4folate reductase [6]

CAS Reg. No.
71822-25-8

2 REACTION AND SPECIFICITY

Catalysed reaction
5-Methyltetrahydrofolate + NADP$^+$ →
→ 5,10-methylenetetrahydrofolate + NADPH (mechanism [5], ping-pong mechanism [7])

Reaction type
Redox reaction

Natural substrates
5,10-Methylenetetrahydrofolate + NADPH (initial enzyme in pathway leading to synthesis of S-adenosylmethionine [2, 7], first step in biosynthesis of methyl groups [9], overview: role in incorporation of methyltetrahydrofolate into cellular metabolism [10]) [1, 2, 7, 9, 10]

Enzyme Handbook © Springer-Verlag Berlin Heidelberg 1994
Duplication, reproduction and storage in data banks are only allowed with the prior permission of the publishers

Substrate spectrum

1 5,10-Methylenetetrahydrofolate + NADPH (specific for (+)-diastereoisomer [1], equilibrium far on the side of methyltetrahydrofolate formation [1])
2 5-Methyltetrahydrofolate + oxidized menadione [1, 9]
3 NADPH + menadione [9]

Product spectrum

1 5-Methyltetrahydrofolate + $NADP^+$
2 5,10-Methylenetetrahydrofolate + reduced menadione [1, 9]
3 $NADP^+$ + reduced menadione [9]

Inhibitor(s)

S-Adenosylmethionine (inhibition partially reversed by S-adenosylhomocysteine [1, 2, 7], effect on equilibrium between active and inactive form of enzyme [8]) [1, 2, 4, 7, 8]; p-Chloromercuribenzoate [1]; Mersalyl [1]; Dicoumarol (inhibits menadione reductase activity only) [1]; Dihydrofolate [4, 7]; Polyglutamate analogues [4]; Phenylglyoxal [6]; Diethyldicarbonate [6]; N-Bromosuccinimide [6]; Dihydropteroyl-polyglutamates [7]; Folylpolyglutamate inhibitors (overview) [12]; More (residues essential for enzyme activity: arginine, histidine, tryptophan) [6]

Cofactor(s)/prostethic group(s)

FAD (flavoprotein [1, 3, 4, 7], each subunit of the dimer or trimer contains bound FAD [3], each subunit (dimer) contains noncovalently bound FAD [7], required [9]) [1, 3, 4, 7, 9]; NADPH (specific for) [1]; NADH (low activity) [1]

Metal compounds/salts

No participation of heavy metal ions in catalytic process [1]

Turnover number (min^{-1})

9600 (NADPH) [7]

Specific activity (U/mg)

153 (pig) [1]; 78 (rat) [1]; 19 [3, 7]; 0.367 [9]; More (assay methods) [7]

K_m-value (mM)

0.033 (NADPH, methylenetetrahydrofolate reductase reaction) [1]; 0.310 (NADH, methylenetetrahydrofolate reductase reaction) [1]; 0.021 (methylenetetrahydrofolate, methylenetetrahydrofolate reductase reaction) [1]; 0.049 (NADPH, reduction of menadione) [1]; 0.085 (NADH, reduction of menadione) [1]; 0.016 (NADPH, methylenetetrahydrofolate reductase reaction, pH 7.2) [7]; 0.088 (methylenetetrahydrofolate, methylenetetrahydrofolate reductase reaction, pH 7.2) [7]; 0.019 (methylenetetrahydrofolate, methylenetetrahydrofolate reductase reaction, pH 6.7) [7]

pH-optimum

6.5 (NADPH-tetrahydrofolate oxidoreductase activity) [9]; 6.6–6.7 (methylenetetrahydrofolate reductase reaction) [1]; 6.6 (methyltetrahydrofolate-menadione oxidoreductase activity) [9]; 7.1 (menadione reductase reaction) [1]; 7.2 (NADPH-menadione oxidoreductase activity) [9]

pH-range

6.2–7.3 (half-maximal activities, methylenetetrahydrofolate reductase reaction) [1]; 5.8–8.6 (5.8: about 60% of activity maximum, 8.6: about 50% of activity maximum, menadione reductase activity) [1]

Temperature optimum (°C)

37 (assay at) [3, 9]

Temperature range (°C)

25–45 (methylenetetrahydrofolate reductase reaction, slow increase of activity between 25°C and 45°C) [1]

3 ENZYME STRUCTURE

Molecular weight

136000 (pig, scanning transmission electron microscopy) [7]
150000 (human, gel filtration) [9]
170000–190000 (pig, gel filtration, single enzyme with methylenetetrahydrofolate reductase and dopamine methyltransferase activity) [2]
210000 (pig, gel filtration) [3]

Subunits

Dimer (2 × 77300, pig, SDS-PAGE, gel filtration [3], dimer or trimer, 2 (or 3) × 77300, pig, SDS-PAGE [7], 2 × 75000, human, SDS-PAGE and inhibition studies [9]) [3, 7, 9]

Glycoprotein/Lipoprotein

–

4 ISOLATION/PREPARATION

Source organism

Rat [1, 11, 13]; Pig (a single enzyme with methylenetetrahydrofolate reductase and dopamine methyltransferase activity [2]) [1–5, 7, 8, 14]; Sheep [6]; Human [9]; Monkey [11]

Source tissue

Liver [1–7, 9, 13]; Brain [13]; Central nervous system (all regions) [13]

Enzyme Handbook © Springer-Verlag Berlin Heidelberg 1994
Duplication, reproduction and storage in data banks are only allowed with the prior permission of the publishers

Localisation in source
Cytoplasm [13]; Soluble [13]

Purification
Pig [1, 3, 7]; Rat [1]; Human [9]

Crystallization
–

Cloned
–

Renaturated
–

---

5 STABILITY

pH

Temperature (°C)
45 (inactivation above) [1]

Oxidation

Organic solvent

General stability information
Inclusion of 10% glycerol during purification is essential for stability [3]; Extremely sensitive to proteolysis [7, 9]

Storage
–20°C, for at least 2 weeks [3]; –70°C, 10% glycerol, several months [7]

---

6 CROSSREFERENCES TO STRUCTURE DATABANKS

PIR/MIPS code

Brookhaven code

7 LITERATURE REFERENCES

[1] Kutzbach, C., Stokstad, E.L.R.: Biochim. Biophys. Acta,250,459–477 (1971)
[2] Pearson, A.G.M., Turner, A.J.: Nature,258,173–174 (1975)
[3] Daubner, S.C., Matthews, R.G.: J. Biol. Chem.,257,140–145 (1982)
[4] Matthews, R.G., Daubner, S.C.: Adv. Enzyme Regul.,20,123–131 (1982)
[5] Vanoni, M.A., Ballou, D.P., Matthews, R.G.: J. Biol. Chem.,258,11510–11514 (1983)
[6] Varalakshmi, K., Savithri, H.S., Rao, N.A.: Biochem. J.,236,295–298 (1986)
[7] Matthews, R.G.: Methods Enzymol.,122,372–381 (1986) (Review)
[8] Jencks, D.A., Matthews, R.G.: J. Biol. Chem.,262,2485–2493 (1987)
[9] Zhou, J., Kang, S.-S., Wong, P.W.K., Fournier, B., Rozen, R.: Biochem. Med. Metab. Biol.,43,234–242 (1990)
[10] Green, J.M., Ballou, D.P., Matthews, R.G.: FASEB J.,2,42–47 (1988) (Review)
[11] Black, K.A., Eells, J.T., Noker, P.E., Hawtrey, C.A., Tephly, T.R.: Proc. Natl. Acad. Sci. USA,82,3854–3858 (1985)
[12] Matthews, R.G., Ghose, C., Green, J.M., Matthews, K. D., Dunlap, R.B.: Adv. Enzyme Regul.,26,157–171 (1987)
[13] Burton, E.G., Sallach, H.J.: Arch. Biochem. Biophys.,166,483–494 (1975)

Enzyme Handbook © Springer-Verlag Berlin Heidelberg 1994
Duplication, reproduction and storage in data banks are only allowed with the prior permission of the publishers

1 NOMENCLATURE

EC number
1.5.1.21

Systematic name
L-Pipecolate:NADP$^+$ 2-oxidoreductase

Recommended name
DELTA^1-Piperideine-2-carboxylate reductase

Synonymes
E.C. 1.5.1.14 (now included with 1.5.1.21)
Reductase, DELTA^1-piperideine-2-carboxylate
1,2-Didehydropipecolate reductase
P2C reductase [1]
1,2-Didehydropipecolic reductase

CAS Reg. No.
52037-88-4

2 REACTION AND SPECIFICITY

Catalysed reaction
DELTA^1-Piperideine-2-carboxylate + NADPH →
→ L-pipecolate + NADP$^+$

Reaction type
Redox reaction

Natural substrates
DELTA^1-Piperideine-2-carboxylate + NADPH (D-lysine metabolism) [1, 2]

Substrate spectrum
1 DELTA^1-Piperideine-2-carboxylate + NADPH (ir [1])
2 More (poor substrates: DELTA^1-pyrroline-5-carboxylate, pyroglutamate) [1]

Product spectrum
1 L-Pipecolate + NADP$^+$ (ir [1])
2 ?

Enzyme Handbook © Springer-Verlag Berlin Heidelberg 1994
Duplication, reproduction and storage in data banks are only allowed with the prior permission of the publishers

Inhibitor(s)
NaF [1]; NADPH (substrate inhibition above 0.15 mM) [1]; Zn^{2+} [1]; Mn^{2+} [1]; Hg^{2+} [1]; Co^{2+} [1]; Ca^{2+} (slight) [1]; Mg^{2+} (slight) [1]; KCN (slight) [1]; p-Chloromercuribenzoate [1]; More (no effect: 0.1 mM L-cysteine, 0.1 mM iodoacetate) [1]

Cofactor(s)/prostethic group(s)
NADPH [1]; NADH (poor coenzyme) [1]

Metal compounds/salts

Turnover number (min^{-1})

Specific activity (U/mg)

K_m-value (mM)
0.23 (DELTA^1-piperideine-2-carboxylate) [1]; 0.13 (NADPH) [1]

pH-optimum
8.0–8.3 [1]

pH-range

Temperature optimum (°C)
30 (assay at) [1]

Temperature range (°C)

---

3 ENZYME STRUCTURE

Molecular weight
120000 (Pseudomonas putida, gel filtration) [1]

Subunits

Glycoprotein/Lipoprotein
–

---

4 ISOLATION/PREPARATION

Source organism
Pseudomonas putida [1, 2]

Source tissue

Localisation in source

Purification

Pseudomonas putida [1]

Crystallization

–

Cloned

–

Renaturated

–

5 STABILITY

pH

Temperature (°C)

Oxidation

Organic solvent

General stability information

NADPH, protection during storage [1]; $NADP^+$, no protection during storage [1]; EDTA, 1 mM, protects during storage and against heat inactivation at 50°C [1]; Sodium diphosphate, 1 mM, protects during storage [1]; Dithiothreitol, 1 mM, protects during storage [1]; Very unstable during purification [1]

Storage

6 CROSSREFERENCES TO STRUCTURE DATABANKS

PIR/MIPS code

Brookhaven code

7 LITERATURE REFERENCES

[1] Payton, C.W., Chang, Y.-F.: J. Bacteriol.,149,864–871 (1982)
[2] Chang, Y.-F., Adams, E.: J. Bacteriol.,117,753–764 (1974)

Enzyme Handbook © Springer-Verlag Berlin Heidelberg 1994
Duplication, reproduction and storage in data banks are only allowed with the prior permission of the publishers

1 NOMENCLATURE

EC number
1.5.1.22

Systematic name
N-(Carboxymethyl)-D-alanine:NAD^+ oxidoreductase (glycine-forming)

Recommended name
Strombine dehydrogenase

Synonymes
Dehydrogenase, strombine
Strombine[N-(carboxymethyl)-D-alanine]dehydrogenase [5]
N-(Carboxymethyl)-D-alanine: NAD^+ oxidoreductase [5]

CAS Reg. No.
79393-84-3

2 REACTION AND SPECIFICITY

Catalysed reaction
Glycine + pyruvate + NADH →
→ N-(carboxymethyl)-D-alanine + NAD^+ + H_2O (enzyme also catalyzes the reaction of EC 1.5.1.17, more slowly [5])

Reaction type
Redox reaction

Natural substrates
Glycine + pyruvate + NADH (functions during periods of elevated energy demand [2], terminal dehydrogenase of glycolysis, role in maintaining energy production under stress of environmental or functional anoxia [3]) [2, 3]

Enzyme Handbook © Springer-Verlag Berlin Heidelberg 1994
Duplication, reproduction and storage in data banks are only allowed with the prior permission of the publishers

Substrate spectrum

1 Glycine + pyruvate + NADH (r [2, 6])
2 More (broad amino acid specificity: glycine (100% activity), L-alanine (78% [2], 76% [3]) [2, 3, 5, 6], serine (82% [3]) [3, 5, 6], L-glutamate (60% [2], not [5]) [2], L-cysteine (20% [6], 67% [2], less than 10% [3]) [2, 3, 5, 6], L-2-aminobutyrate (48% [3]) [3, 6], restricted use of other L-amino acids [3], broad keto-acid specificity: pyruvate (100% activity), oxaloacetate (97% [2], 106% [6]) [2, 5, 6], alpha-ketobutyrate (39%) [2], glyoxylate (27% [2], 38.6% [6], 20% [3], not [5]) [2, 3, 6], alpha-ketovalerate (20%) [6], hydroxypyruvate (30%) [6], not: 2-oxoglutarate [5], L-strombine [5], relative reaction rate with meso-alanopine to D-strombine is 100: 123 [5]) [2, 3, 5, 6]

Product spectrum

1 N-(Carboxymethyl)-D-alanine + NAD^+ + H_2O (r [2, 6])
2 ?

Inhibitor(s)

Succinate (weak) [3]; L-Lactate (weak) [3, 6]; D-Lactate (weak) [3, 6]; meso-Alanopine (weak [3], strong [6]) [3, 6]; Iminodoacetic acid [6]; L-Strombine (weak) [6]

Cofactor(s)/prostethic group(s)

NAD^+; NADH; More (not: $NADP^+$, NADPH) [5]

Metal compounds/salts

Turnover number (min^{-1})

Specific activity (U/mg)

37 [3]; 180 (Glycera alba) [1]; 66 (Scolelepis fuliginosa) [1]; 204 [6]

K_m-value (mM)

180 (glycine, Glycera alba) [1]; 42.4 (glycine, Scolelepis fuliginosa) [1]; 1.2 (pyruvate, Glycera alba) [1]; 0.5 (pyruvate, Scolelepis fuliginosa) [1]; 32.1 (alanine) [2]; 35.7 (glycine) [2]; 0.23 (pyruvate) [2]; 0.66 (oxaloacetate) [2]; 3.35 (alanopine) [2]; 0.009 (NADH) [2]; 0.3 (NAD^+) [2]; 242 (alanine) [3]; 173 (glycine) [3]; 0.61 (pyruvate) [3]; 693 (L-serine) [3]; 44 (L-2-aminobutyrate) [3]; 2.46 (2-oxobutyrate) [3]; 4.18 (glyoxylate) [3]; 0.4 (oxaloacetate) [5]; 2 (2-oxobutyrate) [5]; 0.004 (NADH) [5]; More [5, 6]

pH-optimum

6.0–6.5 (glycine + pyruvate + NADH) [6]; 6.6 (opine formation with glycine or alanine) [2]; 8.8 (increase in velocity up to 8.8, alanopine oxidation) [2]; 9.0 (strombine) [5, 6]; 7.2 (glycine + pyruvate + NADH) [5]

pH-range
6.0–7.5 (opine formation with glycine or alanine, 6.0: about 60% of activity maximum, 7.5: about 50% of activity maximum) [2]

Temperature optimum (°C)
25 (assay at) [1, 2]; 20 (assay at) [5]; 23 (assay at) [6]

Temperature range (°C)

3 ENZYME STRUCTURE

Molecular weight
37000 (Modiolus squamosus, gel filtration) [2]
38200 (Mytilus edulis, gel filtration) [5]
44000 (Aphrodite aculeata, gel filtration) [6]

Subunits
Monomer (Mytilus edulis [5], Aphrodite aculeata [6]) [5, 6]

Glycoprotein/Lipoprotein
–

4 ISOLATION/PREPARATION

Source organism
Arenicola marina (marine invertebrate, EC 1.5.1.7 or EC 1. 5.1.22 ?) [4]; Mytilus edulis (mussel, marine invertebrate, EC 1.5.1.7 or EC 1. 5.1.22 ? [4]) [4, 6]; Nucula nitida (marine invertebrate, EC 1.5.1.7 or EC 1. 5.1.22 ?) [4]; Crassostrea angulata (marine invertebrate, EC 1.5.1.7 or EC 1. 5.1.22 ?) [4]; Aphrodite aculeata (sea mouse) [6]; Glycera alba [1]; Scolelepis fuliginosa [1]; Capitella capitata (low activity) [1]; Modiolus squamosus (marine mussel) [2]; Mercenaria mercenaria (bivalve mollusc) [3]

Source tissue
Pharynx [6]; Intestine [6]; Body-wall [1]; Adductor muscle [2, 3, 5]; Foot muscle [3]; Gill [3]

Localisation in source
Cytoplasm [5, 6]

Purification
Modiolus squamosus (partial) [2]; Mercenaria mercenaria (partial) [3]; Aphrodite aculeata (partial) [6]

Crystallization
–

Enzyme Handbook © Springer-Verlag Berlin Heidelberg 1994
Duplication, reproduction and storage in data banks are only allowed with the prior permission of the publishers

Cloned

–

Renaturated

–

5 STABILITY

pH

Temperature (°C)

Oxidation

Organic solvent

General stability information

Storage

4°C, 1 week [6]

6 CROSSREFERENCES TO STRUCTURE DATABANKS

PIR/MIPS code

Brookhaven code

7 LITERATURE REFERENCES

[1] Blackstock, J., Burdass, M.C.: Biochem. Soc. Trans.,15,383–384 (1987)
[2] Nicchitta, C.V., Ellington, W.R.: Comp. Biochem. Physiol.,77B,233–236 (1984)
[3] Fields, J.H.A., Storey, K.B.: J. Exp. Mar. Biol. Ecol.,105,175–185 (1987)
[4] Siegmund, B., Grieshaber, M.K.: Hoppe-Seyler's Z. Physiol. Chem.,364,807–812 (1983)
[5] Dando, P.R.: Biochem. Soc. Trans.,9,297–298 (1981)
[6] Storey, K.B.: J. Exp. Zool.,225,369–378 (1983)

1 NOMENCLATURE

EC number
1.5.1.23

Systematic name
N^2–(D-1-Carboxyethyl)taurine:NAD$^+$ oxidoreductase (taurine forming)

Recommended name
Tauropine dehydrogenase

Synonymes

CAS Reg. No.
104645-74-1

2 REACTION AND SPECIFICITY

Catalysed reaction
Tauropine + NAD$^+$ + H_2O →
→ taurine + pyruvate + NADH

Reaction type
Redox reaction

Natural substrates
Taurine + pyruvate + NADH [1, 2]

Substrate spectrum
1 Taurine + pyruvate + NADH (r) [1, 2]
2 L-Alanine + pyruvate + NADH [1, 2]
3 Taurine + 2-oxobutyrate + NADH [2]
4 Taurine + 2-oxovalerate + NADH [2]

Product spectrum
1 Tauropine + NAD$^+$ + H_2O [1, 2]
2 ? + NAD$^+$ + H_2O [1, 2]
3 ? + NAD$^+$ + H_2O [2]
4 ? + NAD$^+$ + H_2O [2]

Inhibitor(s)
Succinate [2]; L-Lactate [2]

Cofactor(s)/prostethic group(s)
NAD$^+$ [1, 2]

Enzyme Handbook © Springer-Verlag Berlin Heidelberg 1994
Duplication, reproduction and storage in data banks are only allowed with the prior permission of the publishers

Metal compounds/salts

Turnover number (min^{-1})

Specific activity (U/mg)
463 [2]

K_m-value (mM)
65 (taurine) [1, 2]; 235 (L-alanine) [1, 2]; 0.022 (NADH) [2]; 0.64 (pyruvate) [2]; 0.29 (NAD^+) [2]; 9.04 (tauropine) [2]; 21.1 (2-oxobutyrate) [2]; 3.6 (2-oxovalerate) [2]

pH-optimum
7.0 (taurine + NADH + pyruvate) [2]

pH-range

Temperature optimum (°C)

Temperature range (°C)

---

3 ENZYME STRUCTURE

Molecular weight
42000 (Haliotis lamellosa, SDS-PAGE) [2]
38000 (Haliotis lamellosa, gel filtration) [2]

Subunits
Monomer (1 × 42000, Haliotis lamellosa, SDS-PAGE) [2]

Glycoprotein/Lipoprotein
–

---

4 ISOLATION/PREPARATION

Source organism
Haliotis lamellosa (ormer) [1, 2]

Source tissue
Shell adductor muscle [1, 2]

Localisation in source

Purification
Haliotis lamellosa [1, 2]

Crystallization
–

Cloned
–

Renaturated
–

5 STABILITY

pH

Temperature (°C)

Oxidation

Organic solvent

General stability information

Storage
3–4 weeks, 0–4°C [2]

6 CROSSREFERENCES TO STRUCTURE DATABANKS

PIR/MIPS code

Brookhaven code

7 LITERATURE REFERENCES

[1] Gäde, G.: Biol. Chem. Hoppe-Seyler,368,1519–1523 (1987)
[2] Gäde, G.: Eur. J. Biochem.,160,311–318 (1986)

Enzyme Handbook © Springer-Verlag Berlin Heidelberg 1994
Duplication, reproduction and storage in data banks are only allowed with the prior permission of the publishers

1 NOMENCLATURE

EC number
1.5.1.24

Systematic name
N^5–(L-1-Carboxyethyl)-L-ornithine:NADP$^+$ oxidoreductase (L-ornithine-forming)

Recommended name
N^5–(Carboxyethyl)ornithine synthase

Synonymes
Synthase, N^5–(carboxy ethyl)ornithine

CAS Reg. No.
129070-70-8

2 REACTION AND SPECIFICITY

Catalysed reaction
L-Ornithine + pyruvate + NADPH →
→ N^5–(L-1-carboxyethyl)-L-ornithine + NADP$^+$ + H_2O

Reaction type
Redox reaction
Reductive condensation [1]

Natural substrates

Substrate spectrum
1 L-Ornithine + pyruvate + NADPH (ir) [1]
2 L-Lysine + pyruvate + NADPH [1]

Product spectrum
1 N^5–(L-1-Carboxyethyl)-L-ornithine + NADP$^+$ + H_2O [1]
2 N^6–(L-1-Carboxyethyl)-L-lysine + NADP$^+$ + H_2O [1]

Inhibitor(s)
N^5–(L-1-Carboxyethyl)-L-ornithine (product inhibition) [1]

Cofactor(s)/prostethic group(s)
NADPH [1]

Metal compounds/salts

Enzyme Handbook © Springer-Verlag Berlin Heidelberg 1994
Duplication, reproduction and storage in data banks are only allowed with the prior permission of the publishers

Turnover number (min^{-1})

Specific activity (U/mg)

K_m-value (mM)
0.0066 (NADPH) [1]; 0.150 (pyruvate) [1]; 3.3 (ornithine) [1]; 18.2 (lysine) [1]

pH-optimum
8.0 [1]

pH-range
6.5–9.0 [1]

Temperature optimum (°C)

Temperature range (°C)

3 ENZYME STRUCTURE

Molecular weight
78000 (Streptococcus lactis, gel filtration) [1]
150000 (Streptococcus lactis, HPLC exclusion chromatography, polyacrylamide gradient gel electrophoresis) [1]

Subunits
Dimer or tetramer (2 or 4 × 38000, Streptococcus lactis, SDS-PAGE) [1]

Glycoprotein/Lipoprotein
–

4 ISOLATION/PREPARATION

Source organism
Streptococcus lactis [1, 2]

Source tissue
Cell [1]

Localisation in source

Purification
Streptococcus lactis [1]

Crystallization
–

Cloned
–

Renatured
–

5 STABILITY

pH

Temperature (°C)
25–50 (10 min, stable) [1]; 60 (rapid inactivation) [1]

Oxidation

Organic solvent

General stability information
Glycerol, 20% v/v, 2 months, stable [1]

Storage

6 CROSSREFERENCES TO STRUCTURE DATABANKS

PIR/MIPS code
A33852 (Lactococcus lactis fragment)

Brookhaven code

7 LITERATURE REFERENCES

[1] Thompson, J.: J. Biol. Chem.,264,9592–9601 (1989)
[2] Thompson, J., Nguyen, N.Y., Sackett, D.L., Donkersloot, J.A.: J. Biol. Chem.,266,14573–14579 (1991)

Enzyme Handbook © Springer-Verlag Berlin Heidelberg 1994
Duplication, reproduction and storage in data banks are only allowed with the prior permission of the publishers

1 NOMENCLATURE

EC number
1.5.1.25

Systematic name
Thiomorpholine-3-carboxylate:NAD(P)$^+$ 5,6-oxidoreductase

Recommended name
Thiomorpholine-carboxylate dehydrogenase

Synonymes
Ketimine reductase
Reductase, ketimine
Ketimine-reducing enzyme [1]

CAS Reg. No.
115232-54-7

---

2 REACTION AND SPECIFICITY

Catalysed reaction
Thiomorpholine 3-carboxylate + NAD(P)$^+$ →
→ 3,4-dehydro-1,4-thiomorpholine-3-carboxylate + NAD(P)H (classical ping-pong mechanism [1, 2])

Reaction type
Redox reaction

Natural substrates
Lanthionine ketimine + NAD(P)H [1, 2]
Cystathionine ketimine + NAD(P)H [1, 2]

Substrate spectrum
1 S-Aminoethylcysteine ketimine + NAD(P)H (ir [1]) [1, 2]
2 Lanthionine ketimine + NAD(P)H (ir [1]) [1, 2]
3 Cystathionine ketimine + NAD(P)H (ir [1]) [1, 2]
4 1-Piperidine 2-carboxylate + NAD(P)H [1, 2]
5 ?

Enzyme Handbook © Springer-Verlag Berlin Heidelberg 1994
Duplication, reproduction and storage in data banks are only allowed with the prior permission of the publishers

Product spectrum

1 1,4-Thiomorpholine 3-carboxylic acid + $NAD(P)^+$ [1]
2 1,4-Thiomorpholine 3,5-dicarboxylic acid + $NAD(P)^+$ [1]
3 Cyclothionine + $NADP^+$ [1]
4 ?
5 More (the product is the cyclic imine of the 2-oxoacid corresponding to S-(2-aminoethyl)cysteine, in the reverse direction a number of cyclic unsaturated compounds can act as substrates, more slowly)

Inhibitor(s)

Cofactor(s)/prostethic group(s)

NADH (reduction of lanthionine ketimine is faster by NADPH than by NADH, reduction of S-aminoethylcysteine ketimine is faster by NADPH than by NADH) [1]; NADPH (reduction of lanthionine ketimine is faster by NADPH than by NADH, reduction of S-aminoethylcysteine ketimine is faster by NADPH than by NADH) [1]

Metal compounds/salts

Turnover number (min^{-1})

Specific activity (U/mg)

16.3 [1]; 16.6 [2]

K_m-value (mM)

0.027 (NADH (+ S-aminoethylcysteine ketimine)) [1]; 0.077 (S-aminoethylcysteine ketimine) [1]; 0.47 (lanthionine ketimine (+ NADH)) [1]; 3.0 (cystathionine ketimine (+ NADH)) [1]

pH-optimum

4.5 (S-aminoethylcysteine ketimine [1], lanthionine ketimine (+ NADPH [2]) [1, 2]) [1, 2]; 5.0 (S-aminoethylcysteine ketimine (+ NADH [2]) [1, 2], cystathionine ketimine + NADPH [2], 1-piperidine 2-carboxylate + NADPH [2]) [1, 2]; 6.0 (cystathionine ketimine) [1]

pH-range

Temperature optimum (°C)

Temperature range (°C)

---

3 ENZYME STRUCTURE

Molecular weight

73000–76000 (pig, gel filtration, polyacrylamide gel electrophoresis) [1]
100000 (bovine, polyacrylamide gel electrophoresis) [2]

Subunits
Dimer (2 × 45000, bovine, SDS-PAGE) [2]

Glycoprotein/Lipoprotein
–

4 ISOLATION/PREPARATION

Source organism
Pig [1]; Bovine [2]

Source tissue
Kidney [1]; Brain [2]

Localisation in source
Soluble [2]

Purification
Pig [1]; Bovine [2]

Crystallization
–

Cloned
–

Renaturated
–

5 STABILITY

pH

Temperature (°C)

Oxidation

Organic solvent

General stability information
Irreversible inactivation at protein concentration below 1 mg/ml [1]; Glycerol, 10% v/v, stabilizes [1]; Triton X-100 causes irreversible inactivation [1]

Storage
4°C, 1 month [2]

Enzyme Handbook © Springer-Verlag Berlin Heidelberg 1994
Duplication, reproduction and storage in data banks are only allowed with the prior permission of the publishers

6 CROSSREFERENCES TO STRUCTURE DATABANKS

PIR/MIPS code

Brookhaven code

7 LITERATURE REFERENCES

[1] Nardini, M., Ricci, G., Caccuri, A.M., Solinas, S.P., Vesci, L., Cavallini, D.: Eur. J. Biochem.,173,689–694 (1988)
[2] Nardini, M., Ricci, G., Vesci, L., Pecci, L., Cavallini, D.: Biochim. Biophys. Acta,957,286–292 (1988)

1 NOMENCLATURE

EC number
1.5.1.26

Systematic name
N-(D-1-Carboxyethyl)-beta-alanine:NAD^+ oxidoreductase (beta-alanine-forming)

Recommended name
beta-Alanopine dehydrogenase

Synonymes
Dehydrogenase, beta-alanopine

CAS Reg. No.
113573-64-1

---

2 REACTION AND SPECIFICITY

Catalysed reaction
beta-Alanopine + NAD^+ + H_2O →
→ beta-alanine + pyruvate + NADH

Reaction type
Redox reaction
Reductive condensation [1, 2]

Natural substrates
beta-Alanine + pyruvate + NADH (enzyme functions in metabolism in place of lactate dehydrogenase, which often has very low activity especially in moluscs [1], may play an important physiological role in regulation of cytoplasmic redox balance [2]) [1, 2]

Substrate spectrum
1 beta-Alanine + pyruvate + NADH [1, 2]

Product spectrum
1 beta-Alanopine + NAD^+ + H_2O

Inhibitor(s)

Cofactor(s)/prostethic group(s)
NADH [1, 2]

Metal compounds/salts

Enzyme Handbook © Springer-Verlag Berlin Heidelberg 1994
Duplication, reproduction and storage in data banks are only allowed with the prior permission of the publishers

Turnover number (min^{-1})

Specific activity (U/mg)

K_m-value (mM)

pH-optimum

pH-range

Temperature optimum (°C)

Temperature range (°C)

---

3 ENZYME STRUCTURE

Molecular weight

Subunits

Glycoprotein/Lipoprotein

–

---

4 ISOLATION/PREPARATION

Source organism

Glottidia pyramidata (brachiopoda) [1]; Anadara ovalis (bivalvia) [1]; Crassostrea virginica (bivalvia) [1]; Tectura testudinalis (gastropoda) [1]; Urosalpinx cincerea (gastropoda) [1]; Scapharca broughtonii (shell) [2]

Source tissue

Foot muscle [2]; Adductor muscle [2]; Mantle [2]; Mid-gut gland [2]; Gill [2]

Localisation in source

Purification

Crystallization

–

Cloned

–

Renaturated

–

5 STABILITY

pH

Temperature (°C)

Oxidation

Organic solvent

General stability information

Storage

6 CROSSREFERENCES TO STRUCTURE DATABANKS

PIR/MIPS code

Brookhaven code

7 LITERATURE REFERENCES

[1] Hammen, C.S., Bullock, R.C.: Biochem. Syst. Ecol.,19,263–269 (1991)
[2] Sato, M., Takahara, M., Kanno, N., Sato, Y., Ellington, W.R.: Comp. Biochem. Physiol.,88B,803–806 (1987)

Enzyme Handbook © Springer-Verlag Berlin Heidelberg 1994
Duplication, reproduction and storage in data banks are only allowed with the prior permission of the publishers

1 NOMENCLATURE

EC number
1.5.3.1

Systematic name
Sarcosine:oxygen oxidoreductase (demethylating)

Recommended name
Sarcosine oxidase

Synonymes
Oxidase, sarcosine

CAS Reg. No.
9029-22-5

2 REACTION AND SPECIFICITY

Catalysed reaction
Sarcosine + H_2O + O_2 →
→ glycine + formaldehyde + H_2O_2 (ping-pong bi-bi mechanism [13], mechanism [11, 14])

Reaction type
Redox reaction
Oxidative demethylation [15]

Natural substrates
Sarcosine + H_2O + O_2 (creatinine catabolism [1, 5], sarcosine degradation when sarcosine is sole source of carbon, nitrogen and energy [15]) [1, 5, 15]

Substrate spectrum
1 Sarcosine + O_2 + H_2O [1–17]
2 N-Methyl-L-leucine + O_2 + H_2O [1, 3]
3 N-Methyl-DL-alanine + O_2 + H_2O [1–4, 7]
4 N-Methyl-DL-valine + O_2 + H_2O [1, 3]
5 N-Ethylglycine + O_2 + H_2O [7]
6 Heterocyclic amines + O_2 + H_2O (slowly, e.g.: L-proline, L-pipecolic acid) [17]

Enzyme Handbook © Springer-Verlag Berlin Heidelberg 1994
Duplication, reproduction and storage in data banks are only allowed with the prior permission of the publishers

7 More (very specific for oxygen as acceptor, oxygen can be replaced by: 2,6-dichlorophenolindophenol, phenazine methosulfate, ferricyanide (much smaller V_{max}/K_m values than for O_2) [7], flavin and cytochromes of the c and b (or o) type function as electron carriers [15], not: beta-alanine, N-methylalanine, 1,3-dimethylurea, 1-methylguanidine, methoxyacetate, creatine, creatinine [3], choline [9], N-methyl amino acids [9], dimethylglycine [4, 9], betaine [4, 9], methylene blue [7], cytochrome c [7]) [3, 4, 7, 9, 15]

Product spectrum

1 Glycine + formaldehyde + H_2O_2 [1–17]
2 Formaldehyde + L-Leu + H_2O_2
3 Formaldehyde + L-Val + H_2O_2
4 Acetaldehyde + Gly + H_2O_2
5 ?
6 ?
7 ?

Inhibitor(s)

Iodoacetate [2, 7, 9]; Iodoacetamide [2, 12]; Dithiothreitol [2]; 2-Mercaptoethanol (slight) [2]; o-Phenanthroline (slight [9]) [2, 9]; EDTA (slight [2], not [1, 3]) [2]; N-Ethylmaleimide [2, 7]; Acetate [7, 12, 13, 17]; Propionate [7, 13]; Formaldehyde [7]; Acetaldehyde [7]; N-Methylamino acids [9]; Phenylhydrazine (slight) [9]; Diethyldicarbonate [12]; Methoxyacetate [13]; 2-Furoic acid [17]; 2-Pyrrolecarboxylic acid [17]; 2-Thiophenecarboxylic acid [17]; N-Bromosuccinimide [1, 3]; Hydroxylamine hydrochloride [1]; SDS [1]; Zn^{2+} [1–3, 7, 13]; Ni^{2+} [1]; Cd^{2+} [2, 7, 13]; Fe^{3+} (not [7]) [3]; Pb^{2+} [13]; p-Chloromercuribenzoate (not [1, 3, 7]) [2, 9, 12]; More (chemical modifications) [12]

Cofactor(s)/prostethic group(s)

FAD (flavoprotein: the flavin is both covalently and noncovalently bound in a molar ratio of 1: 1, flavin cofactor associated with the 45000 MW subunit [2], presumably covalently bound and noncovalently bound flavins [2], covalently bound flavin [3], 1 mol of covalently bound flavin per mol of enzyme [6–8, 14, 15], one molecule FAD is covalently bound to subunit B [7], acid-nonextractable flavins [10], 1 mol of covalently bound FAD per mol of enzyme [9]) [2, 3, 6–10, 14, 15]

Metal compounds/salts

Turnover number (min^{-1})

132 (N-ethylglycine) [7]; 204 (N-methyl-L-alanine) [7]; 348 (sarcosine) [2, 7]; 246 (N-methyl-DL-alanine) [2]

Specific activity (U/mg)
25 [1]; 7.3 [2]; 8.93 [7]; 27.6 [9]; More [3]

K_m-value (mM)
0.91 (sarcosine) [1]; 6.4 (sarcosine) [2]; 12.2 (sarcosine) [3]; 6.7 (N-methyl-DL-valine) [1]; 0.58 (N-methyl-L-leucine) [1]; 1.6 (N-methyl-DL-alanine) [1]; 16.5 (N-methyl-DL-alanine) [2]; 6.8 (N-methyl-DL-alanine) [3]; 106 (N-methyl-L-leucine) [3]; 173 (N-methyl-DL-valine) [3]; 0.13 (O_2) [7]; 0.31 (2,6-dichlorophenolindophenol) [7]; 1.67 (phenazine methosulfate) [7]; 2.70 (potassium ferricyanide) [7]

pH-optimum
8 [1, 3]; 8.3 [2]; 8.5–9 [3]

pH-range
6–10 (6: about 65% of activity maximum [1], about 35% of activity maximum [2], about 35% of activity maximum [7], 10: about 80% of activity maximum [1], about 55% of activity maximum [2], about 70% of activity maximum [7]) [1, 2, 7]

Temperature optimum (°C)
37 [3]

Temperature range (°C)

3 ENZYME STRUCTURE

Molecular weight
42000 (Bacillus sp. B-0618, gel filtration) [3]
44000 (Streptomyces sp. KB210–8SY, gel filtration, SDS-PAGE) [1]
45000 (Cylindrocarpon didynum M-1, gel filtration) [9]
168000 (Corynebacterium sp. P-1, gel filtration) [16]
174000 (Corynebacterium sp. U-96, meniscus depletion method) [7]
185000 (Arthrobacter ureafaciens, gel filtration) [2]
190000 (Alcaligenes denitrificans, gel filtration) [5]

Subunits
Momomer (1 × 42000, Bacillus sp. B-0618, 1 × 45000, Cylindrocarpon didynum, gel filtration, SDS-PAGE) [9]
Dimer (or trimer, alpha,beta or alpha,$beta_2$, 100000 (alpha), 55000 (beta), Alcaligenes denitrificans SDS-PAGE) [5]
Tetramer (1 × 96000 + 1 × 45000 + 1 × 23000 + 1 × 14000, Arthrobacter ureafaciens, SDS-PAGE [2], 1 × 110000 (A) + 1 × 44000 (B) + 1 × 21000 (C) + 1 × 10000 (D), Corynebacterium sp. U-96, SDS-PAGE [6], 1 × 100000 + 1 × 42000 + 1 × 20000 + 1 × 6000, Corynebacterium sp. P-1, SDS-PAGE [16]) [2, 6, 16]

Enzyme Handbook © Springer-Verlag Berlin Heidelberg 1994
Duplication, reproduction and storage in data banks are only allowed with the prior permission of the publishers

Glycoprotein/Lipoprotein

–

4 ISOLATION/PREPARATION

Source organism

Arthrobacter sp. (J5 and J11) [5]; Streptomyces sp. KB210–8SY [1]; Arthrobacter ureafaciens [2]; Alcaligenes denitrificans [4, 5]; Bacillus sp. B-0618 [3]; Cylindrocarpon didynum M-1 (fungus) [9]; Pseudomonas sp. [10, 14]; Corynebacterium sp. (U-96 [6–8, 11–14] and P-1 [16, 17]) [6–8, 11–14, 16, 17]; More (straight gram-negative rod) [15]

Source tissue

Localisation in source

Purification

Cylindrocarpon didynum M-1 [9]; Alcaligenes denitrificans [5]; Streptomyces sp. KB210–8SY [1]; Arthrobacter ureafaciens [2]; Bacillus sp. B-0618 [3]; Corynebacterium sp. U-96 [7, 8]; Corynebacterium sp. P-1 [16]; More (straight gram-negative rod) [15]

Crystallization

(Alcaligenes denitrificans) [4]

Cloned

–

Renaturated

–

5 STABILITY

pH

7–9 (stable) [1]; 6.5–9.5 (20°C, 20 h) [2]; 7–10 [3]; 5 (labile below) [3]

Temperature (°C)

45 (10 min, complete loss of activity) [7]; 40 (pH 8.0, 24 h, 50% loss of activity) [2]; 30 (pH 7–9, 10 min) [3]; 20 (pH 6.5–9.5, 20 h) [2]

Oxidation

Organic solvent

General stability information

Storage

Frozen [1]; –70°C, dialyzed enzyme [16]

6 CROSSREFERENCES TO STRUCTURE DATABANKS

PIR/MIPS code

JU0461 (Bacillus sp.); PX0049 (B chain Corynebacterium sp. fragments); JS0671 (precursor Streptomyces sp.)

Brookhaven code

7 LITERATURE REFERENCES

[1] Inouye, Y., Nishimura, M., Matsuda, Y., Hoshika, H., Iwasaki, H., Hujimura, K., Asano, K., Nakamura, S.: Chem. Pharm. Bull.,35,4194–4202 (1987)
[2] Ogushi, S., Nagao, K., Emi, S., Ando, M., Tsuru, D.: Chem. Pharm. Bull.,36,1445–1450 (1988)
[3] Matsuda, Y., Hoshika, H., Inouye, Y., Ikuta, S., Matsuura, K., Nakamura, S.: Chem. Pharm. Bull.,35,711–717 (1987)
[4] Kim, J.M., Shimizu, S., Yamada, H.: Agric. Biol. Chem.,51,1167–1168 (1987)
[5] Kim, J.M., Shimizu, S., Yamada, H.: Agric. Biol. Chem.,50,2811–2816 (1986)
[6] Schuman Jorns, M.: Biochemistry,24,3189–3194 (1985)
[7] Suzuki, M.: J. Biochem.,89,599–607 (1981)
[8] Hayashi, S., Nakamura, S., Suzuki, M.: Biochem. Biophys. Res. Commun.,96,924–930 (1980)
[9] Mori, N., Sano, M., Tani, Y., Yamada, H.: Agric. Biol. Chem.,44,1391–1397 (1980)
[10] Patek, D.R., Dahl, C.R., Frisell, W.R.: Biochem. Biophys. Res. Commun.,46,885–891 (1972)
[11] Hayashi, S.: J. Biochem.,95,1201–1207 (1984)
[12] Hayashi, S., Suzuki, M., Nakamura, S.: J. Biochem.,94,551–558 (1983)
[13] Hayashi, S., Suzuki, M., Nakamura, S.: Biochim. Biophys. Acta,742,630–636 (1983)
[14] Kawamura-Konishi, Y., Suzuki, H.: Biochim. Biophys. Acta,915,346–356 (1987)
[15] Frisell, W.R.: Arch. Biochem. Biophys.,142,213–222 (1971)
[16] Kvalnes-Krick, K., Schuman Jorns, M.: Biochemistry,25,6061–6069 (1986)
[17] Zeller, H.-D., Hille, R., Schuman Jorns, M.: Biochemistry,28,5145–5154 (1989)

Enzyme Handbook © Springer-Verlag Berlin Heidelberg 1994
Duplication, reproduction and storage in data banks are only allowed with the prior permission of the publishers

1 NOMENCLATURE

EC number
1.5.3.2

Systematic name
N-Methyl-L-amino-acid:oxygen oxidoreductase (demethylating)

Recommended name
N-Methyl-L-amino-acid oxidase

Synonymes
Oxidase, N-methylamino acid
N-Methylamino acid oxidase
Demethylase [1]

CAS Reg. No.
9029-23-6

2 REACTION AND SPECIFICITY

Catalysed reaction
An N-methyl-L-amino acid + H_2O + O_2 →
→ an L-amino acid + formaldehyde + H_2O_2

Reaction type
Redox reaction
Oxidative demethylation

Natural substrates

Substrate spectrum
1 N-Methyl-L-amino acid + O_2 + H_2O (e.g.: alpha-N-methyl-L-tryptophan, O, N-dimethyl-DL-tyrosine, N-methyl-DL-phenylalanine) [1]
2 More (susceptibility: N-methyl-L-amino acids with aromatic or heterocyclic substituents > aliphatic N-methylamino acids with long side chains > N-methyl derivatives of beta-hydroxy-amino acids and of basic and acidic amino acids (exception: histidine) > phenylsarcosine, not: N-dimethyl-L-amino acids, L-amino acids, D-amino acids) [1]

Product spectrum
1 L-Amino acid + formaldehyde + H_2O_2
2 ?

Enzyme Handbook © Springer-Verlag Berlin Heidelberg 1994
Duplication, reproduction and storage in data banks are only allowed with the prior permission of the publishers

Inhibitor(s)
More (not: benzoate, atabrine, KCN) [1]

Cofactor(s)/prostethic group(s)
FAD (flavoprotein) [1]

Metal compounds/salts

Turnover number (min^{-1})

Specific activity (U/mg)

K_m-value (mM)
0.262 (alpha-N-methyl-L-tryptophan) [1]

pH-optimum
7.2 (assay at) [1]

pH-range

Temperature optimum (°C)

Temperature range (°C)

---

3 ENZYME STRUCTURE

Molecular weight

Subunits

Glycoprotein/Lipoprotein
–

---

4 ISOLATION/PREPARATION

Source organism
Rabbit [1]

Source tissue
Kidney [1]

Localisation in source

Purification

Crystallization
–

Cloned

–

Renaturated

–

5 STABILITY

pH

Temperature (°C)

Oxidation

Organic solvent

General stability information

Storage

6 CROSSREFERENCES TO STRUCTURE DATABANKS

PIR/MIPS code

Brookhaven code

7 LITERATURE REFERENCES

[1] Moritani, M., Tung, T.-C., Fujii, S., Mito, H., Izumiya, N., Kenmochi, K., Hirohata, R.: J. Biol. Chem.,209,485–492 (1954)

Enzyme Handbook © Springer-Verlag Berlin Heidelberg 1994
Duplication, reproduction and storage in data banks are only allowed with the prior permission of the publishers

1 NOMENCLATURE

EC number
1.5.3.4

Systematic name
N^6–Methyl-L-lysine:oxygen oxidoreductase (demethylating)

Recommended name
N^6–Methyl-lysine oxidase

Synonymes
epsilon-Alkyl-L-lysine:oxygen oxidoreductase [1]
Oxidase, N^6–methyllysine
N^6–Methyllysine oxidase
epsilon-N-Methyllysine demethylase
epsilon-Alkyllysinase [1]
More (identical to histone demethylase) [1]

CAS Reg. No.
37256-28-3

2 REACTION AND SPECIFICITY

Catalysed reaction
N^6–Methyl-L-lysine + H_2O + O_2 →
→ L-lysine + formaldehyde + H_2O_2

Reaction type
Redox reaction
Oxidative demethylation

Natural substrates

Substrate spectrum
1 N^6–Methyl-L-lysine (epsilon-N-monomethyl-L-lysine) + O_2 + H_2O [1, 2]
2 epsilon-N-Monomethyl-D-lysine + H_2O + O_2 (11% the rate of the demethylation of the L-isomer) [1, 2]
3 alpha-Keto-epsilon-methyl-aminocaproic acid + H_2O + O_2 [1]
4 epsilon-N-Ethyl-L-lysine + H_2O + O_2 [2]
5 More (activity restricted to epsilon-N-alkyllysine derivatives [2], epsilon-N-dimethyl-L-lysine is probably oxidized by a different enzyme [2], not: delta-N-monomethyl-L-ornithine [1, 2], [1], not: corresponding benzyl derivatives, alpha-N-methy-L-lysine [2]) [1, 2]

Enzyme Handbook © Springer-Verlag Berlin Heidelberg 1994
Duplication, reproduction and storage in data banks are only allowed with the prior permission of the publishers

Product spectrum

1 L-Lysine + formaldehyde + H_2O_2 [1, 2]
2 D-Lysine + formaldehyde + H_2O_2
3 alpha-Keto-epsilon-aminocaproic acid + formaldehyde + H_2O_2 (alpha-keto-6-amino-1-hexanoic acid is identical with 2-keto-6-amino-1-hexanoic acid)
4 Acetaldehyde + L-lysine + H_2O_2
5 ?

Inhibitor(s)

Ni^{2+} [2]; Zn^{2+} [2]; Co^{2+} [2]; KCN [2]; Mn^{2+} (moderate) [2]; NADH (moderate) [2]; Ascorbic acid (moderate) [2]; FAD (moderate) [2]; FMN (moderate) [2]; Phosphate (inhibits activity in crude homogenate, stimulates enzyme in prepared mitochondrial fraction) [2]; 2,6-Dichlorophenol indophenol [1]; More (not: semicarbazide) [2]

Cofactor(s)/prostethic group(s)

FAD (electron acceptor, not required with whole homogenate, but required for partially purified enzyme) [1]; Phenazine methosulfate (more efficient electron acceptor than FAD) [1]; More (no cofactor requirement) [2]

Metal compounds/salts

Turnover number (min^{-1})

Specific activity (U/mg)

0.0283 [1]; 0.0044 [2]

K_m-value (mM)

1.63 (epsilon-N-methyl-L-lysine) [2]; 1.05 (epsilon-N-monomethyl-L-lysine, crude extract) [1]; 1.9 (epsilon-N-monomethyl-L-lysine, partially purified enzyme) [1]; 0.87 (epsilon-N, epsilon-N-dimethyl-L-lysine) [2]; 5.95 (epsilon-N-ethyl-L-lysine) [2]

pH-optimum

7.0 (partially purified enzyme) [2]; 7.2 [1]

pH-range

6–8 (6: about 35% of activity maximum, 8: about 50% of activity maximum) [1]; 6.0–7.8 (6.0: about 10% of activity maximum, 7.8: about 15% of activity maximum) [2]

Temperature optimum (°C)

38 (assay at) [2]

Temperature range (°C)

3 ENZYME STRUCTURE

Molecular weight

100000–200000 (rat, gel filtration) [1]

Subunits

Glycoprotein/Lipoprotein

–

4 ISOLATION/PREPARATION

Source organism

Rat [1, 2]; Hamster [2]; Chicken [2]; Pigeon [2]; Mouse [2]; Dog [2]; Guinea pig [2]; Cat (low activity) [2]

Source tissue

Kidney [1, 2]; Liver [1, 2]

Localisation in source

Particle-bound [1, 2]

Purification

Rat (partial) [1, 2]

Crystallization

–

Cloned

–

Renaturated

–

5 STABILITY

pH

Temperature (°C)

55 (4 min: 70% loss of activity, 10 min: complete loss of activity) [1]

Oxidation

Organic solvent

General stability information

Storage

Enzyme Handbook © Springer-Verlag Berlin Heidelberg 1994
Duplication, reproduction and storage in data banks are only allowed with the prior permission of the publishers

6 CROSSREFERENCES TO STRUCTURE DATABANKS

PIR/MIPS code

Brookhaven code

7 LITERATURE REFERENCES

[1] Woon Ki Paik, Kim, S.: Arch. Biochem. Biophys.,165,369–378 (1974)
[2] Kim, S., Benoiton, L., Woon Ki Paik: J. Biol. Chem.,239,3790–3796 (1964)

1 NOMENCLATURE

EC number
1.5.3.5

Systematic name
(S)-6-Hydroxynicotine:oxygen oxidoreductase

Recommended name
(S)-6-Hydroxynicotine oxidase

Synonymes
Oxidase, 6-hydroxy-L-nicotine
L-6-Hydroxynicotine oxidase
6-Hydroxy-L-nicotine oxidase [4]
6-Hydroxy-L-nicotine:oxygen oxidoreductase [4]

CAS Reg. No.
37256-29-4

---

2 REACTION AND SPECIFICITY

Catalysed reaction
(S)-6-Hydroxynicotine + H_2O + O_2 →
→ 1-(6-hydroxypyrid-3-yl)-4-(methylamino)butan-1-one + H_2O_2
(mechanism [7, 8])

Reaction type
Redox reaction

Natural substrates
(S)-6-Hydroxynicotine + O_2 + H_2O (inducible enzyme of nicotine catabolism) [6, 9]

Substrate spectrum
1 (S)-6-Hydroxynicotine + O_2 + H_2O (absolute structural [7] and stereochemical specificity [7, 9], other electron acceptors (artificial or natural) than O_2 inactive [7]) [1–8]
2 L-6-Hydroxynornicotine + O_2 + H_2O [8]

Product spectrum
1 1-(6-Hydroxypyrid-3-yl)-4-(methylamino)butan-1-one + H_2O_2
2 1-(6-Hydroxypyrid-3-yl)-4-(amino)butan-1-one + H_2O_2

Enzyme Handbook © Springer-Verlag Berlin Heidelberg 1994
Duplication, reproduction and storage in data banks are only allowed with the prior permission of the publishers

Inhibitor(s)

6-Hydroxy-D-nicotine [1, 9]; Methylene blue [5, 7]; Phenanthroline (9 mM) [6]; [6-Hydroxypyridyl(3)]-(gamma-N-methylaminopropyl)ketone [6]; p-Chloromercuribenzoate [6]; $HgCl_2$ [6]; Urea [6]

Cofactor(s)/prostethic group(s)

FAD (flavoprotein [6], prosthetic group [1, 6], FAD-binding site at the amino-terminus of the polypeptide chain [2], noncovalently bound to apoenzyme [2, 5], 4 mol FAD per mol of enzyme [8], 2 mol FAD per mol of enzyme [5, 6]) [1, 2, 5, 6, 8]

Metal compounds/salts

More (no evidence for participation of metal ions in the electron transport obtained) [6]

Turnover number (min^{-1})

1760 (pH 7.5, 29°C, 6-hydroxynicotine) [5]

Specific activity (U/mg)

More [4, 6, 9]

K_m-value (mM)

0.02 [5, 9]

pH-optimum

7.5 (assay at) [6]

pH-range

Temperature optimum (°C)

22 (assay at) [6]

Temperature range (°C)

---

3 ENZYME STRUCTURE

Molecular weight

93000 (Arthrobacter oxidans, sedimentation studies, diffusion studies) [6]
140000 (Aspergillus oxidans) [8]

Subunits

Dimer (2 × 47000, Aspergillus oxidans, sedimentaion and diffusion studies after denaturation with guanidine hydrochloride) [6]

Glycoprotein/Lipoprotein

–

4 ISOLATION/PREPARATION

Source organism
Arthrobacter oxidans (inducible enzyme) [1–9]

Source tissue

Localisation in source
Cytoplasm [3]

Purification
Arthrobacter oxidans [4, 6, 9]

Crystallization
[6, 8]

Cloned
–

Renaturated
–

5 STABILITY

pH
6–9 [6]

Temperature (°C)
4 (7 days) [6]; 20 (2 h: stable, 2 days: 96% loss of activity) [6]; 40 (5 min: 9% loss of activity, 15 min: 25% loss of activity) [6]; 50 (5 min: 91% loss of activity) [6]; 60 (5 min: 98% loss of activity) [6]

Oxidation

Organic solvent

General stability information
Dilute solutions (when kept in frozen state, loss of activity) [6]; EDTA (no stabilization) [9]; Mercaptoethanol (no stabilization) [9]

Storage
0°C, neutralized saturated ammonium sulfate solution, 2 weeks [6]

Enzyme Handbook © Springer-Verlag Berlin Heidelberg 1994
Duplication, reproduction and storage in data banks are only allowed with the prior permission of the publishers

6 CROSSREFERENCES TO STRUCTURE DATABANKS

PIR/MIPS code

S18264 (Arthrobacter oxydans fragment)

Brookhaven code

7 LITERATURE REFERENCES

[1] Pust, S., Vervoort, J., Decker, K., Bacher, A., Müller, F.: Biochemistry,28,516–521 (1989)
[2] Brandsch, R., Hinkkanen, A.E., Mauch, L., Nagursky, H., Decker, K.: Eur. J. Biochem.,167,315–320 (1987)
[3] Swafford, J.R., Reeves, H.C., Brandsch, R.: J. Bacteriol.,163,792–795 (1985)
[4] Hinkkanen, A., Lilius, E.-M., Nowack, J., Maas, R., Decker, K.: Hoppe-Seyler's Z. Physiol. Chem.,364,801–806 (1983)
[5] Decker, K., Dai, V.D., Möhler, H., Brühmüller, M.: Z. Naturforsch.,27b,1072–1073 (1972)
[6] Dai, V.D., Decker, K., Sund, H.: Eur. J. Biochem.,4,95–102 (1968)
[7] Decker, K., Dai, V.D.: Eur. J. Biochem.,3,132–138 (1967)
[8] Palmer, G., Massey, V. in "Biol. Oxidations" (Singer, T.P., Ed.),263–300 (1968) (Review)
[9] Decker, K., Bleeg, H.: Biochim. Biophys. Acta,105,313–324 (1965)

1 NOMENCLATURE

EC number
1.5.3.6

Systematic name
(R)-6-Hydroxynicotine:oxygen oxidoreductase

Recommended name
(R)-6-Hydroxynicotine oxidase

Synonymes
Oxidase, 6-hydroxy-D-nicotine
D-6-Hydroxynicotine oxidase
6-Hydroxy-D-nicotine oxidase (FAD)

CAS Reg. No.
37233-46-8

2 REACTION AND SPECIFICITY

Catalysed reaction
(R)-6-Hydroxynicotine + H_2O + O_2 →
→ 1-(6-hydroxypyrid-3-yl)-4-(methylamino)butan-1-one + H_2O_2

Reaction type
Redox reaction

Natural substrates
(R)-6-Hydroxynicotine + O_2 + H_2O (inducible enzyme of nicotine catabolism [11])

Substrate spectrum
1 (R)-6-Hydroxynicotine + O_2 + H_2O (highly specific for structural and absolutely specific for stereochemical configuration of its substrate [11]) [1–11]
2 D-6-Aminonicotine + O_2 + H_2O [11]
3 More (methylene blue and 2,6-dichlorophenol-indophenol are able to reoxidize the reduced D-6-hydroxynicotine oxidase [3, 11], no one-electron acceptor can replace O_2 as electron acceptor [11]) [3, 11]

Product spectrum
1 1-(6-Hydroxypyrid-3-yl)-4-(methylamino)butan-1-one + H_2O_2
2 ?
3 ?

Enzyme Handbook © Springer-Verlag Berlin Heidelberg 1994
Duplication, reproduction and storage in data banks are only allowed with the prior permission of the publishers

Inhibitor(s)

Methylene blue [3]; L-6-Hydroxynicotine [4, 11]; o-Phenanthroline (above 1 mM) [11]; D, L-2-Hydroxynicotine [11]; 3-(4-Aminobutyl)pyridine [11]

Cofactor(s)/prostethic group(s)

FAD (flavoprotein, 1 mol FAD per mol of enzyme [3, 5, 6, 10, 11], covalently bound [3, 5, 6, 11], 8alpha-methyl group of FAD is linked to N-3 of a histidyl residue of the polypeptide chain [5, 6, 10], holoenzyme formation from apoenzyme and FAD can be mediated by phosphoenolpyruvate, glyceraldehyde-3-phosphate, glycerate-3-phosphate or glycerol-3-phosphate [6]) [3, 5, 6, 10, 11]

Metal compounds/salts

Turnover number (min^{-1})

1190 (6-hydroxynicotine) [3]

Specific activity (U/mg)

More [2, 4, 11]

K_m-value (mM)

0.05 ((R)-6-hydroxynicotine) [3, 11]; 0.1 ((R)-6-hydroxynicotine) [4]; 0.2 (D-6-aminonicotine) [11]

pH-optimum

8.0 (0.1 M Tris-HCl buffer) [11]; 8.5 (0.1 M glycine-NaOH or phosphate buffer) [11]

pH-range

Temperature optimum (°C)

Temperature range (°C)

---

3 ENZYME STRUCTURE

Molecular weight

50000–53000 (Arthrobacter oxidans [3, 5, 11], sedimentation and diffusion studies [11]) [3, 5, 11]

Subunits

Monomer (1 × 53000, Arthrobacter oxidans, sedimentation and diffusion study after treatment with 6 M guanidine-HCl or 0.1% SDS in 8 M urea [11], 1 × 50000, Arthrobacter oxidans [5]) [5, 11]

Glycoprotein/Lipoprotein

–

4 ISOLATION/PREPARATION

Source organism
Arthrobacter oxidans [1–11]

Source tissue

Localisation in source
Cytoplasm [1]

Purification
Arthrobacter oxidans [2, 4, 11]

Crystallization
–

Cloned
(Arthrobacter oxidans gene, cloned into E. coli) [7, 8]

Renaturated
–

5 STABILITY

pH
6 (unstable below) [11]; 7–9 (stable at neutral or alkaline pH, 4°C) [11]

Temperature (°C)
37 (covalent flavinylation of the apoenzyme with FAD and phosphoenolpyruvate protects the enzyme against degradation) [9]; 45 (inactivated above) [11]

Oxidation

Organic solvent

General stability information
Covalent flavinylation of the apoenzyme with FAD and phosphoenolpyruvate protects against degradation [9]

Storage
4°C, at neutral or alkaline pH, 10 mM mercaptoethanol [11]

6 CROSSREFERENCES TO STRUCTURE DATABANKS

PIR/MIPS code
DEIQHN (Arthrobacter oxydans)

Brookhaven code

Enzyme Handbook © Springer-Verlag Berlin Heidelberg 1994
Duplication, reproduction and storage in data banks are only allowed with the prior permission of the publishers

7 LITERATURE REFERENCES

[1] Swafford, J.R., Reeves, H.C., Bradsch, R.: J. Bacteriol.,163,792–795 (1985)
[2] Hinkkanen, A., Lilius, E.-M., Nowack, J., Maas, R., Decker, K.: Hoppe-Seyler's Z. Physiol. Chem.,364,801–806 (1983)
[3] Decker, K., Dai, V.D., Möhler, H., Brühmüller, M.: : Z. Naturforsch.,27b,1072–1073 (1972)
[4] Decker, K., Bleeg, H.: Biochim. Biophys. Acta,105,313–324 (1965)
[5] Decker, K.: Trends Biochem. Sci.,1,184–185 (1976)
[6] Brandsch, R., Bichler, V.: Eur. J. Biochem.,182,125–128 (1989)
[7] Brandsch, R., Faller, W., Schneider, K.: Mol. Gen. Genet.,202,96–101 (1986)
[8] Brandsch, R., Bichler, V.: FEBS Lett.,192,204–208 (1985)
[9] Brandsch, R., Bichler, V., Krauss, B.: Biochem. J.,258,187–192 (1989)
[10] Möhler, H., Brühmüller, M., Decker, K.: Eur. J. Biochem.,29,152–155 (1972)
[11] Brühmüller, M., Möhler, H., Decker, K.: Eur. J. Biochem.,29,143–151 (1972)

1 NOMENCLATURE

EC number
1.5.3.7

Systematic name
L-Pipecolate:oxygen 1,6-oxidoreductase

Recommended name
L-Pipecolate oxidase

Synonymes
Oxidase, pipecolate
Pipecolate oxidase
L-Pipecolic acid oxidase

CAS Reg. No.
81669-65-0

2 REACTION AND SPECIFICITY

Catalysed reaction
L-Pipecolate + O_2 →
→ H_2O_2 + 2,3,4,5-tetrahydropyridine-2-carboxylate

Reaction type
Redox reaction

Natural substrates
L-Pipecolate + O_2 (human: step in degradation of L-lysine to pipecolic acid, alternative route for lysine utilization [1], Rhodotorula glutinis: biosynthesis of lysine [4]) [1, 4]

Substrate spectrum
1 L-Pipecolate + O_2 [1]

Product spectrum
1 2,3,4,5-Tetrahydropyridine-2-carboxylate + H_2O_2 (i.e. piperidine-6-carboxylic acid, the product reacts with H_2O to form 2-aminoadipate 6-semialdehyde, i.e. 2-amino-6-oxohexanoate) [1]

Enzyme Handbook © Springer-Verlag Berlin Heidelberg 1994
Duplication, reproduction and storage in data banks are only allowed with the prior permission of the publishers

Inhibitor(s)
Phenazine ethosulfate [3]; L-Proline [4]; p-Chloromercuribenzoate [4]; $HgCl_2$ [4]; Hydroxylamine (slight) [4]; KCN (slight) [4]; NaN_3 (slight) [4]

Cofactor(s)/prostethic group(s)
No cofactor requirement, Cynomolgus monkey [3]

Metal compounds/salts

Turnover number (min^{-1})

Specific activity (U/mg)
More [4]

K_m-value (mM)
4.22 (L-pipecolate, Cynomolgus monkey, enzyme in tissue homogenate) [3]; 1.67 (L-pipecolic acid, purified enzyme) [4]

pH-optimum
8.5 [4]

pH-range

Temperature optimum (°C)
37 (assay at) [1, 3]; 25 (assay at) [4]

Temperature range (°C)

---

3 ENZYME STRUCTURE

Molecular weight
43000 (Rhodotorula glutinis, gel filtration) [4]

Subunits
Monomer (1 × 43000, Rhodotorula glutinis, SDS-PAGE) [4]

Glycoprotein/Lipoprotein
–

---

4 ISOLATION/PREPARATION

Source organism
Human [1, 2]; Cynomolgus monkey [3]; Rhodotorula glutinis [4]

Source tissue
Liver [1, 2]; Kidney [3]

Localisation in source
Peroxisomes [1–3]

Purification

Crystallization
–

Cloned
–

Renaturated
–

5 STABILITY

pH

Temperature (°C)
45 (10 min, addition of substrate, without bovine serum albumin: 80% loss of activity, addition of bovine serum albumin: no loss of activity) [4]; 53 (10 min, addition of substrate, without bovine serum albumin: 100% loss of activity) [4]

Oxidation

Organic solvent

General stability information
Bovine serum albumin stabilizes [4]

Storage
4°C: stable for several days, –70°C: stable for several weeks [4]

6 CROSSREFERENCES TO STRUCTURE DATABANKS

PIR/MIPS code

Brookhaven code

7 LITERATURE REFERENCES

[1] Rao, V.V., Chang, Y.-F.: Biochim. Biophys. Acta,1038,295–299 (1990)
[2] Wanders, R.J.A., Romeyn, G.J., Schutgens, R.B.H., Tager, J.M.: Biochem. Biophys. Res. Commun.,164,550–555 (1989)
[3] Mihalik, S.J., Rhead, W.J.: J. Biol. Chem.,264,2509–2517 (1989)
[4] Kinzel, J.J., Bhattacharjee, J.K.: J. Bacteriol.,151,1073–1077 (1982)

Enzyme Handbook © Springer-Verlag Berlin Heidelberg 1994
Duplication, reproduction and storage in data banks are only allowed with the prior permission of the publishers

1 NOMENCLATURE

EC number
1.5.3.8

Systematic name
(S)-Tetrahydroberberine:oxygen 7,14-oxidoreductase

Recommended name
(S)-Tetrahydroprotoberberine oxidase

Synonymes
STOX [4]
(S)-Tetrahydroberberine (THB)oxidase
More (enzyme is now included with EC 1.3.3.8)

CAS Reg. No.

---

2 REACTION AND SPECIFICITY

Catalysed reaction
(S)-7,8,13,13a-Tetrahydrojatrorrhizine + O_2 →
→ 8,13-dihydrojatrorrhizine + H_2O_2 (stereochemistry [2], mechanism [6], enzyme converts tetrahydroprotoberberine alkaloids into protoberberine which by further oxidation of ring C, give the corresponding berberine alkaloids)

Reaction type
Redox reaction

Natural substrates
Tetrahydroprotoberberines + O_2 (final step in protoberberine biosynthesis) [5]
(S)-Tetrahydrocolumbamine (metabolic pathway leading to berberine, jatrorrhizine and palmatine) [6]

Substrate spectrum
1 (S)-7,8,13,13a-Tetrahydrojatrorrhizine + O_2
2 (S)-Tetrahydroprotoberberine + O_2 (specific for tetrahydroberberines of S-configuration) [5]
3 (S)-Tetrahydrocolumbamine + O_2 [2, 6]
4 Scoulerine + O_2
5 Norreticuline + O_2 [6]
6 More (overview [6], not: (R)-enantiomers of tetrahydroprotoberberines, 13,14-dehydrocanadine, canadine methosalt [5]) [5, 6]

Enzyme Handbook © Springer-Verlag Berlin Heidelberg 1994
Duplication, reproduction and storage in data banks are only allowed with the prior permission of the publishers

Product spectrum

1 8,13-Dihydrojatrorrhizine + H_2O_2
2 Protoberberine + H_2O_2
3 Columbamine + H_2O_2 [2, 6]
4 Dehydroscoulerine + H_2O_2 [2]
5 1, 2-Dehydronorreticuline + H_2O_2 [6]
6 ?

Inhibitor(s)

Morine [5, 6]; Protoberberines (above 0.5 mM, end product inhibition) [6]; H_2O_2 (above 0.5 mM, end product inhibition) [6, 7]; Jatrorrhizine [6]; Berberine [6]; (R)-Norreticuline [6]; Dicoumarol [6]; 2,4-Pentanedione [6]; Hg^{2+} (10 mM) [6]; Ag^{2+} (10 mM) [6]; Cd^{2+} (10 mM) [6]; Bathophenanthroline sulphonic acid [7]

Cofactor(s)/prostethic group(s)

FAD (flavoprotein [5, 6], covalently bound flavin [6]) [5, 6]

Metal compounds/salts

Fe^{2+} (firmly bound to enzyme protein, involved in oxidase action) [7]

Turnover number (min^{-1})

112 ((S)-norreticuline) [6]

Specific activity (U/mg)

More [4, 6]

K_m-value (mM)

0.15 (free enzyme [4], (S)-norreticuline) [4, 6]; 0.12 (immobilized enzyme, (S)-norreticuline) [4]; 0.0013 ((S)-tetrahydrojatrorrhizine) [5]; 0.026 ((S)-canadine) [5]; 0.025 ((S)-scoulerine) [6]; 0.007 ((R, S)-corypalmine) [6]; 0.0043 ((R, S)-tetrahydropalmatine) [6]; 0.0133 ((R, S)-canadine) [6]; 0.0065 ((S)-tetrahydroberberine) [7]

pH-optimum

8.8 (Coptis japonica) [7]; 8.9 (free enzyme) [4, 5, 6]; 9.4 (immobilized enzyme) [4]; 9.8 (enzyme in immobilized plant cells [4], Berberis wilsoniae [7]) [4, 7]

pH-range

7–10 (free enzyme, 7: about 25% of activity maximum, 10: about 20% of activity maximum) [4]; 7.6–9.7 (7.6, 9.7: about 50% of activity maximum) [6]

Temperature optimum (°C)

37 (1 h reaction) [7]; 40 (free enzyme) [4, 6]; 50 (immobilized enzyme) [4]; 47 (enzyme in immobilized plant cells) [4]

Temperature range (°C)

3 ENZYME STRUCTURE

Molecular weight

105000 (Berberis wilsoniae, gel filtration) [6]
58000 (Coptis japonica, HPLC-size exclusion chromatography) [7]

Subunits

Dimer (2 × 53000, Berberis wilsoniae, SDS-PAGE [6], 2 × 28000, Coptis japonica, SDS-PAGE [7]) [6, 7]

Glycoprotein/Lipoprotein

–

4 ISOLATION/PREPARATION

Source organism

More (overview) [4]; Annona reticulata [4]; Papaver somniferum [4]; Fumaria parviflora [4]; Coptis japonica [1, 7]; Berberis wilsoniae (var. subcauliata [3–5]) [1–6]

Source tissue

Root [4]; Leaves [4]; Cultured cells [3, 4, 6, 7]

Localisation in source

Membranous vesicles (Golgi-derived smooth vesicles [6], specific gravity: 1.14 g/ml [3, 5, 6], serving only alkaloid biosynthesis [3]) [3, 5, 6]

Purification

Berberis wilsoniae (var. subcauliata, partial [4]) [4, 6]; Coptis japonica [7]

Crystallization

–

Cloned

(Coptis japonica gene, expression in E. coli) [2]

Renaturated

–

5 STABILITY

pH

Temperature (°C)

–20 (12% glycerol, half-life: 102 days) [6]; 4 (half-life: 25 days [6], crude enzyme solution, half-life: 20 h [7]) [6, 7]; 25 (half-life: 4 days (free enzyme) [4, 6], 200 days (immobilized enzyme) [4]) [4, 6]; 45 (1 h, 40% loss of activity) [7]; 50 (1 h, 90% loss of activity) [7]

Enzyme Handbook © Springer-Verlag Berlin Heidelberg 1994
Duplication, reproduction and storage in data banks are only allowed with the prior permission of the publishers

Oxidation

Organic solvent

General stability information

Storage

–20°C, 12% glycerol, half-life: 102 days [6]

6 CROSSREFERENCES TO STRUCTURE DATABANKS

PIR/MIPS code

Brookhaven code

7 LITERATURE REFERENCES

[1] Okada, N., Koizumi, N., Tanaka, T., Ohkubo, H., Nakanishi, S., Yamada, Y.: Proc. Natl. Acad. Sci. USA,86,534–538 (1989)
[2] Frenzel, T., Beale, J.M., Kobayashi, M., Zenk, M.H., Floss, H.G.: J. Am. Chem. Soc.,110,7878–7880 (1988)
[3] Amann, M., Wanner, G., Zenk, M.H.: Planta,167,310–320 (1986)
[4] Amann, M., Zenk, M.H.: Phytochemistry,26,3235–3240 (1987)
[5] Amann, M., Nagakura, N., Zenk, M.H.: Tetrahedron Lett.,25,953–954 (1984) (Review)
[6] Amann, M., Nagakura, N., Zenk, M.H.: Eur. J. Biochem.,175,17–25 (1988)
[7] Okada, N., Shinmyo, A., Okada, H., Yamada, Y.: Phytochemistry,27,979–982 (1988)

1 NOMENCLATURE

EC number
1.5.3.9

Systematic name
(S)-Reticuline:oxygen oxidoreductase (methylene-bridge forming)

Recommended name
Reticuline oxidase

Synonymes
Berberine-bridge-forming enzyme
Tetrahydroprotoberberine synthase
Berberine bridge enzyme [2, 3]
BBE [4]

CAS Reg. No.

2 REACTION AND SPECIFICITY

Catalysed reaction
(S)-Reticuline + O_2 →
→ (S)-scoulerine + H_2O_2 (stereochemistry [3], acts on (S)-reticuline and related compounds, converting the N-methyl group into the methylene bridge ("berberine bridge") of (S)-tetrahydroprotoberberine)

Reaction type
Redox reaction
Methylene-bridge formation

Natural substrates
(S)-Reticuline + O_2 (biosynthesis of protoberberine skeleton [2], biosynthesis of berberine [4]) [2, 4]

Substrate spectrum
1 (S)-Reticuline + O_2 [1, 2]
2 (S)-Protosinomenine + O_2 [1, 2]
3 (S)-Laudonosoline + O_2 [1, 2]
4 More (specific for substrates with S-configuration, not: (S)-reticuline-N-oxide, (R)-reticuline-N-oxide, orientaline, (R, S)-laudanidine) [1]

Enzyme Handbook © Springer-Verlag Berlin Heidelberg 1994
Duplication, reproduction and storage in data banks are only allowed with the prior permission of the publishers

Product spectrum

1 (S)-Scoulerine + H_2O_2 [1]
2 Corresponding substituted (S)-tetrahydroprotoberberine + H_2O_2 [1]
3 Corresponding substituted (S)-tetrahydroberberine + H_2O_2 [1]
4 ?

Inhibitor(s)

EDTA (6 mM) [2]; NaN_3 (50 mM) [2]; NaCN (50 mM) [2]; Protoberberine alkaloids (e.g.: berberine, jatrorrhizine, (S)-norreticuline) [2]; Diethyldithiocarbamate [2]; o-Phenanthroline [1, 2]; Dithioerythritol [1]

Cofactor(s)/prostethic group(s)

Metal compounds/salts

Turnover number (min^{-1})

Specific activity (U/mg)

More [2]

K_m-value (mM)

0.0014 ((S)-reticuline) [1, 2]

pH-optimum

8.9 [1, 2]

pH-range

7–10.8 (7: about 10% of activity maximum, 10.8: about 15% of activity maximum) [2]

Temperature optimum (°C)

40–50 [2]

Temperature range (°C)

3 ENZYME STRUCTURE

Molecular weight

49000–54000 (Berberis beania, gel filtration, SDS-PAGE) [1, 2]

Subunits

Monomer (1 × 54000, Berberis beania, SDS-PAGE) [2]

Glycoprotein/Lipoprotein

–

4 ISOLATION/PREPARATION

Source organism

More (overview) [2]; Papaveraceae (e.g.: Glaucium flavum, Glaucium rubrum) [2]; Fumariaceae (e.g.: Fumaria capreolata, Adlumia fungosa) [2]; Berberis beania [1, 2]; Berberidaceae (e.g.: Berberis vulgaris, Berberis stolonifera) [2]; Ranunculaceae (e.g.: Thallictrum squarrosum, Thalictrum glaucum) [2]; Cissampelos mucronata [2]; Berberis wilsoniae var. subcauliata [4]

Source tissue

Cell cultures [1, 2, 4]

Localisation in source

Membranous vesicles (serving only alkaloid biosynthesis [4], particles with density: 1.14 g/ml [2, 4]) [2, 4]

Purification

Berberis beania [1, 2]

Crystallization

–

Cloned

–

Renaturated

–

5 STABILITY

pH

Temperature (°C)

–20 (half-life: 320 days) [2]; 4 (half-life: 150 days) [2]; 25 (half-life: 18 days) [2]; 37 (half-life: 6 days) [2]

Oxidation

Organic solvent

General stability information

Storage

–20°C, half-life: 320 days [2]

Enzyme Handbook © Springer-Verlag Berlin Heidelberg 1994
Duplication, reproduction and storage in data banks are only allowed with the prior permission of the publishers

6 CROSSREFERENCES TO STRUCTURE DATABANKS

PIR/MIPS code

Brookhaven code

7 LITERATURE REFERENCES

[1] Steffens, P., Nagakura, N., Zenk, M.H.: Tetrahedron Lett.,25,951–952 (1984) (Review)
[2] Steffens, P., Nagakura, N., Zenk, M.H.: Phytochemistry,24,2577–2583 (1985)
[3] Frenzel, T., Beale, J.M., Kobayashi, M., Zenk, M.H., Floss, H.G.: J. Am. Chem. Soc.,110,7878–7880 (1988)
[4] Amann, M., Wanner, G., Zenk, M.H.: Planta,167,310–320 (1986)

1 NOMENCLATURE

EC number
1.5.3.10

Systematic name
N,N-Dimethylglycine:oxygen oxidoreductase (demethylating)

Recommended name
Dimethylglycine oxidase

Synonymes
Oxidase, dimethylglycine

CAS Reg. No.
74870-79-4

2 REACTION AND SPECIFICITY

Catalysed reaction
N,N-Dimethylglycine + H_2O + O_2 →
→ sarcosine + formaldehyde + H_2O_2

Reaction type
Redox reaction

Natural substrates
Dimethylglycine + H_2O + O_2 (reaction in choline metabolism) [1]

Substrate spectrum
1 N,N-Dimethylglycine + H_2O + O_2 [1, 2]
2 More (no substrates: sarcosine, choline, betaine, creatine, creatinine, carnitine, N,N-dimethylaminoethanol, alkylamines) [1]

Product spectrum
1 Sarcosine + formaldehyde + H_2O_2 [1, 2]
2 ?

Inhibitor(s)
Iodoacetate (strong inhibition) [1]; Ag^{2+} [1]; Hg^{2+} [1]; Zn^{2+} [1]; More (no inhibition by Cu^{2+}, Fe^{3+}, PCMB, KCN, EDTA, hydrazine, semicarbazide) [1]

Cofactor(s)/prosthetic group(s)
FAD (2 mol FAD/mol enzyme, covalently bound) [1]

Enzyme Handbook © Springer-Verlag Berlin Heidelberg 1994
Duplication, reproduction and storage in data banks are only allowed with the prior permission of the publishers

Metal compounds/salts

Turnover number (min^{-1})

Specific activity (U/mg)
0.0335 (crude) [2]; 12.3 [1]

K_m-value (mM)
9.1 (dimethylglycine) [1]

pH-optimum
8.0 (assay at) [2]; 9.0 (assay at) [1]

pH-range

Temperature optimum (°C)
30 (assay at) [1, 2]

Temperature range (°C)

---

3 ENZYME STRUCTURE

Molecular weight
170000 (Cylindrocarpon didymum M-1, gel filtration) [1]
180000 (Cylindrocarpon didymum M-1, sedimentation velocity ultracentrifugation) [1]

Subunits
Dimer (2 × 82000, Cylindrocarpon didymum M-1, SDS-PAGE) [1]

Glycoprotein/Lipoprotein
–

---

4 ISOLATION/PREPARATION

Source organism
Cylindrocarpon didymum M-1 [1]; Arthrobacter sp. P1 [2]

Source tissue
Cell [1, 2]

Localisation in source
Cytoplasm [1]

Purification
Cylindrocarpon didymum M-1 [1]

Crystallization

–

Cloned

–

Renaturated

–

5 STABILITY

pH

6.0–7.5 (15 min stable at 40°C) [1]

Temperature (°C)

40 (15 min stable at pH 6.0–7.5) [1]

Oxidation

Organic solvent

General stability information

Storage

6 CROSSREFERENCES TO STRUCTURE DATABANKS

PIR/MIPS code

Brookhaven code

7 LITERATURE REFERENCES

[1] Mori, N., Kawakami, B., Tani, Y., Yamada, H.: Agric. Biol. Chem.,44,1383–1389 (1980)
[2] Levering, P.R., Binnema, D.J., Van Dijken, J.P., Harder, W.: FEMS Microbiol. Lett.,12,19–25 (1981)

Enzyme Handbook © Springer-Verlag Berlin Heidelberg 1994
Duplication, reproduction and storage in data banks are only allowed with the prior permission of the publishers

1 NOMENCLATURE

EC number

1.5.3.11

Systematic name

N^1–Acetylspermidine:oxygen oxidoreductase (deaminating)

Recommended name

Polyamine oxidase

Synonymes

More (undistinguishable from EC 1.4.3.4 in Chemical Abstracts)

CAS Reg. No.

9001-66-5

2 REACTION AND SPECIFICITY

Catalysed reaction

N^1–Acetylspermine + O_2 + H_2O →
→ N^1–acetylspermidine + 3-aminopropanal + H_2O_2

Reaction type

Redox reaction

Natural substrates

Spermine + O_2 + H_2O (reaction in mammalian spermine catabolism) [2]

Substrate spectrum

1 N^1–Acetylspermine + O_2 + H_2O (best substrate) [1]
2 N^1,N^{12}-Diacetylspermine + O_2 + H_2O (second best substrate) [1]
3 Spermine + O_2 + H_2O [1, 2]
4 Spermidine + O_2 + H_2O [1]
5 More (substrates are N-acetyl derivatives of spermine or spermidine, acetylated at the propylamino moiety, no substrates: N^1,N^8-diacetylspermidine, N^8-acetylspermidine, N^1–acetyl-1,3-diaminopropane, N^1–acetylputrescine, putrescine, cadaverine, diaminopropane) [1]

Product spectrum

1 N^1–Acetylspermidine + 3-aminopropanal + H_2O_2 [1]
2 N^1–Acetylspermidine + 3-acetamidopropanal + H_2O_2 [1]
3 Spermidine + 3-aminopropanal + H_2O_2 [1]
4 Putrescine + 3-aminopropanal + H_2O_2 [2]
5 ?

Enzyme Handbook © Springer-Verlag Berlin Heidelberg 1994
Duplication, reproduction and storage in data banks are only allowed with the prior permission of the publishers

Inhibitor(s)

N^8-Acetylspermidine (non-competitive inhibitor to N^1–acetyl-spermidine) [1]; N-(3-Aminopropyl)-1,3-diaminopropane (non-competitive inhibitor to N^1–acetylspermidine) [1]; 3-Amino-propanal (potent inhibitor) [1]; Hg^{2+} [1]; N-Ethylmaleimide [1]; Hydroxylamine [1]; Carbonyl reagents (30 min preincubation inactivates) [1]; Quinacrine [1]; Iron chelators [1]; N^1–Acetyl-putrescine (weak) [1]; Putrescine (weak) [1]; Cadaverine (weak) [1]; Aminopropane (weak) [1]; More (products inhibit, not in presence of benzaldehyde [2], no inhibition: EDTA, NaN_3, NaF [1]) [1, 2]

Cofactor(s)/prostethic group(s)

FAD (tightly bound, flavoprotein, no stimulation by exogenous FAD, not FMN or riboflavin [2]) [1, 2]; Benzaldehyde (activation of spermine and spermidine oxidation at high concentration, not N-acetylspermidine [1]) [1, 2]; Pyridoxal (activation, 30% as effective as benzaldehyde [1]) [1, 2]; Aldehydes (activation by e.g. form-, acet-, propion-, butyl-, hexyl-, salicyl-, o-aminobenz-, p-hydroxybenz-, phenylacetaldehyde. Benz- and anisaldehyde are the best stimulators) [2]; More (no activation by aromatic amines) [2]

Metal compounds/salts

Fe^{2+} (requirement) [1]

Turnover number (min^{-1})

Specific activity (U/mg)

0.141 [2]; 1.38 (presence of benzaldehyde) [2]

K_m-value (mM)

0.0006 (N^1–acetylspermine, cytosolic enzyme) [1]; 0.005 (N^1,N^{12}-diacetylspermine, cytosolic enzyme [1], spermine presence of benzaldehde [1, 2], cytosolic enzyme [1]) [1, 2]; 0.014 (N^1–acetylspermidine, cytosolic enzyme) [1]; 0.015 (spermine, cytosolic enzyme, spermidine presence of benzaldehyde [1, 2], cytosolic enzyme [1]) [1, 2]; 0.02 (spermine) [2]; 0.05 (spermidine [1, 2], cytosolic enzyme [1]) [1, 2]

pH-optimum

More (pI: 4.9) [2]; 10.0 [1, 2]

pH-range

Temperature optimum (°C)

Temperature range (°C)

3 ENZYME STRUCTURE

Molecular weight

55000 (rat, gel filtration) [2]
60000 (rat, gel filtration, sucrose density gradient centrifugation) [1]
61000 (rat, sucrose density gradient centrifugation) [2]

Subunits

Monomer (1 × 60000, rat, SDS-PAGE) [1, 2]

Glycoprotein/Lipoprotein

–

4 ISOLATION/PREPARATION

Source organism

Rat (Wistar strain [2]) [1, 2]

Source tissue

Liver [1, 2]

Localisation in source

Cytosol (highest specific activity [1]) [1]; Peroxisomes (highest proportion of total activity [1]) [1, 2]

Purification

Rat (gel electrophoresis [1]) [1, 2]

Crystallization

–

Cloned

–

Renaturated

–

5 STABILITY

pH

7–9 (quite stable) [2]; 8.9 (more stable during electrophoresis than at pH 7.5) [1]; 10 (marked loss of activity) [2]

Temperature (°C)

60 ($t_{1/2}$: 1 min, 75% loss of activity after 5 min) [2]

Oxidation

Enzyme Handbook © Springer-Verlag Berlin Heidelberg 1994
Duplication, reproduction and storage in data banks are only allowed with the prior permission of the publishers

Organic solvent

General stability information

DTT stabilizes [2]; Freezing, stable to at least once [2]

Storage

–70°C, stable [1, 2]; –20°C, stable [1, 2]; 0–4°C, stable for at least 2 months [1]

6 CROSSREFERENCES TO STRUCTURE DATABANKS

PIR/MIPS code

Brookhaven code

7 LITERATURE REFERENCES

[1] Hölttä, E.: Methods Enzymol.,94,306–311 (1983) (Review)
[2] Hölttä, E.: Biochemistry,16,91–100 (1977)

1 NOMENCLATURE

EC number
1.5.4.1

Systematic name
Pyrimidodiazepine:oxidized-glutathione oxidoreductase (ring-opening, cyclizing)

Recommended name
Pyrimidodiazepine synthase

Synonymes
Synthase, pyrimidodiazepine
PDA synthase [1]

CAS Reg. No.
93586-06-2

2 REACTION AND SPECIFICITY

Catalysed reaction
6-Pyruvoyltetrahydropterin + 2 glutathione →
→ a pyrimidodiazepine + oxidized glutathione (the reduction of 6-pyruvoyltetrahydropterin is accompanied by the opening of the 6-membered pyrazine ring and the formation of the 7-membered diazepine ring)

Reaction type
Redox reaction

Natural substrates
6-Pyruvoyltetrahydropterin + glutathione (enzyme is involved in the formation of the eye pigment drosopterin in Drosophila melanogaster) [1]

Substrate spectrum
1 6-Pyruvoyltetrahydropterin + glutathione [1]

Product spectrum
1 Oxidized glutathione + pyrimidodiazepine (acetyldihydro derivative)

Inhibitor(s)

Cofactor(s)/prostethic group(s)

Metal compounds/salts

Enzyme Handbook © Springer-Verlag Berlin Heidelberg 1994
Duplication, reproduction and storage in data banks are only allowed with the prior permission of the publishers

Turnover number (min^{-1})

Specific activity (U/mg)
0.8 [1]

K_m-value (mM)

pH-optimum

pH-range

Temperature optimum (°C)
30 [1]

Temperature range (°C)

3 ENZYME STRUCTURE

Molecular weight
48000 (Drosophila melanogaster, gel filtration) [1]

Subunits
Dimer (2 × 24000, Drosophila melanogaster, SDS-PAGE) [1]

Glycoprotein/Lipoprotein
–

4 ISOLATION/PREPARATION

Source organism
Drosophila melanogaster [1]

Source tissue
Head [1]

Localisation in source

Purification
Drosophila melanogaster [1]

Crystallization
–

Cloned
–

Renaturated
–

5 STABILITY

pH

Temperature (°C)

Oxidation

Organic solvent

General stability information

Storage

–80°C, stable for at least 1–2 weeks [1]

6 CROSSREFERENCES TO STRUCTURE DATABANKS

PIR/MIPS code

Brookhaven code

7 LITERATURE REFERENCES

[1] Wiederrecht, G.J., Brown, G.M.: J. Biol. Chem.,259,14121–14127 (1984)

Enzyme Handbook © Springer-Verlag Berlin Heidelberg 1994
Duplication, reproduction and storage in data banks are only allowed with the prior permission of the publishers

1 NOMENCLATURE

EC number
1.5.5.1

Systematic name
Electron-transferring-flavoprotein:ubiquinone oxidoreductase

Recommended name
Electron-transferring-flavoprotein dehydrogenase

Synonymes
ETF-QO
ETF:ubiquinone oxidoreductase [2]
Electron transfer flavoprotein dehydrogenase
Electron transfer flavoprotein Q oxidoreductase
Electron transfer flavoprotein-ubiquinone oxidoreductase
Reductase, electron transfer flavoprotein

CAS Reg. No.
86551-03-3

---

2 REACTION AND SPECIFICITY

Catalysed reaction
Reduced electron-transferring flavoprotein + ubiquinone →
→ electron-transferring flavoprotein + ubiquinol (ping-pong mechanism [3])

Reaction type
Redox reaction

Natural substrates

Substrate spectrum
1 Reduced electron-transferring flavoprotein + ubiquinone-1 (electron-transferring flavoprotein from Paracoccus denitrificans or pig liver [7]) [7, 8]

Product spectrum
1 Electron-transferring flavoprotein + ubiquinol-1 [8]

Enzyme Handbook © Springer-Verlag Berlin Heidelberg 1994
Duplication, reproduction and storage in data banks are only allowed with the prior permission of the publishers

Inhibitor(s)

N-Succinimidyl 3-(2-pyridyldithio)propionate (substitute for lysine residues of protein) [2]; N-Ethylmaleimide (inhibitory to thiolated enzyme (with 2-iminothiolane), not inhibitory to unmodified enzyme) [2]; More (not inhibitory: trinitrobenzensulfonic acid) [2]

Cofactor(s)/prostethic group(s)

FAD (1 mol per mol of enzyme) [8]

Metal compounds/salts

4Fe-4S cluster (electron paramagnetic resonance and magnetic circular dichroism studies [4]) [4, 6, 8]

Turnover number (min^{-1})

12000 (electron-transferring flavoprotein, disproportionation) [6]; 4680 (electron-transferring flavoprotein, comproportionation) [6]; More [1, 3]

Specific activity (U/mg)

93.0 (assay method [3], ability for transfer of electrons from electron-transferring flavoprotein to nitroblue tetrazolium used for assay [6]) [3, 6]

K_m-value (mM)

0.00031 (reduced electron-transferring flavoprotein, comproportionation to semiquinone) [6]; 0.00032 (oxidized electron-transferring flavoprotein [5, 6], comproportionation to semiquinone [6], valid for pH range 7.0–8.5, increase of K_m above pH 8.5 [5]) [6]; 0.007 (electron-transferring flavoprotein, disproportionation to oxidized and reduced quinone) [6]; 0.00197 (electron-transferring flavoprotein) [3]; 0.057 (ubiquinone-1) [3]; More (kinetics of partial reactions) [3]

pH-optimum

7.0 [3]; 7.8 (best stability of substrate, value thus chosen for assay) [3]

pH-range

Temperature optimum (°C)

25 (assay at) [3]

Temperature range (°C)

---

3 ENZYME STRUCTURE

Molecular weight

Subunits

? (x × 69000–73000, pig, SDS-PAGE, calculation from flavin content) [6]

Glycoprotein/Lipoprotein
–

4 ISOLATION/PREPARATION

Source organism
Bovine [2, 8]; Pig [3–6]; Paracoccus denitrificans [7]

Source tissue
Heart [2, 8]; Liver [3–6]

Localisation in source
Mitochondria [2, 3]

Purification
Pig [6]; Paracoccus denitrificans (partial) [7]

Crystallization
–

Cloned
–

Renaturated
–

5 STABILITY

pH
7–9 (more than 1 h) [3]; 6 (1 h, 19% loss of activity) [3]; 10 (1 h, 19% loss of activity) [3]; 5.5 (complete inactivation) [3]

Temperature (°C)

Oxidation

Organic solvent

General stability information

Storage
–70°C, protein 1–2 mg/ml, 20% v/v glycerol, several months [6]; 20 mM MOPS/KOH buffer, pH 7.4, 10% v/v ethylene glycol [3]

Enzyme Handbook © Springer-Verlag Berlin Heidelberg 1994
Duplication, reproduction and storage in data banks are only allowed with the prior permission of the publishers

6 CROSSREFERENCES TO STRUCTURE DATABANKS

PIR/MIPS code

Brookhaven code

7 LITERATURE REFERENCES

[1] Frerman, F.: Biochem. Soc. Trans.,16,416–418 (1988)
[2] Steenkamp, D.J.: Biochem. J.,255,869–876 (1988)
[3] Ramsay, R.R., Steenkamp, D.J., Husain, M.: Biochem. J.,241,883–892 (1987)
[4] Johnson, M.K., Morningstar, J.E., Oliver, M., Frerman, F.E.: FEBS Lett.,226,129–133 (1987)
[5] Beckmann, J.D., Frerman, F.E.: Biochemistry,24,3922–3925 (1985)
[6] Beckmann, J.D., Frerman, F.E.: Biochemistry,24,3913–3921 (1985)
[7] Husain, M., Steenkamp, D.J.: J. Bacteriol.,163,709–715 (1985)
[8] Ruzicka, F.J., Beinert, H.: J. Biol. Chem.,252,8440–8445 (1977)

1 NOMENCLATURE

EC number
1.5.99.1

Systematic name
Sarcosine:(acceptor) oxidoreductase (demethylating)

Recommended name
Sarcosine dehydrogenase

Synonymes
Sarcosine N-demethylase
Monomethylglycine dehydrogenase

CAS Reg. No.
37228-65-2

2 REACTION AND SPECIFICITY

Catalysed reaction
Sarcosine + acceptor + H_2O →
→ glycine + formaldehyde + reduced acceptor

Reaction type
Redox reaction
Oxidative deamination

Natural substrates
Sarcosine + acceptor + H_2O [1–16]

Substrate spectrum
1 Sarcosine + acceptor + H_2O (acceptors: 2,6-dichlorophenolindophenol, phenazine methosulfate) [1–16]
2 N-Methyl derivatives of glycine, alanine, valine or leucine + acceptor + H_2O (acceptors: 2,6-dichlorophenolindophenol, phenazine methosulfate, methylene blue, meldola blue, nile blue, potassium ferricyanide) [10]

Product spectrum
1 Glycine + formaldehyde + reduced acceptor [1–16]
2 Amino acid (corresponding) + formaldehyde + reduced acceptor [10]

Inhibitor(s)
Methoxyacetate [4, 15]; Cu^{2+} [10]

Enzyme Handbook © Springer-Verlag Berlin Heidelberg 1994
Duplication, reproduction and storage in data banks are only allowed with the prior permission of the publishers

Cofactor(s)/prostethic group(s)
FAD [2, 4–9, 14]; FMN (Pseudomonas sp.) [13]; Tetrahydropteroyl-pentaglutamate [2, 4, 6–9]

Metal compounds/salts
Nonheme iron [15, 16]

Turnover number (min^{-1})
3.9 (sarcosine) [4]

Specific activity (U/mg)
0.263–0.279 [2, 6]

K_m-value (mM)
0.32–1.0 (sarcosine) [2, 4, 7, 15]; 29 (sarcosine, Pseudomonas putida) [10]; 69 (N-methylalanine, Pseudomonas putida) [10]; 11 (N-methylvaline, Pseudomonas putida) [10]; 20 (N-methylleucine, Pseudomonas putida) [10]; 0.17 (phenazine methosulfate, Pseudomonas putida) [10]; 2.0 (potassium ferricyanide, Pseudomonas putida) [10]

pH-optimum
8.0–9.0 (sarcosine + phenazine methosulfate, Pseudomonas) [10]; 8.0 (sarcosine + phenazine methosulfate) [15, 16]

pH-range
5.5 (not active below, sarcosine + phenazine methosulfate, Pseudomonas putida) [10]; 6.0 (not active below, sarcosine + phenazine methosulfate) [16]

Temperature optimum (°C)

Temperature range (°C)

---

3 ENZYME STRUCTURE

Molecular weight
266000 (Pseudomonas aeruginosa, sedimentation velocity) [11]
170000 (Pseudomonas putida, gel filtration) [10]
99000 (rat, gel filtration) [2]

Subunits
Tetramer (4 × 45000–46000, Pseudomonas putida, SDS-PAGE) [10]
? (x × 105000, rat, SDS-PAGE [8], x × 91000, pig, SDS-PAGE [6]) [6, 8]

Glycoprotein/Lipoprotein
–

4 ISOLATION/PREPARATION

Source organism

Rhizobium meliloti [1]; Pseudomonas putida [3, 10, 12]; Pseudomonas aeruginosa [11]; Pseudomonas reptilovorans [11]; Pseudomonas sp. [13]; Rat [2, 8, 15, 16]; Pig [6]; Monkey [15]

Source tissue

Liver [2, 4, 6–9, 14–16]

Localisation in source

Mitochondria [2, 4, 6–9, 14–16]; Cytoplasmic membranes (Pseudomonas) [11]

Purification

Rat [2, 8, 15, 16]; Pig [6]; Monkey [15]; Pseudomonas putida [10]

Crystallization

–

Cloned

–

Renaturated

–

5 STABILITY

pH

7.5–9.5 (Pseudomonas putida) [10]

Temperature (°C)

Oxidation

Organic solvent

General stability information

Storage

–4°C, 1 week, lyophilized [15]

6 CROSSREFERENCES TO STRUCTURE DATABANKS

PIR/MIPS code

Brookhaven code

Enzyme Handbook © Springer-Verlag Berlin Heidelberg 1994
Duplication, reproduction and storage in data banks are only allowed with the prior permission of the publishers

7 LITERATURE REFERENCES

[1] Smith, L.T., Pocard, J.A., Bernard, T., Le Rudulier, D.: J. Bacteriol.,170,3142–3149 (1988)
[2] Cook, R.J., Wagner, C.: Methods Enzymol.,122,255–260 (1986)
[3] Yamada, H., Shimizu, S., Kim, J.M., Shinmen, Y., Sakai, T.: FEMS Microbiol. Lett.,30,337–340 (1985)
[4] Porter, D.H., Cook, R.J., Wagner, C.: Arch. Biochem. Biophys.,243,396–407 (1985)
[5] Cook, R.J., Misono, K.S., Wagner, C.: J. Biol. Chem.,260,12998–13002 (1985)
[6] Steenkamp, D.J., Husain, M.: Biochem. J.,203,707–715 (1982)
[7] Wittwer, A.J., Wagner, C.: J. Biol. Chem.,256,4109–4115 (1981)
[8] Wittwer, A.J., Wagner, C.: J. Biol. Chem.,256,4102–4108 (1981)
[9] Cook, R.J., Misono, K.S., Wagner, C.: J. Biol. Chem.,259,12475–12480 (1980)
[10] Oka, I., Yoshimoto, T., Rikitake, K., Ogushi, S., Tsur, D.: Agric. Biol. Chem.,43,1197–1203 (1979)
[11] Bater, A.J., Venables, W.A.: Biochim. Biophys. Acta,468,209–226 (1977)
[12] Tsuru, D., Oka, I., Yoshimoto, T.: Agric. Biol. Chem.,40,1011–1018 (1976)
[13] Pinto, J.T., Frisell, W.R.: Arch. Biochem. Biophys.,169,483–491 (1975)
[14] Patek, D.R., Frisell, W.R.: Arch. Biochem. Biophys.,150,347–354 (1972)
[15] Frisell, W.R., MacKenzie, C.G.: Methods Enzymol.,17 A,976–981 (1970)
[16] Frisell, W.R., Mackenzie, C.G.: J. Biol. Chem.,237,94–98 (1962)

1 NOMENCLATURE

EC number
1.5.99.2

Systematic name
N,N-Dimethylglycine:(acceptor) oxidoreductase (demethylating)

Recommended name
Dimethylglycine dehydrogenase

Synonymes
N,N-Dimethylglycine oxidase

CAS Reg. No.
37256-30-7

2 REACTION AND SPECIFICITY

Catalysed reaction
N,N-Dimethylglycine + acceptor + H_2O →
→ sarcosine + formaldehyde + reduced acceptor

Reaction type
Redox reaction
Oxidative deamination

Natural substrates

Substrate spectrum
1 N,N-Dimethylglycine + acceptor + H_2O (acceptors: 2,6-dichlorophenolindophenol, phenazine methosulfate) [1–12]
2 Sarcosine + acceptor + H_2O [2, 7]
3 N-Methyl-L-alanine + acceptor + H_2O [7]
4 epsilon-N-Methyl-L-lysine + acceptor + H_2O [7]

Product spectrum
1 Sarcosine + formaldehyde + reduced acceptor [1–12]
2 Glycine + formaldehyde + reduced acceptor [2]
3 Formaldehyde + L-alanine + reduced acceptor [7]
4 Formaldehyde + L-lysine + reduced acceptor [7]

Inhibitor(s)
Methoxyacetate [3]; Dimethylthetin [7]

Enzyme Handbook © Springer-Verlag Berlin Heidelberg 1994
Duplication, reproduction and storage in data banks are only allowed with the prior permission of the publishers

Cofactor(s)/prostethic group(s)
FAD [2–4, 6–10]; Tetrahydropteroylpentaglutamate [2, 3, 5–10]

Metal compounds/salts
Nonheme iron [11, 12]

Turnover number (min^{-1})
8.4 (N,N-dimethyl glycine) [3]

Specific activity (U/mg)
0.157–0.268 [2, 6]; 82.3 [9]

K_m**-value** (mM)
0.05 (dimethylglycine) [2, 3]; 20.0 (sarcosine) [2]

pH-optimum
8.5–9.0 (dimethylglycine + phenazine methosulfate) [11, 12]

pH-range

Temperature optimum (°C)

Temperature range (°C)

3 ENZYME STRUCTURE

Molecular weight
86000 (rat, gel filtration) [2]

Subunits
? (x × 90000, rat, SDS-PAGE [8, 9], x × 93000, pig, SDS-PAGE [6]) [6, 8, 9]

Glycoprotein/Lipoprotein
–

4 ISOLATION/PREPARATION

Source organism
Rhizobium meliloti [1]; Rat [2, 8, 9, 11, 12]; Pig [6]; Monkey [11]

Source tissue
Liver [2, 3, 5–12]

Localisation in source
Mitochondria [2, 3, 5–12]

Purification
Rat [2, 8, 9, 11, 12]; Pig [6]; Monkey [11]

Crystallization

–

Cloned

–

Renaturated

–

5 STABILITY

pH

Temperature (°C)

Oxidation

Organic solvent

General stability information

Storage

–4°C, 1 week, lyophilized [11]

6 CROSSREFERENCES TO STRUCTURE DATABANKS

PIR/MIPS code

Brookhaven code

7 LITERATURE REFERENCES

[1] Smith, L.T., Pocard, J.A., Bernard, T., Le Rudulier, D.: J. Bacteriol.,170,3142–3149 (1988)
[2] Cook, R.J., Wagner, C.: Methods Enzymol.,122,255–260 (1986)
[3] Porter, D.H., Cook, R.J., Wagner, C.: Arch. Biochem. Biophys.,243,396–407 (1985)
[4] Cook, R.J., Misono, K.S., Wagner, C.: J. Biol. Chem.,260,12998–13002 (1985)
[5] Wagner, C., Briggs, W.T., Cook, R.J.: Arch. Biochem. Biophys.,233,457–461 (1984)
[6] Steenkamp, D.J., Husain, M.: Biochem. J.,203,707–715 (1982)
[7] Wittwer, A.J., Wagner, C.: J. Biol. Chem.,256,4109–4115 (1981)
[8] Wittwer, A.J., Wagner, C.: J. Biol. Chem.,256,4102–4108 (1981)
[9] Wittwer, A.J., Wagner, C.: Proc. Natl. Acad. Sci. USA,77,4484–4488 (1980)
[10] Cook, R.J., Misono, K.S., Wagner, C.: J. Biol. Chem.,259,12475–12480 (1980)
[11] Frisell, W.R., Mackenzie, C.G.: Methods Enzymol.,17 A,976–981 (1970)
[12] Frisell, W.R., Mackenzie, C.G.: J. Biol. Chem.,237,94–98 (1962)

Enzyme Handbook © Springer-Verlag Berlin Heidelberg 1994
Duplication, reproduction and storage in data banks are only allowed with the prior permission of the publishers

1 NOMENCLATURE

EC number
1.5.99.3

Systematic name
L-Pipecolate:(acceptor) 1,6-oxidoreductase

Recommended name
L-Pipecolate dehydrogenase

Synonymes

CAS Reg. No.
9076-63-5

---

2 REACTION AND SPECIFICITY

Catalysed reaction
L-Pipecolate + acceptor →
→ 2,3,4,5-tetrahydropyridine-2-carboxylate + reduced acceptor

Reaction type
Redox reaction

Natural substrates

Substrate spectrum
1 L-Pipecolate + acceptor (acceptors: 2,6-dichlorophenolindophenol, redox dyes) [1, 2]

Product spectrum
1 2,3,4,5-Tetrahydropyridine-2-carboxylate + reduced acceptor (product reacts with H_2O to form 2-aminoadipate 6-semialdehyde, i.e. 2-amino-6-oxohexanoate) [1, 2]

Inhibitor(s)
Quinacrine [1]; Cu^{2+} [1, 2]; Fe^{2+} [2]

Cofactor(s)/prostethic group(s)
FAD [1, 2]; Cytochrome b [1, 2]

Metal compounds/salts

Turnover number (min^{-1})

Enzyme Handbook © Springer-Verlag Berlin Heidelberg 1994
Duplication, reproduction and storage in data banks are only allowed with the prior permission of the publishers

Specific activity (U/mg)

K_m-value (mM)
17 (pipecolate) [1, 2]

pH-optimum
7.4 (pipecolate + 2,6-dichlorophenolindophenol) [1, 2]

pH-range

Temperature optimum (°C)

Temperature range (°C)

---

3 ENZYME STRUCTURE

Molecular weight

Subunits

Glycoprotein/Lipoprotein
–

---

4 ISOLATION/PREPARATION

Source organism
Pseudomonas putida [1, 2]

Source tissue

Localisation in source
Membranes [1, 2]

Purification
Pseudomonas putida (partially) [1, 2]

Crystallization
–

Cloned
–

Renaturated
–

5 STABILITY

pH

Temperature (°C)

Oxidation

Organic solvent

General stability information

Storage

6 CROSSREFERENCES TO STRUCTURE DATABANKS

PIR/MIPS code

Brookhaven code

7 LITERATURE REFERENCES

[1] Rodwell, V.W.: Methods Enzymol.,17 B,174–188 (1971)
[2] Baginsky, M.L., Rodwell, V.W.: J. Bacteriol.,94,1034–1039 (1967)

Enzyme Handbook © Springer-Verlag Berlin Heidelberg 1994
Duplication, reproduction and storage in data banks are only allowed with the prior permission of the publishers

1 NOMENCLATURE

EC number
1.5.99.4

Systematic name
Nicotine:(acceptor) 6-oxidoreductase (hydroxylating)

Recommended name
Nicotine dehydrogenase

Synonymes
Nicotine oxidase
d-Nicotine oxidase

CAS Reg. No.
37256-31-8

2 REACTION AND SPECIFICITY

Catalysed reaction
Nicotine + acceptor + H_2O →
→ (S)-6-hydroxynicotine + reduced acceptor

Reaction type
Redox reaction

Natural substrates

Substrate spectrum
1 Nicotine + acceptor + H_2O (acceptors: 2,6-dichlorophenolindophenol, brilliant cresyl blue, methylene blue, menadione, 5-hydroxy-1,4-naphthoquinone, vitamin K_5) [1–4]
2 Nornicotine + acceptor + H_2O [3]
3 Nicotine-N-oxide + acceptor + H_2O [3]
4 Myosmine + acceptor + H_2O [3]
5 Anabasine + acceptor + H_2O [3]

Product spectrum
1 6-Hydroxynicotine + reduced acceptor [1–4]
2 6-Hydroxynornicotine + reduced acceptor
3 6-Hydroxynicotine-N-oxide + reduced acceptor
4 ? + reduced acceptor
5 6-Hydroxyanabasine + reduced acceptor

Enzyme Handbook © Springer-Verlag Berlin Heidelberg 1994
Duplication, reproduction and storage in data banks are only allowed with the prior permission of the publishers

Inhibitor(s)

Qinacrine [3]; Acriflavine [3]; Cyanide [3]; alpha,alpha'-Dipyridyl [3]; 1,10-Phenanthroline [3]; p-Chloromercuriphenylsulfonate [3]; Hg^{2+} [3]; Iodoacetamide [3]

Cofactor(s)/prostethic group(s)

Bactopterin [1, 2]; FAD [1, 2]; FMN [3]; More (metalloflavoprotein)

Metal compounds/salts

Mo^{6+} [1, 2]; Fe^{2+} [1, 2]

Turnover number (min^{-1})

Specific activity (U/mg)

28.3–29.2 [2, 3]

K_m-value (mM)

0.037 (nicotine, 5-hydroxy-1,4-naphthoquinone) [3]; 0.008 (brilliant cresyl blue) [3]; 0.027 (methylene blue) [3]; 1.24 (2,6-dichlorophenolindophenol) [3]; 0.49 (menadione) [3]; 1.0 (vitamin K_5) [3]

pH-optimum

7.6–8.2 (nicotine + brilliant cresyl blue) [3]

pH-range

Temperature optimum (°C)

Temperature range (°C)

---

3 ENZYME STRUCTURE

Molecular weight

120000 (Arthrobacter oxidans, gel chromatography) [2]

Subunits

Oligomer (Arthrobacter oxidans, SDS-PAGE) [2]

Glycoprotein/Lipoprotein

–

---

4 ISOLATION/PREPARATION

Source organism

Arthrobacter oxidans [1–4]

Source tissue

Localisation in source

Purification

Arthrobacter oxidans [2, 3]

Crystallization

–

Cloned

–

Renaturated

–

5 STABILITY

pH

Temperature (°C)

Oxidation

Organic solvent

General stability information

Storage

6 CROSSREFERENCES TO STRUCTURE DATABANKS

PIR/MIPS code

Brookhaven code

7 LITERATURE REFERENCES

[1] Nagel, M., Koenig, K., Andreesen, J.R.: FEMS Microbiol. Lett.,60,323–326 (1989)
[2] Freudenberg, W., Koenig, K., Andreesen, J.R.: FEMS Microbiol. Lett.,52,13–18 (1988)
[3] Hochstein, L.L., Dalton, B.P.: Biochim. Biophys. Acta,139,56–68 (1967)
[4] Decker, K., Bleeg, H.: Biochim. Biophys. Acta,105,313–324 (1965)

Enzyme Handbook © Springer-Verlag Berlin Heidelberg 1994
Duplication, reproduction and storage in data banks are only allowed with the prior permission of the publishers

1 NOMENCLATURE

EC number
1.5.99.5

Systematic name
N-Methyl-L-glutamate:(acceptor) oxidoreductase (demethylating)

Recommended name
Methylglutamate dehydrogenase

Synonymes
N-Methylglutamate dehydrogenase

CAS Reg. No.
37217-26-8

---

2 REACTION AND SPECIFICITY

Catalysed reaction
N-Methyl-L-glutamate + acceptor + H_2O →
→ L-glutamate + formaldehyde + reduced acceptor

Reaction type
Redox reaction
Oxidative deamination

Natural substrates
N-Methyl-L-glutamate + acceptor + H_2O [1–9]

Substrate spectrum
1 N-Methyl-substituted amino acids + acceptor + H_2O (acceptors: 2,6-dichlorophenolindophenol, cytochrome c, phenazine methosulphate, potassium ferricyanide) [3, 6, 8]

Product spectrum
1 Formaldehyde + amino acids (corresponding) + reduced acceptor [3, 6, 8]

Inhibitor(s)
Formaldehyde [4, 6]; p-Chloromercuribenzoate [6, 8]; Iodoacetamide [6]; 2-Oxoglutarate [6, 8]; Cu^{2+} [6]; Triton X-100 [6, 7]; p-Chloromercuriphenylsulfonate [8]

Enzyme Handbook © Springer-Verlag Berlin Heidelberg 1994
Duplication, reproduction and storage in data banks are only allowed with the prior permission of the publishers

Cofactor(s)/prostethic group(s)
FAD [3]; Cytochrome b [6]

Metal compounds/salts

Turnover number (min^{-1})

Specific activity (U/mg)
0.113 [3]; 0.29 [5]

K_m-value (mM)
0.019–0.2 (2,6-dichlorophenolindophenol) [3, 6–8]; 0.025–0.333 (N-methylglutamate) [3, 6–9]; 0.46–1.3 (N-methylaspartate) [6–8]; 6.0–87 (N-methylalanine) [6–8]; 90.9–200 (sarcosine) [6–8]; 0.011–0.018 (phenazine methosulphate) [6–8]; 0.11 (cytochrome c) [6, 7]; 1.4 (potassium ferricyanide) [8]; 0.20 (N-methylvaline, N-methylisoleucine) [8]; 2.5 (N-methylphenylalanine) [8]; 36 (N-methylserine) [8]

pH-optimum
7.0 (N-methylglutamate + 2,6-dichlorophenolindophenol) [3]; 7.0–8.5 (N-methylglutamate + 2,6-dichlorophenolindophenol) [6, 7]

pH-range

Temperature optimum (°C)

Temperature range (°C)

---

3 ENZYME STRUCTURE

Molecular weight
407000 (Bacterium AT2, gel filtration) [3]
550000 (Pseudomonas aminovorans, gel filtration) [5]

Subunits
Tetramer (4 × 130000, SDS-PAGE) [5]

Glycoprotein/Lipoprotein
–

---

4 ISOLATION/PREPARATION

Source organism
Methylobacterium organophilum [1]; Methylophaga talassica [2]; Methylophaga marina [2]; Bacterium AT2 [3]; Pseudomonas aminovorans [3–7]; Pseudomonas methylica [3]; Pseudomonas sp. AM1 [7]; Pseudomonas sp. MA [8, 9]; Hyphomicrobium vulgare [3]

Source tissue

Localisation in source
Cytosol [3]; Membranes [4, 6, 8, 9]

Purification
Bacterium AT2 [3]; Pseudomonas aminovorans [5–7]; Pseudomonas sp. MA (partially) [8]

Crystallization
–

Cloned
–

Renaturated
–

---

5 STABILITY

pH
6.0 (not stable below) [6]

Temperature (°C)
40 (not stable above) [3, 6]

Oxidation

Organic solvent

General stability information

Storage
Several months, –15°C [6–8]

---

6 CROSSREFERENCES TO STRUCTURE DATABANKS

PIR/MIPS code

Brookhaven code

Enzyme Handbook © Springer-Verlag Berlin Heidelberg 1994
Duplication, reproduction and storage in data banks are only allowed with the prior permission of the publishers

7 LITERATURE REFERENCES

[1] Biville, F., Mazodier, P., Gasser, F., Van Kleef, M.A.G., Duine, J.A.: FEMS Microbiol. Lett.,52,53–58 (1988)

[2] Janvier, M., Frehel, C., Grimont, F., Gasser, F.: Int. J. Syst. Bacteriol.,35,131–139 (1985)

[3] Boulton, C.A., Haywood, G.W., Large, P.J.: J. Gen. Microbiol.,117,293–304 (1980)

[4] Bamforth, C.W., Large, P.J.: Biochem. Soc. Trans.,6,193–195 (1978)

[5] Bamforth, C.W., Large, P.J.: Biochem. J.,167,509–512 (1977)

[6] Bamforth, C.W., Large, P.J.: Biochem. J.,161,357–370 (1977)

[7] Bamforth, C.W., Large, P.J.: Biochem. Soc. Trans.,3,1066–1069 (1975)

[8] Hersh, L.B., Stark, M.J., Worthen, S., Fiero, M.K.: Arch. Biochem. Biophys.,150,219–226 (1972)

[9] Hersh, L.B., Peterson, J.A., Thompson, A.A.: Arch. Biochem. Biophys.,145,115–120 (1971)

1 NOMENCLATURE

EC number
1.5.99.6

Systematic name
Spermidine:(acceptor) oxidoreductase

Recommended name
Spermidine dehydrogenase

Synonymes

CAS Reg. No.
9076-64-6

2 REACTION AND SPECIFICITY

Catalysed reaction
Spermidine + acceptor + H_2O →
→ 1,3-diaminopropane + 4-aminobutanal (condenses non-enzymatically to 1-pyrroline) + reduced acceptor

Reaction type
Redox reaction

Natural substrates
Spermidine + acceptor + H_2O [1–6]

Substrate spectrum
1 Spermidine + acceptor + H_2O (acceptors: potassium ferricyanide, 2,6-dichloroindophenol, phenazine methosulfate, cytochrome c) [1–6]
2 Triamines or tetramines with a 4-aminobutylimino group + acceptor + H_2O [3]
3 Triamines or tetramines with a 3-aminopropylimino group + acceptor + H_2O [3]

Product spectrum
1 1,3-Diaminopropane + 4-aminobutanal (condenses non-enzymatically to 1-pyrroline) + reduced acceptor [1–6]
2 1-Pyrroline + ? + reduced acceptor [3]
3 1,3-Diaminopropane + aminoaldehyde + reduced acceptor [3]

Enzyme Handbook © Springer-Verlag Berlin Heidelberg 1994
Duplication, reproduction and storage in data banks are only allowed with the prior permission of the publishers

Inhibitor(s)
Diamines [3]; p-Chloromercuribenzoate [5, 6]; N-Ethylmaleimide [5, 6]; Quinacrine [5, 6]; Phenylhydrazine [5]

Cofactor(s)/prostethic group(s)
FAD [4–6]; Heme [4–6]

Metal compounds/salts

Turnover number (min^{-1})

Specific activity (U/mg)
420–440 [1, 2, 5, 6]

K_m-value (mM)
0.0002–0.0011 (spermidine) [3, 5, 6]; 0.11 (N-n-propylputrescine) [3]; 0.05 (spermine) [5, 6]; More (triamines and tetramines) [3]

pH-optimum
7.5 (spermidine) [3]; 7.5–10.0 (triamines and tetramines) [3]; 7.2 (spermidine) [5, 6]; 8.8 (spermine) [5, 6]

pH-range

Temperature optimum (°C)

Temperature range (°C)

---

3 ENZYME STRUCTURE

Molecular weight
76000 (Serratia marcescens, sedimentation equilibrium) [5, 6]

Subunits
Monomer (1 × 76000, Serratia marcescens, SDS-PAGE) [6]

Glycoprotein/Lipoprotein
–

---

4 ISOLATION/PREPARATION

Source organism
Serratia marcescens [1–6]

Source tissue

Localisation in source

Purification

Serratia marcescens [1, 2, 5, 6]

Crystallization

–

Cloned

–

Renaturated

–

5 STABILITY

pH

Temperature (°C)

Oxidation

Organic solvent

General stability information

Storage

6 months, –20°C, pH 7.2 [6]

6 CROSSREFERENCES TO STRUCTURE DATABANKS

PIR/MIPS code

Brookhaven code

7 LITERATURE REFERENCES

[1] Okada, M., Kawashima, S., Imahori, K.: Methods Enzymol.,94,303–305 (1983)
[2] Okada, M., Kawashima, S., Imahori, K.: J. Biochem.,85,1225–1233 (1979)
[3] Okada, M., Kawashima, S., Imahori, K.: J. Biochem.,85,1235–1243 (1979)
[4] Tabor, H., Tabor, C.W.: Adv. Enzymol. Relat. Areas Mol. Biol.,36,203–268 (1972) (Review)
[5] Tabor, C.W., Kellog, P.D.: Methods Enzymol.,17 B,746–753 (1971)
[6] Tabor, C.W., Kellog, P.D.: J. Biol. Chem.,245,5424–5433 (1970)

Enzyme Handbook © Springer-Verlag Berlin Heidelberg 1994
Duplication, reproduction and storage in data banks are only allowed with the prior permission of the publishers

1 NOMENCLATURE

EC number
1.5.99.7

Systematic name
Trimethylamine:(acceptor) oxidoreductase (demethylating)

Recommended name
Trimethylamine dehydrogenase

Synonymes

CAS Reg. No.
39307-09-0

2 REACTION AND SPECIFICITY

Catalysed reaction
Trimethylamine + H_2O + acceptor →
→ dimethylamine + formaldehyde + reduced acceptor

Reaction type
Redox reaction

Natural substrates
Trimethylamine + H_2O + FAD-containing flavoprotein [1–3, 9, 10, 12]

Substrate spectrum
1 Trimethylamine + H_2O + acceptor (r, acceptors: an FAD-containing flavoprotein, phenazine methosulphate, 2,6-dichlorophenolindophenol, brilliant cresyl blue, methylene blue) [1–23]
2 Secondary or tertiary amines + H_2O + acceptor [21, 23]

Product spectrum
1 Dimethylamine + formaldehyde + reduced acceptor [1–23]
2 Primary or secondary amines + aldehyde + reduced acceptor

Inhibitor(s)
Tetramethylammonium chloride [9, 20, 21]; Substituted hydrazines [11, 15, 20, 21]; Cyclopropylamines [15]; Ethylamine [18]; Dimethylamine [18]; Acetaldehyde [18]; Trimethylsulphonium chloride [20, 21]; Cu^{2+} [21]; Co^{2+} [21]; Ni^{2+} [21]; Ag^{2+} [21]; Hg^{2+} [21]; Iodoacetamide [21]; N-Ethylmaleimide [21]; p-Chloromercuribenzoate [21]

Enzyme Handbook © Springer-Verlag Berlin Heidelberg 1994
Duplication, reproduction and storage in data banks are only allowed with the prior permission of the publishers

Cofactor(s)/prostethic group(s)
ADP (1 molecule ADP per subunit) [1]; 6-S-Cysteinyl-FMN (and a [4Fe-4S]cluster in the ratio 1: 1 per subunit) [1–6, 8–11, 13–17]

Metal compounds/salts
Iron (4Fe-4S-cluster) [1–6, 8–11, 15–17]

Turnover number (min^{-1})

Specific activity (U/mg)
1.00–1.07 [18, 21]

K_m-value (mM)
0.002 (trimethylamine) [21, 23]; 1.25 (phenazine methosulphate) [21]; 1.7 (diethylamine) [21]; 0.008 (triethylamine) [21]; More (tertiary amines) [21]

pH-optimum
8.5 (trimethylamine + phenazine methosulphate) [21, 23]

pH-range

Temperature optimum (°C)

Temperature range (°C)

---

3 ENZYME STRUCTURE

Molecular weight
166000 (Bacterium W3A1, Hyphomicrobium sp. X [6], SDS-PAGE [6], X-ray analysis [10]) [6, 10]
147000 (Bacterium W3A1, sedimentation equilibrium) [6, 18]
161000 (Bacterium 4B6, gel filtration) [21, 23]

Subunits
Dimer (identical, 2 × 83000, Bacterium W3A1 [2, 4, 6, 10], Bacterium 4B6 [23], X-ray analysis [2, 4, 10], SDS-PAGE [6, 23]) [2, 4, 6, 10, 23]

Glycoprotein/Lipoprotein
–

---

4 ISOLATION/PREPARATION

Source organism
Bacterium W3A1 [1–6, 8–11, 13–16, 18, 19]; Bacterium W6A [19]; Bacterium 4B6 [20–23]; Bacterium C2A1 [22]; Methylophilus methylotrophus [3, 7]; Hyphomicrobium sp. X [5, 6, 8]; Hyphomicrobium vulgare [20]

Source tissue

Localisation in source
Cytoplasm [5, 7]; Cytoplasmic side of membranes [7]

Purification
Bacterium W3A1 [18]; Bacterium 4B6 [21]

Crystallization
[4, 10]

Cloned
–

Renaturated
–

5 STABILITY

pH

Temperature (°C)

Oxidation

Organic solvent

General stability information

Storage
3 months, –20°C [21]

6 CROSSREFERENCES TO STRUCTURE DATABANKS

PIR/MIPS code
A13543 (Bacterium W3A1 fragment)

Brookhaven code
0TMD (METHYLOTROPHIC BACTERIUM W=3=*A*=1=)

Enzyme Handbook © Springer-Verlag Berlin Heidelberg 1994
Duplication, reproduction and storage in data banks are only allowed with the prior permission of the publishers

7 LITERATURE REFERENCES

[1] Lim, L.W., Mathews, F.S., Steenkamp, D.J.: J. Biol. Chem.,263,3075–3078 (1988)
[2] Lim, L.W., Shamala, N., Mathews, F.S., Steenkamp, D.J., Hamlin, R., Xuong, N.: J. Biol. Chem.,261,15140–15146 (1986)
[3] Davidson, V.L., Husain, M., Neher, J.W.: J. Bacteriol.,166,812–817 (1986)
[4] Lim, L.W., Shamala, N., Mathews, F.S., Steenkamp, D.J.: J. Biol. Chem.,259,14458–14462 (1984)
[5] Kasprzak, A.A., Steenkamp, D.J.: J. Bacteriol.,156,348–353 (1983)
[6] Kasprzak, A.A., Papas, E.J., Steenkamp, D.J.: Biochem. J.,211,535–541 (1983)
[7] Burton, S.M., Byrom, D., Carver, M., Jones, G.D.D., Jones, C.W.: FEMS Microbiol. Lett.,17,185–190 (1983)
[8] Steenkamp, D.J., Beinert, H.: Biochem. J.,207,241–252 (1982)
[9] Steenkamp, D.J., Beinert, H.: Biochem. J.,207,233–239 (1982)
[10] Lim, L.W., Mathews, F.S., Steenkamp, D.J.: J. Mol. Biol.,162,869–876 (1982)
[11] Nagy, J., Kenney, W.C., Singer, T.P.: J. Biol. Chem.,254,2684–2688 (1979)
[12] Steenkamp, D.J., Gallup, M.: J. Biol. Chem.,253,4086–4089 (1978)
[13] Steenkamp, D.J., McIntire, W., Kenney, W.C.: J. Biol. Chem.,253,2818–2824 (1978)
[14] Steenkamp, D.J., Kenney, W.C., Singer, T.P.: J. Biol. Chem.,253,2812–2817 (1978)
[15] Steenkamp, D.J., Singer, T.P., Beinert, H.: Biochem. J.,169,361–369 (1978)
[16] Hill, C.L., Steenkamp, D.J., Holm, R.H., Singer, T.P.: Proc. Natl. Acad. Sci. USA,74,547–551 (1977)
[17] Steenkamp, D.J., Singer, T.P.: Biochem. Biophys. Res. Commun.,71,1289–1295 (1976)
[18] Steenkamp, D.J., Mallinson, J.: Biochim. Biophys. Acta,429,705–719 (1976)
[19] Colby, J., Zatman, L.J.: Biochem. J.,148,513–520 (1975)
[20] Large, P.J., McDougall, H.: Anal. Biochem.,64,304–310 (1975)
[21] Colby, J., Zatman, L.J.: Biochem. J.,143,555–567 (1974)
[22] Colby, J., Zatman, L.J.: Biochem. J.,132,101–112 (1973)
[23] Colby, J., Zatman, L.J.: Biochem. J.,121,9–10 (1971)

1 NOMENCLATURE

EC number
1.5.99.8

Systematic name
L-Proline:(acceptor) oxidoreductase

Recommended name
Proline dehydrogenase

Synonymes
L-Proline dehydrogenase

CAS Reg. No.
9050-70-8

---

2 REACTION AND SPECIFICITY

Catalysed reaction
L-Proline + acceptor + H_2O →
→ (S)-1-pyrroline-5-carboxylate + reduced acceptor

Reaction type
Redox reaction

Natural substrates
L-Proline + acceptor + H_2O [1–10]

Substrate spectrum
1 L-Proline + acceptor + H_2O (acceptors: 2,6-dichlorophenolindophenol, phenazine methosulfate, ferricyanide, menadione, cytochrome c) [1–10]

Product spectrum
1 (S)-1-Pyrroline-5-carboxylate + reduced acceptor [1–10]

Inhibitor(s)
Mg^{2+} [2]; Hg^{2+} [3]; Cd^{2+} [3]; Zn^{2+} [3]; L-Azetidine-2-carboxylate [6]; Lactate [6]; Pyruvate [6]

Cofactor(s)/prostethic group(s)
FAD [2, 3, 6]; ADP (insects) [8, 10]

Metal compounds/salts

Enzyme Handbook © Springer-Verlag Berlin Heidelberg 1994
Duplication, reproduction and storage in data banks are only allowed with the prior permission of the publishers

Turnover number (min^{-1})

Specific activity (U/mg)
2.75 [2]; 5.26 [3]; 7.3 [6]

K_m-value (mM)
45–105 (L-proline) [2, 3, 6]; 2–6 (L-proline) [8, 10]; 0.018 (FAD) [3]; 0.0159 (menadione) [8]

pH-optimum
9.5 (L-proline + phenazine methosulfate) [3]; 7.2 (proline + ferricyanide) [4]; 8.0 (L-proline + phenazine methosulfate) [6]; 7.8 (L-proline + menadione) [8]

pH-range

Temperature optimum (°C)

Temperature range (°C)

3 ENZYME STRUCTURE

Molecular weight
260000 (Escherichia coli, gel filtration) [6]
242000 (Pseudomonas aeruginosa, gel filtration, also active as 1-pyrroline-5-carboxylate dehydrogenase) [3]

Subunits
Dimer (2 × 119000–124000, E. coli [6], Pseudomonas aeruginosa [3], SDS-PAGE) [3, 6]

Glycoprotein/Lipoprotein
–

4 ISOLATION/PREPARATION

Source organism
Escherichia coli [1–3, 6]; Salmonella typhimurium [1–3]; Pseudomonas aeruginosa [3]; Zea mays (corn) [4]; Illex illecebrosus (squid) [5]; Insects [7, 9, 10]; Rat [8]

Source tissue
Heart [5]; Flight muscle [7, 9, 10]; Liver [8]

Localisation in source
Cytoplasmic membranes [1, 2]; Mitochondrial membranes [4, 5, 7, 8, 10]

Purification

Escherichia coli [2, 6]; Pseudomonas aeruginosa [3]; Rat [8]

Crystallization

–

Cloned

–

Renaturated

–

5 STABILITY

pH

Temperature (°C)

Oxidation

Organic solvent

General stability information

Storage

1 month, –20°C, pH 7.6, 50% glycerol [3]

6 CROSSREFERENCES TO STRUCTURE DATABANKS

PIR/MIPS code

Brookhaven code

7 LITERATURE REFERENCES

[1] Wood, J.M.: Proc. Natl. Acad. Sci. USA,84,373–377 (1987)
[2] Graham, S.B., Stephenson, J.T., Wood, J.M.: J. Biol. Chem.,259,2656–2661 (1984)
[3] Meile, L., Leisinger, T.: Eur. J. Biochem.,129,67–75 (1982)
[4] Elthon, T.E., Stewart, C.R.: Plant Physiol.,70,567–572 (1982)
[5] Mommsen, T.P., Hochachka, P.W.: Eur. J. Biochem.,120,345–350 (1981)
[6] Scarpulla, R.C., Soffer, R.L.: J. Biol. Chem.,253,5997–6001 (1978)
[7] Balboni, E., Hecht, R.I.: Biochim. Biophys. Acta,462,171–176 (1977)
[8] Kramar, R.: Hoppe-Seyler's Z. Physiol. Chem.,352,1267–1270 (1971)
[9] Crabtree, B., Newsholme, E.A.: Biochem. J.,117,1019–1021 (1970)
[10] Hansford, R.G., Sacktor, B.: J. Biol. Chem.,245,991–994 (1970)

Enzyme Handbook © Springer-Verlag Berlin Heidelberg 1994
Duplication, reproduction and storage in data banks are only allowed with the prior permission of the publishers

1 NOMENCLATURE

EC number
1.5.99.9

Systematic name
5,10-Methylenetetrahydromethanopterin:coenzyme-F420 oxidoreductase

Recommended name
Methylenetetrahydromethanopterin dehydrogenase

Synonymes
N^5, N^{10}–Methylenetetrahydromethanopterin dehydrogenase

CAS Reg. No.
100357-01-5

---

2 REACTION AND SPECIFICITY

Catalysed reaction
5,10-Methylenetetrahydromethanopterin + coenzyme F420 →
→ 5,10-methenyltetrahydromethanopterin + reduced coenzyme F420

Reaction type
Redox reaction

Natural substrates
5,10-Methylenetetrahydromethanopterin + coenzyme F420 [1–4]
More (enzyme is involved in the formation of methane from CO_2)

Substrate spectrum
1 5,10-Methylenetetrahydromethanopterin + coenzyme F420 (r) [1–4]

Product spectrum
1 5,10-Methenyltetrahydromethanopterin + reduced coenzyme F420 [1–4]

Inhibitor(s)

Cofactor(s)/prostethic group(s)
Coenzyme F420 (coenzyme F420 is a 7,8-didemethyl-8-hydroxy-5-deazariboflavin derivative) [1–4]

Metal compounds/salts

Enzyme Handbook © Springer-Verlag Berlin Heidelberg 1995
Duplication, reproduction and storage in data banks are only allowed with the prior permission of the publishers

Turnover number (min^{-1})

Specific activity (U/mg)
197 [3]; 0.111 [4]

K_m-value (mM)
0.004 (coenzyme F420) [2]; 0.05 (methylenetetrahydromethanopterin) [2]

pH-optimum
6.5 [2]; 6.0 [4]

pH-range

Temperature optimum (°C)
90 [2]; 61 [4]

Temperature range (°C)

3 ENZYME STRUCTURE

Molecular weight
32000 (Methanobacterium thermoautotrophicum, SDS-PAGE) [3]

Subunits

Glycoprotein/Lipoprotein
–

4 ISOLATION/PREPARATION

Source organism
Methanosarcina thermophila [1]; Archaeglobus fulgidus [2]; Methanobacterium thermoautotrophicum [3, 4]

Source tissue

Localisation in source

Purification
Methanobacterium thermoautotrophicum [3, 4]

Crystallization
–

Cloned
–

Renaturated

–

5 STABILITY

pH

Temperature (°C)

Oxidation

Organic solvent

General stability information

Storage

6 CROSSREFERENCES TO STRUCTURE DATABANKS

PIR/MIPS code

Brookhaven code

7 LITERATURE REFERENCES

[1] Jablonski, P.E., DiMarco, A.A., Bobik, T.A., Cabell, M.C., Ferry, J.G.: J. Bacteriol.,172,1271–1275 (1990)
[2] Moeller-Zinkhan, D., Boerner, G., Thauer, R.K.: Arch. Microbiol.,152,362–368 (1989)
[3] Mukhopadhyay, B., Daniels, L.: Can. J. Microbiol.,35,499–507 (1988)
[4] Hartzell, P.L., Zvilius, G., Escalante-Semerena, J.C., Donelly, M.I.: Biochem. Biophys. Res. Commun.,133,884–890 (1985)

Enzyme Handbook © Springer-Verlag Berlin Heidelberg 1995
Duplication, reproduction and storage in data banks are only allowed with the prior permission of the publishers

1 NOMENCLATURE

EC number
1.6.1.1

Systematic name
NADPH: NAD^+ oxidoreductase (B-specific)

Recommended name
NAD(P)$^+$ transhydrogenase (B-specific)

Synonymes
Pyridine nucleotide transhydrogenase
Transhydrogenase
Transhydrogenase, nicotinamide adenine dinucleotide (phosphate)
NAD(P) transhydrogenase
Nicotinamide adenine dinucleotide (phosphate) transhydrogenase
NAD transhydrogenase
NADH transhydrogenase
Nicotinamide nucleotide transhydrogenase
NADPH-NAD transhydrogenase
Pyridine nucleotide transferase
NADPH-NAD oxidoreductase
NADH-NADP-transhydrogenase
NADPH:NAD^+ transhydrogenase
H^+-Thase
Non-energy-linked transhydrogenase [2]

CAS Reg. No.
9014-18-0 (not distinguished from EC 1.6.1.2); 9072-60-0 (not distinguished from EC 1.6.1.2)

2 REACTION AND SPECIFICITY

Catalysed reaction
NADPH + NAD^+ →
→ $NADP^+$ + NADH (mechanism [1])

Reaction type
Redox reaction

Natural substrates
NADH + $NADP^+$ [12]

Enzyme Handbook © Springer-Verlag Berlin Heidelberg 1994
Duplication, reproduction and storage in data banks are only allowed with the prior permission of the publishers

Substrate spectrum

1 $NADP^+$ + NADH (degree of reversibility depending on source of enzyme [1], reduction of $NADP^+$ preferred direction of reaction [8, 12], 4B-specific for NAD(P)H [11]) [1, 8, 11, 12, 14, 18, 19]
2 NADPH + thio-NAD^+ [14]
3 NADPH + thio-$NADP^+$ [14, 18, 19]
4 NADH + thio-NAD^+ [14]
5 NADH + thio-$NADP^+$ [14]
6 NADPH + pyridine aldehyde-NAD^+ [1, 18]
7 NADPH + deamino-NAD^+ [1, 18, 19]
8 NADPH + 3-acetylpyridine-NAD^+ [1, 18]
9 More (diaphorase-type reactions with NAD(P)H and $K_3Fe(CN)_6$ and 2,9-dichloroindophenol) [14, 18]

Product spectrum

1 NADPH + NAD^+
2 $NADP^+$ + thio-NADH
3 $NADP^+$ + thio-NADPH
4 NAD^+ + thio-NADH
5 NAD^+ + thio-NADPH
6 $NADP^+$ + pyridine aldehyde-NADH
7 $NADP^+$ + deamino-NADH
8 $NADP^+$ + 3-acetylpyridine-NADH
9 ?

Inhibitor(s)

p-Hydroxymercuribenzoate (dependent on presence of oxidized or reduced substrate [1]) [1, 5, 18]; Phosphate (inhibition of reduction of $NADP^+$ by NADH [5]) [5, 8]; p-Aminophenylarsenoxide [5]; Diphosphate [8]; Arsenate [8]; Pyridoxal phosphate [8]; 5'-AMP [8]; ADP [8]; 2'-AMP [8]; ATP [8]; TTP [8], GTP [8]; CTP [8]; NAD(P)$^+$ (inhibition of 2'-AMP activated reaction [9], uncompetitive to thio-NAD^+ [19], inhibition in absence of Ca^{2+} [12]) [9, 12, 19]; NAD^+ (competitive to thio-NAD^+, uncompetitive with respect to NADPH) [19]; Deoxycholate [14]; More (not inhibitory: palmityl-CoA) [11]

Cofactor(s)/prostethic group(s)

FAD [1, 3, 13]; p-Hydroxymercuribenzoate (activation, depending on presence of oxidized or reduced substrates) [1]; 2'-AMP (activation, partially replaceable by 2', 3'-cyclic AMP or coenzyme A [12], almost no effect [14], activation of reduction of $NADP^+$ or thio-NAD^+ by NADH, no effect on reduction of NAD^+ or thio-NAD^+ by NADPH [19]) [1, 9, 11, 12, 19]

Metal compounds/salts

Ca^{2+} (activation) [1, 5, 7, 11, 12]; Mg^{2+} (activation) [5, 12]; EDTA (slight activating effect at low buffer concentrations) [14]

Turnover number (min^{-1})
14000–20000 (mol NADPH per mol of flavin) [18]

Specific activity (U/mg)
575 [4]; 362 [13]; 251 [19]; More (overview [1]) [1, 6]

K_m-value (mM)
0.015 (NADPH with NAD$^+$ as acceptor) [16]; 0.025 (NADH with NADP$^+$ as acceptor) [16]; 0.03 (thio-NADP$^+$ with NADH as donor) [16]; 0.04 (thio-NAD$^+$ with NADH as donor [19], NADPH with thio-NAD$^+$ as acceptor [16]) [16, 19]; 0.05 (thio-NAD$^+$ with NADH as donor) [16]; 0.06 (NADH with thio-NAD$^+$ as acceptor) [16]; 0.075 (thio-NAD$^+$ with NADPH as donor) [16]; 0.077 (NADH with thio-NAD$^+$ as acceptor) [19]; 0.085 (NADH with thio-NADP$^+$ as acceptor) [16]; 0.11 (NAD$^+$ with NADPH as donor) [16]; 0.38 (NAD$^+$ with NADPH as donor) [19]; 0.25 (thio-NAD$^+$ with NADPH as donor) [19]; 0.4 (deamino-NAD$^+$ with NADPH as donor) [19]; More (dependency on Mg^{2+} concentration [16], kinetic studies [9, 17]) [9, 16, 17]

pH-optimum
7–8 [1, 5]; 7.0 (NADH formation) [8]; 9.6 (NADPH formation) [8]

pH-range
8.4–8.7 (more than half maximal activity at pH 8.4 and 8.7 in absence of Ca^{2+} or Mg^{2+}) [5]; 9.1–9.3 (more than half maximal activity at pH 9.1 and 9.3 in presence of Ca^{2+} or Mg^{2+}) [5]

Temperature optimum (°C)
30–35 [19]

Temperature range (°C)

3 ENZYME STRUCTURE

Molecular weight
6400000 (Pseudomonas aeruginosa, sedimentation equilibrium in presence of NADP$^+$) [10]
1600000 (Pseudomonas aeruginosa, sedimentation equilibrium in presence of 2'-AMP) [10]
421000 (Azotobacter vinelandii, octameric form at pH 8.5–9.0) [6]
More (electron microscopy) [13, 15]

Subunits
Octamer or polymer of octamers (x × 52000–58000, Pseudomonas fluorescens [2], Pseudomonas aeruginosa [2, 10], Azotobacter vinelandii [2, 4, 13], SDS-PAGE [2, 4, 10, 13], amino acid analysis [4], quarternary structure [3, 6], octameric in cell-free extract, polymeric in purified form [5]) [2–6, 10, 13]

Enzyme Handbook © Springer-Verlag Berlin Heidelberg 1994
Duplication, reproduction and storage in data banks are only allowed with the prior permission of the publishers

Glycoprotein/Lipoprotein

–

4 ISOLATION/PREPARATION

Source organism

Pseudomonas fluorescens [1, 2]; Azotobacter vinelandii [2–6, 13–16, 19]; Pseudomonas aeruginosa [2, 7, 9–12, 17, 18]; Azotobacter chroococcum [1]; Azotobacter agile [1]; Beneckea natrigens [8]

Source tissue

Cell

Localisation in source

Purification

Azotobacter vinelandii [4, 13, 14]; Beneckea natrigens (partial) [8]; Pseudomonas aeruginosa [18]; More (overview) [1]

Crystallization

[4]

Cloned

–

Renaturated

–

5 STABILITY

pH

7 (polymeric enzyme stable) [5]

Temperature (°C)

50 (1 h stable, inactivation accelerated by NADH and NADPH) [14]; 51 (25 min, 50% inactivation, accelerated by addition of NADPH, reactivation by FAD) [18]; 65 (15 min, complete inactivation, protection by FAD) [14]; More (rate of thermal inactivation depending on concentration of NAD$^+$, NADP$^+$, NADH, NADPH, free FAD, Mg^{2+}, phosphate, pH) [3]

Oxidation

Organic solvent

General stability information

Urea (8 M, no dissociation [10], 5 min, 50% inactivation [14]) [10, 14]; Bovine serum albumin (0.2%, stabilization of diluted solutions) [14]

Storage

–15°C, 50 mM Tris-HCl buffer, pH 7.5, no loss of activity in 12 weeks [8]; 4°C, 0.1 M phosphate buffer, pH 7.5, 1 mM EDTA, several months stable, storage at –20°C yields a partly insoluble enzyme [14]

6 CROSSREFERENCES TO STRUCTURE DATABANKS

PIR/MIPS code

DEECXA (alpha chain Escherichia coli); DEECXB (beta chain Escherichia coli); DEBOXM (Bovine); A31670 (precursor mitochondrial Bovine fragment)

Brookhaven code

7 LITERATURE REFERENCES

[1] Ryström, J., Heok, J.B., Ernster, L. in "The Enzymes",3rd. Ed. (Boyer, P.D., Ed.) 13,51–88 (1976) (Review)

[2] Voordouw, G., Van Der Vies, S.M., Themmen, A.P.N.: Eur. J. Biochem.,131,527–533 (1983)

[3] Voordouw, G., De Haard, H., Timmermans, J.A.M., Veeger, C., Zabel, P.: Eur. J. Biochem.,127,267–274 (1982)

[4] Voordouw, G., Van Der Vies, S.M., Eweg, J.K., Veeger, C., Van Breemen, J.F.L., Van Bruggen, E.F.J.: Eur. J. Biochem.,111,347–355 (1980)

[5] Voordouw, G., Van Der Vies, S., Scholten, J.W., Veeger, C.: Eur. J. Biochem.,107,337–344 (1980)

[6] Voordouw, G., Veeger, C., Van Breemen, J.F.L., Van Bruggen, E.F.J.: Eur. J. Biochem.,98,447–454 (1979)

[7] Höjeberg, B., Rydström, J.: Eur. J. Biochem.,77,235–241 (1977)

[8] Collins, P.A., Knowles, C.J.: Biochim. Biophys. Acta,480,77–82 (1977)

[9] Widmer, F., Kaplan, N.O.: Biochemistry,15,4693–4699 (1976)

[10] Wermuth, B., Kaplan, N.O.: Arch. Biochem. Biophys.,176,136–143 (1976)

[11] Hoek, J.B., Rydström, J., Höjeberg, B.: Biochim. Biophys. Acta,333,237–245 (1974)

[12] Rydström, J., Hoek, J.B., Höjeberg, B.: Biochem. Biophys. Res. Commun.,52,421–429 (1973)

[13] Middleditch, L.E., Atchison, R.W., Chung, A.E.: J. Biol. Chem.,247,6802–6809 (1972)

[14] Van Den Broek, H.W.J., Santema, J.S., Wassink, J.H., Veeger, C.: Eur. J. Biochem.,24,31–45 (1971)

[15] Van Den Broek, H.W.J., Van Breemen, J.F.L., Van Bruggen, E.F.J., Veeger, C.: Eur. J. Biochem.,24,46–54 (1971)

[16] Van Den Broek, H.W.J., Veeger, C.: Eur. J. Biochem.,24,72–82 (1971)

[17] Cohen, P.T., Kaplan, N.O.: J. Biol. Chem.,245,4666–4672 (1970)

[18] Cohen, P.T., Kaplan, N.O.: J. Biol. Chem.,245,2825–2836 (1970)

[19] Chung, A.E.: J. Bacteriol.,102,438–447 (1970)

Enzyme Handbook © Springer-Verlag Berlin Heidelberg 1994
Duplication, reproduction and storage in data banks are only allowed with the prior permission of the publishers

1 NOMENCLATURE

EC number

1.6.1.2

Systematic name

NADPH: NAD$^+$ oxidoreductase (AB-specific)

Recommended name

NAD(P)$^+$ transhydrogenase (AB-specific)

Synonymes

Pyridine nucleotide transhydrogenase
Transhydrogenase
Transhydrogenase, nicotinamide adenine dinucleotide (phosphate)
NAD(P) transhydrogenase
Nicotinamide adenine dinucleotide (phosphate) transhydrogenase
NAD transhydrogenase
NADH transhydrogenase
Nicotinamide nucleotide transhydrogenase
NADPH-NAD transhydrogenase
Pyridine nucleotide transferase
NADPH-NAD oxidoreductase
NADH-NADP-transhydrogenase
NADPH:NAD$^+$ transhydrogenase
H$^+$-Thase
Energy-linked transhydrogenase

CAS Reg. No.

9014-18-0 (not distinguished from EC 1.6.1.1); 9072-60-0 (not distinguished from EC 1.6.1.1)

2 REACTION AND SPECIFICITY

Catalysed reaction

NADPH + NAD$^+$ →
→ NADP$^+$ + NADH (proposed proton-pump mechanism [33], stereochemistry [35], mechanism [2, 50])

Reaction type

Redox reaction

Enzyme Handbook © Springer-Verlag Berlin Heidelberg 1994
Duplication, reproduction and storage in data banks are only allowed with the prior permission of the publishers

Natural substrates

NADH + NADP$^+$ (forward reaction [9], physiological role [9], coupled to transmembrane transport of protons from cytosol to mitochondria) [2, 9, 59, 60]

Substrate spectrum

1 NADH + NADP$^+$ (r [9, 50, 57], specific for 4A site of NADH (i.e. pro-R hydrogen) and 4B site of NADPH (i.e. pro-S hydrogen) [2, 3, 50], the reaction is coupled to a transmembrane proton translocation from cystosol to mitochondria [1, 9, 33], stereospecificity of NADP$^+$ reduction [35]) [1, 3, 9, 33, 35, 50]
2 NADPH + acetylpyridine adenine dinucleotide [15]
3 More (inactive against 3'-analogs of NADP$^+$ [2, 3], reduction of thio-NAD(P)$^+$ [2, 15, 50], synthesis of diphosphate from inorganic orthophosphate in chromatophores by reverse reaction [19], solubilized and purified enzyme does not catalyze reduction of acetyl pyridine adenine dinucleotide by NADH in absence of NADP$^+$ [5]) [2, 3, 5, 15, 19, 50]

Product spectrum

1 NAD$^+$ + NADPH
2 NADP$^+$ + acetylpyridine adenine dinucleotide (reduced)
3 ?

Inhibitor(s)

5'-AMP [5, 33, 46, 50]; SH-reagents [2, 7]; Triiodothyronine [2, 50]; Mg^{2+} [2, 41, 47, 50]; Ca^{2+} [2, 41, 50]; Mn^{2+} [2, 41, 50]; D_2O [2]; Adenine nucleotides [2, 33, 46, 50]; Palmityl-CoA [2, 47, 54]; N-Ethylmaleimide (degree of inhibition depending on pH and substrate concentration) [7]; Methylmethane thiosulfonate (modification of Cys-893) [7]; Glutathione (with NADPH inhibition of forward and reverse reaction, with NADH no inhibition of forward reaction, 40% inhibition of reverse reaction [15], protection by NADP$^+$ or NAD$^+$, acceleration of inhibition by NADPH [52]) [15, 52]; Phenylarsine oxide [15]; N, N'-Dicylclohexylcarbodiimide [16, 22, 25, 28, 29, 31]; Ethoxyformic anhydride [21]; Dansyl chloride [21]; Pyridoxal phosphate [21]; 5'-[p-(Fluorosulfonyl)benzoyl]-adenosine (structural analog of adenosine) [24]; 2,4-Dinitrophenyl-3'-dephospho-CoA [27]; N-(Ethoxycarbonyl)-2-ethoxy-1,2-dihydroquinoline [22]; S-7-Nitrobenzofuran-4-yl-CoA [27]; S-7-Nitrobenzofuran-4-yl-3'-dephospho-CoA [27]; N-(4-Azido-2-nitrophenyl)-2-aminoethylsulfonate (i.e. NAP-taurine) [30]; Butane-2,3-dione [31, 34]; Diethyldicarbonate [37]; Sr^{2+} [41]; La^{3+} [41]; K^+ [41]; Na^+ [41]; Tl^+ [41]; Pentane-2,4-dione [45]; 5,5'-Dithiobis-(2-nitrobenzoate) [51]; p-Chloromercuribenzoate [56]; p-Chlororomercuriphenyl sulfonate [56]; Phospholipases [53]; More (overview [7], uncouplers of mitochondrial energy transfer [58], substrate and product inhibition [33]) [7, 33, 58]

Cofactor(s)/prostethic group(s)

Lysophosphatidylcholine (activation) [5]; Carbonyl cyanide m-chlorophenylhydrazone (stimulation) [14]; Lysolecithin (stimulation) [48]; Phospholipids (stimulation) [52, 53]; Lipids (activation) [50]; No flavin cofactor (differentiation from EC 1.6.1.1.) [5, 24]

Metal compounds/salts

Ca^{2+} (required for catalytic activity [6], no activation [3]) [6]; Mg^{2+} (required for catalytic activity) [6]

Turnover number (min^{-1})

Specific activity (U/mg)

3.78 [5]; 62.3 [36]; 29.9 [25]; More (assay methods [43]) [18, 23, 29, 43, 44, 46–48, 52]

K_m-value (mM)

0.125 (NAD^+, with NADPH as donor) [33]; 0.02 (NADPH, with NAD^+ as acceptor) [33]; 0.166 (acetylpyridine adenine dinucleotide, with NADPH as donor) [33]; 0.029 (NADPH, with acetylpyridine adenine dinucleotide as acceptor) [33]; 0.01 (NADH, with $NADP^+$ as acceptor) [33]; 0.0017 ($NADP^+$, with NADH as donor) [33]; 0.023 (NADPH) [25]; 0.033 (acetylpyridine adenine dinucleotide) [25]; More (comparison of values for enzyme purified or nonpurified in submitochondrial particles [33], temperature dependency [40], effect of Mg^{2+} and ATP [42, 50]) [2, 33, 40, 42, 50]

pH-optimum

5.5 (reduction of $NADP^+$) [2]; 6.2–6.3 [47]; 7.0 (reduction of NAD^+ [2], ATP-driven proton translocation in reconstituted vesicles [14]) [2, 14]

pH-range

4.7–6.7 (less than 50% of maximal activity above and below) [47]

Temperature optimum (°C)

Temperature range (°C)

---

3 ENZYME STRUCTURE

Molecular weight

220000–280000 (bovine, radiation inactivation, hydrodynamic properties [11], SDS-PAGE after cross-linkage with several bifunctional reagents [38]) [11, 38]

196000 (E. coli) [1]

Enzyme Handbook © Springer-Verlag Berlin Heidelberg 1994
Duplication, reproduction and storage in data banks are only allowed with the prior permission of the publishers

Subunits

Dimer (2 × 109212, bovine, calculation from sequence of cDNA [10], values from SDS-PAGE [1, 16, 47, 49], signal peptide MW 4816, sequence of mRNA [8], cross-linkage experiments [32]) [1, 8, 10, 16, 32, 47, 49]
Tetramer (alpha$_2$, beta$_2$, 2 × 53906 + 2 × 48667, E. coli, calculation from nucleotide sequence) [17]
? (x × 53000 + x × 48000, Rhodobacter capsulatus, SDS-PAGE [5], x × 6784, soluble component, Rhodopseudomonas sphaeroides, amino acid analysis [56]) [5, 56]

Glycoprotein/Lipoprotein

Phospholipoprotein (5 mol loosely bound phospholipids, 9 mol tightly bound phospholipids) [61]

---

4 ISOLATION/PREPARATION

Source organism

Mammals [1, 2, 9, 59, 60]; Bovine [4, 7, 8, 10, 11, 13–15, 18, 21–24, 27–30, 33, 35–44, 47, 49–51, 53, 57, 59–62]; E. coli [3, 17, 20, 25, 26, 46, 52]; Rhodobacter capsulatus (formerly Rhodopseudomonas capsulata) [5, 6]; Rhodospirillum rubrum [19, 34, 45, 48, 50, 54]; Rhodopseudomonas sphaeroides [56]; Rat [12, 16, 31, 50]; Salmonella typhimurium [55]; Micrococcus denitrificans [58]; More (overview) [1, 2]

Source tissue

Heart [2, 4, 8, 13–15, 18, 21–24, 27–33, 35–44, 47, 49, 51, 53, 62]; Liver [2, 12, 16]; Arteries [2]; More (no or low activity in brain, prostate, seminal vesicle, spleen, testis) [2]

Localisation in source

Mitochondrial membrane (membrane topography [4], orientation in membrane [12]) [1, 2, 4, 7, 9–12, 15, 16, 18, 21–24, 27–33, 35–41, 47, 49, 62]; Chromatophores (soluble and insoluble component [56]) [1, 5, 19, 34, 56]; Cytoplasmic membrane [25]

Purification

Rhodobacter capsulatus [5]; Rat [16]; Bovine (affinity chromatography [18, 23, 36], FPLC, comparison of methods [29], immunoexclusion chromatography [39], overview early procedures [50]) [18, 23, 29, 36, 39, 44, 47, 49, 50]; E. coli [25, 46]; Rhodopseudomonas sphaeroides (soluble component) [56]

Crystallization

[5]

Cloned

[10, 17, 20, 26]

Renaturated

(reconstitution) [13, 14, 18, 30, 33, 39, 42, 45, 50]

5 STABILITY

pH

Temperature (°C)

40 (5 min, soluble component complete inactivation) [56]; 44 (2 min, solubilized membrane component, 50% inactivation) [48]; 48 (2 min, membrane particles, 50% inactivation) [48]; More (proteolytic inactivation as a function of temperature [40], protection against thermal inactivation by cations [41], effect of substrates on thermostability in presence and absence of thionitrobenzoate [51]) [40, 41, 51]

Oxidation

Organic solvent

Ethanol (10%, 40°C, inactivation) [62]; n-Butanol (inactivation) [62]; 1,1-Dimethylbutanol (inactivation) [62]; Acetone (used for conversion of mitochondria to an acetone powder causes inactivation) [62]

General stability information

Purified enzyme inactivated at 4°C even in presence of dithiothreitol [5]; Urea, 6 M, inactivation [38]; Cations prevent tryptic inactivation [41]; Inactivation during prolonged column chromatography [44]; Inactivation by refreezing after thawing [48]; Proteolysis stimulated by NADPH and NADP$^+$ [48, 54]

Storage

–70°C or 4°C, Rhodospirillum rubrum membrane component, inactivation in 1 day, stabilization by NADP$^+$ [48]; –15°C, 20 mM sodium Tricine buffer, pH 7.6, 1 mM dithiothreitol, 0.2% Triton X-100, 30% glycerol, 4 weeks, no loss of activity [5]; 4°C, 0.1 M sodium phosphate buffer, pH 7.5, 1 mM dithiothreitol, 0.05% sodium cholate [39]; 4°C, reconstituted with phospholipids, at least 2 months stable [44]; 4°C, unsoluble component: 0.1 M glycyl-glycine buffer, pH 8.0, 10% sucrose, unstable, soluble factor: 120 h with 0.015 mM dithiothreitol stable [56]

6 CROSSREFERENCES TO STRUCTURE DATABANKS

PIR/MIPS code

S02205 (Bovine mitochondrion SGC1 fragment)

Brookhaven code

Enzyme Handbook © Springer-Verlag Berlin Heidelberg 1994
Duplication, reproduction and storage in data banks are only allowed with the prior permission of the publishers

7 LITERATURE REFERENCES

[1] Jackson, J.B., Lever, T.M., Rydström, J., Persson, B., Carlenor, E.: Biochem. Soc. Trans.,19,573–575 (1991) (Reveiw)
[2] Rydström, J., Hoek, J.B., Ernster, L. in "The Enzymes",3rd. Ed. (Boyer, P.D., Ed.) 13,51–88 (1976) (Review)
[3] Hoek, J.B., Rydström, J., Höjeberg, B.: Biochim. Biophys. Acta,333,237–245 (1974)
[4] Yamaguchi, M., Hatefi, Y.: J. Biol. Chem.,266,5728–5735 (1991)
[5] Lever, R.M., Palmer, T., Cunningham, I.J., Cotton, N. P.J., Jackson, J.B.: Eur. J. Biochem.,197,247–255 (1991)
[6] Cotton, N.P.J., Lever, T.M., Nore, B.F., Jones, M.R:, Jackson, J.B.: Eur. J. Biochem.,182,593–603 (1989)
[7] Yamaguchi, M., Hatefi, Y.: Biochemistry,28,6050–6056 (1989)
[8] Yamaguchi, M., Hatefi, Y., Trach, K., Hoch, J.A.: Biochem. Biophys. Res. Commun.,157,24–29 (1988)
[9] Hoek, J.B., Rydström, J.: Biochem. J.,254,1–10 (1988) (Review)
[10] Yamaguchi, M., Hatefi, Y., Trach, K., Hoch, J.A.: J. Biol. Chem.,263,2761–2767 (1988)
[11] Persson, B., Ahnström, G., Rydström, J.: Arch. Biochem. Biophys.,259,341–349 (1987)
[12] Weis, J.K., Wu, L.N.Y., Fisher, R.R.: Arch. Biochem. Biophys.,257,424–429 (1987)
[13] Eytan, G.D., Eytan, E., Rydström, J.: J. Biol. Chem.,262,5015–5019 (1987)
[14] Eytan, G.D., Persson, B., Ekebacke, A., Rydström, J.: J. Biol. Chem.,262,5008–5014 (1987)
[15] Persson, B., Rydström, J.: Biochem. Biophys. Res. Commun.,142,573–578 (1987)
[16] Moody, A.J.: Biochem. Soc. Trans.,14,1210–1212 (1986)
[17] Clarke, D.M., Loo, T.W., Gillam, S., Bragg, P.D.: Eur. J. Biochem.,158,647–653 (1986)
[18] Wu, L.N.Y., Alberta, J.A., Fisher, R.R.: Methods Enzymol.,126,353–360 (1986)
[19] Nore, B.F., Husain, I., Nyren, P., Baltscheffsky, M.: FEBS Lett.,200,133–138 (1986)
[20] Clarke, D.M., Bragg, P.D.: FEBS Lett.,200,23–26 (1986)
[21] Yamaguchi, M., Hatefi, Y.: Arch. Biochem. Biophys.,243,20–27 (1985)
[22] Phelps, D.C., Hatefi, Y.: Arch. Biochem. Biophys.,243,298–304 (1985)
[23] Carlenor, E., Tang, H.-L., Rydström, J.: Anal. Biochem.,148,518–523 (1985)
[24] Phelps, D.C., Hatefi, Y.: Biochemistry,24,3503–3507 (1985)
[25] Clarke, D.M., Bragg, P.D.: Eur. J. Biochem.,149,517–523 (1985)
[26] Clarke, D.M., Bragg, P.D.: J. Bacteriol.,162,367–373 (1985)
[27] Kozlov, I.A., Milgrom, Y.M., Saburova, L.A., Sobolev, A.Y.: Eur. J. Biochem.,145,413–416 (1984)
[28] Phelps, D.C., Hatefi, Y.: Biochemistry,23,4475–4480 (1984)
[29] Persson, B., Enander, K., Tang, H.-L., Rydström, J.: J. Biol. Chem.,259,8626–8632 (1984)
[30] Pennington, R.M., Fisher, R.R.: FEBS Lett.,164,345–349 (1983)
[31] Moody, A.J., Reid, R.A.: Biochem. J.,209,889–892 (1983)
[32] Wu, L.N.W., Fisher, R.R.: J. Biol. Chem.,258,7847–7851 (1983)
[33] Enander, K., Rydström, J.: J. Biol. Chem.,257,14760–14766 (1982)
[34] McFadden, B.J., Fisher, R.R.: Arch. Biochem. Biophys.,190,820–828 (1978)
[35] Wu, L.N.Y., Fisher, R.R.: J. Biol. Chem.,257,11680–11683 (1982)
[36] Wu, L.N.Y., Pennington, R.M., Everett, T.D., Fisher, R.R.: J. Biol. Chem.,257,4052–4055 (1982)
[37] Phelps, D.C., Hatefi, Y.: J. Biol. Chem.,256,8217–8221 (1981)

[38] Anderson, W.M., Fisher, R.R.: Biochim. Biophys. Acta,635,194–199 (1981)
[39] Anderson, W.M., Fowler, W.T., Pennington, R.M., Fisher, R.R.: J. Biol. Chem.,256,1888–1895 (1981)
[40] Blazyk, J.F., Blazyk, J.M., Kline, C.M.: J. Biol. Chem.,256,691–694 (1981)
[41] O'Neal, S.G., Earle, S.R., Fisher, R.R.: Biochim. Biophys. Acta,589,217–230 (1980)
[42] Rydström, J., Fleischer, S.: Methods Enzymol.,55,811–816 (1979)
[43] Rydström, J.: Methods Enzymol.,55,261–273 (1979)
[44] Höjeberg, B., Rydström, J.: Methods Enzymol.,55,275–283 (1979)
[45] Jacobs, E., Fisher, R.R.: Biochemistry,18,4315–4322 (1979)
[46] Hanson, R.L.: J. Biol. Chem.,254,888–893 (1979)
[47] Anderson, W.M., Fisher, R.R.: Arch. Biochem. Biophys.,187,180–190 (1978)
[48] Jacobs, E., Heriot, K., Fisher, R.R.: Arch. Microbiol.,115,151–156 (1977)
[49] Höjeberg, B., Rydström, J.: Biochem. Biophys. Res. Commun.,78,1183–1190 (1977)
[50] Rydström, J.: Biochim. Biophys. Acta,463,155–184 (1977) (Review)
[51] O'Neal, S.G., Fisher, R.R.: J. Biol. Chem.,252,4552–4556 (1977)
[52] Houghton, R.L., Fisher, R.R., Sanadi, D.R.: Biochem. Biophys. Res. Commun.,73,751–757 (1976)
[53] Rydström, J., Heok, J.B., Ericson, B.G., Hundal, T.: Biochim. Biophys. Acta,430,419–425 (1976)
[54] Fisher, R.R., Rampey, S.A., Sadighi, A., Fisher, K.: J. Biol. Chem.,250,819–825 (1975)
[55] Singh, A.P., Bragg, P.D.: J. Gen. Microbiol.,82,237–246 (1974)
[56] Berger, T.J., Orlando, J.A.: Arch. Biochem. Biophys.,159,25–31 (1973)
[57] Fisher, R.R., Kaplan, N.O.: Biochemistry,12,1182–1188 (1973)
[58] Asano, A., Imai, K., Sato, K.: Biochim. Biophys. Acta,143,477–486 (1967)
[59] Fisher, R.R., Earle, S.R. in "The Pyridine Nucleotide Coenzymes" (Everse, J., Andersson, B., You, K.-S., Eds.) pp279–324, Academic Press, N.Y. (1982) (Review)
[60] Rydström, J., Persson, B., Carlenor, E. in "Pyridine Nucleotide Coenzymes, Chemical, Biochemical And Medical Aspects" (Dolphin, D., Poulson, R., Avramovic, O., Eds.) Vol.2B, pp.433–460, Wiley And Sons, N.Y. (1987) (Review)
[61] Rydström, J. in "Mitochondria And Microsomes" (Lee, C.P., Schatz, G., Dallner, G. Eds.) pp.317–335, Addison-Wesley, Reading MA (1981) (Review)
[62] Kaufman, B., Kaplan, N.O.: J. Biol. Chem.,236,2133–2139 (1961)

Enzyme Handbook © Springer-Verlag Berlin Heidelberg 1994
Duplication, reproduction and storage in data banks are only allowed with the prior permission of the publishers

1 NOMENCLATURE

EC number
1.6.2.2

Systematic name
NADH: ferricytochrome-b_5 oxidoreductase

Recommended name
Cytochrome-b_5 reductase

Synonymes
Reductase, cytochrome b_5
Cytochrome b_5 reductase
Dihydronicotinamide adenine dinucleotide-cytochrome b_5 reductase
Reduced nicotinamide adeninedinucleotide-cytochrome b_5 reductase
NADH-ferricytochrome b_5 oxidoreductase
NADH-cytochrome b_5 reductase
NADH 5alpha-reductase [18]
NADH-cytochrome-b_5 reductase

CAS Reg. No.
9032-25-1

---

2 REACTION AND SPECIFICITY

Catalysed reaction
NADH + 2 ferricytochrome b_5 →
→ NAD^+ + H^+ + 2 ferrocytochrome b_5 (proposed mechanism [19])

Reaction type
Redox reaction

Natural substrates
NADH + ferricytochrome b_5
Deoxyhemerythrin + O_2 [9]
5alpha-Dihydrotestosterone + acceptor [18]
More (membrane bound form of somatic cells: essential for lipid metabolism [26], desaturation of fatty acids [40, 43], metabolism of endogenous compounds such as steroids, drugs, carcinogens, environmental pollutants [41], soluble form of erythrocytes: reduction of methemoglobin [6, 42]) [6, 26, 40–43]

Enzyme Handbook © Springer-Verlag Berlin Heidelberg 1994
Duplication, reproduction and storage in data banks are only allowed with the prior permission of the publishers

Substrate spectrum

1 NADH + ferricytochrome b_5 (specific for NADH as electron donor, additional acceptors: ferricyanide [7, 21, 25, 27, 32, 34–37], 2,6-dichlorphenolindophenol [21, 25, 27, 32, 34–37], methemerythrin [9], testosterone [18], methemoglobin-ferrocyanide complex [27, 34], hemin [25], p-benzoquinone [21], 5-hydroxy-1,4-naphthoquinone [21], nitroblue-tetrazolium [21]) [7, 9, 18, 21, 25, 27, 32, 34–37]

2 More (additional electron donor: deamino-NADH, 3-acetylpyridine-NADH [23], poor donor: NADPH [21, 25], poor electron acceptor: methylene blue, ferricytochrome c, O_2, oxidized glutathione, methemoglobin [32], no acceptor: ubiquinone-30, menadione, dihydrofolate, lipoamide [21]) [21, 23, 25, 32]

Product spectrum

1 NAD^+ + H^+ + ferrocytochrome b_5

2 ?

Inhibitor(s)

Benzyl alcohol [3]; Hemin [3]; Inositol hexaphosphate [3]; Wheat germ agglutinin [3]; Phytohemagglutinin [3]; Cl^- [8, 16, 21]; Br^- [16]; F^- [16]; I^- [16]; Acetate [16]; Succinate [16]; Citrate [16]; 5'-(p-Fluorosulfonylbenzoyl)-adenosine [18]; Amytal [18]; Mepacrin [18]; Thenoyltrifluoroacetone [18]; Dicoumarol [18]; Pentachlorophenol [18]; $NADP^+$ [18]; Iodoacetic acid [21]; N-Ethylmaleimide [21, 27, 38]; p-Chloromercuribenzoate [21, 25, 27, 35]; Atebrin [21, 25, 27, 29]; Proflavin [21, 25, 29]; Acrynol [21, 27, 29]; Phosphate [21]; Taurodeoxycholate [21]; Adenine nucleotides [21, 27]; Tris (reduction of cytochrome b_5) [32]; K^+ (reduction of cytochrome b_5 or dichlorphenolindophenol) [32]; High ionic strength [35]; Hg^{2+} [35]; NAD^+ (competitive) [35]; o-Phenanthroline [35]; p-Hydroxymercuribenzoate [34]; alpha,alpha'-Dipyridyl [35]; More (not: NAD^+ [1], ferrocyanide [1], inhibition by halides reversible by dilution [16]) [1, 16]

Cofactor(s)/prostethic group(s)

FAD (1 mol per mol of enzyme) [1, 6, 10, 25, 27, 29, 32]; NADH (NADPH: 7–9% of NADH activity [17]); Detergents (e.g. Triton X-100, activation) [4, 23]; Spermine (activation) [3]; 9-Amino-1,2,3,4-tetrahydroacridine (activation) [3]

Metal compounds/salts

More (rate of reduction dependent on ionic strength) [12, 19, 25]

Turnover number (min^{-1})

40600 (ferricyanide, human red cell membrane) [20]; 42000 (ferricyanide, human, cytosol) [20]; 49600 (ferricyanide, human liver, detergent solubilized) [20]; 5900 (ferricyanide, human liver, protease solubilized) [20]; 1280 (cytochrome b_5) [32]; 30000 (NADH) [38]; More [1, 12, 33]

Specific activity (U/mg)

230.5 (high MW aggregate) [1]; 611.1 (low MW aggregate) [1]; 628 [13]; 790 [27]; More [4, 6, 7, 12, 18, 21, 36, 37]

K_m-value (mM)

0.00016–0.0009 (NADH) [16, 19, 20, 27, 32]; 0.001–0.008 (NADH) [6, 21, 25]; 0.01–0.06 (NADH) [1, 4, 18, 35]; 0.01–0. 04 (ferricytochrome b_5) [4, 6, 19, 20, 34, 38]; 0.002–0.009 (ferricytochrome b_5) [1, 4, 25, 27]; 0.0008 (ferricytochrome b_5) [35]; 0.09 (ferricyanide) [1]; 0.004 (ferricyanide) [21]; 0.01 (ferricyanide) [34]; 0.025–0.089 (testosterone reduction, depending on phosphate concentration) [18]; More (kinetics [9]) [4, 9]

pH-optimum

5.0–5.7 [12]; 5.2 (NADH-methemoglobin-ferrocyanide) [27]; 5.5 (acceptor ferricyanide) [1]; 5.6 [35]; 6.4 (citrate-phosphate buffer) [25]; 6.5 (Tris-phosphate buffer) [34]; 6.5–8.5 (NADH-ferricyanide) [27]; 6.6 (Tris-maleate buffer) [25]; 7.0 (triethanolamine buffer) [34]; 7.0–7.5 [21]; 8.0 (acceptor dichlorphenolindophenol) [1]; 6–8 (acceptor Phascolopsis gouldii cytochrome b_5) [1]; More (no distinct optimum with phosphate or Tris-HCl buffer) [25]

pH-range

5–8 (depending on acceptor) [27]

Temperature optimum (°C)

Temperature range (°C)

3 ENZYME STRUCTURE

Molecular weight

200000–360000 (rabbit [30, 36], Phascolopsis gouldii [1], oligomeric aggregate of detergent-solubilized enzyme in aqueous media, gel filtration) [1, 30, 36]

44000 (potato, FPLC) [13]

30000–34700 (rat erythrocytes, cytosol, gel filtration [24], rabbit erythrocytes, cytosol, gel filtration, calculation from FAD content [25]) [24, 25]

25000 (rabbit liver, cytosol, gel chromatography, sucrose density gradient centrifugation) [34]

27000–28000 (Saccharomyces cerevisiae, gel filtration) [35]

Enzyme Handbook © Springer-Verlag Berlin Heidelberg 1994
Duplication, reproduction and storage in data banks are only allowed with the prior permission of the publishers

Subunits

Monomer (30000–33000, rat [24], rabbit [25], human [27], erythrocytes, soluble form, SDS-PAGE [24, 25, 27], 27000–28000, Saccharomyces cerevisiae, SDS-PAGE [35]) [24, 25, 27, 35]
Oligomer (possibly tetramer [27], membrane bound form, x × 27000–45000, values depending on organism and method of solubilization, Phascolopsis gouldii [1], human [6, 7, 17, 20, 27], rat [6, 24, 33, 37], Pisum sativum [12], bovine [16, 21, 39], rabbit [36], SDS-PAGE [1, 6, 7, 12, 16, 17, 20, 21, 24, 33, 36], gel filtration in presence of 6 M guanidine hydrochloride or sodium deoxycholate [33, 36], x × 22000, Tetrahymena pyriformis, SDS-PAGE [26]) [1, 6, 7, 12, 16, 17, 20, 21, 24, 26, 27, 33, 36, 37, 39]
More (bovine liver: membrane binding domain located at NH_2–terminal site of protein, MW: 6400–6500 [22], rabbit liver: hydrophobic domain of ca. 30 amino acids located at COOH-terminal site of protein [31]) [22, 31]

Glycoprotein/Lipoprotein

Lipoprotein (myristic acid at NH_2–terminus, amide bond, no ester bond) [6, 7]

---

4 ISOLATION/PREPARATION

Source organism

Human [2, 3, 6, 7, 10, 17, 20, 23, 27, 29, 32]; Pig [4, 11, 19, 39]; Rat [5, 14, 18, 24, 33, 37]; Phascolopsis gouldii [1, 9]; Pisum sativum [12]; Potato [13]; Maize [13]; Bovine (calf [38, 39]) [8, 15, 16, 21, 22, 28, 38, 39]; Tetrahymena pyriformis [26]; Rabbit [25, 31, 34, 36, 39]; Saccharomyces cerevisiae (grown anaerobically) [35]

Source tissue

Erythrocytes (soluble form) [1, 3, 6, 9, 10, 20, 24, 25, 27, 29, 32]; Liver [4–6, 11, 14–16, 18–20, 22, 28, 30, 31, 33, 34, 36–39]; Seedlings [12, 13]; Tuber [13]; Neutrophils [17, 23]; Brain [21]

Localisation in source

Membrane bound (cytoplasmic side of erythrocyte membrane [3], membrane of endoplasmic reticulum, outer mitochondrial membrane, tightly bound to cytoplasmic face of membrane [5]) [1, 3, 5, 6, 9, 14, 17, 27]; Microsomes [8, 11–13, 18–21, 26, 28, 30, 32, 33, 35–39]; Cytosol [20, 27, 32, 34]; More (possible conformation of membrane binding domain) [15]

Purification

Human [7, 17, 27, 29, 32]; E. coli (containing human gene) [6]; Phascolopsis gouldii [1, 9]; Pisum sativum [12]; Potato (HPLC, FPLC) [13]; Pig (lysosome- and detergent-solubilization) [19]; Bovine (lysosome-solubilization [21], separation of membrane binding and catalytic domain [22]) [21, 22, 28]; Rabbit [25, 34, 36]; Tetrahymena pyriformis [26]; Rat (detergent-solubilization [33], lysosome-solubilization [37]) [33, 37]; Saccharomyces cerevisiae [35]

Crystallization

(methods of crystallization, preliminary X-ray data) [7, 11]

Cloned

(human gene in E. coli) [2, 6]

Renaturated

(reconstitution) [12]

---

5 STABILITY

pH

7.0–9.0 [38]; 7.5–8.1 [32]

Temperature (°C)

–12 (indefinitely) [38]; 0–5 (several days) [38]; 21 (several h) [38]

Oxidation

O_2 (solubilized enzyme very sensitive to atmospheric oxygen) [1]

Organic solvent

General stability information

Gelatin (stabilization) [35]; Triton X-100 (stabilization) [33]; Phosphate (stabilization) [18]; EDTA (essential for stability) [1]; Dithiothreitol (essential for stability) [1]; Application of HPLC during purification: inactivation [13]; Inactivation caused by solubilization by detergents, activity restored by phosphatidylcholine [18]; FAD (protection against inactivation) [35]; Freezing/thawing (50% inactivation) [33]; Phosphatidylcholine liposomes (stabilization) [33]; Dialysis (inactivation prevented by NADH) [21]

Storage

Liquid N_2 [1]; –90°C, oxidized state, 1 month [26]; –70°C, several months, no loss of activity [33, 36]; –20°C, 25 mM phosphate buffer, pH 7.6, 1 mM EDTA, 0.1 mM dithiothreitol, at least a few months [25]; –20°C, pH 7.5–8.1, 0.5 mM EDTA, protein concentration above 0.02 mg/ml [32]; Frozen, 50 mM potassium phosphate buffer, pH 7.5, 1 mM EDTA, several weeks [37]; 0°C, at least 1 month [16]; 0°C, at least 1 week [21]; 4°C, anaerobic conditions, EDTA, dithiothreitol, several months [9]

Enzyme Handbook © Springer-Verlag Berlin Heidelberg 1994
Duplication, reproduction and storage in data banks are only allowed with the prior permission of the publishers

6 CROSSREFERENCES TO STRUCTURE DATABANKS

PIR/MIPS code

RDHUB5 (Human); B26616 (hepatic Human fragment); A26616 (placental Human fragment); A23896 (Bovine); A26922 (Bovine fragment); A22182 (Bovine fragment); A40495 (Rat); PX0015 (Human fragment); JS0468 (placental Human); A26616 (placental Human fragment); PX0016 (Rat fragment)

Brookhaven code

7 LITERATURE REFERENCES

[1] Bonomi, F., Long, R.C., Kurtz, D.M.: Biochim. Biophys. Acta,999,147–156 (1989)
[2] Rigby, J.S., Bull, P.C., Ashworth, A., Shephard, E.A., Santisteban, I., Phillips, I.R.: Biochem. Soc. Trans.,17,194–195 (1989)
[3] Palmieri, D.A., Rangachari, A., Butterfield, D.A.: Arch. Biochem. Biophys.,280,224–228 (1990)
[4] Tamura, M., Yoshida, S., Tamura, T., Saitoh, T., Takeshita, M.: Arch. Biochem. Biophys.,280,313–319 (1990)
[5] Borgese, N., Longhi, R.: Biochem. J.,266,341–347 (1990)
[6] Shirabe, K., Yubisui, T., Takeshita, M.: Biochim. Biophys. Acta,1008,189–192 (1989)
[7] Murakami, K., Yubisui, T., Takeshita, M., Miyata, T.: J. Biochem.,105,312–317 (1989)
[8] Tamura, M., Yubisui, T., Takeshita, M.: Biochem. J.,251,711–715 (1988)
[9] Utecht, R.E., Kurtz, D.M.: Biochim. Biophys. Acta,953,164–178 (1988)
[10] Takano, T., Ogawa, K., Sato, M., Bando, S., Yubisui, T.: J. Mol. Biol.,195,749–750 (1987)
[11] Miki, K., Kaida, S., Kasai, N., Iyanagi, T., Kobayashi, K., Hayashi, K.: J. Biol. Chem.,262,11801–11802 (1987)
[12] Jollie, D.R., Sligar, S.G., Schuler, M.: Plant Physiol.,85,457–462 (1987)
[13] Galle, A.M., Kader, J.C.: J. Chromatogr.,366,422–426 (1986)
[14] Borgese, N., Pietrini, G.: Biochem. J.,239,393–403 (1986)
[15] Kensil, C.R., Strittmatter, P.: J. Biol. Chem.,261,7316–7321 (1986)
[16] Tamura, M., Yubisui, T., Takeshita, M.: Biochem. J.,230,273–276 (1985)
[17] Tauber, A.I., Wright, J., Higson, F.K., Edelman, S.A., Waxman, D.J.: Blood,66,673–678 (1985)
[18] Golf, S.W., Graf, V., Rempeters, G., Mersdorf, S.: Biol. Chem. Hoppe-Seyler,366,647–653 (1985)
[19] Iyanagi, T., Watanabe, S., Anan, K.F.: Biochemistry,23,1418–1425 (1984)
[20] Kitajima, S., Minakami, S.: J. Biochem.,93,615–620 (1983)
[21] Tamura, M., Yubisui, T., Takeshita, M.: J. Biochem.,94,1547–1555 (1983)
[22] Kensil, C.R., Hediger, M.A., Ozols, J., Strittmatter, P.: J. Biol. Chem.,258,14656–14663 (1983)
[23] Badwey, J.A., Tauber, A.I., Karnovsky, M.L.: Blood,62,152–157 (1983)
[24] Borgese, N., Macconi, D., Parola, L., Pietrini, G.: J. Biol. Chem.,257,13854–13861 (1982)
[25] Yubisui, T., Takeshita, M.: J. Biochem.,91,1467–1477 (1982)

[26] Fukushima, H., Umeki, S., Watanabe, T., Nozawa, Y.: Biochem. Biophys. Res. Commun.,105,502–508 (1982)
[27] Kitajima, S., Yasukochi, Y., Minakami, S.: Arch. Biochem. Biophys.,210,330–339 (1981)
[28] Schafer, D.A., Hultquist, D.E.: Biochem. Biophys. Res. Commun.,95,381–387 (1980)
[29] Yubisui, T., Takeshita, M.: J. Biol. Chem.,255,2454–2456 (1980)
[30] Tajima, S., Mihara, K., Sato, R.: Arch. Biochem. Biophys.,198,137–144 (1979)
[31] Mihara, K., Sato, R., Sakakibara, R., Wada, H.: Biochemistry,17,2829–2834 (1978)
[32] Hultquist, D.E.: Methods Enzymol.,52,463–473 (1978)
[33] Mihara, K., Sato, R.: Methods Enzymol.,52,102–108 (1978)
[34] Lostanlen, D., De Barro, A.V., Leroux, A., Kaplan, J. C.: Biochim. Biophys. Acta,526,42–51 (1978)
[35] Kubota, S., Yoshida, Y., Kumaoka, H.: J. Biochem.,81,187–195 (1977)
[36] Mihara, K., Sato, R.: J. Biochem.,78,1057–1073 (1975)
[37] Takesue, S., Omura, T.: J. Biochem.,67,267–276 (1970)
[38] Strittmatter, P.: Methods Enzymol.,10,561–565 (1967)
[39] Strittmatter, P. in "The Enzymes",2nd. Ed. (Boyer, P.D., Lardy, H., Myrbäck, K., Eds.) 8,113–145 (1963) (Review)
[40] Oshino, N., Sato, R.: J. Biochem.,69,169–180 (1971)
[41] Ortiz De Montellano, P.R. (Ed.) in "Cytochrome P450- Structure, Mechanism And Biochemistry", Plenum New York (1986)
[42] Hultquist, D.E., Passon, P.G.: Nature,229,252–254 (1971)
[43] Kader, J.C.: Biochim. Biophys. Acta,486,429–436 (1977)

Enzyme Handbook © Springer-Verlag Berlin Heidelberg 1994
Duplication, reproduction and storage in data banks are only allowed with the prior permission of the publishers

1 NOMENCLATURE

EC number
1.6.2.4

Systematic name
NADPH:ferrihemoprotein oxidoreductase

Recommended name
NADPH-ferrihemoprotein reductase

Synonymes
NADP-cytochrome reductase
$TPNH_2$ cytochrome c reductase
Ferrihemprotein P450 reductase
Reductase, cytochrome c (reduced nicotinamide adenine dinucleotide phosphate)
NADPH-cytochrome c reductase
TPNH-cytochrome c reductase
Dihydroxynicotinamide adenine dinucleotide phosphate-cytochrome c reductase
Reduced nicotinamide adenine dinucleotide phosphate-cytochrome c reductase
Cytochrome c reductase (reduced nicotinamide adenine dinucleotide phosphate, NADPH, NADPH-dependent)
NADPH-cytochrome c oxidoreductase
NADPH-ferricytochrome c oxidoreductase
NADPH-dependent cytochrome c reductase
NADP-cytochrome c reductase
FAD-cytochrome c reductase
NADPH-cytochrome p-450 reductase [10, 27]
Aldehyde reductase (NADPH-dependent) [10]
EC 1.6.99.2 (formerly) [27]

CAS Reg. No.
9023-03-4

2 REACTION AND SPECIFICITY

Catalysed reaction
NADPH + 2 ferricytochrome →
→ $NADP^+$ + 2 ferrocytochrome

Enzyme Handbook © Springer-Verlag Berlin Heidelberg 1994
Duplication, reproduction and storage in data banks are only allowed with the prior permission of the publishers

Reaction type

Redox reaction

Natural substrates

NADPH + cytochrome P-450 [6]
More (monooxygenase system composed of cytochrome P-450, NADPH-cytochrome c reductase, phospholipids [16, 26], detoxification of drugs, inactivation of procarcinogens [16], biotransformation of airborne compounds [30]) [16, 26, 30]

Substrate spectrum

1 NADPH + ferricytochrome c (NADH less than 5% of NADPH activity [1], additional electron acceptors: 2,6-dichlorophenolindophenol [6, 7, 12, 14, 17, 20, 21, 24, 26, 28, 36], cytochrome P-450 [1, 8, 9, 20, 36, 38], ferricyanide [6, 9, 12, 17, 20, 21, 24, 26, 28], menadione [7, 12, 20, 26], neotetrazolium chloride [20, 26, 36], nitroblue tetrazolium salt [14], vitamin K_3 [36], benzoquinone [36]) [1, 4–9, 12, 14, 17, 20, 21, 24, 26, 28, 36, 38]
2 NADPH + cinnamate [9]
3 NADPH + hexadecanal (hexadecanal replaceable by p-nitroacetophenone, or p-pyridinecarboxaldehyde, or p-nitrobenzaldehyde) [10]
4 NADPH + O_2 (slow reaction, presence of menadione, or duroquinone, or vitamin K_3 essential) [14, 28]
5 17-Hydroxyprogesterone + NADPH [34]
6 More (O-deethylation of 7-ethoxycoumarin [30], N-demethylation of benzphetamine [30, 33], aniline hydroxylase [33], as part of MEOS i.e. microsomal ethanol-oxidizing system composed of NADPH-cytochrome c reductase, cytochrome P-450, phospholipids [33], omega-hydroxylation of fatty acids together with cytochrome P-450 [38]) [30, 33, 38]

Product spectrum

1 $NADP^+$ + ferrocytochrome c
2 $NADP^+$ + p-coumarate [9]
3 $NADP^+$ + hexadecanol [10]
4 $NADP^+$ + O_2^- (superoxide anion) [14, 28]
5 Androstendione + 2-carbon fragment (removal of 2-carbon side chain from 17-position of 21-carbon steroids) [34]
6 ?

Inhibitor(s)

Mersalyl [3, 21]; 3-Aminonicotinamide adenine dinucleotide phosphate [6]; $NADP^+$ [6, 7, 9, 21, 26, 32]; NAD^+ [7]; High ionic strength [9]; 2,6-Dichlorophenolindophenol (formation of superoxide anion) [14]; p-Chloromercuribenzoate [20, 21, 24, 26, 28]; $HgCl_2$ [20, 26, 28]; 2'-AMP [21]; Alizarin [21]; 5,5'-Dithiobis-(2-nitrobenzoate) (in absence of FAD or NADPH) [23]; N_2 [33]; CO [33]; Sodium formate [33]

Cofactor(s)/prostethic group(s)

FAD (ratio FAD:FMN 1:1 [1, 9, 11, 17, 20, 21, 26, 28], 1 mol per subunit [7, 12, 23], tightly bound [20, 28], loosely bound [42]) [1–3, 7–9, 11, 12, 17, 20, 21, 23–26, 28, 31–33, 35, 42]; FMN (ratio FAD:FMN 1:1 [1, 9, 11, 17, 20, 21, 26, 28], tightly bound [20], loosely bound [28, 42], 1 mol per mol of enzyme [17, 20]) [1–3, 8, 9, 11, 17, 20, 21, 23–26, 28, 31, 32, 35, 42]; Nonionic detergent (activation) [17]; Menadione (slight stimulation) [21]; NADPH (not replaceable by NADH [33])

Metal compounds/salts

More (stimulation by increasing ionic strength) [9, 24]

Turnover number (min^{-1})

897 (cytochrome c) [12]; 3870 (ferricyanide) [12]; 458 (2,6-dichlorophenolindophenol) [12]; 87 (menadione) [12]; 6100 (cytochrome c) [26]; More [29, 31, 33]

Specific activity (U/mg)

63.8 [16]; 150–180 [28]; 15.2 [1]; 40 [5]; More [3, 7, 8–11, 13, 14, 20, 24, 25, 27, 30, 34, 36]

K_m-value (mM)

0.0036 (NADPH, similar values [4, 7, 12, 21, 24, 25, 32, 35]) [1]; 0.022 (NADPH, similar values [9, 14, 21, 28]) [11]; 0.013 (cytochrome c, similar value [32]) [5, 21]; 0.006 (cytochrome c, similar values [5, 9]) [35]; 0.0019 (azidonitrophenyl-gamma-aminobutyryl-NADPH) [4]; 0.31 (p-nitrobenzaldehyde) [10]; 1.4 (p-nitroacetophenone) [10]; 2.5 (benzalacetone) [10]; 0.03 (hexadecanol) [10]; 0.0053 (menadione) [14]; 0.077 (2,6-dichlorophenolindophenol) [26]; 7.2 (ethanol, microsomal ethanol oxidizing system) [33]; More (O_2-generation [14]) [11, 12, 14]

pH-optimum

6.9–7.5 (microsomal ethanol-oxidizing system) [33]; 7.0–7. 4 (O_2- generation) [14]; 7.5–9 [21]; 7.7 [18]; 7.8 [20]; 7.8–8.0 [25, 28]; 8–9 [35]

pH-range

6.5–9 [21]; 7.0–8.5 (less than 50% of maximal activity above and below) [26]

Temperature optimum (°C)

Temperature range (°C)

Enzyme Handbook © Springer-Verlag Berlin Heidelberg 1994
Duplication, reproduction and storage in data banks are only allowed with the prior permission of the publishers

3 ENZYME STRUCTURE

Molecular weight

400000 (pig, gel filtration) [1]
100000 (Trypanosoma cruzi, gel filtration) [12]
82000–85000 (Helianthus tuberosus, SDS-PAGE followed by Western blotting [6], Saccharomyces cerevisiae, calculation from FAD content [28], house fly [44], human placenta [45]) [6, 28, 44, 45]
78000–79000 (rabbit, sedimentation equilibrium centrifugation [39], rat liver [46]) [39, 46]
70000 (Nitrobacter winogradskyi, gel filtration [7], pig testis, gel filtration [11], Saccharomyces cerevisiae, gel filtration [28]) [7, 11, 28]
65000–68000 (Saccharomyces cerevisiae, sedimentation equilibrium centrifugation, values depending on pH [23], Candida tropicalis, gel filtration, sedimentation equilibrium centrifugation [26], pig kidney, gel filtration [32, 35], rabbit liver, gel filtration [39]) [23, 26, 32, 35, 39]
More (differences in MW partially due to method of solubilization) [9, 20]

Subunits

Dimer (2 × 36000, Nitrobacter winogradskyi, SDS-PAGE [7], 2 × 52000, Trypanosoma cruzi, SDS-PAGE [12], 2 × 34300–40000, Saccharomyces cerevisiae, SDS-PAGE, sedimentation equilibrium centrifugation after treatment with guanidine-HCl [23]) [7, 12, 23]
Monomer (72000, Saccharomyces cerevisiae, SDS-PAGE [28], 75000, rabbit, Triton-solubilized, SDS-PAGE [39], 68000, rabbit, trypsin-solubilized, SDS-PAGE [39]) [28, 39]
? (x × 72000–87000, pig, SDS-PAGE [1, 11, 14, 29], rabbit, SDS-PAGE [3], horse, SDS-PAGE [5], rat, SDS-PAGE [5, 10, 16, 27, 29], hamster, SDS-PAGE [8], Helianthus tuberosus, SDS-PAGE [9], Lodderomyces elongisporus, SDS-PAGE [17], Candida tropicalis, SDS-PAGE [20], Spodoptera eridania, SDS-PAGE [21]) [1, 3, 5, 8–11, 14, 16, 17, 20, 21, 27, 29]
More (differences in subunit weight partially due to method of solubilization) [9, 20]

Glycoprotein/Lipoprotein

–

4 ISOLATION/PREPARATION

Source organism

Pig [1, 11, 14, 15, 29, 32, 35, 38, 47]; Rat [2, 5, 10, 16, 27, 29, 31, 33, 34, 36, 41, 43, 46]; Rabbit [3, 4, 30, 39]; Human [45]; Helianthus tuberosus (L. var. Blanc commun., Jerusalem artichoke) [6, 9]; Nitrobacter winogradskyi [7]; Hamster [8]; Trypanosoma cruzi [12]; Trichosporon cutaneum [13]; Lodderomyces elongisporus [17]; Aspergillus ochraceus [18]; Saccharomyces cerevisiae (grown anaerobically) [19, 23, 28, 37, 42]; Candida tropicalis (grown on alkanes) [20, 22, 26]; Spodoptera eridania (southern Armyworm) [21]; Catharanthus roseus [24]; Tetrahymena pyriformis [40]; Plants (e.g. maize, potato, avocado, bramble, tulip, leek, Vicia faba, sunflower) [6]; House fly [44]; Horse [5]; Chicken [25]

Source tissue

Polymorphonuclear leukocytes [1, 14]; Liver [2, 5, 8, 16, 27, 29, 31, 33, 34, 36, 38, 39, 43, 46, 47]; Peritoneal neutrophils [3]; Placenta [5, 45]; Tuber [6, 9]; Cell [7]; Brain [10]; Testis [11]; Epimastigotes [12]; Endometrium [15]; Midgut (of larvae) [21]; Seedlings (Vicia faba, leek, sunflower [6]) [6, 24]; Kidney [25, 32, 35]; Lung [30]; Spleen [35]; Bulbs (tulip) [6]; Mesocarp (avocado) [6]

Localisation in source

Membrane bound (outer membrane of mitochondria [25]) [1, 3, 4, 14, 20, 25]; Cytosol [7, 12]; Endoplasmic reticulum [43]; Microsomes [2, 5, 8–10, 13, 15–18, 21, 31, 33, 35–38, 40]; Nucleus (low activity) [19]; Spheroplasts [19]; Nuclear envelope [27]

Purification

Pig [1, 11, 14, 29, 38, 47]; Rat (FAD-depleted enzyme [2]) [2, 5, 10, 16, 29, 33, 36, 41]; Rabbit [3, 30]; Horse [5]; Nitrobacter winogradskyi [7]; Hamster [8]; Helianthus tuberosus [9]; Trypanosoma cruzi [12]; Lodderomyces elongisporus [17]; Candida tropicalis [20, 26]; Spodoptera eridania [21]; Saccharomyces cerevisiae [23, 28]; Catharanthus roseus [24]

Crystallization

(Saccharomyces cerevisiae) [23]

Cloned

–

Renaturated

(reconstitution of O_2-generating system [3], reconstitution of monooxygenase system [9]) [3, 9, 30, 38]

Enzyme Handbook © Springer-Verlag Berlin Heidelberg 1994
Duplication, reproduction and storage in data banks are only allowed with the prior permission of the publishers

5 STABILITY

pH

Temperature (°C)

25–30 (diluted solutions: gradual loss of activity) [41]; 36 (inactivation above) [21]; 40 (50% activity) [21]; 60 (inactivation) [10]; 100 (10 min, inhibition of O_2-formation) [14]; More (FAD and NADPH: protection against thermal inactivation) [23]

Oxidation

Organic solvent

General stability information

FAD, FMN necessary for stabilization during purification [20, 28]; Instable during purification [32]

Storage

–90°C, or –20°C, 24 h, 5–10% loss of activity, reactivation by FAD [12]; –78°C, 30 mM potassium phosphate buffer, pH 7.7, 0.1 mM EDTA, 20% glycerol, 0.4 mM PMSF [16]; –80°C, 50 mM phosphate buffer, pH 7.4, 0.1 mM EDTA, 20% glycerol [8]; –80°C [10]; –70°C, 0.15 mM potassium phosphate buffer, pH 7, 1 mM mercaptoethanol, 1 mM EDTA, 1 micromol FMN, 1 micromol FAD, 0.3% Mulgofen BC-720, 30% glycerol, several months [20]; –20°C, N_2–atmosphere, several weeks [30]; –15°C or –20°C, 10 mM phosphate buffer, pH 7.5, several months [41]; –15°C, more than 1 year [21]; 0–4°C, several months [21]; 0°C, some days [41]; 4°C or room temperature, FAD-depleted enzyme [2]

6 CROSSREFERENCES TO STRUCTURE DATABANKS

PIR/MIPS code

RDRTO4 (Rat); RDPGO4 (Pig fragment); A41447 (Yeast Saccharomyces cerevisiae); A28577 (Brown trout fragments); A25584 (Pig); A25505 (Rabbit); A05233 (Rabbit fragment); S20814 (Yeast Schizosaccharomyces pombe); A37890 (Imperfect fungus Candida tropicalis); A33421 (Human)

Brookhaven code

7 LITERATURE REFERENCES

[1] Kojima, H., Takahashi, K., Sakane, F., Koyama, J.: J. Biochem.,102,1083–1088 (1987)
[2] Kurzban, G.P., Howarth, J., Palmer, G., Strobel, H.W.: J. Biol. Chem.,265,12272–12279 (1990)
[3] Laporte, F., Doussiere, J., Vignais, P.V.: Biochem. Biophys. Res. Commun.,167,790–797 (1990)
[4] Laporte, F., Doussiere, J., Vignais, P.V.: Biochem. Biophys. Res. Commun.,168,78–84 (1990)
[5] Vibet, A., Dintinger, T., Maboundou, J.C., Gaillard, J.L., Divoux, D., Silberzahn, P.: FEBS Lett.,261,31–34 (1990)
[6] Benveniste, I., Lesot, A., Hasenfratz, M.-P., Durst, F.: Biochem. J.,259,847–853 (1989)
[7] Kurokawa, T., Fukumori, Y., Yamanaka, T.: Arch. Microbiol.,148,95–99 (1987)
[8] Ardies, C.M., Lasker, J.M., Bloswick, B.P., Lieber, C. S.: Anal. Biochem.,162,39–46 (1987)
[9] Benveniste, I., Gabriac, B., Durst, F.: Biochem. J.,235,365–373 (1986)
[10] Takahashi, N., Saito, T., Goda, Y., Tomita, K.: J. Biochem.,99,513–519 (1986)
[11] Kuwada, M., Ohsawa, Y., Horie, S.: Biochim. Biophys. Acta,830,45–51 (1985)
[12] Kuwahara, T., White, R.A., Agosin, M.: Arch. Biochem. Biophys.,239,18–28 (1985)
[13] Laurila, H., Käppeli, O., Fiechter, A.: Arch. Microbiol.,140,257–259 (1984)
[14] Sakane, F., Takahashi, K., Koyama, J.: J. Biochem.,96,671–678 (1984)
[15] Sierralta, W.D., Szendro, P.I.: Hoppe-Seyler's Z. Physiol. Chem.,364,1329–1335 (1983)
[16] Shephard, E.A., Pike, S.F., Rabin, B.R., Phillips, I. R.: Anal. Biochem.,129,430–433 (1983)
[17] Honek, H., Schunck, W.-H., Riege, P., Müller, H.-G.: Biochem. Biophys. Res. Commun.,106,1318–1324 (1982)
[18] Jayanthi, C.R., Madyastha, P., Madyastha, K.M.: Biochem. Biophys. Res. Commun.,106,1262–1268 (1982)
[19] Kamimura, K., Tsuchiya, E., Miyakawa, T., Fuki, S., Hirata, A.: Curr. Microbiol.,6,175–180 (1981)
[20] Bertrand, J.C., Gilewicz, M., Bazin, H., Azoulay, E.: Biochem. Biophys. Res. Commun.,94,889–893 (1980)
[21] Crankshaw, D.L., Hetnarski, K., Wilkinson, C.F.: Biochem. J.,181,593–605 (1979)
[22] Bertrand, J.C., Gilewicz, M., Bazin, H., Zacek, M., Azoulay, E.: FEBS Lett.,105,143–146 (1979)
[23] Tyron, E., Cress, M.C., Hamada, M., Kuby, S.A.: Arch. Biochem. Biophys.,197,104–118 (1979)
[24] Madyastha, K.M., Coscia, C.J.: J. Biol. Chem.,254,2419–2427 (1979)
[25] Kulkoski, J.A., Weber, J.L., Ghazarian, J.G.: Arch. Biochem. Biophys.,192,539–547 (1979)
[26] Bertrand, J.C., Bazin, H., Zacek, M., Gilewicz, M., Azoulay, E.: Eur. J. Biochem.,93,237–243 (1979)
[27] Zimmermann, J.J., Kasper, C.B.: Arch. Biochem. Biophys.,190,726–735 (1978)
[28] Kubota, S., Yoshida, Y., Kumaoka, H., Furumichi, A.: J. Biochem.,81,197–205 (1977)
[29] Yasukochi, Y., Masters, B.S.S.: J. Biol. Chem.,251,5337–5344 (1976)
[30] Arinc, E., Philpot, R.M.: J. Biol. Chem.,251,3213–3220 (1976)
[31] Masters, B.S.S., Prough, R.A., Kamin, H.: Biochemistry,14,607–613 (1975)
[32] Fan, L.L., Masters, B.S.S.: Arch. Biochem. Biophys.,165,665–671 (1974)

Enzyme Handbook © Springer-Verlag Berlin Heidelberg 1994
Duplication, reproduction and storage in data banks are only allowed with the prior permission of the publishers

[33] Teschke, R., Hasumura, Y., Lieber, C.S.: Arch. Biochem. Biophys.,163,404–415 (1974)
[34] Betz, G., Roper, M., Tsai, P.: Arch. Biochem. Biophys.,163,318–323 (1974)
[35] Iyanagi, T.: FEBS Lett.,46,51–54 (1974)
[36] Golf, S.W., Graef, V., Staudinger, H.: Hoppe-Seyler' S Z. Physiol. Chem.,355,1063–1069 (1974)
[37] Yoshida, Y., Kumaoka, H., Sato, R.: J. Biochem.,75,1201–1210 (1974)
[38] Ichihara, K., Kusunose, E., Kusunose, M.: Eur. J. Biochem.,38,463–472 (1973)
[39] Iyanagi, T., Mason, H.S.: Biochemistry,12,2297–2308 (1973)
[40] Poole, R.K., Nicholl, W.G., Howells, L., Lloyd, D.: J. Gen. Microbiol.,68,283–294 (1971)
[41] Omura, T., Takesue, S.: J. Biochem.,67,249–257 (1970)
[42] Aoyama, Y., Yoshida, Y., Kubota, S., Kamaoka, H., Furumichi, A.: Arch. Biochem. Biophys.,185,362–369 (1978)
[43] Cooper, M.B., Craft, J.A., Estall, M.R., Rabin, B.R.: Biochem. J.,190,737–746 (1980)
[44] Mayer, R.T., Durrant, J.L.: J. Biol. Chem.,254,4177–4185 (1979)
[45] Osawa, Y., Higashiyama, T., Nakamura, T.: J. Steroid Biochem.,15,449–452 (1981)
[46] Yasukochi, Y., Masters, B.S.S.: J. Biol. Chem.,251,5537–5544 (1976)
[47] Masters, B.S.S., Williams, C.H., Kamin, H.: Methods Enzymol.,10,565–573 (1967)

1 NOMENCLATURE

EC number
1.6.2.5

Systematic name
NADPH:ferricytochrome-c_2 oxidoreductase

Recommended name
NADPH-cytochrome c_2 reductase

Synonymes
Reductase, cytochrome c_2 (reduced nicotinamide adenine dinucleotide phosphate)
Cytochrome c_2 reductase (reduced nicotinamide adenine dinucleotide phosphate, NADPH)

CAS Reg. No.
37256-32-9

2 REACTION AND SPECIFICITY

Catalysed reaction
NADPH + 2 ferricytochrome c_2 →
→ $NADP^+$ + 2 ferrocytochrome c_2

Reaction type
Redox reaction

Natural substrates
NADPH + ferricytochrome c_2 (from Rhodopseudomonas sphaeroides) [1]

Substrate spectrum
1 NADPH + ferricytochrome c_2 (specific for NADPH, additional electron acceptors: 2,6-dichlorophenolindophenol, cytochrome c_2 from Rhodopseudomonas sphaeroides and Rhodospirillum rubrum, $K_3Fe(CN)_6$, cytochrome c of various sources) [1]

Product spectrum
1 $NADP^+$ + ferrocytochrome c_2

Enzyme Handbook © Springer-Verlag Berlin Heidelberg 1994
Duplication, reproduction and storage in data banks are only allowed with the prior permission of the publishers

Inhibitor(s)

Iodoacetate [1]; N-Ethylmaleimide [1]; p-Chloromercuribenzoate [1]; NADH [1]; Thyroxine [1]; More (inhibition by thiol compounds reversed by glutathione, not inhibitory: 8-hydroxyquinoline, o-phenanthroline, EDTA, ATP, NaCN) [1]

Cofactor(s)/prostethic group(s)

FAD (1 mol per mol of enzyme) [1]

Metal compounds/salts

Turnover number (min^{-1})

Specific activity (U/mg)

More [1]

K_m-value (mM)

0.037 (cytochrome c_2) [1]; 0.0125 (2,6-dichlorophenolindophenol) [1]; 0.125 ($K_3Fe(CN)_6$) [1]

pH-optimum

7.5 [1]

pH-range

5.6–8.0 (circa) [1]

Temperature optimum (°C)

Temperature range (°C)

---

3 ENZYME STRUCTURE

Molecular weight

43000 (Rhodopseudomonas sphaeroides, gel filtration) [1]

Subunits

Glycoprotein/Lipoprotein

–

---

4 ISOLATION/PREPARATION

Source organism

Rhodopseudomonas sphaeroides [1]; Human [2]

Source tissue

Cell [1]; 7800 C1 Morris hepatoma cells [2]

Localisation in source

Purification

Rhodopseudomonas sphaeroides [1]

Crystallization

–

Cloned

–

Renaturated

–

5 STABILITY

pH

Temperature (°C)

Oxidation

Organic solvent

General stability information

Storage

4°C, purified enzyme, 0.4 M Tris-HCl buffer, pH 8.0, ca. 2 weeks, addition of glutathione for longer storage, crude enzyme at –15°C, lyophilized, 3 months [1]

6 CROSSREFERENCES TO STRUCTURE DATABANKS

PIR/MIPS code

Brookhaven code

7 LITERATURE REFERENCES

[1] Sabo, D.J., Orlando, J.A.: J. Biol. Chem.,243,3742–3749 (1968)
[2] Norrheim, L., Sorensen, H., Gautvik, K., Bremer, J., Spydevold O.: Biochim. Biophys. Acta,1051,319–323 (1990)

Enzyme Handbook © Springer-Verlag Berlin Heidelberg 1994
Duplication, reproduction and storage in data banks are only allowed with the prior permission of the publishers

1 NOMENCLATURE

EC number
1.6.2.6

Systematic name
NAD(P)H:ferrileghemoglobin oxidoreductase

Recommended name
Leghemoglobin reductase

Synonymes
Ferric leghemoglobin reductase [1]

CAS Reg. No.
60440-35-9

2 REACTION AND SPECIFICITY

Catalysed reaction
NAD(P)H + 2 ferrileghemoglobin →
→ NAD(P)$^+$ + 2 ferroleghemoglobin

Reaction type
Redox reaction

Natural substrates

Substrate spectrum
1 NADH + ferrileghemoglobin

Product spectrum
1 NAD$^+$ + ferroleghemoglobin

Inhibitor(s)
Nicotinate (ferrileghemoglobin-nitrite complex [2]) [1, 2]; Acetate (slight) [1]; Iodoacetamide [2]; p-Hydroxymercuribenzoate [2]; Quinacrine [2]; Amobarbital [2]; Catalase (possibly H_2O_2 as reaction intermediate) [2]

Cofactor(s)/prostethic group(s)
NADH (NADPH: 31% of NADH activity [1], NADPH 54–80% of NADH activity [2]) [1, 2]; FAD [3]

Metal compounds/salts

Enzyme Handbook © Springer-Verlag Berlin Heidelberg 1994
Duplication, reproduction and storage in data banks are only allowed with the prior permission of the publishers

Turnover number (min^{-1})

Specific activity (U/mg)
0.218 (ferrileghemoglobin) [1]; 4.8 (2,6-dichlorophenolindophenol) [3]; 0.5 (ferrileghemoglobin) [3]

K_m-value (mM)
0.0088–0.013 (ferrileghemoglobin, value depending on origin organism) [2]; 0.0095 (leghemoglobin) [1]; 0.0188 (NADH) [1]; 0.051 (NADH) [3]

pH-optimum
5.2 (no measurements below) [1]

pH-range
More (decrease of activity from 100% at pH 5.2 to 30% at pH 6.7, no measurements below 5.2) [1]

Temperature optimum (°C)

Temperature range (°C)

---

3 ENZYME STRUCTURE

Molecular weight
100000–110000 (Glycine max, gel filtration) [1, 3]
83000 (Glycine max, equilibrium ultracentrifugation) [1]

Subunits
Dimer (2 × 54000, Glycine max, SDS-PAGE) [1, 3]

Glycoprotein/Lipoprotein
–

---

4 ISOLATION/PREPARATION

Source organism
Glycine max (soybean) [1–3]; Phaseolus vulgaris (bean) [2]; Vigna unguiculata (cowpea) [2]

Source tissue
Root nodules [1–3]

Localisation in source
Cytosol [1, 3]

Purification
Glycine max [1, 3]

Crystallization

–

Cloned

–

Renaturated

–

5 STABILITY

pH

Temperature (°C)

More (heat labile) [1]

Oxidation

Organic solvent

General stability information

Storage

0°C, 37 days, 96% activity [1]; –70°C [3]

6 CROSSREFERENCES TO STRUCTURE DATABANKS

PIR/MIPS code

Brookhaven code

7 LITERATURE REFERENCES

[1] Saari, L.L., Klucas, R.V.: Arch. Biochem. Biophys.,231,102–113 (1984)
[2] Becana, M., Klucas, R.V.: Proc. Natl. Acad. Sci. USA,87,7295–7299 (1990)
[3] Ji, L., Wood, S., Becana, M., Klucas, R.V.: Plant Physiol.,96,32–37 (1991)

Enzyme Handbook © Springer-Verlag Berlin Heidelberg 1994
Duplication, reproduction and storage in data banks are only allowed with the prior permission of the publishers

1 NOMENCLATURE

EC number
1.6.4.1

Systematic name
NADH:L-cystine oxidoreductase

Recommended name
Cystine reductase (NADH)

Synonymes
Reductase, cystine
NADH-dependent cystine reductase

CAS Reg. No.
9029-18-9

2 REACTION AND SPECIFICITY

Catalysed reaction
NADH + L-cystine →
→ NAD^+ + 2 L-cysteine

Reaction type
Redox reaction

Natural substrates
NADH + L-cystine (enzyme is operating especially while the cell is metabolizing glucose or other metabolites leading to the formation of NADH or reduced triphosphopyridine nucleotides and thus is a connection between carbohydrate metabolism and sulfhydryl group maintenance [1], may provide reduced sulfhydryl groups involved in the transition of mycelium to yeast form in the dimorphic fungus Histoplasma capsulatum [2]) [1, 2]

Substrate spectrum
1 NADH + L-cystine (and alpha-substituted cystines: e.g. alpha-methyl-DL-cystine, alpha-n-propyl-DL-cystine, alpha-isopropyl-DL-cystine, alpha-phenyl-DL-cystine, substrate specificity determined only with crude extract, possibly the reduction of these compounds is catalyzed by a series of different specific enzymes [4]) [1–4]

Enzyme Handbook © Springer-Verlag Berlin Heidelberg 1994
Duplication, reproduction and storage in data banks are only allowed with the prior permission of the publishers

Product spectrum
1 NAD^+ + L-cysteine [1–4]

Inhibitor(s)
More (no effect: streptomycin, fracidin, penicillin) [1]; p-Chloromercuriphenylsulfonic acid (strong) [2]

Cofactor(s)/prostethic group(s)
NADH [1–4]; More (no effect: riboflavin, thiamine, adenylic acid, ribonucleic acid, biotin) [1]

Metal compounds/salts
No effect of divalent metal ions [1]

Turnover number (min^{-1})

Specific activity (U/mg)

K_m-value (mM)
0.9 (L-cystine) [4]; More (K_m is determined with crude enzyme extract only, reduction of alpha-substituted cystines is possibly catalyzed by a series of different specific enzymes: 1.1 mM (DL-cystine), 1.2 mM (alpha-methyl-DL-cystine), 1.4 mM (alpha-n-propyl-DL-cystine), 3.5 mM (alpha-isopropyl-DL-cystine), 1.5 mM (alpha-phenyl-DL-cystine)) [4]

pH-optimum
6.2 (enzyme assay) [1, 4]; 7.0 (enzyme assay) [3]; More (effect of pH on determination of cysteine) [3]

pH-range
More (effect of pH on determination of cysteine) [3]

Temperature optimum (°C)
37 (enzyme assay) [3, 4]

Temperature range (°C)

3 ENZYME STRUCTURE

Molecular weight

Subunits

Glycoprotein/Lipoprotein
–

4 ISOLATION/PREPARATION

Source organism

Pea [1]; Saccharomyces cerevisiae (strain IHM 806, permanently filamentous mutant [1]) [1, 3, 4]; Candida albicans [1]; Histoplasma capsulatum (fungus) [2]; Pig (female) [1]

Source tissue

Seeds (pea) [1]; Cell (yeast) [1]

Localisation in source

Purification

Crystallization

–

Cloned

–

Renaturated

–

5 STABILITY

pH

Temperature (°C)

Oxidation

Organic solvent

General stability information

Storage

6 CROSSREFERENCES TO STRUCTURE DATABANKS

PIR/MIPS code

Brookhaven code

7 LITERATURE REFERENCES

[1] Romano, A.H., Nickerson, W.J.: J. Biol. Chem.,208,409–416 (1954)
[2] Maresca, B., Jacobson, E., Medoff, G., Kobayashi, G.: J. Bacteriol.,135,987–992 (1978)
[3] Thibert, R.J., Sarwar, M., Carroll, J.E.: Mikrochim. Acta,3,615–624 (1969)
[4] Carroll, J.E., Kosicki, G.W., Thibert, R.J.: Biochim. Biophys. Acta,198,601–603 (1970)

Enzyme Handbook © Springer-Verlag Berlin Heidelberg 1994
Duplication, reproduction and storage in data banks are only allowed with the prior permission of the publishers

1 NOMENCLATURE

EC number
1.6.4.2

Systematic name
NADPH:oxidized-glutathione oxidoreductase

Recommended name
Glutathione reductase (NADPH)

Synonymes
Reductase, glutathione
NADPH-glutathione reductase
GSH reductase
GSSG reductase
NADPH-GSSG reductase
Glutathione S-reductase

CAS Reg. No.
9001-48-3

---

2 REACTION AND SPECIFICITY

Catalysed reaction
NADPH + oxidized glutathione →
→ $NADP^+$ + 2 glutathione (mechanism [8], ping-pong mechanism [12])

Reaction type
Redox reaction

Natural substrates
Oxidized glutathione + NADPH (production of substrate for: glutathione peroxidase (E.C. 1.11.1.9), glutathione-homocystine oxidoreductase (E.C. 1.8.4.1), glutathione-protein disulfide oxidoreductase (E.C. 1.8.4.2) et al. [1], maintenance of glutathione/oxidized glutathione ratio is a protective mechanism for intracellular thiols during growth in atmospheric oxygen [2], maintenance of high levels of reduced glutathione in cytoplasm [8], role in cell division cycle [8], role in stress adaption on cellular level [8], enzyme plays a key role in nutrient-induced increase in the thiol content of pancreatic-islet cells and this increase itself participates in the coupling of metabolic to secretory events [30], key enzyme in plants against oxidative stress [31]) [1, 2, 8, 14, 30, 31]

Enzyme Handbook © Springer-Verlag Berlin Heidelberg 1994
Duplication, reproduction and storage in data banks are only allowed with the prior permission of the publishers

Substrate spectrum

1 Oxidized glutathione + NADPH (ir [1, 14], reverse reaction only if very high concentration of $NADP^+$ and glutathione present [39], glutathione or other disulfides, e.g.:
bis-L-gamma-glutamyl-L-cystinyl-bis-beta-alanine [2], bis-N,N'-(gamma-glutamylcystine) (slight) [2], glutathione-S-sulfonate [2], methylene blue (slowly) [42], L-cystine (not [38, 45], slowly [42]) [38, 42], mixed disulfide between coenzyme A and glutathione [2, 40], D, L-lipoate (slight) [2], 5,5'-dithiobis(2-nitrobenzoic acid) [7], S-sulfoglutathione and some mixed disulfides (with the exception of the mixed disulfide of coenzyme A and reduced glutathione) are poor substrates [16], Euglena gracilis enzyme: highly specific for oxidized glutathione [14]) [1–49]

2 More (low transhydrogenase activity with oxidized pyridine nucleotide analogs and diaphorase activity with 2,6-dichlorophenolindophenol as acceptor substrates (NADPH and NADH as donors) [16], branched mechanism [35, 37, 44], bisubstrate mechanism [49], not: cysteamine [38], oxidized lipoic acid [45], oxidized lipoamide [45]) [16, 35, 37, 38, 44, 45, 49]

Product spectrum

1 Glutathione + $NADP^+$ (ir [1, 14])

2 ?

Inhibitor(s)

1,3-Bis-(2-chloroethyl)-1-nitrosourea [2]; 1-(2-Chloroethyl)-3-(2-hydroxyethyl)-1-nitrosourea [2]; N,N'-Bis(trans-4-hydroxychlorohexyl)-N'-nitrosourea [2]; Dinitrosated isomers of N, N'-bis[N(2-chloroethyl)-N-carbonyl]cysteamine [2]; 2-Chloroethylisocyanate [2]; Iodoacetamide [2, 45]; 2,3-Butanedione [2]; 1,2-Cyclohexadione [2]; Phenylglyoxal [2]; N-Alkylmaleimide [2]; Zn^{2+} (root enzyme inhibited, chloroplast enzyme not [26], inhibition increased by NADPH [35]) [2, 14, 35, 39]; Hg^{2+} [2, 7, 14, 42]; Cu^{2+} (root enzyme inhibited, chloroplast enzyme slightly [26]) [2, 7, 14, 26, 45]; Cd^{2+} [2]; Phenylarsenous acid [2]; Arsenite (inhibition increased by NADPH [35]) [2, 7, 35]; Melarsen oxide [2]; p-Hydroxymercuribenzoate [2, 26, 28, 30, 34, 45]; Nitrofurantoin [2, 6]; Nifurtimox [2]; Glutathione (1 mM [14], product inhibition [37, 39, 45], oxidized glutathione, 12 mM, 25% inhibition [49]) [14, 19, 37, 39, 45, 49]; p-Chloromercuribenzoate [14, 42, 49]; NADH [13, 19]; NADPH [13]; Phosphate buffer [19]; Iodide [46]; Fe^{2+} (root enzyme inhibited, chloroplast enzyme slight) [26]; Urea (activation: 0.4–0.6 M, inactivation at higher concentration) [27]; $AgNO_3$ [42, 45]; S-(2,4-Dinitrophenyl)-glutathione [2]; 1-Chloro-2,4-dinitrobenzene [2]; 1-Fluoro-2,4-dinitrobenzene [2]; 2,4,6-Trinitrobenzene sulfonate [2]; Nitrogen mustard [2]; Chromate [2];

Benzylselenosulfate [2]; Iodine [2]; Iodoacetate (in presence but not in absence of reduced coenzyme [3]) [3, 7, 39]; Sulfhydryl reagents (in presence but not in absence of reduced coenzyme) [3]; $NADP^+$ (1 mM [14]) [6, 14, 37, 39, 44]; NaBr (above 0.2 M) [6]; FMN [6]; FAD [6]; Riboflavin [6]; Mg^{2+} [7, 45]; N-Ethylmaleimide ([35], slight [7]) [7, 14, 26, 35, 39, 45, 49]; Mn^{2+} [7, 14, 45]; Ca^{2+} [7, 45]; Co^{2+} [14]; Ni^{2+} [14]; Fe^{3+} [14]; KI [45]; $BaCl_2$ [45]; $Fe(NO_3)_2$ [45]; AMP [45]; ADP [45]; ATP [45]; Cl^- [45, 49]; Br^- [49]; SO_4^{2-} [49]; NO_3^- [49]; Citrate [49]

Cofactor(s)/prostethic group(s)

FAD (FAD enzyme [2, 3, 6, 8, 15, 16, 27, 28, 34–36, 38, 45, 46, 48, 49], 2 mol of FAD per mol of enzyme [5, 11, 13, 15, 16, 36, 44, 48], aminoacid sequence of FAD-binding domain [23], 1 mol of FAD per mol of subunit [34]) [2, 3, 5, 6, 8, 11, 13, 15, 16, 23, 27, 28, 34–36, 38, 44–46, 48]; Flavin (flavoprotein [2, 5, 31, 47], 2 mol of flavin per mol of enzyme [2], 1.06 mol of flavin per mol of subunit [27]) [2, 5, 27, 31, 47]; FAD analogs (properties of glutathione reductase reconstituted with FAD analogues) [24, 29]; NADPH (5% of the activity with NADH [26], can utilize both NADPH and NADH, more active with NADPH [7], best electron donor [14, 15, 16], a single enzyme uses both NADPH and NADH as hydrogen donors [21], absolute specificity for NADPH [27, 28, 34, 47]) [1–43, 45–49]; NADH (activity relative to NADPH: 0.5 % [15], 1.3% [14], 10% [38, 44], 2% [6], less than 1% [46], 3% [49], low activity with NADH [39, 42], enzyme can utilize both NADH and NADPH, more active with NADPH [7], 20 times higher activity with NADH than with NADPH [26], specific for NADH [45]) [6, 7, 14, 15, 26, 38, 39, 42, 44–46, 49]

Metal compounds/salts

Sodium phosphate (activates) [6]; NaCl (activates) [6]; NaBr (activates) [6]; Sodium acetate (activates) [6]; Sodium citrate (activates) [6]; SO_4^{2-} (stimulates) [45]; Na^+ (activates) [45]; NH_4^+ (activates) [45]; Phosphate (activates [45], optimum concentration of potassium phosphate: 0.1 M [46]) [45, 46]; Diphosphate (activates) [45]

Turnover number (min^{-1})

Specific activity (U/mg)

240 [2]; 269 (rat [5]) [5, 33]; 204 (bovine) [5, 36]; 1.18 (Achromobacter starkeyi) [7]; 281.0 (alfalfa) [10]; 273.0 (sainfoin) [10]; 193.3 (sheep) [11]; 61.2 [13]; 145 [25]; 238.0 [27]; 180 [28]; 360 [31]; 249 [35]; 158 [37]; 108 [38]; 246 [39]; 125 [43]; 207 [44]; 28.4 [45]; 320 [46]; 150 [47]; 2.9 [49]; More (assay methods) [4, 5]

Enzyme Handbook © Springer-Verlag Berlin Heidelberg 1994
Duplication, reproduction and storage in data banks are only allowed with the prior permission of the publishers

K_m-value (mM)

More (thermal dependency of K_m [10]) [5, 6, 10, 12, 14, 16, 19, 25–28, 31, 34–39, 44, 45, 48, 49]; 0.0085 (NADPH, human) [2]; 0.065 (glutathione disulfide) [2]; 11.6 (bis-N,N'-(gamma-glutamylcystine)) [2]; 1.0 (mixed disulfide of coenzyme A and glutathione) [2]; 0.250 (glutathione-S-sulfonate, yeast) [2]; 0.0038 (NADH) [2]; 0.070 (oxidized glutathione, E. coli) [9]; 0.025 (NADPH, E. coli) [9]; 0.020 (NADPH, Phycomyces blakesleeanus) [13]; 0.130 (glutathione) [13]; 0.076 (oxidized glutathione) [15]; 0.021 (NADPH) [15]; 3.3 (mixed disulfide of CoA and glutathione) [27]; 0.0005 (FAD) [45]; 0.00078 (FAD) [46]

pH-optimum

4.5 (NADH + oxidized glutathione) [31]; 5.0 (NADH + oxidized glutathione) [15, 16]; 5.0–5.5 (NADH + oxidized glutathione) [44]; 5.7 (NADH + oxidized glutathione) [19]; 6.0 (NADH) [13]; 6.05 (NADH + oxidized glutathione) [21]; 6.6–7.6 [3]; 6.8 (NADPH + oxidized glutathione) [19, 21]; 6.8–7.6 [6]; 6.9 [28]; 7.0 (oxidized glutathione + NADPH [15, 16, 44], NADH + oxidized glutathione [45]) [5, 7, 15, 16, 27, 36, 42, 44, 45]; 7.2–7.3 [46]; 7.5 (NADPH + oxidized glutathione) [13, 31]; 7.5–8.2 [49]; 7.7 (root enzyme) [26]; 7.9 [48]; 8.1 [38]; 8.2 (chloroplast enzyme) [26]; 8.5–9.0 [39]; 9.0 [35]

pH-range

5.2–9.0 (at pH 5.2 and 9.0 about 20% of activity maximum) [28]; 6.1–9.1 (50% of maximal activity at pH 6.1 and 9.1, root enzyme) [26]; 6.3–9.6 (at pH 6.3–9.6 about 50% of activity maximum) [48]; 6.9–9.2 (6.9: about 50% of activity maximum, 9.2: about 83% of activity maximum) [26]; 7–9.5 (7: about 40% of activity maximum, 9.5: about 55% of activity maximum) [39]; 7–11 (7: about 13% of activity maximum, 11: 36% of activity maximum) [34]; 7.3–8.8 (7.3: about 65% of activity maximum, 8.6: about 90% of activity maximum) [38]

Temperature optimum (°C)

25 (assay at) [4]; 28 [7]

Temperature range (°C)

15–40 (15°C: about 25% of activity maximum, 40°C: about 50% of activity maximum) [7]

3 ENZYME STRUCTURE

Molecular weight

56000 (rabbit, SDS-PAGE, gel filtration, monomer, enzyme can also be active as dimer) [28]
63000 (Rhodospirillum rubrum, gel filtration) [49]
79000 (Euglena gracilis Z, gel filtration) [14]
94000 (E. coli, gel filtration) [9]
99800 (Spirulina maxima, pore gradient gel electrophoresis, dimer, enzyme exists predominantly as tetrameric species in equilibrium with a minor dimer fraction) [27]
100000 (bovine [5, 36], analytical ultracentrifugation [36], rat [5], Phycomyces blakesleeanus [13], human, sedimentation equilibrium [20]) [5, 13, 20, 36]
102000 (Hemicentrotus pulcherrimus, gel filtration) [46]
103000 (pig, analytical ultracentrifugation) [15, 16]
104000 (Anabaena sp., PAGE at different acrylamide concentrations, gel filtration) [35]
105000 (E. coli, gel filtration [4], mouse, gel filtration [37]) [4, 37]
106000 (rice, calculation from sedimentation and diffusion coefficients) [48]
108000 (rabbit, ultracentrifugation in glycerol density gradient, dimer, enzyme can also be active as monomer) [28]
109000 (E. coli) [31]
116000 (sheep, gel filtration) [11]
118000 (Saccharomyces cerevisiae, equilibrium sedimentation) [2]
120000 (gerbil, gel filtration) [34]
125000 (spinach, gel filtration [38], rat, gel filtration [44]) [38, 44]
140000 (alfalfa, gel electrophoresis) [10]
145000 (spinach, gel filtration) [39]
149000 (sainfoin, gel electrophoresis) [10]
177000 (Spirulina maxima, sedimentation equilibrium, tetramer, enzyme exists predominantly as a tetrameric species in equilibrium with a minor dimer fraction) [27]
More [25]

Subunits

Monomer (1 × 56000, rabbit, SDS-PAGE, enzyme can also be active as dimer) [28]
Dimer (2 × 60000, rat, SDS-PAGE [44], 2 × 52000, Hemicentrotus pulcherrimus, SDS-PAGE [46], 2 × 52000, rice, SDS-PAGE [48], 2 × 52500, human [2], 2 × 55000, E. coli [31], 2 × 53400, gerbil, SDS-PAGE [34], 2 × 54000, Anabaena sp., SDS-PAGE, gel filtration in presence of urea [35], 2 × 64000, sheep, SDS-PAGE [11], 2 × 49000, E. coli, SDS-PAGE [9], 2 × 50200, Phycomyces blakesleeanus [13], 2 × 40000, Euglena gracilis Z, SDS-PAGE [14], 2 × 59000, pig, SDS-PAGE [15, 16], 2 × 49000, human, SDS-PAGE [20], 2 × 47000, Spirulina maxima, SDS-PAGE, enzyme exists predominantly as a

Enzyme Handbook © Springer-Verlag Berlin Heidelberg 1994
Duplication, reproduction and storage in data banks are only allowed with the prior permission of the publishers

tetrameric species in equilibrium with a minor dimer fraction [27], 2 × 56000, rabbit, SDS-PAGE, enzyme can also be active as monomer [28], 2 × 55000, mouse, SDS-PAGE [37], 2 × 72000, spinach, SDS-PAGE [39], rat [5], bovine [5, 36]) [2, 5, 9, 11, 13–16, 20, 27, 28, 31, 34–37, 39, 44, 46, 48]
Tetramer (4 × 47000, Spirulina maxima, SDS-PAGE, enzyme exists predominantly as a tetrameric species in equilibrium with a minor dimer fraction) [27]
Oligomer (x × 57000 + x × 36000, alfalfa, SDS-PAGE, x × 57000 + x × 37000, sainfoin, SDS-PAGE) [10]
More (three-dimensional structure [8], in absence of thiols the enzyme shows tendency to form aggregates [18], NADH promotes aggregation [19], in absence of thiols, glutathione reductase shows a tendency to form tetramers and larger aggregates (catalytically active) [20]) [8, 18–20]

Glycoprotein/Lipoprotein

–

4 ISOLATION/PREPARATION

Source organism

Human [2, 6, 8, 12, 18–24]; E. coli [2, 4, 9, 31, 32]; Chromatium vinosum [2, 45]; Plasmodium vinckei [2]; Saccharomyces cerevisiae (baker's yeast) [2, 3, 12, 40]; Penicillium chrysogenum [2]; Spinach (Spinacia oleracea) [2, 12, 38, 39]; Mackerel [2]; Mouse [2, 37]; Gerbil [2, 34]; Rabbit [2, 28, 29]; Pig [2, 15, 16, 43]; Bovine [2, 5, 36]; Rat [5, 33, 44]; Achromobacter starkeyi [6]; Medicago sativa (alfalfa) [10]; Onobrychis viciifolia (sainfoin) [10]; Sheep [11]; Phycomyces blakesleeanus [13]; Euglena gracilis Z [14]; Pea (Pisum sativum) [25, 26]; Spirulina maxima (cyanobacterium) [27]; Nicotiana tabacum [31]; Nicotiana glutinosa [31]; Anabaena sp. (strain 7119) [35]; Sea urchin (Hemicentrotus pulcherrimus) [46]; Rice [47, 48]; Rhodospirillum rubrum [49]; Thiobacillus thiooxidans [42]; More (widely distributed among bacteria, fungi, plants, protozoa and animals but not ubiquitous, not in anaerobic eubacteria and 2 species of archaebacteria) [2]

Source tissue

Leaves [2, 10, 26, 31, 38]; Muscle [2]; Liver [2, 5, 28, 29, 33, 34, 37, 44]; Platelets [2, 6]; Leukocytes [2]; Lens fiber [2]; Erythrocytes [2, 8, 15, 16, 18–23, 43]; Cell [7, 14, 27, 31]; Brain [11]; Mycelium [13]; Root [26]; Eggs [46]; Kernel [47, 48]

Localisation in source

Cytoplasm (solely in cytoplasm [14]) [2, 5, 6, 14]; Mitochondria (animals: cytoplasm-like spaces of mitochondria [1]) [1, 2]; Chloroplast (cytoplasm-like spaces of chloroplasts [1], stroma [39]) [1, 2, 25, 26, 39]

Purification

E. coli [4, 9, 31]; Bovine [5, 36]; Rat [5, 33, 44]; Human [6, 18, 21, 22]; Achromobacter starkeyi [7]; Medicago sativa (alfalfa) [10]; Onobrychis viciifolia (sainfoin) [10]; Sheep [11]; Phycomyces blakesleeanus [13]; Pig [15, 43]; Pea (Pisum sativum) [25, 26]; Spirulina maxima (cyanobacterium) [27]; Rabbit [28]; Gerbil [34]; Anabaena sp. (strain 7119, cyanobacterium) [35]; Mouse [37]; Spinach (Spinacia oleracea) [38, 39]; Saccharomyces cerevisiae [3, 40]; Thiobacillus thiooxidans [42]; Hemicentrotus pulcherrimus [46]; Rice [47]; Rhodospirillum rubrum [49]; More (affinity chromatography [17], large scale [41]) [17, 41]

Crystallization

[2, 15, 18, 21, 22, 32]

Cloned

(E. coli gene) [9]

Renaturated

–

---

5 STABILITY

pH

5.6 (30 min, stable) [39]; 5.8–8.4 (67°C: stable, 78 °C: unstable) [14]; 6.0–8.0 (unstable below and above) [42]; 7.5 (unstable below) [31]; 7.5–9.5 (4 °C, stable) [31]

Temperature (°C)

20 (24 h, stable) [49]; 30 (reduced form of the enzyme highly unstable) [26]; 37 (8 h, stable) [37]; 42.5 (complete loss of activity above 42°C, 10 min in presence of NADPH) [13]; 45 (irreversible denaturation above, complete loss of activity after 40 min) [42]; 50 (10 min, complete loss of activity [42], 5 min, 60% loss of activity [45]) [42, 45]; 55 (10 min, 13% loss of activity [13], 5 min, no loss of activity [49], 10 min, pH 7.0, stable [27]) [13, 27, 49]; 60 (10 min, complete loss of activity [39], 3 min, 30% loss of activity [49]) [39, 49]; 60–65 (oxidized form stable) [26]; 62 (unstable above) [49]; 65 (10 min, complete loss of activity [13], 5 min, complete loss of activity [45]) [13, 45]; 67 (pH 5.8–8.4, stable) [14]; 70 (highly stable, significant loss of activity only after 10 min) [31]; 75 (1 h, 25% loss of activity [28], 10 min, stable [37]) [28, 37]; 78 (pH 5. 8–8.4, unstable) [14]; 80 ($NADP^+$, oxidized glutathione, glutathione and dichloroindophenol protect against thermal inactivation at 80°C) [37]; More (root enzyme more stable than chloroplast enzyme, oxidized form more stable than reduced form) [26]

Enzyme Handbook © Springer-Verlag Berlin Heidelberg 1994
Duplication, reproduction and storage in data banks are only allowed with the prior permission of the publishers

Oxidation

Organic solvent

General stability information

Complete inactivation after incubation with NADPH [37, 39]; Complete inactivation after incubation with NADH [37]; Partial inactivation after incubation with ATP [38]; Repeated freezing and thawing: stable to [28, 45], no effect [27], inactivation [2]; Concentrated solution, stable for years [2]; Crystals and amorphous protein obtained by precipitation with ammonium sulfate are stable [2]; Freezing denaturates [5]; Urea: quite stable to [13, 28], activation from 0.4–0.6 M, inactivation at higher concentration [27]; $NADP^+$: stabilizes [31], stabilizes against thermal inactivation at 80°C [37]; FAD: stabilizes [31], no stabilization against thermal inactivation at 80°C [37]; Dithiothreitol stabilizes [31]; Glycerol stabilizes [31]; Pure enzyme heated in presence of bovine serum albumin and 0.020 mM NADPH: loss of activity beyond 30°C, significant reduction even at 0°C [31]; Glutathione protects against thermal inactivation at 80 °C [37]; Oxidized glutathione protects against thermal inactivation at 80°C [37]; Dichloroindophenol protects against thermal inactivation at 80°C [37]

Storage

Concentrated solution, years [2]; 4°C, activation after storage over days, months or years [2]; –20°C, pH 6.8, 0.05 M potassium phosphate buffer, indefinitely stable [3]; –20°C, 2–3 months [7]; 4°C, pH 7.5, 50 mM sodium phosphate buffer, 1 mM 2-mercaptoethanol, 1 mM EDTA [13]; –20°C, several weeks [14]; 0–4°C, several months [18]; –20°C, 1 year [27]; –20°C, with several freeze-thawing cycles, more than 2 years, stable [28]; –18°C, in the dark, 10% glycerol, 5 mM $NADP^+$, for at least 1 year [31]; 25°C, in the dark [35]; In refrigerator, several months [36]; –70°C, pH 7.0, 1 mM EDTA, 1 mM dithiothreitol, 0.020 mM FAD, 5 mM $NADP^+$ [37]; 4°C, several weeks [39]; –20°C, 2 years [45]; –28°C, 50 mM phosphate buffer, a few months [46]; 0°C, 6 days [49]

6 CROSSREFERENCES TO STRUCTURE DATABANKS

PIR/MIPS code

RDHUU (Human); RDECU (Escherichia coli); S08979 (Human); S15236 (Pseudomonas aeruginosa); PX0017 (Spirulina sp. fragments); S18973 (Garden pea)

Brookhaven code

3GRS (HUMAN (HOMO SAPIENS) ERYTHROCYTE)

7 LITERATURE REFERENCES

[1] Williams, C.H. in "The Enzymes",3rd. Ed. (Boyer, P.D., Ed.) 13,89–173 (1976) (Review)
[2] Schirmer, R.H., Krauth-Siegel, R.L., Schulz, G.E. in "Coenzymes Cofactors" (Pt. A) 3,553–596 (1989) (Review)
[3] Colman, R.F.: Methods Enzymol.,17,500–503 (1971) (Review)
[4] Williams, C.H., Arscott, L.D.: Methods Enzymol.,17,503–509 (1971) (Review)
[5] Carlberg, I., Mannervik, B.: Methods Enzymol.,113,484–490 (1985) (Review)
[6] Moroff, G., Kosow, D.P.: Biochim. Biophys. Acta,527,327–336 (1978)
[7] Ruiz-Herrera, J., Amezcua-Ortega, R., Trujillo, A.: J. Biol. Chem.,243,4083–4088 (1968)
[8] Schulz, G.E., Schirmer, R.H., Pai, E.F. in "Flavins Flavoproteins" Proc. Int. Symposium,6th Meeting (Yagi, K., Yamano, T., Eds.),557–567 (1978) (Review)
[9] Scrutton, N.S., Berry, A., Perham, R.N.: Biochem. J.,245,875–880 (1987)
[10] Kidambi, S.P., Mahan, J.R., Matches, A.G.: Plant Physiol.,92,363–367 (1990)
[11] Acan, N.L., Tezcan, E.F.: FEBS Lett.,250,72–74 (1989)
[12] Wong, K.K., Vanoni, M.A., Blanchard, J.S.: Biochemistry,27,7091–7096 (1988)
[13] Montero, S., De Arriaga, D., Soler, J.: Biochim. Biophys. Acta,952,56–66 (1988)
[14] Shigeoka, S., Onishi, T., Nakano, Y., Kitaoka, S.: Biochem. J.,242,511–515 (1987)
[15] Boggaram, V., Brobjer, T., Larson, K., Mannervik, B.: Anal. Biochem.,98,335–340 (1979)
[16] Boggaram, V., Larson, K., Mannervik, B.: Biochim. Biophys. Acta,527,337–347 (1978)
[17] Harding, J.J.: J. Chromatogr.,77,191–199 (1973)
[18] Worthington, D.J., Rosemeyer, M.A.: Eur. J. Biochem.,48,167–177 (1974)
[19] Worthington, D.J., Rosemeyer, M.A.: Eur. J. Biochem.,67,231–238 (1976)
[20] Worthington, D.J., Rosemeyer, M.A.: Eur. J. Biochem.,60,459–466 (1975)
[21] Nakashima, K., Miwa, S., Yamauchi, K.: Biochim. Biophys. Acta,445,309–323 (1976)
[22] Krohne-Ehrich, G., Schirmer, R.H., Untucht-Grau, R.: Eur. J. Biochem.,80,65–71 (1971)
[23] Untucht-Grau, R., Schirmer, R.H., Schirmer, I., Krauth-Siegel, R.L.: Eur. J. Biochem.,120,407–419 (1981)
[24] Krauth-Siegel, R.L., Schirmer, R.H., Ghisla, S.: Eur. J. Biochem.,148,335–344 (1985)
[25] Connell, J.P., Mullet, J.E.: Plant Physiol.,82,351–356 (1986)
[26] Bielawski, W., Joy, K.W.: Phytochemistry,25,2261–2265 (1986)
[27] Rendon, J.L., Calcagno, M., Mendoza-Hernandez, G., Ondarza, R.N.: Arch. Biochem. Biophys.,248,215–223 (1986)
[28] Zanetti, G.: Arch. Biochem. Biophys.,198,241–246 (1979)
[29] Zanetti, G., Beretta, C., Malandra, D.: Arch. Biochem. Biophys.,244,831–837 (1986)
[30] Malaisse, W.J., Dufrane, S.P., Mathias, P.C.F., Carpinelli, A.R., Malaisse-Lagae, F., Garcia-Morales, P., Valverde, I., Sener, A.: Biochim. Biophys. Acta,844,256–264 (1985)
[31] Mata, A.M., Pinto, M.C.: Z. Naturforsch.,39c,908–915 (1984)
[32] Ermler, U., Schulz, G.E.: Proteins Struct. Funct. Genet.,9,174–179 (1991)
[33] Carlberg, I., Altmejd, B., Mannervik, B.: Biochim. Biophys. Acta,677,146–152 (1981)
[34] Le Trang, N., Bhargava, K.K., Cerami, A.: Anal. Biochem.,133,94–99 (1983)

Enzyme Handbook © Springer-Verlag Berlin Heidelberg 1994
Duplication, reproduction and storage in data banks are only allowed with the prior permission of the publishers

[35] Serrano, A., Rivas, J., Losada, M.: J. Bacteriol.,158,317–324 (1984)
[36] Carlberg, I., Mannervik, B.: Anal. Biochem.,116,531–536 (1981)
[37] Lopez-Barea, J., Lee, C.-Y.: Eur. J. Biochem.,98,487–499 (1979)
[38] Wirth, E., Latzko, E.: Z. Pflanzenphysiol.,89,69–75 (1978)
[39] Halliwell, B., Foyer, C.H.: Planta,139,9–17 (1978)
[40] Carlberg, I., Mannervik, B.: Biochim. Biophys. Acta,484,268–274 (1977)
[41] Pigiet, V.P., Conley, R.R.: J. Biol. Chem.,252,6367–6372 (1977)
[42] Tano, T., Ishii, K., Sugio, T., Imai, K.: Agric. Biol. Chem.,40,1879–1880 (1976)
[43] Mannervik, B., Jacobsson, K., Boggaram, V.: FEBS Lett.,66,221–224 (1976)
[44] Carlberg, I., Mannervik, B.: J. Biol. Chem.,250,5475–5480 (1975)
[45] Chung, Y.C., Hurlbert, R.E.: J. Bacteriol.,123,203–211 (1975)
[46] Ii, I., Sakai, H.: Biochim. Biophys. Acta,350,141–150 (1974)
[47] Ida, S., Morita, Y.: Agric. Biol. Chem.,35,1542–1549 (1971)
[48] Ida, S., Morita, Y.: Agric. Biol. Chem.,35,1550–1557 (1971)
[49] Boll, M.: Arch. Mikrobiol.,66,374–388 (1969)

1 NOMENCLATURE

EC number
1.6.4.4

Systematic name
NAD(P)H:protein-disulfide oxidoreductase

Recommended name
Protein-disulfide reductase (NAD(P)H)

Synonymes
Reductase, protein disulfide
Protein disulfide reductase
Insulin-glutathione transhydrogenase
Disulfide reductase

CAS Reg. No.
9029-19-0

2 REACTION AND SPECIFICITY

Catalysed reaction
NAD(P)H + protein disulfide →
→ $NAD(P)^+$ + protein dithiol

Reaction type
Redox reaction

Natural substrates

Substrate spectrum
1 NAD(P)H + protein disulfide (reaction rate with NADPH is approximately twice that with NADH) [1]

Product spectrum
1 $NAD(P)^+$ + protein dithiol [1]

Inhibitor(s)
Iodoacetate [1]; Hg^{2+} [1]; Ag^+ [1]; Cu^{2+} [1]; Arsenite [1]; Cd^{2+} [1]; Zn^{2+} [1]; p-Chloromercuribenzoate [1]

Enzyme Handbook © Springer-Verlag Berlin Heidelberg 1994
Duplication, reproduction and storage in data banks are only allowed with the prior permission of the publishers

Cofactor(s)/prostethic group(s)

NADH (reaction rate with NADPH is approximately twice that with NADH) [1]; NADPH (reaction rate with NADPH is approximately twice that with NADH) [1]

Metal compounds/salts

Turnover number (min^{-1})

Specific activity (U/mg)

More [1]

K_m-value (mM)

0.002 (NADPH) [1]

pH-optimum

6.9–7.4 [1]

pH-range

6.2–7.8 (6.2: about 50% of activity maximum, 7.8: about 20% of activity maximum) [1]

Temperature optimum (°C)

30 (enzyme assay) [1]

Temperature range (°C)

---

3 ENZYME STRUCTURE

Molecular weight

Subunits

Glycoprotein/Lipoprotein

–

---

4 ISOLATION/PREPARATION

Source organism

Pea [1]; Oat [1]; Broad bean [1]; Wheat [1]; Barley [1]; Maize [1]

Source tissue

Seeds [1]; Roots [1]; Shoots [1]

Localisation in source

Purification

Pea [1]

Crystallization

–

Cloned

–

Renaturated

–

5 STABILITY

pH

5.5 (5 min at 5°C or at 22°C, irreversible inactivation) [1]

Temperature (°C)

60 (10 min, stable) [1]; 70 (10 min, complete inactivation) [1]

Oxidation

Organic solvent

General stability information

Storage

–15°C, 8 weeks, stable [1]

6 CROSSREFERENCES TO STRUCTURE DATABANKS

PIR/MIPS code

Brookhaven code

7 LITERATURE REFERENCES

[1] Hatch, M.D., Turner, J.F.: Biochem. J.,76,556–562 (1960)

Enzyme Handbook © Springer-Verlag Berlin Heidelberg 1994
Duplication, reproduction and storage in data banks are only allowed with the prior permission of the publishers

1 NOMENCLATURE

EC number
1.6.4.5

Systematic name
NADPH:oxidized-thioredoxin oxidoreductase

Recommended name
Thioredoxin reductase (NADPH)

Synonymes
Reductase, thioredoxin
NADP-thioredoxin reductase
NADPH-thioredoxin reductase

CAS Reg. No.
9074-14-0

2 REACTION AND SPECIFICITY

Catalysed reaction
NADPH + oxidized thioredoxin →
→ $NADP^+$ + reduced thioredoxin (mechanism [11])

Reaction type
Redox reaction

Natural substrates
NADPH + oxidized thioredoxin (metabolic function of thioredoxin reductase-thioredoxin system: supplies reducing equivalents for a wide variety of acceptors (e.g.: ribonucleotide reductase, nonspecific protein disulfide reductase, methionine sulfoxide reductase, D-proline reductase) [11], reduction of free radicals at the surface of the epidermis, enzyme may play a role in physiology of pancreatic beta-cells [19]) [11, 19]

Substrate spectrum
1 NADPH + oxidized thioredoxin (protein of 12000 MW, containing a single disulfide [11], mutant thioredoxins [2], thioredoxin C-2 [3]) [2, 3, 11]
2 Disulfide isomerase + NADPH [1]
3 5,5'-Dithiobis(2-nitrobenzoic acid) + NADPH [18, 20]
4 Alloxan + NADPH (rat [18], bovine [21], not: E. coli [21]) [18, 21]
5 Dithiothreitol + NADPH [24]

Enzyme Handbook © Springer-Verlag Berlin Heidelberg 1994
Duplication, reproduction and storage in data banks are only allowed with the prior permission of the publishers

6 More (reduction of thioredoxin by NADPH is virtually complete, equilibrium constant is 48 at pH 7 [11], specific for NADPH (B side of nicotinamide ring) [1, 27], highly specific for NADPH and thioredoxin [13], catalyzes reaction: E. coli thioredoxin (reduced) + T4-thioredoxin (oxidized) --> E. coli thioredoxin (oxidized) + T4 thioredoxin (reduced) [25]) [1, 11, 13, 25, 27]

Product spectrum

1 $NADP^+$ + reduced thioredoxin
2 $NADP^+$ + disulfide isomerase (with reduced disulfides)
3 $NADP^+$ + ?
4 $NADP^+$ + ?
5 $NADP^+$ + ?
6 ?

Inhibitor(s)

Ca^{2+} [3, 5, 8]; 13-cis-Retinoic acid [6]; $NADP^+$ (product inhibition) [11]; Arsenite [12, 17, 18]; N-Ethylmaleimide [15]; 5,5'-Dithiobis(2-nitrobenzoic acid) (above 0.1 mM) [20]; p-Chloromercuribenzoate + NADPH [26]

Cofactor(s)/prostethic group(s)

NADPH (strictly dependent on NADPH, inactive with NADH [13]) [1–28]; FAD (flavoprotein [8, 9, 11–13, 17, 18, 20, 26, 27], 2 FAD per dimer [9, 13, 26]) [8, 9, 11–13, 17, 18, 20, 26, 28]

Metal compounds/salts

Turnover number (min^{-1})

30000 (protein disulfide-isomerase [1], rat reductase, rat thioredoxin, bovine thioredoxin) [18]; 1365 (wild type thioredoxin, pH 8.0) [2]; 620 (mutant thioredoxin-R and thioredoxin-LAC, pH 8.0) [2]; 10000 (alloxan, bovine enzyme) [21]

Specific activity (U/mg)

More [4, 9, 13, 16]; 35.0 [18]

K_m-value (mM)

0.035 (protein disulfide-isomerase) [1]; 0.002 (wild type thioredoxin, pH 8.0) [2]; 0.125 (mutant thioredoxins: thioredoxin R, thioredoxin K 36 R, pH 8.0) [2]; 0.000470 (thioredoxin C-2) [4]; 0.000760 (E. coli thioredoxin) [4]; 0.003 (NADPH) [9]; 0.008 (NADPH) [11]; 0.0017 (thioredoxin) [11]; 0.00237 (E. coli reductase, E. coli thioredoxin) [13]; 0.00958 (E. coli reductase, Rhodobacter sphaeroides thioredoxin) [13]; 0.00327 (Rhodobacter sphaeroides reductase, Rhodobacter sphaeroides thioredoxin) [13]; 0.0146 (Rhodobacter sphaeroides, E. coli thioredoxin) [13]; 0.006 (NADPH) [18]; 0.0025 (rat reductase, rat thioredoxin, bovine thioredoxin) [18]; 0.330 (alloxan, bovine enzyme) [21]; 0.0032 (NADPH) [22]

pH-optimum

7.7 [11]

pH-range

Temperature optimum (°C)

25 (assay at) [22, 26]

Temperature range (°C)

---

3 ENZYME STRUCTURE

Molecular weight

116000 (rat, gel filtration) [18]
105000 (human, HeLa cells, gel filtration) [20]
75000 (Saccharomyces cerevisiae, gel filtration) [26]
73000–75000 (E. coli, amino acid analysis based on 2 mol FAD per molecule of enzyme) [11]
68000 (Spinacia oleracea, gel filtration [9], Rhodobacter sphaeroides, native polyacrylamide gel electrophoresis [13]) [9, 13]
65000 (wheat, gel filtration) [22]
62000 (Daucus carota, gel filtration) [12]
60000 (Chromatium vinosum, gel filtration) [17]
More [24]

Subunits

Dimer (2 × 33000, Spinacia oleracea, SDS-PAGE [9], 2 × 34000, Daucus carota, SDS-PAGE [12], 2 × 34000, Rhodobacter sphaeroides, SDS-PAGE [13], 2 × 33500, Chromatium vinosum, SDS-PAGE [17], 2 × 58000, rat, SDS-PAGE [18], 2 × 58000, human, HeLa cells, SDS-PAGE [20], 2 × 38000, Saccharomyces cerevisiae, SDS-PAGE [26]) [9, 12, 13, 17, 18, 20, 26]
? (x × 58000, human melanoma metastases, SDS-PAGE) [8]

Glycoprotein/Lipoprotein

–

---

4 ISOLATION/PREPARATION

Source organism

Bovine [1, 19, 21]; E. coli (enzyme with active site mutation [7]) [1, 5, 7, 10, 11, 15, 19, 23, 25, 28]; Human (HeLa cells [20]) [3, 5, 6, 8, 14, 20]; Corynebacterium nephridii [4]; Spinacia oleracea (spinach) [9]; Daucus carota [12]; Rhodobacter sphaeroides Y [13]; Guinea pig [14]; Saccharomyces cerevisiae (involvement of thioredoxin reductase in the dimethyl sulphoxide reductase system [16]) [16, 26]; Chromatium vinosum [17]; Rat [18, 27]; Mouse [19]; Wheat [22]; Euglena gracilis [24]

Enzyme Handbook © Springer-Verlag Berlin Heidelberg 1994
Duplication, reproduction and storage in data banks are only allowed with the prior permission of the publishers

Source tissue

Thymus [1, 21]; Keratinocytes [5, 14]; Metastatic melanotic melanoma [3, 5, 6, 8]; Amelanotic melanoma [3]; Leaves [9]; Cell [16]; Liver [18, 27]; Pancreatic islets [19]; Cultured cells (HeLa cells) [20]

Localisation in source

Plasma membrane (associated) [8]; Cytoplasm [8]

Purification

E. coli [23, 28]; Corynebacterium nephridii [4]; Human (HeLa cells [20]) [8, 20]; Spinacia oleracea (spinach) [9]; Daucus carota (partial) [12]; Rhodobacter sphaeroides [13]; Saccharomyces cerevisiae [16, 26]; Chromatium vinosum [17]; Rat [18, 27]

Crystallization

–

Cloned

[10]

Renaturated

–

5 STABILITY

pH

Temperature (°C)

70 (10 min, complete loss of activity) [22]

Oxidation

Organic solvent

General stability information

Storage

–20°C, several months [27]

6 CROSSREFERENCES TO STRUCTURE DATABANKS

PIR/MIPS code

RDECT (Escherichia coli)

Brookhaven code

1SRX (ESCHERICHIA COLI B); 1TRX (ESCHERICHIA COLI); 2TRX (ESCHERICHIA COLI); 3TRX (HUMAN (HOMO SAPIENS) RECOMBINANT IN (ESCHERICHIA COLI)); 4TRX (HUMAN (HOMO SAPIENS) RECOMBINANT IN (ESCHERICHIA COLI))

7 LITERATURE REFERENCES

[1] Lundström, J., Holmgren, A.: J. Biol. Chem.,265,9114–9120 (1990)
[2] Gleason, F.K., Lim, C.-J., Gerami-Nejad, M., Fuchs, J. A.: Biochemistry,29,3701–3709 (1990)
[3] Schallreuter, K.U., Wood, J.M.: Biochim. Biophys. Acta,997,242–247 (1989)
[4] McFarlan, S.C., Hogenkamp, H.P.C., Eccleston, E.D., Howard, J.B., Fuchs, J.A.: Eur. J. Biochem.,179,389–398 (1989)
[5] Schallreuter, K.U., Pittelkow, M.R., Wood, J.M.: Biochem. Biophys. Res. Commun.,162,1311–1316 (1989)
[6] Schallreuter, K.U., Wood, J.M.: Biochem. Biophys. Res. Commun.,160,573–579 (1989)
[7] Prongay, A.J., Engelke, D.R., Williams, C.H.: J. Biol. Chem.,264,2656–2664 (1989)
[8] Schallreuter, K.U., Wood, J.M.: Biochim. Biophys. Acta,967,103–109 (1988)
[9] Florencio, F.J., Yee, B.C., Johnson, T.C., Buchanan, B.: Arch. Biochem. Biophys.,266,496–507 (1988)
[10] Russel, M., Model, P.: J. Bacteriol.,163,238–242 (1985)
[11] Williams, C.H. in "The Enzymes",3rd. Ed. (Boyer, P.D., Ed.) 13,89–173 (1976) (Review)
[12] Johnson, T.C., Cao, R.Q., Kung, J.E., Buchanan, B.B.: Planta,171,321–331 (1987)
[13] Clement-Metral, J.D., Höög, J.-O., Holmgren, A.: Eur. J. Biochem.,161,119–126 (1986)
[14] Schallreuter, K.U., Wood, J.M.: Biochem. Biophys. Res. Commun.,136,630–637 (1986)
[15] O'Donnell, M.E., Williams, C.H.: Biochemistry,24,7617–7621 (1985)
[16] Gibson, R.M., Large, P.J.: FEMS Microbiol. Lett.,26,89–94 (1985)
[17] Johnson, T.C., Crawford, N.A., Buchanan, B.B.: J. Bacteriol.,158,1061–1069 (1984)
[18] Luthman, M., Holmgren, A.: Biochemistry,21,6628–6633 (1982)
[19] Grankvist, K., Holmgren, A., Luthman, M., Täljedal, I.-B.: Biochem. Biophys. Res. Commun.,107,1412–1418 (1982)
[20] Lik-Shing Tsang, M., Weatherbee, J.A.: Proc. Natl. Acad. Sci. USA,78,7478–7482 (1981)
[21] Holmgren, A., Lyckeborg, C.: Proc. Natl. Acad. Sci. USA,77,5149–5152 (1980)
[22] Suske, G., Wagner, W., Follmann, H.: Z. Naturforsch.,34c,214–221 (1979)
[23] Pigiet, V.P., Conley, R.R.: J. Biol. Chem.,252,6367–6372 (1977)
[24] Munavalli, S., Parker, D.V., Hamilton, F.D.: Proc. Natl. Acad. Sci. USA,72,4233–4237 (1975)
[25] Berglund, O., Holmgren, A.: J. Biol. Chem.,250,2778–2782 (1975)
[26] Speranza, M.L., Ronchi, S., Minchiotti, L.: Biochim. Biophys. Acta,327,274–281 (1973)
[27] Larsson, A.: Eur. J. Biochem.,35,346–349 (1973)
[28] Moore, E.C., Reichard, P., Thelander, L.: J. Biol. Chem.,239,3445–3452 (1964)

Enzyme Handbook © Springer-Verlag Berlin Heidelberg 1994
Duplication, reproduction and storage in data banks are only allowed with the prior permission of the publishers

1 NOMENCLATURE

EC number
1.6.4.6

Systematic name
NADPH:CoA-glutathione oxidoreductase

Recommended name
CoA-glutathione reductase (NADPH)

Synonymes
Coenzyme A glutathione disulfide reductase
Coenzyme A disulfide-glutathione reductase

CAS Reg. No.
37256-33-0

2 REACTION AND SPECIFICITY

Catalysed reaction
NADPH + CoA-glutathione →
→ $NADP^+$ + CoA + glutathione

Reaction type
Redox reaction

Natural substrates
CoA-glutathione + NADPH [1–6]

Substrate spectrum
1 CoA-glutathione + NADPH (ir) [1–6]
2 More (substrate is a mixed disulfide)

Product spectrum
1 Glutathione + CoA + $NADP^+$ [1–6]
2 ?

Inhibitor(s)

Cofactor(s)/prostethic group(s)
NADPH [1–6]

Metal compounds/salts

Enzyme Handbook © Springer-Verlag Berlin Heidelberg 1994
Duplication, reproduction and storage in data banks are only allowed with the prior permission of the publishers

Turnover number (min^{-1})

Specific activity (U/mg)

K_m-value (mM)
0.20–0.23 (CoA-glutathione) [4, 6]

pH-optimum
5.75 (CoA-glutathione + NADPH) [4]; 5.5 (CoA-glutathione + NADPH) [6]

pH-range

Temperature optimum (°C)

Temperature range (°C)

---

3 ENZYME STRUCTURE

Molecular weight
108000 (Saccharomyces cerevisiae, sucrose gradient ultracentrifugation) [6]
42500 (rat, gel filtration, also active as glutathione reductase) [4]

Subunits

Glycoprotein/Lipoprotein
–

---

4 ISOLATION/PREPARATION

Source organism
Rat [1, 3–5]; Escherichia coli [2]; Saccharomyces cerevisiae (yeast) [6]

Source tissue
Heart [1]; Kidney [1]; Lung [1]; Brain [1]; Liver [1, 3–5]; Muscle [1]; Spleen [1]; Testis [1]; Intestine [1]

Localisation in source

Purification
Rat [4]; Saccharomyces cerevisiae [6]

Crystallization
–

Cloned
–

Renaturated

–

5 STABILITY

pH

Temperature (°C)

Oxidation

Organic solvent

General stability information

Storage

6 CROSSREFERENCES TO STRUCTURE DATABANKS

PIR/MIPS code

Brookhaven code

7 LITERATURE REFERENCES

[1] Acuna, R., Vargas, E., Ondarza, R.N. in "Temas Bioquim. Actual." (Pina, E., Pena, A., Chagoya De Sanchez, V., Eds.) 355–359 (1978)
[2] Loewen, P.C.: Can. J. Biochem.,55,1019–1027 (1977)
[3] Eriksson, S., Guthenberg, C., Mannervik, B.: FEBS Lett.,39,296–300 (1974)
[4] Ondarza, R.N., Escamilla, E., Gutiérrez, J., De La Chica, G.: Biochim. Biophys. Acta,341,162–171 (1974)
[5] Dyar, R.E., Wilken, D.R.: Arch. Biochem. Biophys.,153,619–626 (1972)
[6] Ondarza, R.N., Abney, R., López-Colomé, A.M.: Biochim. Biophys. Acta,191,239–248 (1969)

Enzyme Handbook © Springer-Verlag Berlin Heidelberg 1994
Duplication, reproduction and storage in data banks are only allowed with the prior permission of the publishers

1 NOMENCLATURE

EC number
1.6.4.7

Systematic name
NADH:asparagusate oxidoreductase

Recommended name
Asparagusate reductase (NADH)

Synonymes
Asparagusate dehydrogenase
Asparagusic dehydrogenase

CAS Reg. No.
56126-52-4

2 REACTION AND SPECIFICITY

Catalysed reaction
NADH + asparagusate →
→ NAD^+ + 3-mercapto-2-mercaptomethylpropanoate

Reaction type
Redox reaction

Natural substrates
Asparagusate + NADH [1–4]
Lipoate + NADH [1–4]

Substrate spectrum
1 Asparagusate + NADH (r) [1–4]
2 Lipoate + NADH (r) [1–4]

Product spectrum
1 Dihydroasparagusate + NAD^+ [1–4]
2 Dihydrolipoate + NAD^+ [1–4]

Inhibitor(s)
Cetyltrimethylammonium bromide [1]; p-Chloromercuribenzoate [2, 3]; Hg^{2+} [2, 3]; N-Ethylmaleimide [2, 3]

Cofactor(s)/prostethic group(s)
FAD [1–3]; NAD^+ [1–4]

Enzyme Handbook © Springer-Verlag Berlin Heidelberg 1994
Duplication, reproduction and storage in data banks are only allowed with the prior permission of the publishers

Metal compounds/salts

Turnover number (min^{-1})

Specific activity (U/mg)
0.215 [1–3]

K_m-value (mM)
20 (asparagusate) [1–3]; 3 (lipoate) [1–3]

pH-optimum
5.9 [1–3]

pH-range
5.0–7.0 [3]

Temperature optimum (°C)
40–45 [2, 3]

Temperature range (°C)

---

3 ENZYME STRUCTURE

Molecular weight
110000–112000 (Asparagus officinalis, gel filtration, sedimentation equilibrium) [1–3]

Subunits

Glycoprotein/Lipoprotein
–

---

4 ISOLATION/PREPARATION

Source organism
Asparagus officinalis [1–4]

Source tissue

Localisation in source
Mitochondria [1–4]

Purification
Asparagus officinalis [1–4]

Crystallization
–

Cloned

–

Renaturated

–

5 STABILITY

pH

Temperature (°C)

70 (not stable above) [3]

Oxidation

Organic solvent

General stability information

Storage

1 month, –80°C [1, 2, 4]

6 CROSSREFERENCES TO STRUCTURE DATABANKS

PIR/MIPS code

Brookhaven code

7 LITERATURE REFERENCES

[1] Yanagawa, H.: Methods Enzymol.,143,516–521 (1987)
[2] Yanagawa, H.: Methods Enzymol.,62,172–181 (1979)
[3] Yanagawa, H., Egami, F.: J. Biol. Chem.,251,3637–3644 (1976)
[4] Yanagawa, H., Egami, F.: Biochim. Biophys. Acta,384,342–352 (1975)

Enzyme Handbook © Springer-Verlag Berlin Heidelberg 1994
Duplication, reproduction and storage in data banks are only allowed with the prior permission of the publishers

1 NOMENCLATURE

EC number
1.6.4.8

Systematic name
NADPH:trypanothione oxidoreductase

Recommended name
Trypanothione reductase

Synonymes

CAS Reg. No.
102210-35-5

2 REACTION AND SPECIFICITY

Catalysed reaction
NADPH + trypanothione →
→ $NADP^+$ + reduced trypanothione

Reaction type
Redox reaction

Natural substrates

Substrate spectrum
1 Trypanothione + NADPH (r, trypanothione is the oxidized form of N^1,N^8-bis(glutathionyl)spermidine)
2 Glutathionylspermidine disulfide + NADPH (r) [5, 8, 9]

Product spectrum
1 Reduced trypanothione + $NADP^+$ [1–10]
2 2-Glutathionylspermidine + $NADP^+$ [5, 8, 9]

Inhibitor(s)
Melarsen-trypanothione adduct [3]; Nifurtimox [4–6]; 1,3-Bis(2-chloroethyl)-1-nitrosourea [5]; 2-(5-Nitro-2-furanylmethylidene)-N,N'-(1,4-piperazinediylbis(1,3-propanediyl))bishydrazinecarboximidamide tetrahydrobromide [5]; 2,3-Bis(3-(2-amidinohydrazono)-butyl)-1,4-naphthoquinone dihydrochloride [5]; Naphthoquinone derivatives [6]; Nitrofuran derivatives [6]

Enzyme Handbook © Springer-Verlag Berlin Heidelberg 1994
Duplication, reproduction and storage in data banks are only allowed with the prior permission of the publishers

Cofactor(s)/prostethic group(s)

FAD [2, 4, 9, 10]; $NADP^+$; More (activity is dependent on a redox-active cysteine at the active centre)

Metal compounds/salts

Turnover number (min^{-1})

9600 (trypanothione) [2]; 14200 (trypanothione) [9]; 31000 (trypanothione) [10]

Specific activity (U/mg)

28.9 [2]; 284 [9]; 295 [10]

K_m-value (mM)

0.018–0.058 (trypanothione) [2, 3, 5, 8–10]; 0.022–0.275 (glutathionylspermidine disulfide) [5, 8, 9]; 0.005–0.007 (NADPH) [2, 5, 9, 10]

pH-optimum

7.5 (glutathionylspermidine disulfide + NADPH) [5]; 7.5–8.0 (trypanothione + NADPH) [10]

pH-range

Temperature optimum (°C)

Temperature range (°C)

---

3 ENZYME STRUCTURE

Molecular weight

110000 (E. coli with Trypanosoma congolense gene, gel filtration [2], Crithidia fasciculata, SDS-PAGE [10]) [2, 10]
105000 (Trypanosoma cruzi, gel filtration) [9]

Subunits

Dimer (2 × 53443, Trypanosoma congolense, DNA sequence [7], 2 × 50000–53800, Trypanosoma congolense [2], Trypanosoma cruzi [4, 9], Crithidia fasciculata [10], SDS-PAGE) [2, 4, 7, 9, 10]

Glycoprotein/Lipoprotein

–

4 ISOLATION/PREPARATION

Source organism

Crithidia fasciculata (insect-parasitic trypanosomatid) [1–10]; Leishmania donovani [1]; Leishmania aethiopica [1]; Trypanosoma brucei [1, 3]; Trypanosoma cruzi [1, 2, 4–7, 9]; Trypanosoma congolense [1, 2, 4, 5, 7]

Source tissue

Localisation in source

Purification

Trypanosoma congolense (gene cloned and expressed in E. coli) [2]; Trypanosoma cruzi [9]; Crithidia fasciculata [10]

Crystallization

[4, 5, 9]

Cloned

[1, 2, 7]

Renaturated

–

5 STABILITY

pH

Temperature (°C)

Oxidation

Organic solvent

General stability information

Storage

6 CROSSREFERENCES TO STRUCTURE DATABANKS

PIR/MIPS code

A27727 (Trypanosoma congolense)

Brookhaven code

Enzyme Handbook © Springer-Verlag Berlin Heidelberg 1994
Duplication, reproduction and storage in data banks are only allowed with the prior permission of the publishers

7 LITERATURE REFERENCES

[1] Taylor, M.C., Chapman, C.J., Kelly, J.M., Fairlamb, A.H., Miles, M.A.: Biochem. Soc. Trans.,17,579–580 (1989)
[2] Sullivan, F.X., Shames, S.L., Walsh, C.T.: Biochemistry,28,4986–4992 (1989)
[3] Fairlamb, A.H., Henderson, G.B., Cerami, A.: Proc. Natl. Acad. Sci. USA,86,2607–2611 (1989)
[4] Krauth-Siegel, R.L., Jockers-Scheruebl, M.C., Becker, K., Schirmer, R.H.: Biochem. Soc. Trans.,17,315–317 (1989)
[5] Jockers-Scheruebl, M.C., Schirmer, R.H., Krauth-Siegel, R.L.: Eur. J. Biochem.,180,267–272 (1989)
[6] Henderson, G.B., Ulrich, P., Fairlamb, A.H., Rosenberg, I., Pereira, M., Sela, M., Cerami, A.: Proc. Natl. Acad. Sci. USA,85,5374–5378 (1988)
[7] Shames, S.L., Kimmel, B.E., Peoples, O.P., Agabian, N., Walsh, C.T.: Biochemistry,27,5014–5019 (1988)
[8] Henderson, G.B., Fairlamb, A.H., Ulrich, P., Cerami, A.: Biochemistry,26,3023–3027 (1987)
[9] Krauth-Siegel, R.L., Enders, B., Henderson, G.B., Fairlamb, A.H., Schirmer, R.H.: Eur. J. Biochem.,164,123–128 (1987)
[10] Shames, S.L., Fairlamb, A.H., Cerami, A., Walsh, C.T.: Biochemistry,25,3519–3526 (1986)

1 NOMENCLATURE

EC number
1.6.4.9

Systematic name
NADPH:bis-gamma-glutamylcysteine oxidoreductase

Recommended name
Bis-gamma-glutamylcystine reductase (NADPH)

Synonymes
Reductase, bis-gamma-glutamylcystine
More (not identical to EC 1.6.4.2 or EC 1.6.4.10)

CAS Reg. No.
117056-54-9

2 REACTION AND SPECIFICITY

Catalysed reaction
NADPH + bis-gamma-glutamylcystine →
→ $NADP^+$ + 2 gamma-glutamylcysteine

Reaction type
Redox reaction

Natural substrates
NADPH + bis-gamma-glutamylcystine (highly specific) [2]

Substrate spectrum
1 NADPH + bis-gamma-glutamylcystine (highly specific) [1, 2]
2 NADPH + 5,5'-dithiol-bisnitrobenzoate (in vitro, not found in Halobacterium halobium, reduction at 23% the rate of bis-gamma-glutamylcystine reduction) [2]
3 More (no activity with reduced glutathione or other biological disulfides [1], no substrates are: thiols found in Halobacterium halobium: tetrathionate, cystine, CoA, thiols not found in Halobacterium halobium: bisacetylcystine, cystinylbisglycine, homocystine [2]) [1, 2]

Product spectrum
1 $NADP^+$ + gamma-glutamylcysteine [1]
2 ?
3 ?

Enzyme Handbook © Springer-Verlag Berlin Heidelberg 1994
Duplication, reproduction and storage in data banks are only allowed with the prior permission of the publishers

Inhibitor(s)
Hg^{2+} (0.0025 mM, strong) [2]; Cu^{2+} (0.1 mM, strong) [2]; Zn^{2+} (0.1 mM) [2]; AsO_2- (25 mM) [2]; Tris-buffer [2]; More (no inhibition: AsO_2^- (5 mM), Ni^{2+} (0.1 mM), Mn^{2+} (1.0 mM), Mg^{2+} (100 mM)) [2]

Cofactor(s)/prostethic group(s)
FAD (flavoprotein, 1 mol FAD/mol subunit) [2]; NADPH [1, 2]; NADH (7% as effective as NADPH) [1]; More (no cofactors are lipoic acid disulfide with either NADPH or NADH and dihydro-lipoamide with either $NADP^+$ or NAD^+ [1]) [1]

Metal compounds/salts
Sodium phosphate (activation) [2]

Turnover number (min^{-1})
1700 (FAD)

Specific activity (U/mg)
25.0 [1]; 28 [2]

K_m-value (mM)
0.29 (NADH) [2]; 0.81 (bis-gamma-glutamylcystine) [2]

pH-optimum
7.5 [2]

pH-range
6.3–8.2 (about half-maximal activity at pH 6.3 and 8.2) [2]

Temperature optimum (°C)
30 (assay at)

Temperature range (°C)

---

3 ENZYME STRUCTURE

Molecular weight
122000 (Halobacterium halobium, gel filtration) [1]

Subunits
Dimer (2 × 61000, Halobacterium halobium, SDS- or urea-PAGE) [1]

Glycoprotein/Lipoprotein
–

4 ISOLATION/PREPARATION

Source organism
Halobacterium halobium [1, 2]

Source tissue
Cell [1, 2]

Localisation in source

Purification
Halobacterium halobium (metal-chelate affinity chromatography with Cu^{2+} or Ni^{2+}) [1]

Crystallization
–

Cloned
–

Renaturated
–

---

5 STABILITY

pH

Temperature (°C)
60 (stable in high (4.3 M) and intermediate (2.2 M) ionic strength buffers, inactivation in low (0.5 M) ionic strength buffers) [1]; 80 (inactivation) [1]

Oxidation

Organic solvent

General stability information
High ionic strength, above 3.5 M, increases thermal stability [1]; High ionic strength, above 2 M, stabilizes during purification and storage [1]; Low ionic strength, below 1.0 M, rapid denaturation, at 0.3 M, $t_{1/2}$: 1 min, at 0.8 M, $t_{1/2}$: 5 min [2]

Storage
20°C, stable for over a year in high ionic strength buffer, above 3.5 M [1]

Enzyme Handbook © Springer-Verlag Berlin Heidelberg 1994
Duplication, reproduction and storage in data banks are only allowed with the prior permission of the publishers

6 CROSSREFERENCES TO STRUCTURE DATABANKS

PIR/MIPS code

Brookhaven code

7 LITERATURE REFERENCES

[1] Sundquist, A.R., Fahey, R.C.: J. Bacteriol.,170,3459–3467 (1988)
[2] Sundquist, A.R., Fahey, R.C.: J. Biol. Chem.,264,719–725 (1989)

1 NOMENCLATURE

EC number
1.6.4.10

Systematic name
NADH:CoA-disulfide oxidoreductase

Recommended name
CoA-disulfide reductase (NADH)

Synonymes
More (not identical to EC 1.6.4.1, EC 1.6.4.2, EC 1.6.4.9)

CAS Reg. No.

---

2 REACTION AND SPECIFICITY

Catalysed reaction
NADH + CoA-disulfide →
→ NAD^+ + 2 CoA

Reaction type
Redox reaction

Natural substrates
NADH + CoA-disulfide [1]

Substrate spectrum
1 NADH + CoA-disulfide [1]

Product spectrum
1 NAD^+ + CoA [1]

Inhibitor(s)

Cofactor(s)/prostethic group(s)
NADH [1]

Metal compounds/salts

Turnover number (min^{-1})

Specific activity (U/mg)
More [1]

Enzyme Handbook © Springer-Verlag Berlin Heidelberg 1994
Duplication, reproduction and storage in data banks are only allowed with the prior permission of the publishers

K_m-value (mM)

pH-optimum
7.4 (assay at) [1]

pH-range

Temperature optimum (°C)
37 (assay at) [1]

Temperature range (°C)

3 ENZYME STRUCTURE

Molecular weight

Subunits

Glycoprotein/Lipoprotein
–

4 ISOLATION/PREPARATION

Source organism
Bacillus megaterium QM [1]

Source tissue
Cell (growing and sporulating) [1]; Spore (dormant, highest activity) [1]

Localisation in source

Purification

Crystallization
–

Cloned
–

Renaturated
–

5 STABILITY

pH

Temperature (°C)

Oxidation

Organic solvent

General stability information

Storage

6 CROSSREFERENCES TO STRUCTURE DATABANKS

PIR/MIPS code

Brookhaven code

7 LITERATURE REFERENCES

[1] Setlow, B., Setlow, P.: J. Bacteriol.,132,444–452 (1979)

Enzyme Handbook © Springer-Verlag Berlin Heidelberg 1994
Duplication, reproduction and storage in data banks are only allowed with the prior permission of the publishers

1 NOMENCLATURE

EC number
1.6.5.3

Systematic name
NADH:ubiquinone oxidoreductase

Recommended name
NADH dehydrogenase (ubiquinone)

Synonymes
Ubiquinone reductase
Type I dehydrogenase
Complex I dehydrogenase
Reductase, ubiquinone
Coenzyme Q reductase
Complex I (electron transport chain)
Complex I (mitochondrial electron transport)
Complex I (NADH:Q1 oxidoreductase)
Dihydronicotinamide adenine dinucleotide-coenzyme Q reductase
DPNH-Coenzyme Q reductase
DPNH-ubiquinone reductase
Mitochondrial electron transport complex 1
Mitochondrial electron transport complex I
NADH coenzyme Q1 reductase
NADH-coenzyme Q oxidoreductase
NADH-coenzyme Q reductase
NADH-CoQ oxidoreductase
NADH-CoQ reductase
NADH-ubiquinone reductase
NADH-ubiquinone oxidoreductase
NADH-ubiquinone-1 reductase
Reduced nicotinamide adenine dinucleotide-coenzyme Q reductase
NADH:ubiquinone oxidoreductase complex [1]
NADH-Q6 oxidoreductase [4]
Electron transfer complex I [13]
More (the complex, present in mitochondria, can be degraded to form EC 1.6.99.3)

CAS Reg. No.
9028-04-0

Enzyme Handbook © Springer-Verlag Berlin Heidelberg 1994
Duplication, reproduction and storage in data banks are only allowed with the prior permission of the publishers

2 REACTION AND SPECIFICITY

Catalysed reaction

NADH + ubiquinone →
→ NAD^+ + ubiquinol (mechanism [16])

Reaction type

Redox reaction

Natural substrates

NADH + ubiquinone (the enzyme complex is a segment of the respiratory chain responsible for electron transfer from NADH to ubiquinone-1 and ATP-energized transhydrogenation from NADH to $NADP^+$) [2]

Substrate spectrum

1 NADH + ubiquinone (specific for NADH (above pH 6.8 [12]) [2, 4, 12], specific for ubiquinone isoprenologs [2, 7, 8], electron acceptors: ubiquinone-1 [2, 9, 12, 16], ubiquinone-2 [4], ubiquinone-6 [4, 16], ubiquinone-10 [4, 9], duroquinone [12], ferricyanide [2, 4, 7–9, 16], 2,6-dichloroindophenol [4, 7, 12], 2-methylnaphthoquinone [7], methylene blue [7], cytochrome c [7, 9], menadione [9]) [1–16]

2 More (overview: dehydrogenase and transhydrogenase activity [6], transhydrogenation from NADH and NADPH to NAD^+ or 3-acetylpyridine dinucleotide [7], ferricyanide reductase reaction: ping pong bi bi mechanism [8]) [6–8]

Product spectrum

1 NAD^+ + ubiquinol

2 ?

Inhibitor(s)

Piericidin A (not [4], inhibition of complex I, no effect on soluble NADH dehydrogenase [7], inhibition of reduction of ubiquinones, no inhibition of ferricyanide reduction [2, 8]) [2, 6–8]; Rotenoids (inhibition of reduction of ubiquinones, no inhibition of ferricyanide reduction) [2]; Mercurials (inhibition of reduction of ubiquinones, no inhibition of ferricyanide reduction [2, 8], inhibition of complex I, no effect on soluble NADH dehydrogenase [7]) [2, 7, 8]; Amytal (inhibition of reduction of ubiquinones, no inhibition of ferricyanide reduction) [2, 16]; Seconal (inhibition of reduction of ubiquinones, no inhibition of ferricyanide reduction) [2]; Demerol (inhibition of reduction of ubiquinones, no inhibition of ferricyanide reduction [2]) [2, 6]; Flavone (NADH-ubiquinone reductase not inhibited, NADH-duroquinone reductase inhibited [12]) [4, 12]; Rhein [6]; Barbiturates (inhibition of reduction of ubiquinones, no inhibition of ferricyanide reduction [8]) [6, 8];

p-Chloromercuribenzoate [7]; p-Chloromercuriphenylsulfonate [7]; Mersalyl [7]; NADH (inhibition of quinone reductase activity) [7]; Rotenone (not [4, 9, 12, 15], NADH-ubiquinone reductase not inhibited, NADH-duroquinone reductase inhibited [12], inhibition of reduction of ubiquinones, no inhibition of ferricyanide reduction [8, 16]) [6–8, 12, 16]

Cofactor(s)/prostethic group(s)

FMN (1.4–1.5 nmol per mg protein [2, 6], 1 mol FMN per 650000–700000 MW protein [16]) [2, 6, 16]; Ubiquinone-10 (4.2–4.5 nmol per mg protein [2, 6], 3 mol per 650000–700000 MW protein [16]) [2, 6, 16]; FAD (non-covalently linked) [4]; Guanidine (activates) [7]; Arginine (activates) [7]; Arginine methyl ester (activates) [7]; Alkylguanidines (activate) [7]

Metal compounds/salts

Iron (iron-sulfur clusters [3, 5, 8], no iron-sulfur clusters [4], 16–18 gatom of iron per 650000–700000 MW protein [16], 23–26 ng of nonheme iron per mg protein [2, 6], characterization of iron-sulfur clusters by electron paramagnetic resonance spectroscopy [3, 5] and magnetic circular dichroism [10], 15 binuclear and 3 tetranuclear iron-sulfur (FeS) clusters, 1 binuclear [2Fe-2S] and 1 tetranuclear [4Fe-4S]cluster each in FP and HP and 3 binuclear and 1 tetranuclear in IP [8], contains 4 major iron-sulfur centers [16]) [2–6, 8, 10, 16]

Turnover number (min^{-1})

150000 (ubiquinone-2) [4]; 30000 (ubiquinone-6) [4]

Specific activity (U/mg)

43 (NADH + cytochrome c) [16]; 61.9 (ubiquinone-6) [4]; 100 (NADH + 2,6-dichlorophenol indophenol) [16]; 150–160 (NADH + ubiquinone-1) [16]; 160–170 (NADH + menadione) [16]; 215 (NADH + ferricyanide) [16]; 1671 (ubiquinone-2) [4]

K_m-value (mM)

0.007 (NADH (+ ferricyanide)) [16]; 0.012 (cytochrome c (+ NADH)) [16]; 0.014 (NADH [8], NADH (+ cytochrome c) [16]) [8, 16]; 0.019 (NADH (+ ubiquinone)) [9]; 0.031 (NADH) [4]; 0.055 (NADH(+ ferricyanide)) [9]; 0.040 (ubiquinone-1) [8]; 0.143 (ubiquinone-1) [9]; 0.6 (cytochrome c) [7]; 4.0 (ferricyanide (+ NADH)) [16]; More [12]

pH-optimum

More (dichlorophenol indophenol reduction, pH-independent) [12]; 4.5–9.0 (broad, ubiquinone-6) [4]; 5.0 (NADPH + menadione) [7]; 5.5 (NADPH + ferricyanide, NADPH + 3-acetylpyridine adenine dinucleotide) [7]; 6.2 (ubiquinone-2) [4]; 7.1 (NADH + ubiquinone-1) [12]; 7.5 (NADH + ferricyanide) [7]; 8.0 (NADH + menadione, ubiquinone-1, cytochrome c, dichlorophenol indophenol or 3-acetylpyridine adenine dinucleotide) [7]

Enzyme Handbook © Springer-Verlag Berlin Heidelberg 1994
Duplication, reproduction and storage in data banks are only allowed with the prior permission of the publishers

pH-range
4.5–8.0 (50% of activity maximum at pH 4.5 and 8.0, ubiquinone-2) [4]

Temperature optimum (°C)
30 (assay at) [2]; 38 (assay at) [2, 16]

Temperature range (°C)

---

3 ENZYME STRUCTURE

Molecular weight
53000 (Saccharomyces cerevisiae, SDS-PAGE) [4]
400000 (Beta vulgaris, gel filtration) [9]
610000 (Neurospora crassa, sedimentation equilibrium data) [13]

Subunits
Monomer (1 × 53000, Saccharomyces cerevisiae, SDS-PAGE) [4]
More (mammalian NADH:ubiquinone oxidoreductase complex contains 25 polypeptides and 8 iron-sulfur (FeS) clusters of the mitochondrial respiratory chain [1], structurally composed of 3 distinct fragments: 1. FP water-soluble Fe-S flavoprotein composed of 3 polypeptides (MW: 51000, 24000, 9000 [8]) containing FMN and 2 FeS clusters, NADH dehydrogenase activity in presence of quinones or ferric complexes as electron acceptors, 2. IP: water-soluble Fe-S-protein, composed of 5–6 polypeptides (MW: 75000, 49000, 30000, 18000, 15000, 13000 [8]), contains 4 FeS clusters, 3. HP: water-insoluble fraction containing phospholipids and hydrophobic polypeptides, contains about 16 polypeptides and 2 FeS clusters, estimations of MW of polypeptides of complex I [6], contains 25 unlike polypeptides [8], Beta vulgaris: contains 14 major polypeptides [9], Phaseolus aureus: contains at least 5 major polypeptides of MW 76000, 46000, 39000, 33000 and 27000 [11], Arum maculatum: numerous polypeptides (MW: 39200, 33000) [12], Neurospora crassa: 25 different subunits [13], at least 22 subunits (MW between 70000 and 11000) [14], bovine complex I composed of at least 10 polypeptides ranging in MW from 10000 to 70000 [16]) [1, 6, 8, 9, 11–14, 16]

Glycoprotein/Lipoprotein
Lipoprotein (contains 0.22 mg of lipids per mg of protein [2, 6, 16], contains phospholipid [8]) [2, 6, 8, 16]

---

4 ISOLATION/PREPARATION

Source organism
E. coli [3]; Saccharomyces cerevisiae [4, 15]; Paracoccus denitrificans [5, 8]; Bovine [8, 10, 16]; Beta vulgaris [9]; Phaseolus aureus [11]; Arum maculatum [12]; Neurospora crassa [13, 14]

Source tissue

Heart [8, 10, 16]

Localisation in source

Mitochondria [4, 8–16]; Membrane (inner membrane of mitochondria [4, 9, 16]) [4, 9, 15, 16]

Purification

Saccharomyces cerevisiae [4]; Phaseolus aureus [11]; Bovine [10]; More (methodology for the stepwise fragmentation of Complex I and isolation of subunits containing FeS clusters) [1]

Crystallization

–

Cloned

–

Renaturated

–

---

5 STABILITY

pH

Temperature (°C)

Oxidation

Organic solvent

General stability information

Storage

---

6 CROSSREFERENCES TO STRUCTURE DATABANKS

PIR/MIPS code

FEYBQI (chain ndhI Synechocystis sp. PCC 6803); DNHUN1 (chain 1 Human mitochondrion SGC1); QXBO1M (chain 1 Bovine mitochondrion SGC1); QXMS1M (chain 1 Mouse mitochondrion SGC1); QQRT1M (chain 1 Rat mitochondrion SGC1 fragment); QXXL1M (chain 1 African clawed frog mitochondrion SGC1); DNWTU1 (chain 1 Wheat mitochondrion); DNOBU1 (chain 1 Bertero's evening primrose mitochondrion); DENTN1 (chain 1 Common tobacco chloroplast); DELVN1 (chain 1 Liverwort Marchantia polymorpha chloroplast); DERZN1 (chain 1 Rice chloroplast); QXYB1 (chain 1 Synechocystis sp. PCC 6803); DNHUN2 (chain 2 Human mitochondrion SGC1); QXBO2M (chain 2 Bovine mitochondrion SGC1);

Enzyme Handbook © Springer-Verlag Berlin Heidelberg 1994
Duplication, reproduction and storage in data banks are only allowed with the prior permission of the publishers

QXMS2M (chain 2 Mouse mitochondrion SGC1); QXXL2M (chain 2 African clawed frog mitochondrion SGC1); QXFF2Y (chain 2 Fruit fly Drosophila yakuba mitochondrion SGC4); QXFF2M (chain 2 Fruit fly Drosophila melanogaster mitochondrion SGC4 fragment); DNETU2 (chain 2 Sugar beet mitochondrion); DERZN2 (chain 2 Rice chloroplast); DENTN2 (chain 2 Common tobacco chloroplast); DELVN2 (chain 2 Liverwort Marchantia polymorpha chloroplast); DNHUN3 (chain 3 Human mitochondrion SGC1); QXBO3M (chain 3 Bovine mitochondrion SGC1); QXMS3M (chain 3 Mouse mitochondrion SGC1); QXXL3M (chain 3 African clawed frog mitochondrion SGC1); DNWTU3 (chain 3 Wheat mitochondrion); DNZMU3 (chain 3 Maize mitochondrion); DNOBU3 (chain 3 Bertero's evening primrose mitochondrion); DENTN3 (chain 3 Common tobacco chloroplast); DERZN3 (chain 3 Rice chloroplast); DELVN3 (chain 3 Liverwort Marchantia polymorpha chloroplast); DNHUNL (chain 4L Human mitochondrion SGC1); QXBO4L (chain 4L Bovine mitochondrion SGC1); QXMS4L (chain 4L Mouse mitochondrion SGC1); QXXL4L (chain 4L African clawed frog mitochondrion SGC1); DENTNL (chain 4L Common tobacco chloroplast); DERZNL (chain 4L Rice chloroplast); QXZM4L (chain 4L Maize chloroplast); DELVNL (chain 4L Liverwort Marchantia polymorpha chloroplast); QXYB4L (chain 4L Synechocystis sp. PCC 6803); DNHUN4 (chain 4 Human mitochondrion SGC1); QXGI4M (chain 4 Common gibbon mitochondrion SGC1 fragment); QXBO4M (chain 4 Bovine mitochondrion SGC1); QXMS4M (chain 4 Mouse mitochondrion SGC1); QXXL4M (chain 4 African clawed frog mitochondrion SGC1); QXASM4 (chain 4 Aspergillus amstelodami mitochondrion SGC3); QXAS4M (chain 4 Emericella nidulans mitochondrion SGC3 fragment); DENTN4 (chain 4 Common tobacco chloroplast); DELVN4 (chain 4 Liverwort Marchantia polymorpha chloroplast); DERZN4 (chain 4 Rice chloroplast); QXZM4 (chain 4 Maize chloroplast); DNHUN5 (chain 5 Human mitochondrion SGC1); QXBO5M (chain 5 Bovine mitochondrion SGC1); QXMS5M (chain 5 Mouse mitochondrion SGC1); QXXL5M (chain 5 African clawed frog mitochondrion SGC1); DNMUU5 (chain 5 Mouse-ear cress mitochondrion); DNOBU5 (chain 5 Bertero's evening primrose mitochondrion); DENTN5 (chain 5 Common tobacco chloroplast); DERZN5 (chain 5 Rice chloroplast); DELVN5 (chain 5 Liverwort Marchantia polymorpha chloroplast); DEHUN6 (chain 6 Human mitochondrion SGC1); DEBON6 (chain 6 Bovine mitochondrion SGC1); DEMSN6 (chain 6 Mouse mitochondrion SGC1); DERTN6 (chain 6 Rat mitochondrion SGC1 fragment); DEXLN6 (chain 6 African clawed frog mitochondrion SGC1); DELVN6 (chain 6 Liverwort Marchantia polymorpha chloroplast); DERZN6 (chain 6 Rice chloroplast); QXYB6 (chain 6 Synechocystis sp. PCC 6803); DERZ49 (49K protein Rice chloroplast); S19921 (chain 2 Synechococcus sp. PCC 7942); S04435 (chain 3 Synechocystis sp. PCC 6803); S01650 (chain 1 Chlamydomonas rein-hardtii mitochondrion); JQ0385 (chain 4 Chlamydomonas reinhardtii mitochondrion); S01523 (chain 1 Liverwort

Marchantia polymorpha chloroplast); S01569 (chain 2 Liverwort Marchantia polymorpha chloroplast); S01599 (chain 3 Liverwort Marchantia polymorpha chloroplast); S01522 (chain 392 Liverwort Marchantia polymorpha chloroplast); S01528 (chain 4 Liverwort Marchantia polymorpha chloroplast); S01526 (chain 4L Liverwort Marchantia polymorpha chloroplast); S01512 (chain 5 Liverwort Marchantia polymorpha chloroplast); S01525 (chain 6 Liverwort Marchantia polymorpha chloroplast); S05719 (chain 2 Soybean chloroplast); S08445 (chain 5 Fava bean chloroplast); C34879 (chain 1 Fava bean mitochondrion fragment); S04586 (chain 1 Common tobacco mitochondrion fragment); JV0006 (chain 3 Garden petunia mitochondrion); JQ1374 (chain 3 Wheat mitochondrion); S06835 (chain 4 Wheat mitochondrion fragment); S04434 (chain 3 Maize chloroplast); S04583 (chain 1 Maize mitochondrion fragment); A34051 (29/21K chain precursor mitochondrial Neurospora crassa); A23431 (chain 1 Neurospora crassa mitochondrion SGC3); S02154 (chain 2 Podospora anserina mitochondrion SGC3); S02155 (chain 3 Podospora anserina mitochondrion SGC3); S02153 (chain 4 Podospora anserina mitochondrion SGC3); S09132 (chain 4L Podospora anserina mitochondrion SGC3); S09133 (chain 5 Podospora anserina mitochondrion SGC3); S06058 (chain I Podospora anserina mitochondrion SGC3); S05466 (chain 1 Leishmania major mitochondrion SGC6 fragment); D26696 (polypeptide 1 Sauroleishmania tarentolae mitochondrion SGC6); A26696 (polypeptide 5 Sauroleishmania tarentolae mitochondrion SGC6 fragment); S07752 (chain 1 Paramecium tetraurelia mitochondrion SGC6); JS0233 (chain 2 Paramecium tetraurelia mitochondrion SGC6); S07734 (chain 2 Paramecium tetraurelia mitochondrion SGC6); S07727 (chain 3 Paramecium tetraurelia mitochondrion SGC6); S07754 (chain 4 Paramecium tetraurelia mitochondrion SGC6); S07733 (chain 400 Paramecium tetraurelia mitochondrion SGC6); S07744 (chain 5 Paramecium tetraurelia mitochondrion SGC6); S01210 (chain 1 Brine shrimp mitochondrion SGC4 fragments); S01220 (chain 2 Brine shrimp mitochondrion SGC4 fragments); S01213 (chain 3 Brine shrimp mitochondrion SGC4 fragments); S01211 (chain 4 Brine shrimp mitochondrion SGC4 fragment); S01209 (chain 4L Brine shrimp mitochondrion SGC4); S01877 (chain 5 Brine shrimp mitochondrion SGC4 fragment); S01212 (chain 6 Brine shrimp mitochondrion SGC4 fragment); S01191 (chain 1 Fruit fly Drosophila melanogaster mitochondrion SGC4); S01185 (chain 3 Fruit fly Drosophila melanogaster mitochondrion SGC4); S01187 (chain 4 Fruit fly Drosophila melanogaster mitochondrion SGC4); S01188 (chain 4L Fruit fly Drosophila melanogaster mitochondrion SGC4); S01186 (chain 5 Fruit fly Drosophila melanogaster mitochondrion SGC4); S01189 (chain 6 Fruit fly Drosophila melanogaster mitochondrion SGC4); S01277 (chain 2 Fruit fly Drosophila virilis mitochondrion SGC4 fragment Contains: NADH dehydrogenase EC 1.6.99.3); D30020 (chain 1 Fruit fly Drosophila yakuba mitochondrion SGC4); A25797 (chain 2 Fruit fly

zyme Handbook © Springer-Verlag Berlin Heidelberg 1994
plication, reproduction and storage in data banks are only
owed with the prior permission of the publishers

Drosophila yakuba mitochondrion SGC4); G25797 (chain 3 Fruit fly Drosophila yakuba mitochondrion SGC4); I25797 (chain 4 Fruit fly Drosophila yakuba mitochondrion SGC4); A30020 (chain 4L Fruit fly Drosophila yakuba mitochondrion SGC4); H25797 (chain 5 Fruit fly Drosophila yakuba mitochondrion SGC4); S01499 (chain 1 Sea urchin Strongylocentrotus purpuratus mitochondrion SGC8); S01500 (chain 2 Sea urchin Strongylocentrotus purpuratus mitochondrion SGC8); S01507 (chain 3 Sea urchin Strongylocentrotus purpuratus mitochondrion SGC8); S01508 (chain 4 Sea urchin Strongylocentrotus purpuratus mitochondrion SGC8); S01502 (chain 4L Sea urchin Strongylocentrotus purpuratus mitochondrion SGC8); S01509 (chain 5 Sea urchin Strongylocentrotus purpuratus mitochondrion SGC8); S01510 (chain 6 Sea urchin Strongylocentrotus purpuratus mitochondrion SGC8); S04688 (chain 3 Rainbow trout mitochondrion SGC1 fragment); S04690 (chain 4 Rainbow trout mitochondrion SGC1 fragment); S04689 (chain 4L Rainbow trout mitochondrion SGC1 fragment); S10187 (chain 1 Chicken mitochondrion SGC1); S10188 (chain 2 Chicken mitochondrion SGC1); S10194 (chain 3 Chicken mitochondrion SGC1); S10196 (chain 4 Chicken mitochondrion SGC1); S10195 (chain 4L Chicken mitochondrion SGC1); S10197 (chain 5 Chicken mitochondrion SGC1); S10199 (chain 6 Chicken mitochondrion SGC1); A00449 (chain 5 Common gibbon mitochondrion SGC1 fragment); A00436 (chain 4 Gorilla mitochondrion SGC1 fragment); A00447 (chain 5 Gorilla mitochondrion SGC1 fragment); A00435 (chain 4 Chimpanzee mitochondrion SGC1 fragment); A00437 (chain 4 Orangutan mitochondrion SGC1 fragment); A00448 (chain 5 Orangutan mitochondrion SGC1 fragment); A30113 (24K chain precursor mitochondrial Human); PX0053 (20K polypeptide Bovine fragments); S06180 (24K chain Bovine fragment Contains: NADH dehydrogenase 24K chain EC 1.6.99.3); B30113 (24K chain precursor mitochondrial Bovine); S04104 (49K chain Bovine); JX0186 (9K polypeptide Bovine); A31910 (29K chain Bovine mitochondrion fragment); A31868 (24K chain precursor mitochondrial Rat); S04747 (chain 1 Rat mitochondrion SGC1); S04748 (chain 2 Rat mitochondrion SGC1); S04754 (chain 3 Rat mitochondrion SGC1); S04756 (chain 4 Rat mitochondrion SGC1); S04755 (chain 4L Rat mitochondrion SGC1); S04757 (chain 5 Rat mitochondrion SGC1); S04758 (chain 6 Rat mitochondrion SGC1); A40296 (25k chain NQO2 Paracoccus denitrificans); A39588 (50k chain Paracoccus denitrificans); D40296 (NADH-binding chain NQO1 Paracoccus denitrificans fragment); S14966 (Synechocystis sp. PCC 6803); S16447 (Wheat mitochondrion); S16448 (Wheat mitochondrion); S19057 (Yeast Saccharomyces cerevisiae); S05628 (Emericella nidulans mitochondrion SGC3); S04724 (Emericella nidulans mitochondrion SGC3); S14277 (Neurospora crassa); A35935 (31K chain precursor Neurospora crassa); S17663 (Neurospora crassa); S17664 (Neurospora crassa); S17647 (Neurospora crassa); S07320 (Neurospora crassa mitochondrion SGC3); S06214 (Migratory locust mitochondrion SGC4); S08469 (Starfish Asterina pectinifera mitochondrion SGC8); S08468 (Starfish Asterias forbesii mitochondrion SGC8); S14204 (Starfish Pisaster ochraceus mitochondrion); S14203 (Starfish Pisaster ochraceus mitochondrion); S14211 (Starfish Pisaster ochraceus mitochondrion); S14213 (Starfish Pisaster ochraceus mitochondrion); S14212 (Starfish Pisaster ochraceus mitochondrion); S14206 (Starfish Pisaster ochraceus mitochondrion); S14214 (Starfish Pisaster ochraceus mitochondrion); F30396 (chain 3 Pink salmon mitochondrion SGC1); D30401 (chain 4L Pink salmon mitochondrion SGC1 fragment); G30396 (chain 3 Coho salmon mitochondrion SGC1); E30401 (chain 4L Coho salmon mitochondrion SGC1 fragment); H30396 (chain 3 Sockeye salmon mitochondrion SGC1); F30401 (chain 4L Sockeye salmon mitochondrion SGC1 fragment); A30401 (chain 3 Chinook salmon mitochondrion SGC1); G30401 (chain 4L Chinook salmon mitochondrion SGC1 fragment); B30401 (chain 3 Cutthroat trout mitochondrion SGC1); H30401 (chain 4L Cutthroat trout mitochondrion SGC1 fragment); E30396 (chain 3 Rainbow trout mitochondrion SGC1); C30401 (chain 4L Rainbow trout mitochondrion SGC1 fragment); S08425 (Atlantic cod mitochondrion); S08426 (Atlantic cod mitochondrion); B41451 (chain 1 Bullfrog mitochondrion SGC1 fragment); A41451 (chain 2 Bullfrog mitochondrion SGC1); S17854 (Human); S16382 (Human); B30863 (24K precursor Human mitochondrion SGC1); S15107 (Bovine); A37941 (51k chain precursor Bovine); A39362 (51k chain precursor Bovine); A33552 (precursor Bovine); S17676 (Bovine); S17677 (Bovine); A30863 (24K precursor Bovine mitochondrion SGC1); S16157 (Rat mitochondrion); S16887 (Rat mitochondrion)

Brookhaven code

7 LITERATURE REFERENCES

[1] Ragman, C.I., Hatefi, Y.: Methods Enzymol.,126,360–369 (1986) (Review)
[2] Hatefi, Y.: Methods Enzymol.,53,11–14 (1978) (Review)
[3] Meinhardt, S.W., Matsushita, K., Kaback, H.R., Ohnishi, T.: Biochemistry,28,2153–2160 (1989)
[4] De Vries, S., Grivell, L.A.: Eur. J. Biochem.,176,377–384 (1988)
[5] Meinhardt, S.W., Kula, T., Yagi, T., Lillich, T., Ohnishi, T.: J. Biol. Chem.,262,9147–9153 (1987)
[6] Hatefi, Y., Galante, Y.M., Stiggall, D.L., Ragan, C.I.: Methods Enzymol.,56,577–602 (1979) (Review)
[7] Galante, Y.M., Hatefi, Y.: Methods Enzymol.,53,15–21 (1978) (Review)
[8] Hatefi, Y.: Annu. Rev. Biochem.,54,1015–1069 (1985) (Review)
[9] Soole, K.L., Dry, I.B., Wiskich, J.T.: Plant Physiol.,98,588–594 (1992)
[10] Kowal, A.T., Morningstar, J.E., Johnson, M.K., Ramsay, R.R., Singer, T.P.: J. Biol. Chem.,261,9239–9245 (1986)

Enzyme Handbook © Springer-Verlag Berlin Heidelberg 1994
Duplication, reproduction and storage in data banks are only allowed with the prior permission of the publishers

[11] Cottingham, I.R., Moore, A.L.: Biochem. J.,254,303–305 (1988)
[12] Cook, N.D., Cammack, R.: Biochim. Biophys. Acta,827,30–35 (1985)
[13] Leonard, K., Haiker, H., Weiss, H.: J. Mol. Biol.,194,277–286 (1987)
[14] Ise, W., Haiker, H., Weiss, H.: EMBO J.,4,2075–2080 (1985)
[15] De Vries, S., Grivell, L.A.: Eur. J. Biochem.,176,377–384 (1988)
[16] Hatefi, Y. in "The Enzymes Of Biological Membranes" (Martonosi, A. Ed.) 4,3–41 (1976) (Review)

1 NOMENCLATURE

EC number
1.6.5.4

Systematic name
NADH:monodehydroascorbate oxidoreductase

Recommended name
Monodehydroascorbate reductase (NADH)

Synonymes
NADH:semidehydroascorbic acid oxidoreductase [1]
MDHA [2]
Semidehydroascorbate reductase [9]
AFR [4, 5]
AFR-reductase [20]
Ascorbic free radical reductase [5]
Ascorbate free radical reductase [6]
SOR [13]
MDAsA reductase (NADPH) [8]
SDA reductase [9]
NADH:ascorbate radical oxidoreductase [10]
NADH-semidehydroascorbate oxidoreductase [1]
Ascorbate free-radical reductase [10]
NADH:AFR oxidoreductase [20]

CAS Reg. No.
9029-26-9

2 REACTION AND SPECIFICITY

Catalysed reaction
NADH + 2 monodehydroascorbate →
→ NAD^+ + 2 ascorbate

Reaction type
Redox reaction

Enzyme Handbook © Springer-Verlag Berlin Heidelberg 1994
Duplication, reproduction and storage in data banks are only allowed with the prior permission of the publishers

Natural substrates

NAD(P)H + monodehydroascorbate (enzyme sustains glyoxysomal NAD^+ during beta-oxidation, the glyoxylate cycle and gluconeogenesis in the endosperm during germination [4], key enzyme for maintaining the ascorbic acid system in the reduced state [5, 20], regeneration of ascorbate from monodehydroascorbate [8, 17], physiologically NADPH is the electron donor [8], involvement in oxidation of NADH by lipid peroxide in mitochondria and microsomes [16], functions to provide cytoplasmic reducing equivalents to intramitochondrial cytochrome P-450 [21]) [4, 5, 8, 16, 17, 20, 21]

Substrate spectrum

1 NADH + monodehydroascorbate
2 More (other electron acceptors: 2,6-dichlorophenolindophenol [1, 22], cytochrome b_5 [1], cytochrome c (+ ascorbate [22]) [1, 22], monodehydroisoascorbate [22], D-isoascorbic acid [19], ferricyanide [22], 1,2-naphthoquinone-4-sulfonate [22], methylene blue (+ ascorbate) [22], semidehydro-D(-)-ascorbic acid reduced more rapidly than semidehydro-L(+)-ascorbic acid [18], Fe^{3+} in complex form is reduced by NADH in presence of the enzyme [19]) [1, 18, 19, 22]

Product spectrum

1 NAD^+ + ascorbate
2 ?

Inhibitor(s)

Citrate (slight) [1]; Sodium diethyldithiocarbamate [1]; Quinaldic acid [1]; alpha,alpha'-Bipyridyl [1]; 1-(2-Thenoyl)-3,3,3-trifluoroacetone [1, 9]; Cu^{2+} [8]; Zn^{2+} [8]; Tris-buffer (2-amino-2-(hydroxymethyl)-1,3-propanediol) [10]; Tes-buffer (2-([2-hydroxy-1,1-bis(hydroxymethyl)ethyl]amino)ethanesulfonic acid) [10]; Imidazole buffer [10]; Phosphate buffer [10]; Mn^{2+} [19]; p-Chloromercuribenzoate [1, 8, 10, 19]; 2-Iodoacetamide [1]; N-Ethylmaleimide (inhibition partially reversed by thiol-containing compounds [6]) [1, 6, 8, 10]; Mersalyl (inhibition partially reversed by thiol-containing compounds [6], pyridine nucleotides protect [6]) [6, 10]; More (inhibition of membrane-bound enzyme by lectins [7], insulin inhibits enzyme in plasma membrane [11]) [7, 11]

Cofactor(s)/prostethic group(s)

NADH (activity threefold higher with NADPH than with NADH [8]) [1, 2, 4, 8, 10, 11, 13, 17, 18, 22]; NADPH (not active with NADPH [1, 19], activity threefold higher with NADPH than with NADH [8]) [8]; o-Phenanthroline (stimulation) [1]; Cytochrome b_5 (contains cytochrome b_5) [19]; Flavin (contains a flavin) [19]; FAD (contains 1 mol of FAD per mol of enzyme) [22]

Metal compounds/salts
Diphosphate (stimulation) [1]

Turnover number (min^{-1})
More [1]; 9000 (NADPH + monodehydroascorbate) [22]; 12000 (NADH + monodehydroascorbate) [22]

Specific activity (U/mg)
39.5 [1]; 0.264 [6]; 61.0 [10]; 256 [22]

K_m-value (mM)
0.0012 (semidehydroascorbate) [1]; 0.12 (NADH) [1]; 0.210 (NADPH) [8]; 0.007 (NADH) [8]; 0.077 (NADH) [10]; 0.30 (NADPH) [10]; 0.05 (NADH) [18]; 0.005 (semidehydroascorbate) [18]; 0.0046 (NADH) [22]; 0.023 (NADPH) [22]; 0.0014 (monodehydroascorbate) [22]

pH-optimum
7.0 [8]; 7.0–7.2 [1]; 7.4 [18]; 7.5–8.5 [22]; 8 [10]

pH-range
5.5–9.5 (5.5: about 60% of activity maximum, 9.5: about 80% of activity maximum) [22]; 6.6–7.9 (at pH 6.6 and 7.9 about 50% of activity maximum) [1]

Temperature optimum (°C)
39 [1]; 41 [8]

Temperature range (°C)

3 ENZYME STRUCTURE

Molecular weight
42000 (Solanum tuberosum, gel filtration) [10]
47000 (Cucumis sativus, gel filtration) [22]
52000 (Euglena gracilis) [8]
66000 (Neurospora crassa, gel filtration) [1]

Subunits
Monomer (1 × 42000, Solanum tuberosum, SDS-PAGE [10], 1 × 47000, Cucumis sativus, gel filtration [22]) [10, 22]

Glycoprotein/Lipoprotein
–

Enzyme Handbook © Springer-Verlag Berlin Heidelberg 1994
Duplication, reproduction and storage in data banks are only allowed with the prior permission of the publishers

4 ISOLATION/PREPARATION

Source organism

Cuscuta reflexa (parasitic plant, activity is ten times lower than in etiolated and green plants) [5]; Neurospora crassa [1]; Glycine max (soybean) [1]; Nicotiana tabacum (tobacco) [3]; Ricinus communis (castor bean) [4]; Rat [7, 9, 11, 13, 15, 18, 21]; Euglena gracilis Z [8]; Solanum tuberosum (potato) [10, 20]; Human [12]; Pig [14, 19]; Jerusalem artichoke [20]; Onion [20]; Broad bean [20]; Cauliflower [20]; Cucumber (Cucumis sativus [22]) [20, 22]; Pea [20]; Pterocladia sp. [20]; Gigartina sp. [20]; Hypnea sp. [20]; Gracilaria sp. [20]; Mouse [10, 20]

Source tissue

Adrenal gland [9, 21]; Heart [9]; Brain [9]; Lung [9]; Spleen [9]; Polymorphonuclear leukocytes [12]; Kidney [19]; Cell [1]; Root nodules [2]; Root [20]; Cultured cells [3]; Callus [3]; Leaf [3]; Seedlings [5]; Liver [7, 9, 11, 13, 14, 15]; Tubers [10, 20]; Bud [20]; Etiolated hook [20]; Fibroblasts [20]; Virus-transformed mouse fibroblasts [20]; Fruit [22]

Localisation in source

Membrane (glyoxysomal membrane [4], plasma membrane [7, 11], outer mitochondrial membrane [9, 21]) [4, 7, 9, 11, 21]; Mitochondria [9, 21]; Cytoplasm [1]; Cytosol [8]; Microsomes [14–16, 18, 19]; More (localized in a hitherto not identified vesicle fraction) [13]

Purification

Neurospora crassa [1]; Solanum tuberosum [10]; Euglena gracilis Z (partial) [8]; Cucumis sativus [22]

Crystallization

–

Cloned

–

Renaturated

–

---

5 STABILITY

pH

Temperature (°C)

43 (pH 6.0–7.4, stable) [8]; 50 (pH 6.0–7.4, unstable) [8]; 60 (1 min, 20% loss of activity [10], 14 min, complete inactivation [1]) [1, 10]

Oxidation

Organic solvent

General stability information

Storage

4°C, Tricine-sodium hydroxide buffer, pH 8, 30 days, complete loss of activity, addition of 1–10 mM $MgSO_4$: 62% loss of activity [10]

6 CROSSREFERENCES TO STRUCTURE DATABANKS

PIR/MIPS code

Brookhaven code

7 LITERATURE REFERENCES

[1] Schulze, H.-U., Schott, H.-H., Staudinger, H.: Hoppe-Seyler's Z. Physiol. Chem.,353,1931–1942 (1972)
[2] Dalton, D.A., Post, C.J., Langeberg, L.: Plant Physiol.,96,812–818 (1991)
[3] Tanaka, K., Masuda, R., Sugimoto, T., Kawamura, Y., Kuboi, T.: Agric. Biol. Chem.,54,2003–2008 (1990)
[4] Bowditch, M.I., Donaldson, R.P.: Plant Physiol.,94,531–537 (1990)
[5] Tommasi, F., De Gara, L., Liso, R., Arrigoni, O.: J. Plant Physiol.,135,766–768 (1990)
[6] Borraccino, G., Dipierro, S., Arrigoni, O.: Phytochemistry,28,715–717 (1989)
[7] Navas, P., Estevez, A., Villalba, J.M., Buron, M.I., Crane, F.L.: Biochem. Biophys. Res. Commun.,154,1029–1033 (1988)
[8] Shigeoka, S., Yasumoto, R., Onishi, T., Nakano, Y., Kitaoka, S.: J. Gen. Microbiol.,133,227–232 (1987)
[9] Nishino, H., Ito, A.: J. Biochem.,100,1523–1531 (1986)
[10] Borraccino, G., Dipierro, S., Arrigoni, O.: Planta,167,521–526 (1986)
[11] Goldenberg, H.: Biochem. Biophys. Res. Commun.,94,721–726 (1980)
[12] Stankova, L., Bigley, R., Wyss, S.R., Aebi, H.: Experientia,35,852–853 (1979)
[13] Geiß, D., Schulze, H.-U.: FEBS Lett.,60,374–379 (1975)
[14] Weber, H., Weis, W., Wolf, B.: Hoppe-Seyler's Z. Physiol. Chem.,355,595–599 (1974)
[15] Schulze, H.-U., Gallenkamp, H., Staudinger, H.: Hoppe-Seyler's Z. Physiol. Chem.,354,391–406 (1973)
[16] Green, R.C., O'Brien, P.J.: Biochim. Biophys. Acta,293,334–342 (1973)
[17] Schulze, H.-U., Gallenkamp, H., Staudinger, H.: Hoppe-Seyler's Z. Physiol. Chem.,351,809–817 (1970)
[18] Oehler, G., Weis, W., Staudinger, H.: Hoppe-Seyler's Z. Physiol. Chem.,353,495–496 (1972)
[19] Kersten, H., Kersten, W., Staudinger, H.: Biochim. Biophys. Acta,27,598–608 (1958)
[20] Arrigoni, O., Dipierro, S., Borraccino, G.: FEBS Lett.,125,242–244 (1981)
[21] Natarajan, R.D., Harding, B.W.: J. Biol. Chem.,260,3902–3905 (1985)
[22] Hossain, M.A., Asada, K.: J. Biol. Chem.,260,12920–12926 (1985)

Enzyme Handbook © Springer-Verlag Berlin Heidelberg 1994
Duplication, reproduction and storage in data banks are only allowed with the prior permission of the publishers

1 NOMENCLATURE

EC number
1.6.6.1

Systematic name
NADH:nitrate oxidoreductase

Recommended name
Nitrate reductase (NADH)

Synonymes
Assimilatory nitrate reductase
Reductase, nitrate
NADH-nitrate reductase
NADH-dependent nitrate reductase
Assimilatory NADH: nitrate reductase [3]

CAS Reg. No.
9013-03-0

2 REACTION AND SPECIFICITY

Catalysed reaction
NADH + nitrate →
→ NAD^+ + nitrite + H_2O (mechanism: random bi bi [3], NADH-cytochrome c reductase (diaphorase), a functional moiety of the spinach NADH-nitrate reductase complex: hexa uni ping pong mechanism [4])

Reaction type
Redox reaction

Natural substrates
Nitrate + NADH (key enzyme involved in the first step of nitrate assimilation in plants)
More (possible role in iron assimilation (NADH: Fe(III)-citrate reductase activity)) [21]

Enzyme Handbook © Springer-Verlag Berlin Heidelberg 1994
Duplication, reproduction and storage in data banks are only allowed with the prior permission of the publishers

Substrate spectrum

1 Nitrate + NADH

2 More (NADH also reduces: ferricyanide [8, 17, 65], methylene blue [8], cytochrome c [17, 65, 75], dichlorophenolindophenol [17], benzoquinone [8], menadione [8], chlorate [37], bromate [37], iodate [37], ferric citrate [21], also nitrate reduction with reduced bromophenol blue [42], no NADPH nitrate reductase activity [5, 57, 72], molecular oxygen as electron acceptor in NADH-nitrate reductase system [61], $FADH_2$–nitrate reductase activity [72]) [5, 8, 17, 21, 37, 42, 57, 61, 65, 72, 75]

Product spectrum

1 Nitrate + NAD^+ + H_2O

2 ?

Inhibitor(s)

Aminooxyacetate [29]; O-Methoxylamine [29]; Thiols (inactivation by NAD^+ in presence of thiol compounds [30]) [30, 49]; NADH (no inactivation in presence of NADH alone [32, 33], but by simultaneous presence of a low concentration of cyanide [32, 33], or superoxide [33], in presence of dithiothreitol and/or FAD but not with cysteine [33], reaction by incubation with oxidant systems [47], inactivation in absence of nitrate, which prevents inactivation, cyanide enhances degree of inactivation [48]) [32, 33, 47–49]; Bromophenol blue [42]; NAD(P)H [50]; NADPH [49]; Dithioerythritol [49]; p-Chloromercuribenzoate [50, 58]; Atabrine [58]; Carbamoyl phosphate [72]; NAD^+ (inactivation by NAD^+ in presence of thiol compounds [30], product inhibition [3, 12]) [3, 4, 12, 30]; Nitrite (product inhibition [3, 12]) [3, 12, 17]; Adenosine 5'-diphosphoribose (dead-end inhibition) [3]; Thiocyanate (dead-end inhibition [3]) [3, 17]; Ferrocytochrome c [10]; p-Hydroxymercuribenzoate (NADH-nitrate reductase inhibited, bromophenol blue activity not affected [42], disappearance of NADH-diaphorase activity and appearance of FAD-requirement for the inactivation of NAD(P)H of $FMNH_2$–nitrate reductase [64]) [12, 21, 31, 42, 56, 64, 75]; Cyanide (reactivation by flavins in light [20], reactivation by incubation with oxidant systems [47]) [12, 17, 20, 47, 75, 76]; Hydroxylamine (photoinactivation in presence of flavins [27]) [17, 21, 27, 29]; Azide [17, 21, 31, 58, 76]; Cyanate [17, 58]; Pyridoxal phosphate [17, 72]; Hydrazine (slight) [17]; Semicarbazide (slight) [17]; Glycine hydroxamate (slight) [17]; Hypoxanthine (slight) [17]; Oxidized cytochrome c [17]; Fe^{2+} [18]; Cu^{2+} [18, 72]; Zn^{2+} [18, 72]; Pb^{2+} [18]; Fe^{3+} [18]; Potassium ferricyanide [22]; Cl^- [26]; NaCl [54]; ADP [72]; Co^{2+} [72]; Mn^{2+} [72]; NO_2^- [72]; MoO_4^{2-} [72]; VO_3^- [72]; More (leupeptin inhibits thiol-dependent acid endoproteinase, responsible for degradation of enzyme in barley primary leaf extract [52], trypsin, Staphylococcus aureus V8 protease or natural inactivator protease from corn cause loss of activity [65], radiation inactivation analysis [44], inactivation by maize root inactivation enzyme of maize and pea leaf enzyme [50], effect of trypsin and maize root proteinase [43], NADH-nitrate reductase inhibitor from young soybean leaves [11], corn root inactivating protein [45], inhibitor from roots of rice seedlings (protein, NADH protects against inhibition [46], 2 inhibitors: MW 460000 and 240000 [56]) [14, 46, 56], inhibition of partial activities [21], inhibition of NADH: Fe(III)-citrate reductase activity by monospecific anti-nitrate reductase rabbit sera [21], inhibition by increased ionic strength [37], effect of ionic strength on partial activities [37]) [11, 14, 21, 37, 43–46, 50, 52, 56, 65]

Cofactor(s)/prostethic group(s)

FAD (2–3 mol per mol of enzyme (MW 280000) [7], minimum of 2 mol per mol of enzyme (MW 356000) [13], 1 FAD per subunit (MW 100000) of the tetramer [28], 1 FAD per 115000 MW subunit [71], constituent of the soybean enzyme [58], exogenous supply of flavin is necessary for assay with dithionite [58], prosthetic group [65], no stimulation of activity [76]) [7, 13, 28, 58, 65, 71, 76]; Heme (1 heme iron per 115000 MW subunit [71], 2–3 mol per mol of enzyme (MW 280000) [7], minimum of 2 mol per mol of enzyme (MW 356000) [13], 1 heme per subunit (MW 100000) of the tetramer [28], prosthetic group [65]) [7, 13, 28, 65, 71]; Calmodulin (activation) [62]; Cytochrome (partially purified enzyme contains a cytochrome [16], enzyme associated with a cytochrome of the b-type [17], contains cytochrome b_{557} [17, 29], Neurospora enzyme contains cytochrome b_{557} [58], no cytochrome involvement demonstrated in higher plants [58], 1 cytochrome b_{557} per subunit [71]) [16, 17, 29, 58, 71]; L-Cysteine (activation) [31]; L-Cysteic acid (activation) [31]; L-Cysteine methyl ester (activation) [31]; 2-Mercaptoethylamine (activation) [31]; Glutathione (activation) [31]; Ferricyanide (inactive but ferricyanide-activatable form of enzyme) [35]

Metal compounds/salts

Ca^{2+} (activation) [62]; Molybdenum (2 mol per mol of enzyme (MW 280000) [7], minimum of 2 mol of molybdenum per mol of enzyme (MW 356000) [13], 1 Mo^{6+} per subunit (MW 100000) of the tetramer [28], oxidation-reduction midpoint potentials of the molybdenum center in spinach enzyme [39], plant enzyme contains molybdenum [58], Chlorella enzyme contains molybdenum [65], 1 molybdenum + pterin per subunit (MW 115000) [71]) [7, 13, 28, 39, 58, 65, 71]; Phosphate (stimulation [3, 41], increases activity [40], Neurospora enzyme has a requirement for phosphate [58]) [3, 40, 41, 58]

Turnover number (min^{-1})

Specific activity (U/mg)

100 (nitrate reductase) [21]; 24.8 [66]; 10 (NADH: Fe(III)-citrate reductase) [21]; 8 [5]; 23 [9]; 12 (Cucurbita pepo) [12]; 6.9 (Zea mays) [12]; 21 [34]; 60–70 [67]; 6. 48 [68]; 80–130 [69]; 51 [70]; 95 [71]; 37 [72]; 80 [73]; More (in vivo assay method [53, 58], in vitro assay method [58]) [53, 58]

Enzyme Handbook © Springer-Verlag Berlin Heidelberg 1994
Duplication, reproduction and storage in data banks are only allowed with the prior permission of the publishers

K_m-value (mM)

0.24 (nitrate) [5]; 0.003 (NADH) [12]; 0.070 (nitrate) [12]; 0.13 (dichlorophenolindophenol (+ NADH)) [17]; 0.02 (Fe(III)-citrate (+ NADH)) [21]; 0.05 (nitrate) [21]; 0.007 (NADH, ionic strength 50 mM) [37]; 0.013 (nitrate, ionic strength 50 mM) [37]; 0.221 (ClO_3^-, ionic strength 50 mM) [37]; 0.739 (BrO_3^-, ionic strength 50 mM) [37]; 10.751 (IO_3^-, ionic strength 50 mM) [37]; 0.006 (NADH, ionic strength 200 mM) [37]; 0.018 (nitrate, ionic strength 200 mM) [37]; 0.060 (reduced bromophenol blue) [42]; 0.5 (nitrate (+ reduced bromophenol blue)) [42]; 0.009 (NADH) [57]; 0.0026 (FAD) [57]; 0.23 (nitrate) [57]; 0.270 (nitrate) [72]; 0.0038 (NADH) [72]; 0.000008 (FAD) [72]; 0.11 (nitrate) [76]; 0.0081 (NADH) [76]; More (inorganic phosphate increases K_m for nitrate [40], effect of ionic strength on K_m [26, 37], of partial enzyme activities [37]) [9, 26, 37, 40]

pH-optimum

6.3 (NADH: Fe(III)-citrate reductase activity) [21]; 6.5 [76]; 7.0 (soybean, in vivo assay [53]) [37, 53]; 7.5 (maize, spinach, in vivo assay [53]) [5, 12, 31, 53, 72]; More (pH-optima of partial activities range from 5.5–8.1) [41]

pH-range

5.5–7.5 (5.5: about 70% of activity maximum, 7.5: about 20% of activity maximum, NADH: Fe(III)-citrate reductase activity) [21]

Temperature optimum (°C)

30 [22]

Temperature range (°C)

–10 – 40 (–10°C: about 10% (wheat), about 30% (maize) of activity maximum, 40°C: about 20% (wheat), about 60% (maize) of activity maximum) [22]; 10–30 (Q_{10} is 2.77, between 10°C and 20°C, reaction rate at 30°C only a little faster than at 20°C) [16]

---

3 ENZYME STRUCTURE

Molecular weight

205000 (Hordeum vulgare, gel filtration) [68]
221000 (Hordeum vulgare, gel filtration, sucrose density gradient centrifugation) [5]
270000 (Spinacia oleracea, gel filtration, sucrose density gradient centrifugation) [69]
280000 (Chlorella vulgaris, sedimentation equilibrium study) [7]
330000 (Glycine max, gel filtration) [76]
356000 (Chlorella vulgaris, gel filtration, sucrose density gradient centrifugation) [13]
360000 (Chlorella vulgaris, sedimentation equilibrium study) [2]

Subunits

Dimer (2 × 110000, Hordeum vulgare, SDS-PAGE [1], 2 × 100000, Hordeum vulgare, SDS-PAGE [5], 2 × 103000, Hordeum vulgare, SDS-PAGE [68], 2 × 110000–120000, Spinacia oleracea, SDS-PAGE [69], 1 × 110000 + 1 × 120000, Spinacia oleracea, SDS-PAGE [70], 2 × 115000, Cucurbita maxima, SDS-PAGE [71], 2 × 130000, Hordeum vulgare, SDS-PAGE [72]) [1, 5, 68–72]

Tetramer (4 × 100000, Chlorella vulgaris, at low protein concentration the tetramer dissociates to a fully active dimer [28], 4 × 90000, Chlorella vulgaris, SDS-PAGE, homotetramer with dihedral symmetry [2], 200000 MW higher-plant NADH-nitrate reductase complex contains 4 types of components: 1. alpha (MW 40000, FAD containing), 2. beta (cytochrome b_{557} subunit), 3. gamma (MW 40000), 4. molybdenum-containing component (MW 1000, carried on the gamma subunit) [6]) [2, 6]

More (Chlorella vulgaris: active monomer (MW 90000) contains 3 subunits [7], composed of at least 3 subunits [13], spinach: molybdenum, heme and FAD components are located in distinct domains (covalently linked by exposed hinge regions) molybdenum domain is important in maintenance of subunit interactions in enzyme complex [38]) [7, 13, 38]

Glycoprotein/Lipoprotein

–

4 ISOLATION/PREPARATION

Source organism

Hordeum vulgare (barley) [1, 5, 6, 40, 52, 55, 68, 72]; Chlorella fusca [59]; Chlorella vulgaris [2, 3, 7, 13, 28, 44, 45, 65]; Spinacia oleracea (spinach) [4, 10, 19, 20, 26, 29, 37–39, 47, 49, 51, 53, 58, 64, 66, 68–70]; Cucurbita pepo (squash [8, 12, 42], marrow [58]) [8, 12, 42, 58]; Triticum aestivum (wheat) [9, 22, 30, 33, 34, 48, 58]; Zea mays (corn) [12, 22, 25, 32, 36, 43, 45, 50, 53, 58, 60, 61, 67, 74]; Oryza sativa (rice) [14]; Agrostemma githago [15]; Chlorella pyrenoidosa [16, 57]; Chlorella sp. (Berlin strain [17]) [17, 41]; Cucurbita maxima [18, 21, 71]; Lycopersicon esculentum (tomato) [23, 58]; Nicotiana tabacum [24, 35]; Ankistrodesmus braunii [27]; Thalassiosira pseudonana [31]; Pisum arvense (pea) [50]; Lemna minor (duckweed) [58]; Glycine max (soybean, 3 forms [75], contains EC 1.6.6.1 and 1.6.6.2 [78, 79], 1.6.6.1 most active in young cotyledons, as cotyledons age 1.6. 6.2 becomes more active [79]) [53, 75, 76, 78, 79]; Suaeda maritima (var. macrocarpa) [54]; Chenopodium album (lambsquarter) [58]; Amaranthus hybrides (pigweed) [58]; Brassica oleracea italica (broccoli) [58]; Raphanus sativus (radish) [58]; Cucumis sativus (cucumber) [58]; Capsicum frutescens (green pepper) [58]; Amaranthus sp. [62]; Sinapis alba [63]; Nicotiana plumbaginifolia [73]; Gossypium hirsutum (cotton) [77]

Enzyme Handbook © Springer-Verlag Berlin Heidelberg 1994
Duplication, reproduction and storage in data banks are only allowed with the prior permission of the publishers

Source tissue

Leaves (of seedlings [9, 60], primary leaf [45, 52, 67, 76]) [1, 4, 5, 9, 12, 14, 19, 20, 30, 32, 33, 38, 40, 45, 47, 48, 52, 55, 58, 60–62, 67, 68, 70, 72–76]; Seedlings [9, 55, 60, 67]; Cotyledons [12, 18, 21, 58, 63, 71]; Embryos [15]; Root [19, 36, 58]; Shoots [22, 58, 68]; Suspension culture (of XO cell line of Nicotiana tabacum) [35]; Scutellum (of very young maize [60]) [43, 58, 60]; Petioles [58]; Stem [58]; Aleurone layer [58]; More (enzymes from leaf and roots have different antigenic behaviour [19], higher activities in chlorophyllous than in nonchlorophyllous tissue [58]) [19, 58]

Localisation in source

Cytoplasm [58]; Chloroplast [58]

Purification

Hordeum vulgare [5, 40, 72]; Chlorella vulgaris [7, 13]; Cucurbita pepo [8, 12]; Triticum aestivum [9]; Zea mays (2 forms with different MW [67]) [12, 67]; Chlorella pyrenoidosa (partial) [16, 57]; Spinacia oleracea [19, 66, 68, 70]; Cucurbita maxima [21, 71]; Thalassiosira pseudonana [31]; Nicotiana plumbaginifolia [73]; Glycine max [76]

Crystallization

–

Cloned

(gene from : Zea mays [25], Lycopersicon esculentum [23], Nicotiana tabacum [24], flavin domain of corn leaf [74]) [23–25, 74]

Renaturated

–

5 STABILITY

pH

6.5–8.5 (1 h, stable) [51]; 7.5 (highest stability [9], 0°C, half-life: 42 h [72]) [9, 72]

Temperature (°C)

10 (pH 7.5, $t_{1/2}$ 28 min) [9]; 25 (pH 7.5, $t_{1/2}$ 28 min [9], 20 min, stable [51]) [9, 51]; 30 (20 min, about 20% loss of activity) [51]; 35 (20 min, about 45% loss of activity) [51]; 40 (pH 7.5, $t_{1/2}$ 1 min [9], 40 min, 50% loss of NADH-nitrate reductase activity, slight increase of bromophenol blue activity [42], 20 min, about 90% loss of activity [51]) [9, 42, 51]; More (temperature stability of component enzymatic acitivities) [9]

Oxidation

Organic solvent

General stability information

FAD: increases stability at 25°C from $t_{1/2}$ 30 min (minus FAD) to 70 min, no effect on stability at 10°C [9], stabilization against heat inactivation [51, 59], urea inactivation [51], guanidine hydrochloride inactivation [51], some protection at acid pH [51]; NO_3^- stabilizes [77]; NADH: stabilization at 0°C and 25°C [22], stabilizes [77]; Decay characteristics in crude, partially-purified and highly purified preparations [34]; Chymostatin, stabilizes [36]; Glycerol, 20%, stabilizes [57]; Urea, 4 M, inactivates [51]; Guanidine hydrochloride inactivates [51]; SDS, 0.03%, inactivates [51]; Tris-HCl, 0.25 M, pH 8.5, 3 mM dithiothreitol, 0.005 mM FAD, 0.001 mM molybdate, 1 mM EDTA: stabilzes at 0°C and 30°C [55]; Unstable both in vivo and in vitro, stability in vitro and vivo varies greatly with species, plant age and tissue [58]

Storage

–20°C, phosphate buffer, pH 6.9, 40% glycerol, stable for more than 2 months [13]; 0–5°C, partially purified enzyme, stable for at least a week [51]; –15°C, partially purified enzyme, + 20% glycerol, 25% loss of activity after 48 h [57]; Crude extract, stable for several days at 0°C or several months at –80°C [67]

6 CROSSREFERENCES TO STRUCTURE DATABANKS

PIR/MIPS code

RDMUNH (Mouse-ear cress); RDTONH (Tomato); RDNTNT (Tobacco Nicotiana tomentosiformis); RDNTNS (Wood tobacco); RDSPNH (Spinach); RDBHNH (Barley cv. Himalaya); RDBHNS (Barley cv. Steptoe fragment); S01640 (1 Mouse-ear cress fragment); S01641 (2 Mouse-ear cress fragment); S07554 (Rice); JH0182 (Emericella nidulans); S04349 (Escherichia coli); S17197 (Chlorella vulgaris); A41667 (Winter squash); A41085 (Curled-leaved tobacco fragment); S19254 (Maize); A35499 (flavin chain Maize)

Brookhaven code

Enzyme Handbook © Springer-Verlag Berlin Heidelberg 1994
Duplication, reproduction and storage in data banks are only allowed with the prior permission of the publishers

7 LITERATURE REFERENCES

[1] Min Kuo, T., Somers, D.A., Kleinhofs, A., Warner, R.L.: Biochim. Biophys. Acta,708,75–81 (1982)
[2] Howard, W.D., Solomonson, L.P.: J. Biol. Chem.,257,10243–10250 (1982)
[3] Howard, W.D., Solomonson, L.P.: J. Biol. Chem.,256,12725–12730 (1981)
[4] De La Rosa, F.F., Palacian, E.: Plant Sci. Lett.,21,1–8 (1981)
[5] Kuo, T., Kleinhofs, A., Warner, R.L.: Plant Sci. Lett.,17,371–381 (1980)
[6] Wray, J.L., Small, I.S., Brown, J.: Biochem. Soc. Trans.,7,739–741 (1979)
[7] Giri, L., Ramadoss, C.S.: J. Biol. Chem.,254,11703–11712 (1979)
[8] Smarrelli, J., Campbell, W.H.: Plant Sci. Lett.,16,139–147 (1979)
[9] Sherrard, J.H., Dalling, M.J.: Plant Physiol.,63,346–353 (1979)
[10] Maldonado, J.M., Notton, B.A., Hewitt, E.J.: Plant Sci. Lett.,13,143–150 (1978)
[11] Jolly, S.O., Tolbert, N.E.: Plant Physiol.,62,197–203 (1978)
[12] Campbell, W.H., Smarrelli, J.: Plant Physiol.,61,611–616 (1978)
[13] Solomonson, L.P., Lorimer, G.H., Hall, R.L., Borchers, R., Bailey, J.L.: J. Biol. Chem.,250,4120–4127 (1975)
[14] Kadam, S.S., Gandhi, A.P., Sawhney, S.K., Naik, M.S.: Biochim. Biophys. Acta,350,162–170 (1974)
[15] Kende, H., Shen, T.C.: Biochim. Biophys. Acta,286,118–125 (1972)
[16] Vennesland, B., Jetschmann, C.: Biochim. Biophys. Acta,229,554–564 (1971)
[17] Solomonson, L.P., Vennesland, B.: Biochim. Biophys. Acta,267,544–557 (1972)
[18] Smarrelli, J., Campbell, W.H.: Biochim. Biophys. Acta,742,435–445 (1983)
[19] Ferrario, S., Hirel, B., Gadal, P.: Biochem. Biophys. Res. Commun.,113,733–737 (1983)
[20] Jawali, N., Sane, P.V.: FEBS Lett.,158,213–216 (1983)
[21] Redinbaugh, M.G., Campbell, W.H.: Biochem. Biophys. Res. Commun.,114,1182–1188 (1983)
[22] Datta, N., Rao, L.V.M., Guha-Mukherjee, S., Sopory, S. K.: Phytochemistry,22,821–824 (1983)
[23] Daniel-Vedele, F., Dorbe, M.-F., Caboche, M., Rouze, P.: Gene,85,371–380 (1989)
[24] Vaucheret, H., Vincentz, M., Kronenberger, J., Caboche, M., Rouze, P.: Mol. Gen. Genet.,216,10–15 (1989)
[25] Gowri, G., Campbell, W.H.: Plant Physiol.,90,792–798 (1989)
[26] Barber, M.J., Notton, B.A., Kay, C.J., Solomonson, L. P.: Plant Physiol.,90,70–94 (1989)
[27] Balanidin, T., Fernandez, V.M., Aparicio, P.J.: Plant Physiol.,82,65–70 (1986)
[28] Solomonson, L.P., McCreery, M.J.: J. Biol. Chem.,261,806–810 (1986)
[29] Jawali, N., Sane, P.V.: Phytochemistry,23,225–228 (1984)
[30] Aryan, A.P., Wallace, W., Nicholas, D.J.D.: Phytochemistry,23,719–721 (1984)
[31] Smarrelli, J., Campbell, W.H.: Phytochemistry,19,1601–1605 (1980)
[32] Echevarria, C., Maurino, S.G., Maldonado, J.M.: Phytochemistry,23,2155–2158 (1984)
[33] Aryan, A.P., Wallace, W.: Biochim. Biophys. Acta,827,215–220 (1985)
[34] Jones, P.W., Mhuimhneachain, M.N.: Phytochemistry,24,385–392 (1985)
[35] Trinity, P.M., Filner, P.: Phytochemistry,30,69–71 (1991)
[36] Long, D.M., Oaks, A.: Plant Physiol.,93,846–850 (1990)
[37] Barber, M.J., Notton, B.A.: Plant Physiol.,93,537–540 (1990)
[38] Kubo, Y., Ogura, N., Nakagawa, H.: J. Biol. Chem.,263,19684–19689 (1988)

[39] Barber, M.J., Notton, B.A., Solomonson, L.P.: FEBS Lett.,213,372–374 (1987)
[40] Oji, Y., Ryoma, Y., Wakiuchi, N., Okamoto, S.: Plant Physiol.,83,472–474 (1987)
[41] Kay, C.J., Barber, M.J.: J. Biol. Chem.,261,14125–14129 (1986)
[42] Campbell, W.H.: Plant Physiol.,82,729–732 (1986)
[43] Batt, R.G., Wallace, W.: Biochim. Biophys. Acta,744,205–211 (1983)
[44] Solomonson, L.P., Mc Creery, M.J., Kay, C.J., Barber, M.J.: J. Biol. Chem.,262,8934–8939 (1987)
[45] Poulle, M., Oaks, A., Bzonek, P., Goodfellow, V.J., Solomonson, L.P.: Plant Physiol.,85,375–378 (1987)
[46] Leong, C.C., Shen, T.-C.: Biochim. Biophys. Acta,612,245–252 (1980)
[47] Maldonado, J.M., Notton, B.A., Hewitt, E.J.: Planta,156,289–294 (1982)
[48] Aryan, A.P., Batt, R.G., Wallace, W.: Plant Physiol.,71,582–587 (1983)
[49] Palacian, E., De La Rosa, F., Castillo, F., Gomez-Moreno, C.: Arch. Biochem. Biophys.,161,441–447 (1974)
[50] Wallace, W.: Biochim. Biophys. Acta,377,239–250 (1975)
[51] De La Rosa, F.F., Castillo, F., Palacian, E.: Phytochemistry,16,875–879 (1977)
[52] Wray, J.L., Kirk, D.W.: Plant Sci. Lett.,23,207–213 (1981)
[53] Maurino, S.G., Echevarria, C., Mejias, J.A., Vargas, M.A., Maldonado, J.M.: J. Plant Physiol.,124,123–130 (1986)
[54] Billard, J.P., Boucaud, J.: Phytochemistry,21,1225–1228 (1982)
[55] Kuo, T.-M., Warner, R.L., Kleinhofs, A.: Phytochemistry,21,531–533 (1982)
[56] Leong, C.C., Shen, T.-C.: Biochim. Biophys. Acta,703,129–133 (1982)
[57] Schloemer, R.H., Garrett, R.H.: Plant Physiol.,51,591–593 (1973)
[58] Hageman, R.H., Hucklesby, D.P.: Methods Enzymol.,23 A,491–503 (1971) (Review)
[59] Zumft, W.G., Aparicio, P.J., Paneque, A., Losada, M.: FEBS Lett.,9,157–160 (1970)
[60] Sorger, G., Gooden, D.O., Earle, E.D., McKinnon, J.: Plant Physiol.,82,473–478 (1986)
[61] Ruoff, P., Lillo, C.: Biochem. Biophys. Res. Commun.,172,1000–1005 (1990)
[62] Sane, P.V., Kumar, N., Baijal, M., Singh, K.K., Kochhar, V.K.: Phytochemistry,26,1289–1291 (1987)
[63] Schuster, C., Schmidt, S., Mohr, H.: Planta,177,74–83 (1989)
[64] Castillo, F., De La Rosa, F.F., Calero, F., Palacian, E.: Biochem. Biophys. Res. Commun.,69,277–284 (1976)
[65] Solomonson, L.P., Barber, M.J., Robbins, A.P., Oaks, A.: J. Biol. Chem.,261,11290–11294 (1986)
[66] Fido, R.J.: Plant Sci.,50,111–115 (1987)
[67] Nakagawa, H., Poulle, M., Oaks, A.: Plant Physiol.,75,285–289 (1984)
[68] Campbell, J.M., Wray, J.L.: Phytochemistry,22,2375–2382 (1983)
[69] Nakagawa, H., Yonemura, Y., Yamamoto, H., Sato, T., Ogura, N., Sato, R.: Plant Physiol.,77,124–128 (1985)
[70] Fido, R.J., Notton, B.A.: Plant Sci. Lett.,37,87–91 (1984)
[71] Redinbaugh, M.G., Campbell, W.H.: J. Biol. Chem.,260,3380–3385 (1985)
[72] Oji, Y., Mamano, T., Ryoma, Y., Miki, Y., Okamoto, S. : J. Plant Physiol.,119,247–256 (1985)
[73] Moureaux, T., Leydecker, M.-T., Meyer, C.: Eur. J. Biochem.,179,617–620 (1989)
[74] Hyde, G.E., Campbell, W.H.: Biochem. Biophys. Res. Commun.,168,1285–1291 (1990)
[75] Nelson, R.S., Streit, L., Harper, J.E.: Plant Physiol.,80,72–76 (1986)
[76] Jolly, S.O., Campbell, W., Tolbert, N.E.: Arch. Biochem. Biophys.,174,431–439 (1976)
[77] Tischler, C.R., Purvis, A.C., Jordan, W.R.: Plant Physiol.,61,714–717 (1978)
[78] Streit, L., Nelson, R.S., Harper, J.E.: Plant Physiol.,78,80–84 (1985)
[79] Orihuel-Iranzo, B., Campbell, W.H.: Plant Physiol.,65,595–599 (1980)

Enzyme Handbook © Springer-Verlag Berlin Heidelberg 1994
Duplication, reproduction and storage in data banks are only allowed with the prior permission of the publishers

1 NOMENCLATURE

EC number
1.6.6.2

Systematic name
NAD(P)H:nitrate oxidoreductase

Recommended name
Nitrate reductase (NAD(P)H)

Synonymes
Assimilatory nitrate reductase
Assimilatory NAD(P)H-nitrate reductase [8]
NAD(P)H bispecific nitrate reductase [14]
Reductase, nitrate (reduced nicotinamide adenine dinucleotide (phosphate))
Nitrate reductase NAD(P)H
NAD(P)H-nitrate reductase

CAS Reg. No.
9029-27-0

---

2 REACTION AND SPECIFICITY

Catalysed reaction
NAD(P)H + nitrate →
→ $NAD(P)^+$ + nitrite + H_2O

Reaction type
Redox reaction

Natural substrates
NAD(P)H + nitrate (plays a role in first step of nitrate assimilation) [5]

Substrate spectrum
1 NAD(P)H + nitrate
2 More (other electron donors: combination of dithionite with flavin nucleotides or viologens [12], $FMNH_2$ [2], reduced methyl viologen [2, 12], also reduction of ClO_3^-, $K_3Fe(CN)_6$, cytochrome c and 2,6-dichlorophenolindophenol with NADPH or NADH [12]) [2, 12]

Enzyme Handbook © Springer-Verlag Berlin Heidelberg 1994
Duplication, reproduction and storage in data banks are only allowed with the prior permission of the publishers

Product spectrum

1 $NAD(P)^+$ + nitrite + H_2O
2 ?

Inhibitor(s)

Dithionite [14]; Nitrite [12]; Cyanate [12]; Azide [12]; NAD(P)H (preincubation with NAD(P)H and cyanide: inactivation [1], preincubation with NAD(P)H alone: activation [1], hydroxyl radical is involved in in vitro irreversible inactivation [5], effect of temperature on inactivation [5], FAD enhances inactivation [5], in vitro an active diaphorase moiety is required for inactivation by reduced pyridine nucleotides, in vivo the absence of nitrate rather than the presence of ammonium is the triggering event for inactivation [9]) [1, 5, 7, 9, 14]; Cyanide (preincubation with NAD(P)H and cyanide: inactivation [1]) [1, 7, 12]; p-Chloromercuribenzoate (inhibition of diaphorase activity [7], NADH or NADPH protect [7]) [1, 7, 12]; p-Hydroxymercuribenzoate [19]

Cofactor(s)/prostethic group(s)

NADH (utilizes both NADH and NADPH, more active with NADH [1, 3, 4], at high nitrate concentration: more active with NADPH [12], at low nitrate concentration: more active with NADH [12], NADPH to NADH activity ratio is 1.8 [19]) [1, 3, 4, 7, 12, 19]; NADPH (utilizes both NADH and NADPH, more active with NADH [1, 3, 4], at high nitrate concentration: more active with NADPH [12], at low nitrate concentration: more active with NADH [12], NADPH to NADH activity ratio is 1.8 [19]) [1, 3, 4, 7, 12, 19]; Cytochrome b_{557} (contains cytochrome b_{557} [1, 3, 12], 1 molecule per subunit [3], 4 mol per mol of enzyme (Ankistrodesmus braunii), 1 mol per mol of enzyme (Rhodotorula glutinis) [20]) [1, 3, 12, 20]; FAD (prosthetic group [17], essential for NAD(P)H-dependent activity [17], maximum activity requires addition of flavin nucleotides, FAD more effective than FMN [1], complete reduction of cytochrome by NADPH takes place only when added FAD is present in the enzyme preparation before addition of NADPH [12], structure of FAD binding site [16], enzyme contains 4 molecules of FAD [18], activation [4]) [1, 4, 12, 16–18]; FMN (maximum activity requires addition of flavin nucleotides, FAD is more effective than FMN [1], activation [4]) [1, 4]; Heme (enzyme contains 4 heme groups [18], minimum of 4 molecules of heme per molecule of native enzyme [19]) [18, 19]

Metal compounds/salts

Molybdenum (molybdoprotein [3, 4, 8, 12], contains one molecule of molybdenum per subunit [3], enzyme contains 2 atoms of molybdenum [18]) [3, 4, 8, 12, 18]

Turnover number (min^{-1})

28175 [19]; More (above 6000 [1]) [1, 10]

Specific activity (U/mg)
217.4 [1]; 148.2 [12]; 0.718 (NADH-nitrate reductase) [7]; 0.766 (NADPH-nitrate reductase) [7]; 72–80 [19]; 61.25 [19]

K_m-value (mM)
0.045 (NADPH [4], nitrate (+ NADH) [12]) [4, 12]; 0.10 (NADH) [4]; 0.017 (NADH) [1]; 0.030 (NADPH) [1]; 0.120 (nitrate (+ NADH)) [1]; 0.110 (nitrate (+ NADPH)) [1]; 0.0115 (NADH) [7]; 0.0145 (NADPH) [7]; 0.13 (nitrate (+ NADH)) [7]; 0.14 (nitrate (+ NADPH)) [7]; 0.125 (nitrate (+ NADPH)) [12]; 0.020 (NADPH) [12]; 0.160 (NADH) [12]; 0.000004 (FAD, as a protector of NADH-cytochrome c reductase activity) [17]; More [8, 14, 20]

pH-optimum
7.0 [3, 8]; 7.4–7.7 [7]; 7.5 [12, 20]; 7.8–8.2 [13]

pH-range
7.1–8.2 (at pH 7.1 and 8.2: about 50% of activity maximum) [7]; 7–9 (at pH 7 and 9: about 60% of activity maximum) [13]

Temperature optimum (°C)

Temperature range (°C)

3 ENZYME STRUCTURE

Molecular weight
460000–467400 (Ankistrodesmus braunii, gel filtration, density gradient centrifugation, disc gel electrophoresis, pore gradient electrophoresis, sucrose-density gradient centrifugation) [18, 19]
365000 (Candida nitratophila, sedimentation equilibrium centrifugation) [1]
230000 (Rhodotorula glutinis, sucrose density gradient centrifugation, gel filtration) [12]

Subunits
Dimer (2 × 118000, Rhodotorula glutinis, SDS-PAGE) [12]
Tetramer (4 × 95000, Candida nitratophila, SDS-PAGE) [1]
Octamer (8 × 58750–59000, Ankistrodesmus braunii, SDS-PAGE) [18, 19]

Glycoprotein/Lipoprotein
–

4 ISOLATION/PREPARATION

Source organism
Candida nitratophila [1, 3]; Hansenula anomala (ascomycetous yeast, enzyme exists in 2 interconvertible forms, active and inactive, depending on *the state of the* molybdenum center [5]) [4–6]; Sphagum magellanicum [2];

Enzyme Handbook © Springer-Verlag Berlin Heidelberg 1994
Duplication, reproduction and storage in data banks are only allowed with the prior permission of the publishers

Sphagum palustre [2]; Sphagum pappilosum [2]; Sphagum squarrosum [2]; Sphagum subsecundum [2]; Sphagum cuspidatum [2]; Sphagum fallox [2]; Sphagum fimbriatum [2]; Sphagum molle [2]; Sphagum rubellum [2]; Sphagum subnitens [2]; Chlorella variegata [7]; Chlamydomonas reinhardtii (mutant 104 [8], mutant 305 [8], complementation from mutants [8]) [8, 9]; Glycine max (contains EC 1.6.6.1 and EC 1.6.6.2 [10], NADH: nitrate oxidoreductase is more active in young cotyledons, as cotyledons age NAD(P)H: nitrate oxidoreductase becomes more active [11]) [10, 11]; Rhodotorula glutinis [12, 20]; Cyanidium caldarium (alga) [13]; Hordeum vulgare (barley, mutant nar la) [14]; Monoraphidium braunii [15]; Ankistrodesmus braunii [16–20]

Source tissue

Cell [4]; Cotyledons [11]

Localisation in source

Soluble [8, 17]; Chloroplast (specifically located in the pyrenoid region of chloroplast) [15]; More (not membrane-bound) [4]

Purification

Candida nitratophila [1, 3]; Hansenula anomala [4]; Sphagum magellanicum [2]; Sphagum palustre [2]; Sphagum pappilosum [2]; Sphagum squarrosum [2]; Sphagum subsecundum [2]; Sphagum cuspidatum [2]; Sphagum fallox [2]; Sphagum fimbriatum [2]; Sphagum molle [2]; Sphagum rubellum [2]; Sphagum subnitens [2]; Chlorella variegata [7]; Chlamydomonas reinhardtii [8]; Glycine max [10]; Rhodotorula glutinis [12]; Hordeum vulgare (partial) [14]; Ankistrodesmus braunii [18, 19]

Crystallization

–

Cloned

–

Renaturated

–

---

5 STABILITY

pH

Temperature (°C)

30 (inactivation) [4]; 60 (3 min, complete loss of NADPH-nitrate reductase activity) [13]; More (heat activation of benzyl viologen: nitrate reductase activity) [13]

Oxidation

Organic solvent

General stability information

Labile to dilution with protein-free buffer (2.5 mg bovine serum albumin, stabilization) [1]

Storage

4°C, 50% loss of activity after 7 days [1]

6 CROSSREFERENCES TO STRUCTURE DATABANKS

PIR/MIPS code

RDBJNH (European white birch); RDBHNP (Barley)

Brookhaven code

7 LITERATURE REFERENCES

[1] Hipkin, C.R., Ali, A.H., Cannons, A.: J. Gen. Microbiol.,132,1997–2003 (1986)
[2] Deising, H.: Z. Naturforsch.,42c,653–656 (1987)
[3] Hipkin, C.R., Kau, D.A., Cannons, A.C., Jones, D.H., Gallon, J.R., Kay, C.J., Barber, M.J., Solomonson, L.P.: Biochem. Soc. Trans.,17,928–929 (1989)
[4] Minagawa, N., Yoshimoto, A.: Agric. Biol. Chem.,47,125–127 (1983)
[5] Minagawa, N., Yoshimoto, A.: Agric. Biol. Chem.,49,2217–2219 (1985)
[6] Minagawa, N., Yoshimoto, A.: Agric. Biol. Chem.,49,2213–2215 (1985)
[7] Hipkin, C.R., Al-Bassam, B.A., Syrett, P.J.: Planta,144,137–141 (1979)
[8] Fernandez, E., Cardenas, J.: Biochim. Biophys. Acta,657,1–12 (1981)
[9] Cordoba, F., Cardenas, J., Fenandez, E.: Biochim. Biophys. Acta,827,8–13 (1985)
[10] Streit, L., Nelson, R.S., Harper, J.E.: Plant Physiol.,78,80–84 (1985)
[11] Orihuel-Iranzo, B., Campbell, W.H.: Plant Physiol.,65,595–599 (1980)
[12] Guerrero, M.G., Gutierrez, M.: Biochim. Biophys. Acta,482,272–285 (1977)
[13] Rigano, C.: Arch. Mikrobiol.,76,265–276 (1971)
[14] Harker, A.R., Narayanan, K.R., Warner, R.L., Kleinhofs, A.: Phytochemistry,25,1275–1279 (1986)
[15] Lopez-Ruiz, A., Roldan, J.M., Verbelen, J.P., Diez, J.: Plant Physiol.,78,614–618 (1985)
[16] Marquez, A.J., De La Rosa, M.A, Vega, J.M.: J. Chromatogr.,235,435–443 (1982)
[17] De La Rosa, M.A., Marquez, A.J., Vega, J.M.: Z. Naturforsch.,37c,24–30 (1982)
[18] De La Rosa, M.A., Vega, J.M., Zumft, W.G.: J. Biol. Chem.,256,5814–5819 (1981)
[19] De La Rosa, M.A., Diez, J., Vega, J.M., Losada, M.: Eur. J. Biochem.,106,249–256 (1980)
[20] Guerrero, M.G., Vega, J.M., Losada, M.: Annu. Rev. Plant Physiol.,32,169–204 (1981) (Review)

Enzyme Handbook © Springer-Verlag Berlin Heidelberg 1994
Duplication, reproduction and storage in data banks are only allowed with the prior permission of the publishers

1 NOMENCLATURE

EC number
1.6.6.3

Systematic name
NADPH:nitrate oxidoreductase

Recommended name
Nitrate reductase (NADPH)

Synonymes
Assimilatory nitrate reductase [1]
Assimilatory reduced nicotinamide adenine dinucleotide phosphate-nitrate reductase [4]
NADPH-nitrate reductase [4, 6, 7]
Assimilatory NADPH-nitrate reductase [5]
Triphosphopyridine nucleotide-nitrate reductase [9]
NADPH:nitrate reductase [12]

CAS Reg. No.
9029-28-1

---

2 REACTION AND SPECIFICITY

Catalysed reaction
NADPH + nitrate →
→ $NADP^+$ + nitrite + H_2O (random order rapid-equilibrium mechanism [2])

Reaction type
Redox reaction

Natural substrates
Nitrate + NADPH
More (physiological electron flow: NADPH --> sulfhydryls --> FAD --> cytochrome b_{557} --> Mo --> NO_3^-) [3]

Substrate spectrum
1 Nitrate + NADPH (ir [9])
2 More (associated activities: NADPH-cytochrome c reductase [2, 4, 5, 14, 16], reduced methyl viologen dye: nitrate reductase [2, 4, 5, 14, 15], $FADH_2$–nitrate reductase [4, 5, 14, 15]) [2, 4, 5, 14–16]

Enzyme Handbook © Springer-Verlag Berlin Heidelberg 1994
Duplication, reproduction and storage in data banks are only allowed with the prior permission of the publishers

Product spectrum

1 $NADP^+$ + nitrite
2 ?

Inhibitor(s)

8-Hydroxyquinoline (slight [2, 4]) [2, 4, 9, 14]; 1,10-Phenanthroline (slight [2]) [2, 9, 14]; Potassium chlorate (slight) [2]; 2-Amino-pyridine adenine dinucleotide ($NADP^+$ or FAD protect against inactivation) [3]; 3-Aminopyridine adenine dinucleotide [3]; p-Chloromercuribenzoate (reversed by glutathione or cysteine [9]) [5, 9, 13]; Thiourea [9, 14]; Potassium ethyl xanthate [9]; Hydrazine [14]; Hydroxylamine hydrochloride [14]; NADPH (substrate inhibition when both FAD and nitrate are unsaturating) [16]; $NADP^+$ (product inhibition) [2, 16]; Nitrite (product inhibition [16], slight [2]) [2, 16]; Azide [2, 4, 5, 9, 14]; Phenylglyoxal [4]; EDTA [4]; p-Hydroxymercuribenzoate (reversed by sulfhydryl reagents [4]) [2, 4, 14]; Iodoacetamide [2]; Cyanide (not [13]) [2, 4, 5, 9, 14]

Cofactor(s)/prostethic group(s)

FAD (flavoprotein [2, 5, 9, 14], activation of NADPH-nitrate reductase activity [5], contains 5.91–7.78 nmol FAD per mg of protein [5], acts as electron carrier [14]) [2, 5, 9, 14]; Haem (haemoprotein component) [2]; Cytochrome b_{557} (i. e. protoporphyrin IX heme [14], 2 mol of cytochrome b_{557} per mol of protein [4], involved in intracellular electron transport from NADPH to nitrate [5], associated with the enzyme [14, 15]) [4, 14, 15]; NADPH (3-fold greater affinity than with NADH) [14]; FMN (acts as electron carrier) [14]; More (NADH: no effect) [5]

Metal compounds/salts

Molybdenum (enzyme contains molybdenum [2, 4, 7, 8, 14], 1 mol of molybdenum per mol of protein [4], indispensible role in nitrate reductase complex [7]) [2, 4, 7, 8, 14]; Phosphate (required for maximal activity) [2]

Turnover number (min^{-1})

Specific activity (U/mg)

More [4, 14, 15]; 2.5 [2]; 15.5 [5]; 16.0 [13]; 200 [16]

K_m-value (mM)

More [16]; 0.080 (nitrate (+ NADPH), pH 6.5) [2]; 0.060 (nitrate (+ NADPH), pH 7.5) [2]; 0.009 (NADPH (+ nitrate), pH 6.5, NADPH (+ cytochrome c)) [2]; 0.010 (NADPH (+ nitrate), pH 7.5) [2]; 0.029 (cytochrome c (+ NADPH), pH 7.5) [2]; 0.024 (cytochrome c (+ NADPH), pH 8.5) [2]; 0.015 (NADPH (+ cytochrome c)) [2]; 2.5 ($FADH_2$ (+ NO_3^-)) [5]; 3.0 ($FMNH_2$ (+ NO_3^-)) [5]; 0.20 (nitrate (+ NADPH) [5], NADH [14]) [5, 14]; 0.045 (NADPH (+ nitrate)) [5]; 0.012–0.015 (nitrate) [10]; 0.062 (NADPH) [14]; 0.000017 (FAD (electron carrier)) [14]; 0.0038 (FMN (electron carrier)) [14]

pH-optimum
7.0 [9]; 7–8 [5, 14]; 7.5–8.0 [2]; More (optima of associated activities) [14, 15]

pH-range
6–8.5 (6: about 55% of activity maximum, 8.5: about 50% of activity maximum) [14]

Temperature optimum (°C)
25 (assay at) [13]; 30 (assay at) [5]

Temperature range (°C)

3 ENZYME STRUCTURE

Molecular weight
180000 (Aspergillus nidulans, gel filtration, sucrose density gradient centrifugation) [5]
190000 (Aspergillus nidulans, gel filtration, sucrose density gradient centrifugation) [2]
197000 (Aspergillus nidulans, gel filtration, sucrose density gradient centrifugation) [13]
199000 (Penicillium chrysogenum, gel filtration, glycerol density gradient centrifugation) [16]
228000 (Neurospora crassa, gel filtration, sucrose density gradient centrifugation) [14]

Subunits
Dimer (2 × 91000, Aspergillus nidulans, SDS-PAGE [6], 1 × 97000 + 1 × 98000, Penicillium chrysogenum, SDS-PAGE [16], 1 × 115000 + 1 × 130000, Neurospora crassa, SDS-PAGE [4], 2 × 145000, Neurospora crassa (wild-type), SDS-PAGE [12]) [4, 6, 12, 16]
Tetramer (2 × 59000 + 2 × 38000, Aspergillus nidulans, SDS-PAGE [5], 4 × 116000, Funaria hygrometrica [10]) [5, 10]
More (nit-1 mutant enzyme is the apoprotein of nitrate reductase) [12]

Glycoprotein/Lipoprotein
–

4 ISOLATION/PREPARATION

Source organism
Penicillium chrysogenum [16]; Aspergillus nidulans [1, 2, 5–7, 13]; Neurospora crassa (wild-type and nit-1 mutant [12], nit-3 mutant [15]) [3, 4, 8, 9, 12, 14, 15]; Marchantia polymorpha (bryophyte) [11]

Enzyme Handbook © Springer-Verlag Berlin Heidelberg 1994
Duplication, reproduction and storage in data banks are only allowed with the prior permission of the publishers

Source tissue
Mycelium [9]

Localisation in source
Soluble [9, 14]; Cytoplasm [14]

Purification
Aspergillus nidulans (partial [2]) [2, 5, 13]; Neurospora crassa (nit-3 mutant [15]) [4, 9, 12, 14, 15]; Penicillium chrysogenum [16]

Crystallization
[14]

Cloned
–

Renaturated
–

5 STABILITY

pH

Temperature (°C)
4 (half-life: 3–5 days) [2]; 25 (half-life: 120 min) [1]; 30 (half-life: 66 min) [1]; 35 (half-life: 18.8 min (wild type) [1], 3.8–18.4 min (various temperature sensitive mutant strains)) [1]; 37 (half-life: 20 min [13], 14.5 min [1], less than 20 min (absence of FAD) [5]) [1, 5, 13]; 40 (50% loss of activity after 5 min) [9]; 49 (more than 90% loss of activity after 3 min) [14]; 50 (complete loss of activity after 5 min) [9]; More (heat lability of associated activities) [14, 15]

Oxidation

Organic solvent

General stability information

Storage
–15°C, purified enzyme, overnight storage, about 50% loss of activity [9]

6 CROSSREFERENCES TO STRUCTURE DATABANKS

PIR/MIPS code
S16292 (Neurospora crassa)

Brookhaven code

7 LITERATURE REFERENCES

[1] MacDonald, D.W., Cove, D.J.: Eur. J. Biochem.,47,107–110 (1974)
[2] McDonald, D.W., Coddington, A.: Eur. J. Biochem.,46,169–178 (1974)
[3] Klein Amy, N., Garrett, R.H., Anderson, B.M.: Biochim. Biophys. Acta,480,83–95 (1977)
[4] Pan, S.-S., Nason, A.: Biochim. Biophys. Acta,523,297–313 (1978)
[5] Minagawa, N., Yoshimoto, A.: J. Biochem.,91,761–774 (1982)
[6] Cooley, R.N., Tomsett, A.B.: Biochim. Biophys. Acta,831,89–93 (1985)
[7] Downey, R.J.: Biochem. Biophys. Res. Commun.,50,920–925 (1973)
[8] Nicholas, D.J.D., Nason, A.: J. Biol. Chem.,207,353–360 (1954)
[9] Nason, A., Evans, H.J.: J. Biol. Chem.,202,655–673 (1953)
[10] Johri, M.M., Henriques, B.M.A. in "Chemistry And Chemical Taxonomy Of Bryophytes" (Zinsmeister, H.D., Mues, R., Eds.),265–273, Oxford University Press, Oxford (1990)
[11] Takio, S., Takami, S., Hino, S.: J. Hattori Bot. Lab.,58,131–147 (1985)
[12] Horner, R.D.: Biochim. Biophys. Acta,744,7–15 (1983)
[13] Downey, R.J.: J. Bacteriol.,105,759–768 (1971)
[14] Garrett, R.H., Nason, A.: J. Biol. Chem.,244,2870–2882 (1969)
[15] Antoine, A.D.: Biochemistry,13,2289–2294 (1974)
[16] Renosto, F., Ornitz, D.M., Peterson, D., Segel, I.H.: J. Biol. Chem.,256,8616–8625 (1981)

Enzyme Handbook © Springer-Verlag Berlin Heidelberg 1994
Duplication, reproduction and storage in data banks are only allowed with the prior permission of the publishers

1 NOMENCLATURE

EC number
1.6.6.4

Systematic name
NAD(P)H:nitrite oxidoreductase

Recommended name
Nitrite reductase (NAD(P)H)

Synonymes
Reductase, nitrite (reduced nicotinamide adenine dinucleotide (phosphate))
NADH-nitrite oxidoreductase [1]
NADPH-nitrite reductase [1]
Assimilatory nitrite reductase [12]

CAS Reg. No.
9029-29-2

2 REACTION AND SPECIFICITY

Catalysed reaction
3 NAD(P)H + nitrite →
→ 3 $NAD(P)^+$ + NH_4OH + H_2O

Reaction type
Redox reaction

Natural substrates
NAD(P)H + nitrite (nitrite assimilation [16]) [12, 16]

Substrate spectrum
1 NAD(P)H + nitrite
2 More (other electron donors: dithionite [9, 12, 14, 15], $FADH_2$ [15], $FMNH_2$ [15], no reaction with (electron acceptors): nitrous oxide [6], hyponitrite [6], nitrate [6], no reaction with (electron donors): FMN [7], FAD [7], benzyl viologen [7], enzyme-hydroxylamine complex occurs as an intermediate in nitrite reductase reaction and hydroxylamine reductase reaction [6], enzyme also has hydroxylamine reductase activity [1, 4, 5, 7, 9, 12, 15], enzyme also catalyzes: NADH-dependent reduction of horse heart cytochrome c [5], 2,6-dinitrophenol-indophenol [5], $K_3Fe(CN)_6$ [5]) [1, 4–7, 9, 12, 14, 15]

Enzyme Handbook © Springer-Verlag Berlin Heidelberg 1994
Duplication, reproduction and storage in data banks are only allowed with the prior permission of the publishers

Product spectrum

1 $NAD(P)^+ + NH_4OH + H_2O$
2 ?

Inhibitor(s)

Cyanide [1, 7, 9, 12–15]; p-Chloromercuribenzoate [1, 7, 9, 13, 15, 16]; Sulphite [1, 9]; Arsenite [1, 7, 9]; 2,2'-Bipyridine [1, 13, 15]; 1,10-Phenanthroline [1, 12, 13]; NaN_3 [1, 14]; NAD^+ (substrate inhibition at low concentration [4], substrate inhibition at high concentration, at low concentration required for full activity, maximum at 1 mM [6]) [4, 6]; $NADP^+$ [7]; CO [9]; 8-Hydroxyquinoline [9, 13, 15]; Antimycin A [15]; Sodium sulfide [12]; Hydroxylamine (competitive to nitrite [15]) [12, 15]; Sodium bisulfite [12]; EDTA [13]; Sodium diethyldithiocarbamate [13]; Salicylic acid [13]; Dinitrophenol [15]; Iodine [15]; Rose bengal [15]; 1-Cyclohexyl-3-(2-morpholinoethyl)-carbodiimide metho-p-toluenesulphonate [15]; Nitrite (competitive to hydroxylamine) [15]; Atabrine [13]; Thiourea [13]; Diquinolyl [13]; KCN [16]; Hydrazine sulphate [13]; 2-Heptyl-4-hydroxyquinoline-N-oxide [13]; Naphthoquinone [13]; Iodosobenzoate [13]; Cyanate [14]; Carbamoyl phosphate [14]; p-Hydroxymercuribenzoate [14]; N-Ethylmaleimide [15]; Rotenone [15]; More (inactivation by preincubation with reduced pyridine nucleotide plus FAD, NO_2^- [11, 12], hydroxylamine [11, 12], cyanide [11], sulfite [11] or arsenite [11] protect against inactivation [11, 12], inactivation by preincubation with NADH, nitrite can protect against inactivation and reverse the process once it has occured [14], NADH inhibits in absence of appropriate substrate (nitrite or hydroxylamine) [15]) [11, 12, 14, 15]

Cofactor(s)/prostethic group(s)

Flavin (contains 0.4 mol of flavin per mol of enzyme [1], E. coli: no covalently bound flavin, enzyme contains one non-covalently bound FAD molecule [5], little or no flavin associated with purified protein [8]) [1, 5, 8]; FAD (E. coli: enzyme contains one non-covalently bound FAD molecule [5], exogenously added FAD is required for activity in vitro [8, 10], increases activity [13], stimulation [14], addition required for maximal activity [16]) [5, 8, 10, 13, 14, 16]; FMN (not detected [5], increases activity [13], no effect [14]) [13]; Siroheme [5, 8–10]; NADH (enzyme is specific for NADPH, reaction with NADH is approximately 1% of the rate with NADPH [7]) [1, 2, 3–6, 9, 12–15]; NADPH (enzyme is specific for NADPH, reaction with NADH is approximately 1% of the rate with NADPH [7]) [7–10, 12, 15]; More (heme not detectable) [1]

Metal compounds/salts

Fe ([2Fe-2S]-iron sulphur clusters [3], enzyme contains 5 Fe atoms per subunit [5], 9–10 mol Fe per mol of enzyme, heme iron associated with siroheme moieties, nonheme iron involved in iron-sulfur centers [8], deficiency of Fe derepresses activity [13]) [3, 5, 8, 13]; Fe^{3+} (stimulation) [13]; Cu (deficiency derepresses activity) [13]; Zn (deficiency derepresses activity) [13]; Mg (deficiency derepresses activity) [13] More (Mo not detectable [5], Azotobacter agile: enzyme contains an essential metal component) [5, 16]

Turnover number (min^{-1})

Specific activity (U/mg)

5.1 [1]; 0.18 [7]; 26.9 [9]; 0.730 (NADPH + NO_2^-) [12]; 0.520 (NADH + NO_2^- [12]) [12, 13]

K_m-value (mM)

0.016 (NADH (+ 2 mM NO_2^-, in presence of 1 mM NAD^+)) [6]; 0.005–0.0055 (NO_2^-) [6, 14]; 5.3 (hydroxylamine) [6, 15]; 0.4 (NO_2^-) [7]; 0.015 (NADPH [8], NADH [14]) [8, 14]; 0.0075 (NO_2^-) [8]; 0.00002 (FAD) [8]; 0.01 (NO_2^- (+ NADPH or NADH)) [12]; 2.5 (hydroxylamine) [12]; 0.0001 (FAD) [12]; 0.8 (NO_2^- (+ dithionite)) [12]; 0.05 (FAD, FMN) [13]; 0.0048 (NO_2^-) [15]; 0.0063 (NADH (+ NO_2^-)) [15]; 0.150 (NADH (+ hydroxylamine)) [15]; More (effect of NAD^+ on K_m for NADH [6]) [6, 13]

pH-optimum

7.1 [16]; 7.5–8.5 [7]; 7.5 [12]; 7.6 [13]

pH-range

6–8 (6: about 35% of activity maximum) [16]; 6.5–9.5 (6.5: about 50% of activity maximum, 9.5: about 15% of activity maximum) [7]; 4.0–11.0 (4.0: about 40% of activity maximum, 11.0: about 70% of activity maximum) [13]

Temperature optimum (°C)

Temperature range (°C)

3 ENZYME STRUCTURE

Molecular weight

67000 (Azotobacter chroococcum, sucrose density gradient centrifugation) [14]

190000 (E. coli, glycerol density gradient centrifugation) [1]

290000 (Neurospora crassa, gel filtration, sucrose density gradient centrifugation) [12]

Enzyme Handbook © Springer-Verlag Berlin Heidelberg 1994
Duplication, reproduction and storage in data banks are only allowed with the prior permission of the publishers

Subunits

Dimer (2 × 88000, E. coli, SDS-PAGE [1], 2 × 140000, Neurospora crassa, SDS-PAGE [8]) [1, 8]

Glycoprotein/Lipoprotein

–

4 ISOLATION/PREPARATION

Source organism

E. coli (strain OR75Ch15 [1], K12 [2, 4–6], strain Bn [7]) [1–7]; Neurospora crassa [8–13]; Azotobacter chroococcum [14]; Azotobacter agile [16]

Source tissue

Cell [1, 5, 7]; Mycelium [9]

Localisation in source

Purification

E. coli (strain OR75Ch15 [1], K12 [5], strain Bn [7]) [1, 5 , 7]; Neurospora crassa [8–13]

Crystallization

–

Cloned

[2]

Renaturated

–

5 STABILITY

pH

Temperature (°C)

50 (5 min, complete loss of activity) [16]

Oxidation

Organic solvent

General stability information

Nitrite stabilizes [5]; Hydroxylamine stabilizes [5]; FAD increases stability [5]; Dialysis: for 15 h against 1 mM cyanide, 1 mM 8-hydroxyquinoline or 1 mM EDTA followed by dialysis against 0.05 M phosphate buffer, no decrease in activity [7], against water or a variety of buffers, deactivation [13], stable to dialysis for 6 h against 0.1 M K_2HPO_4 and 1 mM glutathione [16]

Storage

4°C, addition of 1 mM NO_2^- or 50 mM hydroxylamine, 10% loss of activity after 24 h [5]; –17°C, overnight, 50% loss of activity [13]; –15°C, 0.1 M phosphate buffer, pH 7.5, 1 mM glutathione, partially purified enzyme, stable [16]

6 CROSSREFERENCES TO STRUCTURE DATABANKS

PIR/MIPS code

S00529 (Escherichia coli fragment); JH0181 (long form Emericella nidulans); PS0299 (short form Emericella nidulans)

Brookhaven code

7 LITERATURE REFERENCES

[1] Coleman, K.J., Cornish-Bowden, A., Cole, J.A.: Biochem. J.,175,483–493 (1978)
[2] Macdonald, H., Cole, J.: Mol. Gen. Genet.,200,328–334 (1985)
[3] Cammack, R., Jackson, R.H., Cornish-Bowden, A., Cole, J.A.: Biochem. J.,207,333–339 (1982)
[4] Jackson, R.H., Cole, J.A., Cornish-Bowden, A.: Biochem. J.,199,171–178 (1981)
[5] Jackson, R.H., Cornish-Bowden, A., Cole, J.A.: Biochem. J.,193,861–867 (1981)
[6] Coleman, K.J., Cornish-Bowden, A., Cole, J.A.: Biochem. J.,175,495–499 (1978)
[7] Lazzarini, R.A., Atkinson, D.E.: J. Biol. Chem.,236,3330–3335 (1961)
[8] Prodouz, K.N., Garrett, R.H.: J. Biol. Chem.,256,9711–9717 (1981)
[9] Greenbaum, P., Prodouz, K.N., Garrett, R.H.: Biochim. Biophys. Acta, 526, 52–64 (1978)
[10] Vega, J.M., Garrett, R.H.: J. Biol. Chem.,250,7980–7989 (1975)
[11] Vega, J.M., Greenbaum, P., Garrett, R.H.: Biochim. Biophys. Acta,377,251–257 (1975)
[12] Lafferty, M.A., Garrett, R.H.: J. Biol. Chem.,249,7555–7567 (1974)
[13] Nicholas, D.J.D., Medina, A., Jones, O.T.G.: Biochim. Biophys. Acta,37,468–476 (1960)
[14] Vega, J.M., Guerrero, M.G., Leadbetter, E., Losada, M.: Biochem. J.,133,701–708 (1973)
[15] Wang, R., Nicholas, D.J.D.: Phytochemistry,25,2463–2469 (1986)
[16] Spencer, D., Takahashi, H., Nason, A.: J. Bacteriol.,73,553–562 (1957)

Enzyme Handbook © Springer-Verlag Berlin Heidelberg 1994
Duplication, reproduction and storage in data banks are only allowed with the prior permission of the publishers

1 NOMENCLATURE

EC number
1.6.6.6

Systematic name
NADH:hyponitrite oxidoreductase

Recommended name
Hyponitrite reductase

Synonymes

CAS Reg. No.
9029-30-5

2 REACTION AND SPECIFICITY

Catalysed reaction
2 NADH + hyponitrite →
→ 2 NAD^+ + 2 NH_2OH

Reaction type
Redox reaction

Natural substrates
Hyponitrite + NADH [1]

Substrate spectrum
1 Hyponitrite + NADH [1]

Product spectrum
1 NH_2OH + NAD^+ [1]

Inhibitor(s)
Cyanide [1]; Hydrazine sulfate [1]; p-Chloromercuribenzoate [1]; 2,6-Dinitrophenol [1]; 2,2'-Diquinolyl [1]

Cofactor(s)/prostethic group(s)
Flavin (flavoprotein) [1]; NADH [1]

Metal compounds/salts
Metalloprotein [1]

Enzyme Handbook © Springer-Verlag Berlin Heidelberg 1994
Duplication, reproduction and storage in data banks are only allowed with the prior permission of the publishers

Turnover number (min^{-1})

Specific activity (U/mg)

K_m-value (mM)

pH-optimum

pH-range

Temperature optimum (°C)

Temperature range (°C)

3 ENZYME STRUCTURE

Molecular weight

Subunits

Glycoprotein/Lipoprotein
–

4 ISOLATION/PREPARATION

Source organism
Neurospora crassa [1]

Source tissue

Localisation in source

Purification

Crystallization
–

Cloned
–

Renaturated
–

5 STABILITY

pH

Temperature (°C)

Oxidation

Organic solvent

General stability information

Storage

6 CROSSREFERENCES TO STRUCTURE DATABANKS

PIR/MIPS code

Brookhaven code

7 LITERATURE REFERENCES

[1] Medina, A., Nicholas, D.J.D.: Nature,179,533–534 (1957)

Enzyme Handbook © Springer-Verlag Berlin Heidelberg 1994
Duplication, reproduction and storage in data banks are only allowed with the prior permission of the publishers

1 NOMENCLATURE

EC number
1.6.6.7

Systematic name
NADPH:4(dimethylamino)azobenzene oxidoreductase

Recommended name
Azobenzene reductase

Synonymes
New Coccine (NC)-reductase [1]
NC-Reductase
Azo-dye reductase [15]
Orange II azoreductase [4]
NAD(P)H:1-(4'-sulfophenylazo)-2-naphthol oxidoreductase
Orange I azoreductase [5]
Azo reductase [9, 12]
Azoreductase [7, 10, 11]
Nicotinamide adenine dinucleotide (phosphate) azoreductase [14]
$NADPH_2$–dependent azoreductase [14]
Reductase, azobenzene
Dimethylaminobenzene reductase
p-Dimethylaminoazobenzene azoreductase
Dibromopropylaminophenylazobenzoic azoreductase
N, N-Dimethyl-4-phenylazoaniline azoreductase
p-Aminoazobenzene reductase
Azo dye reductase
Methyl red azoreductase

CAS Reg. No.
9029-31-6

2 REACTION AND SPECIFICITY

Catalysed reaction
NADPH + 4-(dimethylamino)azobenzene →
→ $NADP^+$ + N, N-dimethyl-1,4-phenylenediamine + aniline

Reaction type
Redox reaction

Enzyme Handbook © Springer-Verlag Berlin Heidelberg 1994
Duplication, reproduction and storage in data banks are only allowed with the prior permission of the publishers

Natural substrates

NADPH + 4-(dimethylamino)azobenzene (key enzyme of azo dye degradation, selective agent during experimental evolution in continuous cultures) [5]

Substrate spectrum

1 NADPH + 4-(dimethylamino)azobenzene
2 Methyl red + NAD(P)H (Methyl red is identical with 2'-carboxy-4-N, N-dimethylazobenzene) [2]
3 New Coccine + NADPH [1]
4 Orange II + NAD(P)H (Orange II is identical with 4-[(2-hydroxy-1-naphthalenyl)azo]benzene sulfonic acid) [4]
5 Orange I + NAD(P)H (Orange I is identical with 4-[(4-hydroxy-1-naphthalenyl)azo]benzene sulfonic acid) [5]
6 Direct blue 15 + NAD(P)H [6]
7 1,2-Dimethyl-4(p-carboxyphenylazo)-5-hydroxybenzene + NADPH [13]
8 Amaranth + NADPH [8, 9]
9 More (specific for reduction of Methyl red [2], electron transport protein [3], Orange I azoreductase: absolute requirement for a hydroxyl group in 4'-position of the naphthol ring of the substrate molecule [5], Orange II azoreductase: requirement for substrates with a 2-naphthol moiety [5]) [2, 3, 5]

Product spectrum

1 $NADP^+$ + N, N-1,4-phenylenediamine + aniline
2 2-Carboxy-1-phenylamine + N, N-dimethyl-1,4-phenylenediamine + $NADP^+$
3 ?
4 Sulfanilic acid + $NAD(P)^+$ + aminonaphthol (sulfanilic acid is identical with 4-amino-1-benzenesulfonic acid, aminonaphthol is identical with 1-amino-2-hydroxynaphthalene) [4]
5 4-Amino-1-benzenesulfonic acid + 1-amino-4-hydroxynaphthalene + $NADP^+$
6 ?
7 1,2-Dimethyl-4-amino-5-hydroxybenzene + p-carboxyphenylamine + $NADP^+$
8 alpha-Naphthylamine-4-sulfonic acid + beta-naphthol-3,6-disulfonic acid + $NADP^+$
9 ?

Inhibitor(s)

SDS [1]; Orange II (substrate inhibition at high concentration) [4]; Orange I (substrate inhibition) [5]; CO [8]; Ag^+ [1]; Hg^{2+} [1]; Cu^{2+} [1]; Fe^{2+} [1]; Fe^{3+} [1]; 4-Dimethylaminoazobenzene [15]; Fatty acids (oxidized unsaturated) [15]

Cofactor(s)/prostethic group(s)

NADPH (specific for [1]) [1, 2, 4, 5, 8, 14, 15]; NADH [2, 4, 5, 14]; FAD (increases activity [1, 8, 9, 13, 14], one mol of enzyme contains 2 mol of FAD [2], prosthetic group [15]) [1, 2, 8, 9, 13–15]; FMN (increases activity [1, 8, 9], enzyme contains no FMN [2]) [1, 2, 8, 9]; Riboflavin (no activation [1], activation [8, 9]) [8, 9]; Methylviologen (activation) [8]; Flavin (prosthetic group seems to be flavin [1], typical flavoprotein absorption spectrum [2]) [1, 2]; More (no activity with NADH [1], effect of different cofactors on azoreductase activity of intestinal anaerobic bacteria [6], apparent non-identity of cytochrome c reductase and flavin-dependent azoreductase activity [10], P-450 type cytochromes are responsible for azoreductase activity [11]) [1, 6, 10, 11]

Metal compounds/salts

Mg^{2+} (slight activation) [1]; Dicumarol [3, 14]

Turnover number (min^{-1})

Specific activity (U/mg)

0.566 [1]; 17.833 [2]; 2.87 [5]; More (continuous assay for azo reductase) [12]

K_m-value (mM)

0.25 (NADPH) [2]; 0.005 (NADPH) [4]; 0.4 (NADH) [2]; 0.180 (NADH) [4]; 0.32 (Methylene red) [2]; 1.0 (1-(4'-sulfophenylazo)-2-naphthol) [4]; 0.8 (1-(4'-carboxyphenylazo)-2-naphthol) [4]; 1.5 (1-(3'-carboxyphenylazo)-2-naphthol) [4]; 1.3 (1-(2'-methyl-4-sulfophenylazo)-2-naphthol) [4]; 0.9 (1-phenylazo-2-hydroxy-6-sulfonaphthalene) [4]; 0.0026 (1-(4'-sulfophenylazo)-4-naphthol) [5]; 0.0024 (1-(4'-carboxyphenylazo)-4-naphthol) [5]; 0.204 (1-(4'-sulfophenylazo)-4-anthranol) [5]; 0.040 (1-(4'-sulfophenyl-N-methylhydrazo)-4-naphthol) [5]; 0.0011 (NADPH) [5]; 0.054 (NADH) [5]; 0.034 (amaranth) [9]; 0.29 (1,2-dimethyl-4-(p-carboxyphenylazo)-5-hydroxy-benzene) [13]; 0.12 ($NADP^+$) [13]

pH-optimum

5.5–6.0 [2]; 5.0–8.0 (Orange II + NADH) [4]; 6.2 (Orange I) [5]; 6.5 (carboxy-Orange II + NADH) [4]; 6.8 (carboxy-Orange I) [5]; 7.0 [1]

pH-range

4.0–10.0 (4.0: about 10% of activity maximum, 10.0: about 20% of activity maximum) [1]; 4.0–7.0 (4.0: about 25% of activity maximum, 7.0: about 20% of activity maximum) [2]

Temperature optimum (°C)

40 [1]; 41 [5]; 45 [4]

Enzyme Handbook © Springer-Verlag Berlin Heidelberg 1994
Duplication, reproduction and storage in data banks are only allowed with the prior permission of the publishers

Temperature range (°C)

5–70 (at 5°C and 70°C: about 55% of activity maximum) [1]; 20–50 (20°C: about 25% of activity maximum (carboxy-Orange II), about 30% of activity maximum (Orange II), 50 °C: about 95% of activity maximum (carboxy-Orange II), about 80% of activity maximum (Orange II)) [4]

3 ENZYME STRUCTURE

Molecular weight

21500 (Pseudomonas sp. K24, gel filtration, Orange I azoreductase) [5]
32000 (Pseudomonas sp. strain KF 46, gel filtration, Orange II azoreductase) [4]
52000 (rat, gel filtration) [2]

Subunits

Monomer (1 × 20200, Pseudomonas sp. K24, SDS-PAGE, Orange I azoreductase [5], 1 × 30000, Pseudomonas sp. KF46, SDS-PAGE, Orange II azoreductase [4]) [4, 5]
Dimer (?, 2 × 30000, rat, SDS-PAGE) [2]

Glycoprotein/Lipoprotein

–

4 ISOLATION/PREPARATION

Source organism

Rat [2, 3, 7–10, 13–15]; Bacillus cereus T-105 [1]; Pseudomonas sp. (strain KF46 (Orange II azoreductase induced by both Orange II and carboxy-Orange II) [4], strain K24 [5]) [4, 5]; Mouse [11]; Eubacterium hadrum [6]; Eubacterium sp. [6]; Clostridium clostridiiforme [6]; Butyrivibrio sp. [6]; Bacteroides sp. [6]; Clostridium paraputrificum [6]; Clostridium nexile [6]; Clostridium sp. [6]; Clostridium perfringens [6]; Guinea pig [13]; Chicken [13]; Dog [13]; Meladow vole [13]

Source tissue

Cell [1]; Liver [2, 3, 7–9, 11, 13–15]

Localisation in source

Intracellular [1]; Extracellular [6]; Soluble [7]; Mitochondria (activity higher in crude outer mitochondrial membrane fraction than in mitoplast and intact mitochondria) [7]

Purification

Bacillus cereus T-105 [1]; Rat [2]; Pseudomonas sp. strain KF 46 [4]; Pseudomonas sp. strain K24 [5]

Crystallization

–

Cloned

–

Renaturated

–

5 STABILITY

pH

6.0–8.0 (5°C, 7 days, stable) [1]

Temperature (°C)

35 (stable below) [4]; 60 (10 min, stable up to) [1]; 80 (10 min, 45% loss of activity) [1]

Oxidation

Inactivation by oxygen [6]

Organic solvent

General stability information

Storage

–20°C, 50% glycerol, several months, stable [4]

6 CROSSREFERENCES TO STRUCTURE DATABANKS

PIR/MIPS code

Brookhaven code

7 LITERATURE REFERENCES

[1] Matsudomi, N., Kobayashi, K., Akuta, S.: Agric. Biol. Chem.,41,2323–2329 (1977)
[2] Huang, M.-T., Miwa, G.T., Lu, A.Y.H.: J. Biol. Chem.,254,3930–3934 (1979)
[3] Huang, M.-T., Miwa, G.T., Cronheim, N., Lu, A.Y.H.: J. Biol. Chem.,254,11223–11227 (1979)
[4] Zimmermann, T., Kulla, H.G., Leisinger, T.: Eur. J. Biochem.,129,197–203 (1982)
[5] Zimmermann, T., Gasser, F., Kulla, H.G., Leisinger, T.: Arch. Microbiol.,138,37–43 (1984)
[6] Rafii, F., Franklin, W., Cerniglia, C.E.: Appl. Environ. Microbiol.,56,2146–2151 (1990)
[7] Moreno, S.N.J., Mason, R.P., Docampo, R.: J. Biol. Chem.,259,14609–14616 (1984)

Enzyme Handbook © Springer-Verlag Berlin Heidelberg 1994
Duplication, reproduction and storage in data banks are only allowed with the prior permission of the publishers

[8] Fujita, S., Peisach, J.: Biochim. Biophys. Acta,719,178–189 (1982)
[9] Mallett, A.K., King, L.J., Walker, R.: Biochem. J.,201,589–595 (1982)
[10] Mallett, A.K., Walker, R., King, L.J.: Biochem. Soc. Trans.,6,1302–1305 (1978)
[11] Fujita, S., Peisach, J.: J. Biol. Chem.,253,4512–4513 (1978)
[12] Mallett, A.K., King, L.J., Walker, R.: Biochem. Soc. Trans.,5,1522–1524 (1977)
[13] Smith, E.J., Van Loon, E.J.: Anal. Biochem.,31,315–320 (1969)
[14] Daniel, J.W.: Biochem. J.,11,19P-20P (1969)
[15] Ketterer, B., Ross-Mansell, P., Davidson, H.: Biochem. J.,107,15P-16P (1968)

1 NOMENCLATURE

EC number
1.6.6.8

Systematic name
NADPH:guanosine-5'-phosphate oxidoreductase (deaminating)

Recommended name
GMP reductase

Synonymes
Guanosine 5'-monophosphate reductase [1]
NADPH:GMP oxidoreductase (deaminating) [11]
Reductase, guanylate
Guanosine monophosphate reductase
Guanylate reductase

CAS Reg. No.
9029-32-7

2 REACTION AND SPECIFICITY

Catalysed reaction
NADPH + guanosine 5'-phosphate →
→ $NADP^+$ + inosine 5'-phosphate (mechanism: ordered sequential with GMP binding first [9], first order reaction [12])

Reaction type
Redox reaction
Reductive deamination [7]

Natural substrates
NADPH + guanosine 5'-phosphate (enzyme of salvage pathway of purine synthesis [3], enzyme involved in interconversion of purine ribonucleotides [7]) [3, 7]

Substrate spectrum
1 NADPH + guanosine 5'-phosphate (ir [7, 14], r [11], the rate of the reverse reaction is 6% of the forward reaction [11])
2 More (arabinosylGMP, 2'-dGMP and 8-azaGMP reductively deaminated to their corresponding IMP analog at rates 1–2% the rate with GMP [9], less than 10% of the NADPH rate: thionicotinamide-NADPH, deamino-NADPH, 3-acetylpyrimidine-NADPH [9], specific for GMP [14]) [9, 14]

Enzyme Handbook © Springer-Verlag Berlin Heidelberg 1994
Duplication, reproduction and storage in data banks are only allowed with the prior permission of the publishers

Product spectrum

1 $NADP^+$ + inosine 5'-phosphate (ir [7, 14], r [11])
2 ?

Inhibitor(s)

6-Chloro-9-beta-D-ribofuranosylpurine 5'-phosphate [13]; Iodoacetamide [13]; 6-Thio-IMP [9, 13]; 6-Thio-GMP [9, 13]; ADP [13]; AMP (weak) [13, 14]; Adenosine 2'-phosphate [13]; ATP (no inhibition without Mg^{2+}, MgATP: no inhibition at high GMP concentration (0.010 mM-0.100 mM), strong inhibition at lower concentration (below 0.010 mM) [8], no effect [2, 11]) [8]; 8-Aza-XMP [9]; 6-Thio-XMP [9]; Arabinosyl-XMP [9]; 8-Aza-7-deaza-XMP [9]; 2'-dXMP [9]; 6-Chloropurine ribonucleotide [9]; p-Hydroxymercuribenzoate [11, 12]; 5,5'-Dithiobis(2-nitrobenzoic acid) [12]; Mg^{2+} [12]; Fe^{2+} [12]; Zn^{2+} [12]; Mn^{2+} [12]; Ni^{2+} [12]; Cu^{2+} [12]; Ca^{2+} [12]; XMP (GTP and diguanosine tetraphosphate counteract inhibition [10]) [2, 9, 10, 12]; IMP [2, 11–14]; GTP (slight) [13]

Cofactor(s)/prostethic group(s)

NADPH (specific for [8, 11, 14], markedly preferred as coenzyme [9]); $NADP^+$ [11]; GTP (nonessential activator) [9]; 2-Mercaptoethanol (in absence of mercaptoethanol 50% of maximal velocity [11], activation [12]) [11, 12]; Dithiothreitol (activation) [12]; Cysteine (activation) [12]; Thioglycolic acid (activation) [12]; More (NADH: no cofactor function) [2]

Metal compounds/salts

No requirement for monovalent or divalent cations [14]

Turnover number (min^{-1})

Specific activity (U/mg)

More [13]; 0.35 [9]; 78 [11]; 0.564 [10]; 0.0052 [12]

K_m-value (mM)

0.031 (NADPH) [2]; 0.0026 (GMP) [9]; 0.0014 (GMP) [2]; 0.010 (less than, NADPH) [8]; 0.0094 (GMP) [8]; 0.0169 (NADPH) [9]; More [10–12, 14]

pH-optimum

7.5 [2, 12]; 8.3 [8]; 8.5 [11]; 7.5–8.2 [14]

pH-range

7–9.5 (7: about 25% of activity maximum, 9.5: about 30% of activity maximum) [11]; 6–9 (6: about 50% of activity maximum, 9: about 70% of activity maximum) [12]

Temperature optimum (°C)

37 (assay at) [9]

Temperature range (°C)

3 ENZYME STRUCTURE

Molecular weight

36000 (E. coli) [1]
37437 (E. coli, determination of nucleotide sequence) [4]
90000 (bovine, gel filtration) [2]
170000 (human, gel filtration) [9]
250000 (Artemia salina, gel filtration) [11]

Subunits

Tetramer (4 × 45000, Salmonella typhimurium) [6]

Glycoprotein/Lipoprotein

–

4 ISOLATION/PREPARATION

Source organism

Mycoplasma mycoides [8]; Human [9, 12]; Aerobacter aerogenes [13, 14]; E. coli [1, 4, 14]; Bovine [2]; Bacillus subtilis (mutants containing mutations in the salvage pathway of purine synthesis) [3]; Leishmania donovani [5]; Salmonella typhimurium [6, 7, 14]; Artemia salina [10, 11]

Source tissue

Thymus [2]; Erythrocytes [9, 12]; Cysts [10]

Localisation in source

Purification

Bovine [2]; Human [9]; Artemia salina (partial) [10, 11]; Aerobacter aerogenes [13]

Crystallization

–

Cloned

(E. coli K-12 guaC gene) [1]

Renaturated

–

5 STABILITY

pH

Temperature (°C)

67 (15 min, 40% loss of activity) [12]

Enzyme Handbook © Springer-Verlag Berlin Heidelberg 1994
Duplication, reproduction and storage in data banks are only allowed with the prior permission of the publishers

Oxidation

Organic solvent

General stability information

Thawing and freezing, many times with less than 15% loss of activity over 6 months [9]

Storage

–70°C [9]; –70°C, 2 months, less than 20% loss of activity [12]

6 CROSSREFERENCES TO STRUCTURE DATABANKS

PIR/MIPS code

S01671 (Escherichia coli)

Brookhaven code

7 LITERATURE REFERENCES

[1] Moffat, K.G., Mackinnon, G.: Gene,40,141–143 (1985)
[2] Stephens, R.W., Whittaker, V.K.: Biochem. Biophys. Res. Commun.,53,975–981 (1973)
[3] Endo, T., Uratani, B., Freese, E.: J. Bacteriol.,155,169–179 (1983)
[4] Andrews, S.C., Guest, J.R.: Biochem. J.,255,35–43 (1988)
[5] Spector, T., Jones, T.E.: Biochem. Pharmacol.,31,3891–3897 (1982)
[6] Neuhard, J., Nygaard, P. in "Escherichia Coli And Salmonella Typhimurium Cellular And Molecular Biology" (Neidhardt, F.C., Ingraham, J.L., Low, K.B., Magasanik, B., Schaechter, M., Umbarger, E., Eds.) 1,445–473, American Society For Microbiology, Washington (1987)
[7] Garber, B.B., Jochimsen, B.U., Gots, J.S.: J. Bacteriol.,143,105–111 (1980)
[8] Mitchell, A., Sin, I.L., Finch, L.R.: J. Bacteriol.,134,706–712 (1978)
[9] Spector, T., Jones, T.E., Miller, R.L.: J. Biol. Chem.,254,2308–2315 (1979)
[10] Renart, M.F., Renart, J., Sillero, M.A.G., Sillero, A.: Biochemistry,15,4962–4966 (1976)
[11] Renart, M.F., Sillero, A.: Biochim. Biophys. Acta,341,178–186 (1974)
[12] Mackenzie, J.J., Sorensen, L.B.: Biochim. Biophys. Acta,327,282–294 (1973)
[13] Brox, L.W., Hampton, A.: Biochemistry,7,398–405 (1968)
[14] Mager, J., Magasanik, B.: J. Biol. Chem.,235,1474–1478 (1960)

1 NOMENCLATURE

EC number
1.6.6.9

Systematic name
NADH:trimethylamine-N-oxide oxidoreductase

Recommended name
Trimethylamine-N-oxide reductase

Synonymes
Reductase, trimethylamine N-oxide
Trimethylamine N-oxide reductase
Trimethylamine oxide reductase
TMAO reductase [1, 2]
TOR [1]

CAS Reg. No.
37256-34-1

2 REACTION AND SPECIFICITY

Catalysed reaction
NADH + trimethylamine-N-oxide →
→ NAD^+ + trimethylamine + H_2O

Reaction type
Redox reaction

Natural substrates
Trimethylamine-N-oxide + ? (cytochrome 554,557 may be the physiological electron donor [15], trimethylamine-N-oxide acts as a terminal electron acceptor for an anaerobic respiratory chain [5, 14, 15], anaerobic respiration [8–10], cytochrome c-556 may be the physiological electron donor [12]) [5, 8–10, 12, 14, 15]

Enzyme Handbook © Springer-Verlag Berlin Heidelberg 1994
Duplication, reproduction and storage in data banks are only allowed with the prior permission of the publishers

Substrate spectrum

1 Trimethylamine-N-oxide + NADH (other electron donors: benzyl viologen [1, 4, 7, 8], FMN (17% of the activity with methyl viologen) [1], reduced methylene blue [4], reduced N-methylphenazonium methosulfate [4], NADH (no effect [1, 4]) [5, 6], no effect: NADPH [1, 4], ascorbate [1], cytochrome c [1], $FADH_2$ [1, 4]) [1, 4–8]
2 N, N-Dimethyldodecylamine N-oxide + electron donor [4]
3 Hydroxylamine N-oxide + electron donor (r [3]) [3, 4]
4 gamma-Picoline N-oxide + electron donor [3, 4]
5 Adenosine N'-oxide + electron donor [4]
6 Chlorate + electron donor [1, 3, 12]
7 More (a single enzyme responsible for both trimethylamine-N-oxide and dimethylsulphoxide reductase [12, 13]) [1, 3–8, 12, 13]

Product spectrum

1 Trimethylamine + NAD^+ + H_2O
2 N, N-Dimethyldodecylamine + H_2O + oxidized electron donor
3 Hydroxylamine + H_2O + oxidized electron donor
4 gamma-Picoline + H_2O + oxidized electron donor
5 Adenosine + H_2O + oxidized electron donor
6 ?
7 ?

Inhibitor(s)

Hg^{2+} [1]; Cu^{2+} [1]; FMN (inhibition, when added to the standard assay mixture containing benzyl viologen as electron acceptor) [1]; FAD (inhibition, when added to the standard assay mixture containing benzyl viologen as electron acceptor) [1]; KCN (not [6], partial [4], E. coli enzyme inhibited, Salmonella typhimurium enzyme not [2]) [2–4]; N-Ethylmaleimide [2]; NaN_3 (10 mM, 5% inhibition of inducible enzyme, 50–60% inhibition of constitutive enzyme [2], E. coli enzyme inhibited, Salmonella typhimurium enzyme not [3], not [6]) [2, 3]; p-Chloromercuribenzoate (5 mM, 2% inhibition of inducible enzyme, 50–60% inhibition of constitutive enzyme [2]) [2, 4, 6]; EDTA (E. coli enzyme inhibited, Salmonella typhimurium enzyme not [3], slight [6]) [3, 6]; o-Phenanthroline (slight [3, 4]) [3, 4, 6]; Na_2MoO_4 [4]; Na_2WO_4 [4]; 5,5-Dithiobis-(2-nitrobenzoic acid) [4]; Iodoacetate (partial) [4, 6]; Sodium chlorate (partial) [4]; 2,2-Dipyridyl (slight [4]) [4, 6]; NADH (slight inhibition at 14 mM) [6]; Semicarbazide [6]; Cupferron [6]; 8-Hydroxyquinoline (slight) [6]; $CuSO_4$ [6]; Fe^{2+} [6]; Mo^{6+} [6]; Mn^{2+} [6]; Ni^{2+} [6]; Co^{2+} [6]

Cofactor(s)/prostethic group(s)

NADH; NAD^+; FAD (stimulation) [4, 6]; FMN (stimulation) [4, 6]

Metal compounds/salts

Molybdenum (enzyme contains 1.96 atoms of molybdenum [3], contains molybdopterin [3], enzyme contains 1.3 atoms of molybdenum [4]) [3, 4]; Iron (enzyme contains 0.96 atoms of iron) [3]; Zinc (enzyme contains 1.52 atoms of zinc) [3]; Fe^{3+} (slight stimulation) [6]

Turnover number (min^{-1})

Specific activity (U/mg)

211.53 [4]; 237 [11]; More [1]

K_m-value (mM)

0.89 (trimethylamine-N-oxide) [1]; 2.2 (chlorate) [1]; 0.67 (trimethylamine-N-oxide (+ methyl viologen), pH 6.9) [3]; 0.15 (methyl viologen (+ trimethylamine-N-oxide), pH 6.9) [3]; 0.95 (trimethylamine-N-oxide (+ benzyl viologen), pH 5.5) [3]; 0.33 (benzyl viologen (+ trimethylamine-N-oxide)) [3]; 0.02 (trimethylamine N-oxide) [4]; 2.41 (N, N-dimethyldodecylamine N-oxide) [4]; 6.92 (gamma-picoline N-oxide) [4]; 0.3 (trimethylamine-N-oxide) [13]; 6 (dimethylsulphoxide) [13]

pH-optimum

5.5 (trimethylamine-N-oxide + benzyl viologen [3]) [2, 3]; 5.65 [1]; 6.9 (trimethylamine-N-oxide + methyl viologen) [3]; 7.0–7.5 [6]

pH-range

6.0–8.0 (6.0: about 35% of activity maximum, 8.0: about 80% of activity maximum) [6]

Temperature optimum (°C)

45 [1]

Temperature range (°C)

---

3 ENZYME STRUCTURE

Molecular weight

47000 (Shewanella sp., 2 MW forms: 47000 (SDS-PAGE in absence of reducing agents) and 84000 (gel filtration, SDS-PAGE under reducing conditions)) [4]

80000 (Rhodobacter capsulatus, SDS-PAGE) [12]

84000 (Shewanella sp., 2 MW forms: 47000 (SDS-PAGE in absence of reducing conditions) and 84000 (gel filtration, SDS-PAGE under reducing conditions)) [4]

95000 (Proteus vulgaris, gradient gel electrophoresis, gel isoelectric focusing) [13]

Enzyme Handbook © Springer-Verlag Berlin Heidelberg 1994
Duplication, reproduction and storage in data banks are only allowed with the prior permission of the publishers

110000 (E. coli K12, a single inducible trimethylamine-N-oxide reductase which can exist as a dimer (MW 230000) or a monomer (MW 110000)) [11]
200000 (E. coli, TMAO reductase I, the major enzyme among inducible TMAO reductases, SDS-PAGE) [3]
230000 (E. coli K12, a single inducible trimethylamine-N-oxide reductase which can exist as a dimer (MW 230000) or a monomer (MW 110000)) [11]
332000 (Salmonella typhimurium, inducible enzyme, gel filtration) [1]

Subunits

Monomer (1 × 84000, Shewanella sp., gel filtration, SDS-PAGE under reducing conditions [4], E. coli K12, a single inducible trimethylamine-N-oxide reductase which can exist as a dimer (MW 230000) or a monomer (MW 110000) [11]) [4, 11]
Dimer (2 × 95000, E. coli, TMAO reductase I, the major enzyme among inducible TMAO reductases, SDS-PAGE [3], a single inducible trimethylamine-N-oxide reductase which can exist as a dimer (MW 230000) or a monomer (MW 110000) [11]) [3, 11]
Tetramer (4 × 84000, Salmonella typhimurium, inducible enzyme, gel filtration) [1]

Glycoprotein/Lipoprotein

–

4 ISOLATION/PREPARATION

Source organism

Erythrobacter sp. (aerobic photosynthetic bacterium) [10]; Vibrio parahaemolyticus [6]; E. coli (a constitutive enzyme and an inducible enzyme with different properties [2, 8], TMAO reductase I, the major enzyme among inducible TMAO reductases [3], reduction of trimethylamine-N-oxide is catalyzed by at least 2 enzymes in E. coli: trimethylamine-N-oxide reductase (anaerobically induced by trimethylamine N-oxide) and the constitutive enzyme dimethyl sulfoxide reductase [7], inducible enzyme [11]) [2, 3, 5, 7, 8, 11, 15]; Shewanella sp. NCMB 400 (marine bacterium, 2 enzyme forms) [4]; Rhodobacter capsulatus (a single enzyme responsible for both trimethylamine-N-oxide and dimethylsulphoxide reductase [12]) [9, 12]; Proteus vulgaris (a single enzyme responsible for both trimethylamine-N-oxide and dimethylsulphoxide reductase) [13]; Alteromonas sp. NCMB (inducible) [14]; Salmonella typhimurium (inducible enzyme [1], mutants defective in trimethylamine oxide reduction [16]) [1, 16]

Source tissue

Localisation in source

Cytoplasm (low activity) [10]; Soluble (inducible enzyme) [2]; Periplasm (inducible enzyme [8]) [4, 8, 10, 12, 14]; Membrane (constitutive enzyme [2, 8]) [2, 6, 8]

Purification

Salmonella typhimurium (inducible enzyme) [1]; E. coli (TMAO reductase I, the major enzyme among inducible TMAO reductases [3]) [3, 11]; Shewanella sp. (2 enzyme forms) [4]

Crystallization

–

Cloned

–

Renaturated

–

---

5 STABILITY

pH

6–8.5 (stable in pH-range 6–8.5, stability peak at pH 7.3) [1]

Temperature (°C)

60 (20 min, 10% loss of activity) [2]; 65 (5 min, 90% loss of constitutive activity, 10% loss of inducible activity) [8]; 75 (10 min, pH 7.3, quite stable) [1]; 80 (rapid loss of activity) [1]; 90 (10 min, pH 7.3, complete loss of activity) [1]

Oxidation

Aeration for 10 min causes 10% inactivation [2]

Organic solvent

General stability information

2-Mercaptoethanol protects against inactivation during preparation of cell extract, ineffective as stabilizer during storage [6]

Storage

0°C, 50 mM sodium phosphate buffer, pH 7.3, 30% loss of activity after 17 days [1]; –80°C, 10% loss of activity after 15 days [1]

Enzyme Handbook © Springer-Verlag Berlin Heidelberg 1994
Duplication, reproduction and storage in data banks are only allowed with the prior permission of the publishers

6 CROSSREFERENCES TO STRUCTURE DATABANKS

PIR/MIPS code

Brookhaven code

7 LITERATURE REFERENCES

[1] Kwan, H.S., Barrett, E.L.: J. Bacteriol.,155,1455–1458 (1983)
[2] Violet, M., Medani, C.-L., Giordano, G.: FEMS Microbiol. Lett.,27,85–91 (1985)
[3] Yamamoto, I., Okubo, N., Ishimoto, M.: J. Biochem.,99,1773–1779 (1986)
[4] Clarke, G.J., Ward, F.B.: J. Gen. Microbiol.,13,379–386 (1988)
[5] Cox, J.C., Knight, R.: FEMS Microbiol. Lett.,12,249–252 (1981)
[6] Unemoto, T., Hayashi, M., Miyaki, K., Hayashi, M.: Biochim. Biophys. Acta,110,319–328 (1965)
[7] Yamamoto, I., Hinakura, M., Seki, S., Seki, Y., Kondo, H.: Curr. Microbiol.,20,245–249 (1990)
[8] Silvestro, A., Pommier, J., Pascal, M.-C., Giordano, G.: Biochim. Biophys. Acta,999,208–216 (1989)
[9] McEwan, A.G., Richardson, D.J., Hudig, H., Ferguson, S.J., Jackson, J.B.: Biochim. Biophys. Acta,973,308–314 (1989)
[10] Arata, H., Serikawa, Y., Takamiya, K.-i.: J. Biochem.,103,1011–1015 (1988)
[11] Silvestro, A., Pommier, J., Giordano, G.: Biochim. Biophys. Acta,954,1–13 (1988)
[12] McEwan, A.G., Richardson, D.J., Jackson, J.B., Ferguson, S.J.: Biochem. Soc. Trans.,16,182–183 (1988)
[13] Styrvold, O.B., Strom, A.R.: Arch. Microbiol.,140,74–78 (1984)
[14] Easter, M.C., Gibson, D.M., Ward, F.B.: J. Gen. Microbiol.,129,3689–3696 (1983)
[15] Bragg, P.D., Hackett, N.R.: Biochim. Biophys. Acta,725,168–177 (1983)
[16] Kwan, H.S., Barrett, E.L.: J. Bacteriol.,155,1147–1155 (1983)

1 NOMENCLATURE

EC number
1.6.6.10

Systematic name
NAD(P)H:4-nitroquinoline-N-oxide oxidoreductase

Recommended name
Nitroquinoline-N-oxide reductase

Synonymes
Reductase, nitroquinoline N-oxide
Nitroquinoline N-oxide reductase
4-Nitroquinoline 1-oxide reductase
More (in reference 1–3 the enzyme is named 4-nitroquinoline 1-oxide reductase (nitric oxide reductase, EC 1.6.99.2)) [1–3]

CAS Reg. No.
37256-35-2

---

2 REACTION AND SPECIFICITY

Catalysed reaction
2 NAD(P)H + 4-nitroquinoline-N-oxide →
→ 2 NAD(P)$^+$ + 4-hydroxyquinoline-N-oxide

Reaction type
Redox reaction

Natural substrates

Substrate spectrum
1 NAD(P)H + 4-nitroquinoline-N-oxide

Product spectrum
1 NAD(P)$^+$ + 4-hydroxyaminoquinoline-N-oxide

Inhibitor(s)

Cofactor(s)/prostethic group(s)
NADH [3]; NADPH [3]

Metal compounds/salts

Enzyme Handbook © Springer-Verlag Berlin Heidelberg 1994
Duplication, reproduction and storage in data banks are only allowed with the prior permission of the publishers

Turnover number (min^{-1})

Specific activity (U/mg)

K_m-value (mM)

pH-optimum
6.4 (enzyme assay) [1]

pH-range

Temperature optimum (°C)
25 (enzyme assay) [1]

Temperature range (°C)

3 ENZYME STRUCTURE

Molecular weight

Subunits

Glycoprotein/Lipoprotein
–

4 ISOLATION/PREPARATION

Source organism
Guinea pig [3]; Hamster [3]; Dog [1]; Rat [2, 3]; Rabbit [3]; Cat [3]

Source tissue
Digestive tract (distribution in mucosa of digestive tract [1], of dog [1], of various animals [3]) [1, 3]

Localisation in source

Purification

Crystallization
–

Cloned
–

Renaturated
–

5 STABILITY

pH

Temperature (°C)

Oxidation

Organic solvent

General stability information

Storage

6 CROSSREFERENCES TO STRUCTURE DATABANKS

PIR/MIPS code

Brookhaven code

7 LITERATURE REFERENCES

[1] Suzuki, K.: Gann,69,229–235 (1978)
[2] Booth, D.R.: Cell Tissue Kinet.,23,331–340 (1990)
[3] Suzuki, K.: Nihon Univ. J. Med.,20,345–357 (1978)

Enzyme Handbook © Springer-Verlag Berlin Heidelberg 1994
Duplication, reproduction and storage in data banks are only allowed with the prior permission of the publishers

1 NOMENCLATURE

EC number
1.6.6.11

Systematic name
NADH:hydroxylamine oxidoreductase

Recommended name
Hydroxylamine reductase (NADH)

Synonymes
Reductase, hydroxylamine
Ammonium dehydrogenase
NADH-hydroxylamine reductase
N-Hydroxy amine reductase [3]

CAS Reg. No.
9032-06-8

2 REACTION AND SPECIFICITY

Catalysed reaction
NADH + hydroxylamine →
→ NAD^+ + NH_3 + H_2O

Reaction type
Redox reaction

Natural substrates
Hydroxylamine + NADH (Neurospora crassa pathway of nitrate reduction) [8]

Substrate spectrum
1 Hydroxylamine + NADH
2 N-Methylhydroxylamine + NADH [3, 4]
3 N, N-Dimethylhydroxylamine + NADH [3]
4 N-Methyl-N-benzyl-hydroxylamine + NADH [3]
5 N-Methyl-N-n-octyl-hydroxylamine + NADH [3]
6 Phenylhydroxylamine + NADH

Enzyme Handbook © Springer-Verlag Berlin Heidelberg 1994
Duplication, reproduction and storage in data banks are only allowed with the prior permission of the publishers

7 More (enzyme also reduces anthranilohydroxamate and nicotinohydroxamate (affinity is about twenty times lower than for hydroxylamine) [5], possible identity of hydroxylamine reductase and the enzyme which reduces anthranilic hydroxamate to the amide [6], not: O-methylhydroxylamine, hydrazine, nitrophenylhydroxylamine, salicylhydroxamic acid [8], no other substances beside nitrite and dinitrobenzene stimulate oxidation of NADPH [8], no reduction of N,N-diethyl or O-methyl compounds [4], phenylhydroxylamine (very slowly) [4]) [4–6, 8]

Product spectrum

1 NH_3 + NAD^+ + H_2O
2 Methylamine + NAD^+ + H_2O
3 Dimethylamine + NAD^+ + H_2O
4 N-Methyl-N-benzylamine + NAD^+ + H_2O
5 N-Methyl-N-n-octylamine + NAD^+ + H_2O
6 Phenylamine + NAD^+ + H_2O
7 ?

Inhibitor(s)

Nitrite (nitrite and hydroxylamine reductase are associated with the same enzyme, nitrite and hydroxylamine are competitive for the same binding site) [1]; p-Chloromercuribenzoate [1, 6, 7]; p-Hydroxymercuribenzoate [5]; N-Ethylmaleimide (not [6]) [1, 5]; Rotenone [1]; Antimycin 1 [1]; 8-Hydroxyquinoline [1, 7, 8]; 2,2'-Dipyridyl [1, 7, 8]; Potassium cyanide [1]; Dinitrophenol [1]; Potassium thiocyanate [1]; $HgCl_2$ [5, 6]; KCN [5, 7, 8]; NADH (inhibits in absence of hydroxylamine) [1]; Salicylaldoxime [7, 8]; NaN_3 [7, 8]; Potassium ethyl xanthate [7, 8]; Sodium diethyl dithiocarbamate [7, 8]; o-Phenanthroline [7, 8]; Hydrazine [7]; Sulfite [8]; Cu^{2+} [8]; Uranyl acetate (ATP, ADP, sodium diphosphate restores activity) [4]; Inorganic phosphate [4–6]; Sodium arsenate [4, 6]; $CaCl_2$ (reactivation by washing) [4]; NaCl (reactivation by washing) [4]; KCl (reactivation by washing) [4]; Deoxycholic acid [4]; Duponol C [4]; Ethyl alcohol [4]; Propyl alcohol [4]; Butyl alcohol [4]; N-Methyl hydroxylamine [4]; More (hydrolytic enzymes, e.g. murine phospholipase, subtilopeptidase A, trypsin, papain, bromelain inhibit) [4]

Cofactor(s)/prostethic group(s)

NADH; NADPH (about 65% of the activity with NADH [5, 7]) [5, 7, 8]; FAD (stimulates) [8]; More (purified enzyme: absolute requirement for an unidentified, dissociable, heat-stable, organic factor obtained from soybean leaf extract, no flavin requirement) [7]

Metal compounds/salts

Mn^{2+} (stimulates) [7]

Turnover number (min^{-1})

Specific activity (U/mg)
More [1]

K_m-value (mM)
0.07 (NADH) [8]; 0.1 (NADPH) [8]; 5.3 (hydroxylamine) [1]; 0.150 (NADH) [1]; 0.5 (hydroxylamine) [6]; 3.8 (hydroxylamine) [8]; More [7]

pH-optimum
6.3 [3]; 6.6–7.1 [7]; 6.8–7.6 [6]; 7.5–8.5 (tris(hydroxymethyl)aminomethane or diphosphate buffer) [8]; 7.6 [5]; 8.5 [1]

pH-range
5.4–8.4 (5.4: considerable activity in Tris-maleate buffer, 8.4: about 50% of activity maximum) [5]; 5.5–7.2 (5.5: about 70% of activity maximum, 7.2: 50% of activity maximum) [5]; More (little or no activity in phosphate buffer unless 1 mM EDTA is included in the reaction mixture) [8]

Temperature optimum (°C)
37 (assay at) [3]

Temperature range (°C)

3 ENZYME STRUCTURE

Molecular weight

Subunits
More (3 protein fractions are required to reconstitute NADH-hydroxylamine reductase activity: detergent-extracted cytochrome b_5 and its flavoprotein and a third microsomal protein) [2]

Glycoprotein/Lipoprotein
–

4 ISOLATION/PREPARATION

Source organism
Derxia gummosa (nitrite and hydroxylamine reductase are associated with the same enzyme) [1]; Pig (multicomponent enzyme system catalyzes reduction of hydroxylamine and a number of its mono- and disubstituted derivatives [2, 3]) [2–4]; Rat [4–6]; Mouse [4]; Sheep [4]; Rabbit [4]; Neurospora crassa (3 hydroxylamine reductases: one is identical with sulfite reductase and is absolutely specific for NADPH (accepts hydroxylamine

Enzyme Handbook © Springer-Verlag Berlin Heidelberg 1994
Duplication, reproduction and storage in data banks are only allowed with the prior permission of the publishers

and sulfate as substrates at a common binding site), a second hydroxylamine reductase uses either NADH or NADPH (stimulated by FAD), a third hydroxylamine reductase is also able to accept NADH and NADPH) [9]; Glycine max (soybean) [7]; Opossum (very little activity) [4]; More (overview [4], absent from reptiles, amphibians and fish) [4]

Source tissue

Liver [2–6]; Kidney [3]; Leaves [7]; More (not: brain, serum (rat, cat, bovine), low activity: lung, spleen) [3]

Localisation in source

Microsomes [2, 3]; Mitochondria (firmly attached to membrane) [4–6]; Soluble [7]; More (not: spinach chloroplast) [4]

Purification

Derxia gummosa (nitrite and hydroxylamine reductase are associated with the same enzyme) [1]; Glycine max (enzyme not seperated from pyridine nucleotide-nitrite enzyme) [7]; Neurospora crassa (partial) [8]

Crystallization

–

Cloned

–

Renaturated

–

5 STABILITY

pH

5.4 (complete loss of activity after brief exposure) [6]; 7.5–9.5 (highest stability) [8]; 8.4 (no loss of activity after brief exposure) [6]

Temperature (°C)

27 (stable for more than 1 h) [5]; 38 (1 h, inactivation) [5]; 40 (membrane bound enzyme, 10 min, pH 7.4, dithiothreitol, no loss of activity) [6]; 50 (membrane bound enzyme, 10 min, pH 7.4, dithiothreitol, 50% loss of activity [6], 5 min, stable [8]) [6, 8]; 60 (membrane bound enzyme, pH 7.4, dithiothreitol, complete loss of activity) [6]; More (NADH prevents heat damage) [6]

Oxidation

Organic solvent

General stability information

Freezing partially inactivates membrane bound enzyme [6]; Solubilization with deoxycholate, n-butanol or acetone destroys membrane bound enzyme [6]; Dithiothreitol stabilizes during storage [6]; NADH prevents heat damage [6]; Dialysis: against a sodium diphosphate (3 mM) - cysteine (1 mM) solution, pH 8, 2 h, less than 50% loss of activity, against 10 mM phosphate, pH 8, 1 h, complete loss of activity [8]

Storage

4°C [7]; 4°C, overnight [8]; –15°C, slow loss of activity over several weeks [8]; –20°C [7]

6 CROSSREFERENCES TO STRUCTURE DATABANKS

PIR/MIPS code

Brookhaven code

7 LITERATURE REFERENCES

[1] Wang, R., Nicholas, D.J.D.: Phytochemistry,25,2463–2469 (1986)
[2] Kadlubar, F.F., Ziegler, D.M.: Arch. Biochem. Biophys.,162,83–92 (1974)
[3] Kadlubar, F.F., Mckee, E.M., Ziegler, D.M.: Arch. Biochem. Biophys.,156,46–57 (1973)
[4] Bernheim, M.L.C.: Enzymologia,43,167–176 (1972)
[5] Bernheim, M.L.C., Hochstein, P.: Arch. Biochem. Biophys.,124,436–442 (1968)
[6] Bernheim, M.L.C.: Arch. Biochem. Biophys.,134,408–413 (1969)
[7] Roussos, G.G., Nason, A.: J. Biol. Chem.,235,2997–3007 (1960)
[8] Zucker, M., Nason, A.: J. Biol. Chem.,213,463–478 (1955)
[9] Siegel, L.M., Leinweber, F.-J., Monty, K.J.: J. Biol. Chem.,240,2705–2711 (1965)

Enzyme Handbook © Springer-Verlag Berlin Heidelberg 1994
Duplication, reproduction and storage in data banks are only allowed with the prior permission of the publishers

1 NOMENCLATURE

EC number
1.6.6.12

Systematic name
NADPH:4-(dimethylamino)phenylazobenzene oxidoreductase

Recommended name
4-(Dimethylamino)phenylazobenzene reductase

Synonymes
N, N'-Dimethyl-p-aminoazobenzene oxide reductase [1]
Dimethylaminoazobenzene N-oxide reductase [1]
NADPH-dependent DMAB N-oxide reductase [1]

CAS Reg. No.

2 REACTION AND SPECIFICITY

Catalysed reaction
NADPH + 4-(dimethylamino)phenylazoxybenzene →
→ $NADP^+$ + 4-(dimethylamino)phenylazobenzene

Reaction type
Redox reaction

Natural substrates

Substrate spectrum
1 NADPH + 4-(dimethylamino)phenylazoxybenzene (high specificity for NADPH and 4-(dimethylamino) phenylazoxybenzene) [1]

Product spectrum
1 $NADP^+$ + 4-(dimethylamino)phenylazobenzene

Inhibitor(s)

Cofactor(s)/prostethic group(s)
NADPH (specific for) [1]

Metal compounds/salts

Turnover number (min^{-1})
150 (NADPH) [1]

Enzyme Handbook © Springer-Verlag Berlin Heidelberg 1994
Duplication, reproduction and storage in data banks are only allowed with the prior permission of the publishers

Specific activity (U/mg)
0.270 [1]

K_m-value (mM)
0.021 (NADPH) [1]; 0.700 (4-(dimethylamino) phenylazoxybenzene) [1]

pH-optimum
6.5 (enzyme assay) [1]

pH-range

Temperature optimum (°C)
37 (enzyme assay) [1]

Temperature range (°C)

---

3 ENZYME STRUCTURE

Molecular weight
370000 (rat, gel filtration) [1]

Subunits
Tetramer (4 × 92000, rat, SDS-PAGE) [1]

Glycoprotein/Lipoprotein
–

---

4 ISOLATION/PREPARATION

Source organism
Rat [1]

Source tissue
Liver [1]; Kidney [1]; Heart [1]; Spleen [1]; Lung [1]

Localisation in source
Cytosol [1]

Purification
Rat [1]

Crystallization
–

Cloned
–

Renaturated
–

5 STABILITY

pH

Temperature (°C)

Oxidation

Organic solvent

General stability information

Storage
  −20°C, several weeks [1]

6 CROSSREFERENCES TO STRUCTURE DATABANKS

PIR/MIPS code

Brookhaven code

7 LITERATURE REFERENCES

[1] Lasmet Johnson, P.R., Ziegler, D.M.: J. Biochem. Toxicol.,1,15–27 (1986)

Enzyme Handbook © Springer-Verlag Berlin Heidelberg 1994
Duplication, reproduction and storage in data banks are only allowed with the prior permission of the publishers

1 NOMENCLATURE

EC number
1.6.6.13

Systematic name
NAD(P)H:N-hydroxy-2-acetamidofluorene N-oxidoreductase

Recommended name
N-Hydroxy-2-acetamidofluorene reductase

Synonymes
N-Hydroxy-2-acetylaminofluorene reductase

CAS Reg. No.
99890-08-1

2 REACTION AND SPECIFICITY

Catalysed reaction
NAD(P)H + N-hydroxy-2-acetamidofluorene →
→ $NAD(P)^+$ + 2-acetamidofluorene

Reaction type
Redox reaction

Natural substrates

Substrate spectrum
1 N-Hydroxy-2-acetamidofluorene + NAD(P)H (other electron donors: cysteine, glutathione, dithiothreitol, 2-mercaptoethanol) [1]
2 N-Hydroxy-4-acetylaminobiphenyl + NAD(P)H (much slower reaction) [1]

Product spectrum
1 2-Acetamidofluorene + $NAD(P)^+$ [1]
2 4-Acetylaminobiphenyl + $NAD(P)^+$ [1]

Inhibitor(s)
N-Ethylmaleimide [1]; $CuSO_4$ [1]; Disulfiram [1]; p-Chloromercuribenzoate [1]

Enzyme Handbook © Springer-Verlag Berlin Heidelberg 1994
Duplication, reproduction and storage in data banks are only allowed with the prior permission of the publishers

Cofactor(s)/prostethic group(s)
NADH [1]; NADPH [1]; Cysteine (enhances activity, can replace NADH) [1]; Gluthatione (enhances activity, can replace NADH) [1]; Dithiothreitol (enhances activity, can replace NADH) [1]; 2-Mercaptoethanol (enhances activity, can replace NADH) [1]

Metal compounds/salts

Turnover number (min^{-1})

Specific activity (U/mg)
910 [1]

K_m-value (mM)

pH-optimum

pH-range

Temperature optimum (°C)

Temperature range (°C)

---

3 ENZYME STRUCTURE

Molecular weight
34000 (rabbit, gel filtration) [1]

Subunits
Monomer (1 × 34000, rabbit, SDS-PAGE) [1]

Glycoprotein/Lipoprotein
–

---

4 ISOLATION/PREPARATION

Source organism
Rabbit [1]

Source tissue
Liver [1]

Localisation in source
Cytosol [1]; Microsomes [1]

Purification
Rabbit [1]

Crystallization

–

Cloned

–

Renaturated

–

5 STABILITY

pH

Temperature (°C)

Oxidation

Organic solvent

General stability information

Storage

6 CROSSREFERENCES TO STRUCTURE DATABANKS

PIR/MIPS code

Brookhaven code

7 LITERATURE REFERENCES

[1] Kitamura, S., Tatsumi, K.: Biochem. Biophys. Res. Commun.,133,67–74 (1985)

Enzyme Handbook © Springer-Verlag Berlin Heidelberg 1994
Duplication, reproduction and storage in data banks are only allowed with the prior permission of the publishers

1 NOMENCLATURE

EC number
1.6.8.1

Systematic name
NAD(P)H:FMN oxidoreductase

Recommended name
NAD(P)H dehydrogenase (FMN)

Synonymes
FMN reductase
NAD(P)H:flavin oxidoreductase [1]
Reductase, riboflavin mononucleotide (reduced nicotinamide adenine dinucleotide (phosphate))
Flavin mononucleotide reductase
Flavine mononucleotide reductase
NAD(P)H-FMN reductase
NAD(P)H:FMN oxidoreductase
Riboflavin mononucleotide reductase
Riboflavine mononucleotide reductase

CAS Reg. No.
64295-83-6

2 REACTION AND SPECIFICITY

Catalysed reaction
NAD(P)H + FMN →
→ $NAD(P)^+$ + $FMNH_2$ (sequential mechanism [1], 2 distinct enzymes: 1. NADH-specific FMN reductase [4, 5], 2. NADPH-specific FMN reductase [4, 5])

Reaction type
Redox reaction

Natural substrates
NAD(P)H + FMN (production of $FMNH_2$, one of the substrates of the luciferase reaction in bioluminescent bacteria) [1, 6]

Enzyme Handbook © Springer-Verlag Berlin Heidelberg 1994
Duplication, reproduction and storage in data banks are only allowed with the prior permission of the publishers

Substrate spectrum

1 NAD(P)H + FMN (2 distinct enzymes: 1. NADH-specific FMN reductase, 2. NADPH-specific FMN reductase [4, 5])
2 NAD(P) H + riboflavin [1, 7]
3 NAD(P)H + FAD [1]
4 1-Deazariboflavin + NAD(P)H [7]
5 Lumiflavin + NAD(P)H [7]
6 Alloxazine + NADPH (no reaction with NADH) [7]
7 More (higher specificity for riboflavin and FMN than FAD [1], overview: reaction with various flavin isosteres [2]) [1, 2]

Product spectrum

1 $NAD(P)^+$ + $FMNH_2$
2 $NAD(P)^+$ + reduced riboflavin
3 $NAD(P)^+$ + $FADH_2$
4 $NAD(P)^+$ + reduced 1-deazariboflavin
5 $NAD(P)^+$ + reduced lumiflavin
6 $NADP^+$ + reduced alloxazine
7 ?

Inhibitor(s)

Lumichrome [7]; Alloxazine [7]; AMP [7]; Erythritol [7]

Cofactor(s)/prostethic group(s)

NADH (higher specificity for NADH than for NADPH [1], 2 distinct enzymes: 1. NADH-specific FMN reductase, 2. NADPH-specific FMN reductase [4, 5]) [1, 4, 5]; NADPH (higher specificity for NADH than for NADPH [1], 2 distinct enzymes: 1. NADH-specific FMN reductase, 2. NADPH-specific FMN reductase [4, 5]) [1, 4, 5]

Metal compounds/salts

Turnover number (min^{-1})

930 (NADH, NADH:FMN oxidoreductase) [5]; 2040 (NADPH, NADPH:FMN oxidoreductase) [5]

Specific activity (U/mg)

1.85 (NADPH reductase) [4]; 2.05 (NADH reductase) [4]; 31.0 (NADH:FMN oxidoreductase) [5]; 51.0 (NADPH:FMN oxidoreductase) [5]; 86 [1]

K_m-value (mM)

More (overview: K_m of various flavin isosteres [2]) [1, 2, 6, 7]; 0.000016 (lumiflavin, NADH:FMN oxidoreductase, Tris buffer) [7]; 0.0002 (lumiflavin, NADH:FMN oxidoreductase, phosphate buffer) [7]; 0.00027 (riboflavin, NADH:FMN oxidoreductase, Tris buffer) [7]; 0.0005 (FMN, NADPH-specific activity [4], riboflavin, NADH:FMN oxidoreductase, phosphate buffer [7]) [4, 7]; 0.001 (FMN, NADH:FMN oxidoreductase, phosphate buffer) [7]; 0.0011 (FMN, NADH-specific activity [4], Tris buffer [7]) [4, 7]; 0.0022 (alloxazine, NADPH:FMN oxidoreductase, citrate buffer) [7]; 0.0059 (lumiflavin, NADPH:FMN oxidoreductase, phosphate buffer) [7]; 0.0066 (alloxazine, NADPH:FMN oxidoreductase, phosphate buffer) [7]; 0.036 (riboflavin, phosphate buffer) [7]

pH-optimum

pH-range

Temperature optimum (°C)

23 (assay at) [1, 4]

Temperature range (°C)

---

3 ENZYME STRUCTURE

Molecular weight

19000 (Beneckea harveyi, NADH-specific enzyme, gel filtration) [4]
24000 (Beneckea harveyi, gel filtration) [1]
30000 (Beneckea harveyi, gel filtration, NADH:FMN oxidoreductase) [5]
31000 (Beneckea harveyi, NADH-specific enzyme, linear sucrose gradient centrifugation) [4]
40000 (Beneckea harveyi, gel filtration, NADPH:FMN oxidoreductase) [4, 5]
45000 (Photobacterium fischeri, gel filtration) [5]
63000 (Beneckea harveyi, NADPH-specific enzyme, linear sucrose gradient centrifugation) [4]

Subunits

Monomer (1 × 24000, Beneckea harveyi, SDS-PAGE) [1]

Glycoprotein/Lipoprotein

–

---

4 ISOLATION/PREPARATION

Source organism

Beneckea harveyi (2 distinct enzymes: 1. NADH-specific FMN reductase, 2. NADPH-specific FMN reductase [4, 5]) [1–7]; Photobacterium fischeri [5]

Enzyme Handbook © Springer-Verlag Berlin Heidelberg 1994
Duplication, reproduction and storage in data banks are only allowed with the prior permission of the publishers

Source tissue
Cell [1, 4, 5]

Localisation in source

Purification
Beneckea harveyi [1, 4, 5]

Crystallization
–

Cloned
–

Renaturated
–

---

5 STABILITY

pH

Temperature (°C)
55 (5 min, 60% loss of activity of NADH-specific enzyme, complete loss of activity of NADPH-specific enzyme) [4]

Oxidation

Organic solvent

General stability information

Storage
4°C, 10% glycerol, 1 mM dithiothreitol, 0.05 M phosphate, pH 7.0, 50% loss of activity after 2 months [1]

---

6 CROSSREFERENCES TO STRUCTURE DATABANKS

PIR/MIPS code

Brookhaven code

7 LITERATURE REFERENCES

[1] Michaliszyn, G.A., Wing, S.S., Meighen, E.A.: J. Biol. Chem.,252,7495–7499 (1977)
[2] Walsh, C., Fisher, J., Spencer, R., Graham, D.W., Ashton, W.T., Brown, J.E., Brown, R.D., Rogers, E.F.: Biochemistry,17,1942–1951 (1978)
[3] Spencer, R., Fisher, J., Walsh, C.: Biochemistry,16,3594–3602 (1977)
[4] Gerlo, E., Charlier, J.: Eur. J. Biochem.,57,461–467 (1975)
[5] Jablonski, E., DeLuca, M.: Biochemistry,16,2932–2936 (1977)
[6] Jablonski, E., DeLuca, M.: Biochemistry,17,672–678 (1978)
[7] Nefsky, B., DeLuca, M.: Arch. Biochem. Biophys.,216,10–16 (1982)

Enzyme Handbook © Springer-Verlag Berlin Heidelberg 1994
Duplication, reproduction and storage in data banks are only allowed with the prior permission of the publishers

1 NOMENCLATURE

EC number
1.6.8.2

Systematic name
NADPH:riboflavin oxidoreductase

Recommended name
NADPH dehydrogenase (flavin)

Synonymes
NADPH:flavin oxidoreductase [1]
Flavin reductase [1]
Reductase, riboflavin mononucleotide (reduced nicotinamide adenine dinucleotide phosphate)
Flavin mononucleotide reductase
Flavine mononucleotide reductase
FMN reductase (NADPH)
NADPH-dependent FMN reductase
NADPH-flavin reductase
NADPH-FMN reductase
NADPH-specific FMN reductase
Riboflavin mononucleotide reductase
Riboflavine mononucleotide reductase

CAS Reg. No.
56626-29-0

2 REACTION AND SPECIFICITY

Catalysed reaction
NADPH + riboflavin →
→ $NADP^+$ + reduced riboflavin

Reaction type
Redox reaction

Natural substrates

Enzyme Handbook © Springer-Verlag Berlin Heidelberg 1994
Duplication, reproduction and storage in data banks are only allowed with the prior permission of the publishers

Substrate spectrum

1 NADPH + riboflavin [1]
2 NADPH + galactoflavin [1]
3 NADPH + FMN [1]
4 NADPH + FAD [1]
5 NADPH + menadione [1]
6 NADPH + p-nitrobluetetrazolium [1]
7 NADPH + 2,6-dichlorophenol indophenol [1]
8 NADPH + methylene blue [1]

Product spectrum

1 $NADP^+$ + reduced riboflavin
2 $NADP^+$ + reduced galactoflavin
3 $NADP^+$ + $FMNH_2$
4 $NADP^+$ + $FADH_2$
5 $NADP^+$ + reduced menadione
6 $NADP^+$ + reduced p-nitrobluetetrazolium
7 $NADP^+$ + reduced 2, 6-dichlorophenol indophenol
8 NADPH + reduced methylene blue

Inhibitor(s)

Cofactor(s)/prostethic group(s)

NADPH (20-fold selectivity for NADPH over NADH) [1]; More (does not contain firmly bound flavin) [1]

Metal compounds/salts

Turnover number (min^{-1})

Specific activity (U/mg)

22 [1]

K_m-value (mM)

0.006 (FMN) [1]; 0.01 (FAD) [1]; 0.04 (riboflavin, galactoflavin) [1]; 0.05 (NADPH (+ FMN)) [1]

pH-optimum

8 (assay at) [1]

pH-range

Temperature optimum (°C)

36 (assay at) [1]

Temperature range (°C)

3 ENZYME STRUCTURE

Molecular weight
38000–40000 (Entamoeba histolytica, gel filtration, SDS-PAGE) [1]

Subunits
Monomer (1 × 38000, Entamoeba histolytica, SDS-PAGE) [1]

Glycoprotein/Lipoprotein
–

4 ISOLATION/PREPARATION

Source organism
Entamoeba histolytica [1]

Source tissue
Cell [1]

Localisation in source

Purification
Entamoeba histolytica [1]

Crystallization
–

Cloned
–

Renaturated
–

5 STABILITY

pH

Temperature (°C)

Oxidation

Organic solvent

General stability information

Storage

Enzyme Handbook © Springer-Verlag Berlin Heidelberg 1994
Duplication, reproduction and storage in data banks are only allowed with the prior permission of the publishers

6 CROSSREFERENCES TO STRUCTURE DATABANKS

PIR/MIPS code

Brookhaven code

7 LITERATURE REFERENCES

[1] Lo, H.-S., Reeves, R.E.: Mol. Biochem. Parasitol.,2,23–30 (1980)

1 NOMENCLATURE

EC number
1.6.99.1

Systematic name
NADPH:(acceptor) oxidoreductase

Recommended name
NADPH dehydrogenase

Synonymes
NADPH diaphorase
OYE
Dehydrogenase, reduced nicotinamide adenine dinucleotide phosphate
Diaphorase
Dihydronicotinamide adenine dinucleotide phosphate dehydrogenase
NADPH-dehydrogenase
NADPH-diaphorase
$NADPH_2$–dehydrogenase
Old yellow enzyme
Reduced nicotinamide adenine dinucleotide phosphate dehydrogenase
TPNH dehydrogenase
TPNH-diaphorase
Triphosphopyridine diaphorase
Triphosphopyridine nucleotide diaphorase

CAS Reg. No.
9001-68-7

2 REACTION AND SPECIFICITY

Catalysed reaction
NADPH + acceptor →
→ $NADP^+$ + reduced acceptor (mechanism [1, 6])

Reaction type
Redox reaction

Natural substrates
NADPH + acceptor (physiological role as electron acceptor) [5]

Enzyme Handbook © Springer-Verlag Berlin Heidelberg 1994
Duplication, reproduction and storage in data banks are only allowed with the prior permission of the publishers

Substrate spectrum

1 NADPH + acceptor (acceptor: O_2 [1], ferricytochrome c [1], $Fe(CN)_6^{3-}$ [1, 2], leuco-2,3',6-trichlorophenol indophenol [2], methylene blue [2], nitroblue tetrazolium hydrochloride [9], menadione [2], FMN [2], benzoquinone [2], ferric chloride with o-phenanthroline [2], acetylpyridine analog of NADPH [2])

Product spectrum

1 $NADP^+$ + reduced acceptor

Inhibitor(s)

Co^{2+} [2]; Iodoacetate [2]; 2'-AMP [2]; Fe^{3+} [1]; Mercuric chloride [2]; p-Chloromercuribenzoate [2]; Zn^{2+} [2]; $NADP^+$ [2]; N-Ethylmaleimide [8]; p-Hydroxymercuribenzoate [8]

Cofactor(s)/prostethic group(s)

Flavin (flavoprotein [1], FMN in yeast [1], FAD in plants [1], 2 FMN per enzyme molecule [1], 2 FMN per dimer [4], 1 FAD per molecule of 35000 MW [2]) [1, 2, 4]; NADPH (reduction of enzyme with alpha- and beta-NADPH, alpha-NADPH is more effective [7]) [1–8]; NADH (at a rate of one-tenth as fast as with NADPH [1], no activity [2]) [1]

Metal compounds/salts

No metal content [1]

Turnover number (min^{-1})

Specific activity (U/mg)

28.1 [1]

K_m-value (mM)

0.0011 (NADPH) [9]; 0.006 (NADPH) [2]; 0.0073 (nitroblue tetrazolium hydrochloride) [9]; 0.2 (NADPH) [8]

pH-optimum

7.0 [8]

pH-range

Temperature optimum (°C)

Temperature range (°C)

3 ENZYME STRUCTURE

Molecular weight

30000–40000 (spinach, ultracentrifugal analysis) [2]
106000 (yeast, low speed sedimentation without reaching equilibrium) [1]
More [1]

Subunits

Dimer (2 × 50000, Saccharomyces carlsbergensis) [4]

Glycoprotein/Lipoprotein

–

4 ISOLATION/PREPARATION

Source organism

Euglena gracilis var. bacillaris [8]; Yeast [1]; Spinach [2]; Saccharomyces carlsbergensis (brewer's yeast) [3–7]; Rat [9]

Source tissue

Leaf [2]; Brain [9]

Localisation in source

Chloroplast [2]

Purification

Spinach [2]

Crystallization

[1, 4]

Cloned

[3]

Renaturated

–

5 STABILITY

pH

Temperature (°C)

Oxidation

Organic solvent

Enzyme Handbook © Springer-Verlag Berlin Heidelberg 1994
Duplication, reproduction and storage in data banks are only allowed with the prior permission of the publishers

General stability information

Storage

–5°C, 15% loss of activity after 96 h [8]

6 CROSSREFERENCES TO STRUCTURE DATABANKS

PIR/MIPS code

Brookhaven code

7 LITERATURE REFERENCES

[1] Akeson, A., Ehrenberg, A., Theorell, H. in "The Enzymes",2nd Ed. (Boyer, P.D., Lardy, H., Myrbäck, K. Eds.) 7,477–494, Academic Press, New York (1963) (Review)
[2] Jagendorf, A.T.: Methods Enzymol.,6,430–434 (1963) (Review)
[3] Saito, K., Davio, M., Lockridge, O., Massey, V. in " Flavins Flavoproteins Proc. Int. Symp." (Curti, B., Rochi, S., Zanetti, G. Eds.) 10th Meeting,349–352 (1990)
[4] Fox, K.M., Jacques, S.M., Karplus, P.A. in "Flavins Flavoproteins Proc. Int. Symp." (Curti, B., Rochi, S., Zanetti, G. Eds.) 10th Meeting,353–356 (1990)
[5] Miura, R., Yamaichi, K., Tagawa, K., Miyake, Y.: J. Biochem.,102,1311–1320 (1987)
[6] Honma, T., Ogura, Y.: Biochim. Biophys. Acta,484,9–23 (1977)
[7] Massey, V., Schopfer, L.M.: J. Biol. Chem.,261,1215–1222 (1986)
[8] Mohanty, M.K., Hunter, F.R., Myers, J.B.: J. Protozool.,24,335–340 (1977)
[9] Kemp, M.C., Kuonene, D.R., Roberts, P.J.: Biochem. Soc. Trans.,15,501 (1987)

1 NOMENCLATURE

EC number
1.6.99.2

Systematic name
NAD(P)H:(quinone-acceptor) oxidoreductase

Recommended name
NAD(P)H dehydrogenase (quinone)

Synonymes
Quinone reductase
Menadione reductase
Phylloquinone reductase
1.6.5.2 (formerly)
Dehydrogenase, reduced nicotinamide adenine dinucleotide (phosphate, quinone)
DT-Diaphorase
Flavoprotein NAD(P)H-quinone reductase
Menadione oxidoreductase
NAD(P)H menadione reductase
NAD(P)H-quinone dehydrogenase
NAD(P)H-quinone oxidoreductase
NAD(P)H:(quinone-acceptor)oxidoreductase (EC 1.6.99.2)
NAD(P)H: menadione oxidoreductase
NADH-menadione reductase
Naphthoquinone reductase
p-Benzoquinone reductase
Reduced NAD(P)H dehydrogenase
Viologen accepting pyridine nucleotide oxidoreductase
Vitamin K reductase
Diaphorase [1]
Reduced nicotinamide-adenine dinucleotide (phosphate) dehydrogenase [3]
Vitamin-K reductase [4]

CAS Reg. No.
9032-20-6

Enzyme Handbook © Springer-Verlag Berlin Heidelberg 1994
Duplication, reproduction and storage in data banks are only allowed with the prior permission of the publishers

2 REACTION AND SPECIFICITY

Catalysed reaction

NAD(P)H + acceptor →
→ $NAD(P)^+$ + reduced acceptor (ordered bi-bi mechanism [10], ping-pong mechanism [5, 11])

Reaction type

Redox reaction

Natural substrates

Vitamin K + NAD(P)H [4]
More (protection of cells against damage by reactive oxygen species generated during oxidative cycling of quinones and semiquinone radicals [15], possible role in seed germination [20]) [15, 20]

Substrate spectrum

1 NAD(P)H + acceptor ((reduced) acceptor: menadione (2-methyl-1,4-naphthoquinone) [1, 2, 4, 8, 11, 15, 16, 18, 19], methylene blue (not [3]) [1], ferricyanide (low activity [11]) [1, 3, 4, 11], 2,6-dichlorophenol indophenol (slowly [3, 5, 11]) [1, 3, 4, 15, 16, 18], vitamin K (2-methyl-3-phythyl-1,4-naphthoquinone, not [3–5]) [2, 8, 15], coenzyme Q_{10} (not [1, 11]) [2], benzoquinones (not p-benzoquinone [3]) [2, 4, 5, 16, 19], vitamin K_2 [4], naphthoquinones [4, 5, 16, 19], benzo(a) pyrene [8], 3,6-quinone [8], cyclized-dopamine ortho-quinone [8], coenzyme Q_1 [11], 2-methoxyquinone [12], 2-hydroxymethyl-5-methoxy-2,5-cyclohexadiene-1,4-dione [22], 2-methoxy-2,5-cyclohexadiene-1,4-dione [22], 4,5-dimethoxy-3,5-cyclohexadiene-1,2-dione [22], 2-methoxy-5-methoxymethyl-2,5-cyclohexadiene [22])
2 More (no reaction with cytochromes [1, 3, 4, 20], lipoic acid [1, 3], O_2 [3], vitamin K_3 [3], ubiquinone-30 [3], 2-p-iodophenyl-3-p-nitrophenyl-5-phenyltetrazolium chloride [3], coenzyme Q_6 [5]) [1, 3–5, 20]

Product spectrum

1 $NAD(P)^+$ + reduced acceptor
2 ?

Inhibitor(s)

Dicoumarol (slight [22]) [1–6, 11, 14–16, 22]; 2,4-Dinitrophenol [1, 2, 20]; o-Phenanthroline (not [5]) [1, 2]; Amytal [1]; Tromexan [2, 4]; Phenindione [2]; Salicylate [2]; alpha,alpha-Dipyridyl [2]; 8-Hydroxyquinoline [2]; 2,4-Dinitroaniline [2]; 2,4-Dinitronaphthol [2]; 4-Hydroxycoumarin [4]; 4-Methoxycoumarin [4]; Dicoumarol dimethylether [4]; p-Chloromercuribenzoate [5]; 2-Pivaloyl-1,3-indanedione [6]; Cibacron blue (and related anthraquinone dyes [7]) [7, 14]; (A)-2-Azido-NAD^+ (photodependent inhibition) [9]; (A)-8-Azido-(NAD^+) (photodependent inhibition) [9]; 1,4-Naphthohydroquinone [10]; $NADP^+$ [10]; Iodoacetic acid [20]; EDTA [20]; Thenoyltrifluoroacetone [20]; DTT [20]; 5,5'-Dithiobis(2-nitrobenzoate) [20]

Cofactor(s)/prostethic group(s)

FAD (enzyme contains FAD [1, 4, 6, 14], contains 1 mol of FAD per mol of enzyme [6], stimulates [1, 2], requires FMN or FAD as cofactor, FMN is more effective [11], required [15]) [1, 2, 4, 6, 11, 14, 15]; FMN (enzyme requires FMN or FAD as cofactor, FMN is more effective [11], stimulation [1, 2], contains FNN as prosthetic group [3, 5]) [1–3, 5, 11]; NAD^+ [1]; $NADP^+$ [1]; NADH [2, 4, 5, 11, 12, 15, 20]; NADPH (not [11]) [2, 4, 5, 12, 15, 20]; Deamino-NADH [2]; Glutathione (stimulates) [3]; Cysteine (stimulates) [3]; Mercaptoethanol (stimulates) [3]

Metal compounds/salts

Turnover number (min^{-1})

700000 (menadione) [4]

Specific activity (U/mg)

More [1, 4, 19]; 2.8 [3]; 2.9 [2]; 46.2 [5]; 371 [6]; 710 (cytosol) [13]; 830 (microsomes) [13]; 1212 (dichloroindophenol) [18]; 1770 [16]; 2452 (2-methyl-1,4-naphthoquinone) [18]

K_m-value (mM)

0.0000025 (FMN) [5]; 0.002 (menadione) [11, 15]; 0.0025 (menadione (+ NADH)) [1]; 0.22 (NADPH) [3]; 0.024 (methylene blue (+ NADPH)) [1]; 0.034 (NADH) [3]; 0.035 (menadione (+ NADH)) [5]; 0.040 (menadione (+ NADPH)) [1]; 0.115 (2-methoxy-5-(methoxymethyl)-2,5-cyclohexadiene) [22]; 0.223 (NADH(+ menadione)) [5]; 0.3 (vitamin K_1 (+ NADPH)) [1]; 0.33 (vitamin K_1 (+ NADH)) [1]; 0.34 (NADPH (+ menadione)) [5]; 1.4 (potassium ferricyanide (+ NADPH)) [1]; 1.7 (potassium ferricyanide (+ NADH)) [1]; More [5, 10, 11, 16, 18, 21–23]

Enzyme Handbook © Springer-Verlag Berlin Heidelberg 1994
Duplication, reproduction and storage in data banks are only allowed with the prior permission of the publishers

pH-optimum

4.4 [5]; 5.7 (menadione) [4]; 7–8.5 [20]; 7.0–8.0 [11]; 8.0 (2-methoxy-5-(methoxymethyl)-2,5-cyclohexadiene) [22]; 8.5 (2-hydroxymethyl-5-methoxy-2,5-cyclohexadiene-1,4-dione) [22]; 9–9.5 [19]; More [2]

pH-range

Temperature optimum (°C)

25 (assay at) [13, 19]

Temperature range (°C)

3 ENZYME STRUCTURE

Molecular weight

36000 (Halobacterium cutirubrum, gel filtration) [21]
39700–41700 (Microcystis aeruginosa, gel filtration, amino acid composition) [19]
47000–69000 (Phanerochaete chrysosporium, gel filtration, SDS-PAGE) [22]
50000 (E. coli, gel filtration) [11]
52000 (bovine, sedimentation equilibrium, sedimentation velocity [4], mouse, gel filtration [16]) [4, 16]
55000 (rat, gel filtration) [6]
60000 (Clostridium tyrobutyricum, gel filtration) [5]
60200–67000 (castor bean, gel filtration, SDS-PAGE, amino acid analysis) [20]
61000 (rat, gel filtration) [13]

Subunits

Monomer (1 × 60000, Clostridium tyrobutyricum, SDS-PAGE [5], 1 × 63000–67000, castor bean, SDS-PAGE [20]) [5, 20]
Dimer (2 × 24000, E. coli, SDS-PAGE [1], 2 × 27000 (each subunit consists of 2 polypeptide chains of MW 18000 and 8000–9000), rat, SDS-PAGE [6, 23], 2 × 30000, mouse, SDS-PAGE [16], 2 × 32000, rat, SDS-PAGE [13]) [1, 6, 11, 13, 16, 23]
Octamer (8 × 5100, Microcystis aeruginosa, SDS-PAGE) [19]

Glycoprotein/Lipoprotein

Glycoprotein (not a glycoprotein [23]) [8]

4 ISOLATION/PREPARATION

Source organism

Phanerochaete chrysosporium [22]; Halobacterium cutirubrum [21]; Bovine (ox [1]) [1, 4, 18]; Dog [2]; Octopus vulgaris [3]; Clostridium tyrobutyricum [5]; Rat [6, 8, 9, 13–15, 18, 23]; Microcystis aeruginosa [10, 19]; E. coli [11]; Castor bean [20]; Sporotrichum pulverulentum [12]; Mouse [16, 17]; More (widely distributed in animal kingdom) [10]

Source tissue

Brain [1, 18]; Liver [2, 4, 6, 8, 9, 13–16, 18, 23]; Hepatopancreas [3]; Mycelium [12, 22]; Seed [20]; Heart [18]; Lung [18]; Kidney [18]; Small intestine [18]; Skeletal muscle [18]; Mammary gland [18]

Localisation in source

Soluble [19]; Mitochondria [2, 18]; Cytosol [8, 13, 16, 18, 23]; Intracellular [12, 22]; Microsomes [13, 18]; Membrane vesicles (interior surface) [21]

Purification

Clostridium tyrobutyricum [5]; Bovine [1, 4, 18]; Dog [2]; Octopus vulgaris [3]; Rat (partial, isoforms [8]) [6, 8, 13, 14, 18, 23]; Phanerochaete chrysosporium [22]; Mouse [16, 17]; Microcystis aeruginosa [19]; Castor bean (2 isoenzymes: D-I, D-II) [20]

Crystallization

–

Cloned

–

Renaturated

–

5 STABILITY

pH

Temperature (°C)

–20 (some weeks) [4]; 0 (some days) [2]; 20 (several h) [2]; 57 (5 min, 90% loss of activity) [2]

Oxidation

Organic solvent

Enzyme Handbook © Springer-Verlag Berlin Heidelberg 1994
Duplication, reproduction and storage in data banks are only allowed with the prior permission of the publishers

General stability information

Sucrose, 0.25 M, essential for stabilization during storage at −20°C and purification [18]; Stable to dialysis against 0.01 M potassium phosphate buffer, pH 7.5 and 0.01 M Tris buffer, pH 8.2 and 8.9, 16 h or more at 3°C [2]

Storage

−20°C, 0.25 M sucrose, several months [18]; 5°C, 0.05 M sodium phosphate, pH 8, enzyme D-II is stable for 1 month, enzyme D-1 loses 30% of its activity after 1 week [20]

6 CROSSREFERENCES TO STRUCTURE DATABANKS

PIR/MIPS code

A30879 (Human); A26140 (Rat); A28950 (Rat); A41135 (Human); A32667 (Human); A35607 (Rat); A34162 (Rat)

Brookhaven code

7 LITERATURE REFERENCES

[1] Giuditta, A., Strecker, H.J.: Biochim. Biophys. Acta,48,10–19 (1961)
[2] Wosilait, W.D.: J. Biol. Chem.,235,1196–1201 (1960)
[3] Di Prisco, G., Casola, L., Giuditta, A.: Biochem. J.,105,455–460 (1967)
[4] Märki, F., Martius, C.: Biochem. Z.,333,111–135 (1960)
[5] Petitdemange, H., Marczak, R., Raval, G., Gay, R.: Can. J. Microbiol.,26,324–329 (1980)
[6] Rase, B., Bartfai, T., Ernster, L.: Arch. Biochem. Biophys.,172,380–386 (1976)
[7] Prestera, T., Prochaska, H.J., Talalay, P.: Biochemistry,31,824–833 (1992)
[8] Segura-Aguilar, J., Kaiser, R., Lind, C.: Biochim. Biophys. Acta,1120,33–42 (1992)
[9] Deng, P.S.K., Zhao, S.-H., Iyanagi, T., Chen, S.: Biochemistry,30,6942–6948 (1991)
[10] Cloete, F., Viljoen, C.C., Scott, W.E., Oosthuizen, M.M.J.: Biochim. Biophys. Acta,870,279–291 (1986)
[11] Hayashi, M., Hasegawa, K., Oguni, Y., Unemoto, T.: Biochim. Biophys. Acta,1035,230–236 (1990)
[12] Buswell, J.A., Eriksson, K.-E.: Methods Enzymol.,161,271–274 (1988)
[13] Sharkis, D.H., Swenson, R.P.: Biochem. Biophys. Res. Commun.,161,434–441 (1989)
[14] Prochaska, H.J.: Arch. Biochem. Biophys.,267,529–538 (1988)
[15] Fasco, M.J., Principe, L.M.: Biochem. Biophys. Res. Commun.,104,187–192 (1982)
[16] Prochaska, H.J., Talalay, P.: J. Biol. Chem.,261,1372–1378 (1986)
[17] Amzel, L.M., Bryant, S.H., Prochaska, H.J., Talalay, P.: J. Biol. Chem.,261,1379 (1986)
[18] Lind, C., Cadenas, E., Hochstein, P., Ernster, L.: Methods Enzymol.,186,287–301 (1990)
[19] Viljoen, C.C., Cloete, F., Scott, W.E.: Biochim. Biophys. Acta,827,247–259 (1985)
[20] Viljoen, C.C., Cloete, F., Botes, D.P., Kruger, H.: Phytochemistry,22,365–370 (1983)
[21] Lanyi, J.K.: J. Biol. Chem.,247,3001–3007 (1972)
[22] Constam, D., Muheim, A., Zimmermann, W., Fiechter, A. : J. Gen. Microbiol.,137,2209–2214 (1991)
[23] Wallin, R.: Biochem. J.,181,127–135 (1979)

1 NOMENCLATURE

EC number
1.6.99.3

Systematic name
NADH:(acceptor) oxidoreductase

Recommended name
NADH dehydrogenase

Synonymes
Cytochrome c reductase
Type I dehydrogenase
EC 1.6.2.1 (formerly)
beta-NADH dehydrogenase
Dehydrogenase, reduced nicotinamide adenine dinucleotide
Diaphorase
Dihydrocodehydrogenase I dehydrogenase
Dihydronicotinamide adenine dinucleotide dehydrogenase
Diphosphopyridine diaphorase
DPNH dehydrogenase
DPNH diaphorase
NADH diaphorase
NADH hydrogenase
NADH oxidoreductase
NADH-menadione oxidoreductase
Proteins, specific or class, gene MURF3
Reduced diphosphopyridine nucleotide diaphorase
More (enzyme is present in a mitochondrial complex as EC 1.6.5.3)

CAS Reg. No.
9079-67-8

---

2 REACTION AND SPECIFICITY

Catalysed reaction
NADH + acceptor →
→ NAD^+ + reduced acceptor (ping-pong mechanism [10], with ferricyanide as substrate: ordered bi-bi reaction mechanism [6])

Reaction type
Redox reaction

Enzyme Handbook © Springer-Verlag Berlin Heidelberg 1994
Duplication, reproduction and storage in data banks are only allowed with the prior permission of the publishers

Natural substrates

NADH + acceptor (enzyme connected with the respiratory chain) [8]

Substrate spectrum

1 NADH + acceptor (acceptors: ubiquinone-1 [16, 20], 2,6-dichlorophenol indophenol [2, 4, 5, 7–10, 14–17, 19, 20, 22, 23], menadione (not [22]) [2, 9, 10, 14, 16, 19, 20], juglone [2, 16], FMN [4], cytochrome c (not [16]) [4, 7, 9, 10, 14, 15, 20, 22], ferricyanide [6–10, 14–17, 19, 20, 22, 23], coenzyme Q_1 [23], methemoglobin-ferrocyanide complex [7], cytochrome b_5 [7, 15], methemoglobin (via cytochrome b_5, when NADH is used as electron donor) [7], duroquinone [16], benzoquinone [17], naphthoquinone [17], caldariellaquinone [17], 3-(4', 5'-dimethyl-thiazol-2-yl)-2, 4-diphenyl tetrazolium bromide [21])

2 More (no reaction as acceptor: potassium ferricyanide [2], lawsone (2-hydroxy-1,4-naphthoquinone) [2], vitamin K_1 (2-methyl-3-phythyl-1,4-naphthoquinone) [2], vitamin K_5 (4-amino-2-methyl-1-naphthol) [2], coenzyme Q_6 [2], coenzyme Q_{10} [2, 22], lipoic acid [2, 23], lipoamide [2], horse heart cytochrome c [2], after preparation has been subject to certain treatments cytochrome c may act as acceptor [3]) [2, 3, 22, 23]

Product spectrum

1 NAD^+ + reduced acceptor

2 ?

Inhibitor(s)

FMN [5]; Riboflavin [5]; NAD^+ [6, 14, 15, 19]; Ferrocyanide [6]; p-Chloromercuribenzoate [7, 15]; N-Ethylmaleimide [7, 15]; Atebrin [7]; Acrinol [7]; ADP [7, 19]; SCN^- [10]; NO_3^- [10]; Cl^- [10]; p-Hydroxymercuribenzoate [14, 22]; AMP [18, 19]; Adenine [19]; Deoxyadenosine [19]; Deoxy-AMP [19]; cAMP [19]; ATP [19]; p-Chloromercuriphenyl sulfonic acid [23]

Cofactor(s)/prostethic group(s)

Flavin (flavoprotein [14, 16, 17, 20, 23], flavin component is probably FAD [23], contains no flavin [1, 15, 19], contains 13.5 nmol per mg of protein [20], 85% of the total flavin is FAD, 15% is FMN [2], FMN, FAD or riboflavin required (soluble enzyme) [4], FMN required [8], contains 1 mol FAD per subunit (MW 65000) [9], FAD as prosthetic group [10], 1.4 mol of flavin per mol of enzyme [17], FAD stimulates [17, 23], FMN stimulates [23], absolute requirement for FAD [19]) [2, 4, 8–10, 14–17, 19, 20]; Guanidine (stimulates) [20]; Alkylguanidines (stimulate) [20]; 3-Acetylpyridine-NADH (can replace NADH) [2]; Thionicotinamide-NADH (can replace NADH) [2]; Arginine methyl ester (stimulates) [20]; NADH (relative dehydrogenation activity of NADH and NADPH is 2.9:1.0 at pH 7.5

[1], specific for NADH as electron donor [4, 5, 8, 10, 16–19, 21], specific for NADH above pH 7.0, below pH 7.0 low oxidation of NADPH [16]) [1, 4, 5, 8, 10, 16–19, 21]; NADPH (in absence of guanidine, NADPH is essentially unoxidized at pH values above 6.0 [20], less effective than NADH [7], relative dehydrogenation activity of NADH and NADPH is 2.9:1.0 at pH 7.5 [1], cannot replace NADH [2, 4, 5, 15], specific for NADH above pH 7.0, below pH 7.0 low oxidation of NADPH [16], oxidized at a rate less than 6% the rate of NADH oxidation [20], no activity with NADPH [23]) [1, 7, 16, 20]; Phospholipid (inactive in absence of) [12]; Detergent (inactive in absence of detergent) [8, 12]; Triton X-100 (slightly activates) [17]; More (3-pyridine aldehyde deamino NADH and 3-pyridine aldehyde NADH cannot replace NADH) [2]

Metal compounds/salts

K^+ (stimulates) [9, 10, 19]; Na^+ (stimulates) [2, 9, 10]; Li^+ (activates) [10]; Cs^+ (activates) [10]; Rb^+ (activates) [10]; Iron (iron-sulfur protein [20], contains non-heme iron [14, 20], 9.3 nmol per mg of protein [14], no iron-sulfur centres detected [16], 74.2 nmol per mg protein [20], contains approximately 0.06 mol of iron per mol of enzyme [23]) [14, 16, 20, 23]; Sodium arsenate (increases activity) [5]; More (kinetic properties in relation to salt activation) [13]

Turnover number (min^{-1})

Specific activity (U/mg)

2.74 [1]; 376 [2]; 790 [7]; 0.43 [8]; 24.2 [10]; 27.5 [14]; 69.2 [16]; 24.7 [21]; More [4, 9]

K_m-value (mM)

0.002 (cytochrome c) [10]; 0.0029 (2,6-dichlorophenol indophenol) [4]; 0.005 (ferricyanide) [7]; 0.0079 (FMN) [4]; 0.012 (cytochrome c) [4]; 0.017 (NADH (+ juglone)) [2]; 0.02 (cytochrome b_5) [7]; 0.021 (menadione (+ NADH)) [2]; 0.022 (NADH (+ 2,6-dichlorophenol indophenol)) [2]; 0.030 (2,6-dichlorophenol indophenol (+ NADH)) [2]; 0.034 (NADH (+ menadione)) [2]; 0.035 (menadione) [10]; 0.049 (ferricyanide) [6]; 0.05 (naphthoquinone) [17]; 0.063 (2,6-dichlorophenol indophenol) [5]; 0.098–0.11 (2,6-dichlorophenol indophenol) [7, 10]; 0.1 (benzoquinone) [17]; 0.133 (juglone (+ NADH)) [2]; 2.1 (ferricyanide) [10]; More [4–7, 10, 11, 14–16, 18, 19, 21–23]

pH-optimum

4.5 (NADH + ferricyanide) [17]; 5.2 (NADH + methemoglobin-ferrocyanide) [7]; 5.6 [11]; 6.5–7.0 [19]; 6.5–8.5 (NADH + ferricyanide) [7]; 7.2 [16]; 7.4 (2,6-dichlorophenol indophenol + NADH) [22]; 7.5 (2,6-dichlorophenol indophenol + NADH [17, 23]) [4, 5, 17, 23]; 7.5–8.0 [21]; 8.5–9.0 (2,6-dichlorophenol indophenol + NADH) [9]

Enzyme Handbook © Springer-Verlag Berlin Heidelberg 1994
Duplication, reproduction and storage in data banks are only allowed with the prior permission of the publishers

pH-range

3.5–7 (50% of activity maximum at pH 3.5 and 7, NADH + ferricyanide) [17]; 6.6–9.2 (about 50% of activity maximum at pH 6.6 and 9.2) [4]; 6.5–9.0 (6.5: about 50% of activity maximum, 9.0: about 75% of activity maximum) [5]

Temperature optimum (°C)

37 [21]

Temperature range (°C)

---

3 ENZYME STRUCTURE

Molecular weight

29000 (bovine, sedimentation velocity) [1]
37000 (Rhodopseudomonas capsulata, soluble enzyme, gel filtration) [4]
64000 (halophilic bacterium, gel filtration) [2]
69000 (bovine, gel filtration) [20]
74000 (bovine, calculation on the basis of FMN content, minimum molecular weight) [20]
79000 (Drosophila hydei, gel filtration) [11]
87000 (Candida utilis, polyacrylamide gel electrophoresis) [23]
95000 (Sulfolobus acidocaldarius, HPLC gel filtration) [17]
97000 (Rhodopseudomonas capsulata, membrane-bound enzyme, gel filtration) [5]
138000 (alkalophilic bacillus, gel filtration) [9]
144000 (human, erythrocyte membrane, gel filtration) [7]
215000 (Propionibacterium shermanii, gel filtration) [14]
480000 (bovine, polyacrylamide gel electrophoresis) [15]

Subunits

Monomer (1 × 38000, Rhodopseudomonas capsulata, soluble enzyme, SDS-PAGE) [4]
Dimer (2 × 50000, Sulfolobus acidocaldarius, SDS-PAGE [17], 2 × 65000, alkalophilic bacillus, SDS-PAGE [9]) [9, 17]
Hexamer (6 × 15000, Rhodopseudomonas capsulata, membrane-bound, SDS-PAGE) [4]
? (x × 63000, Bacillus subtilis, SDS-PAGE [21], x × 51000 + x × 24000 + x × 37000, Candida utilis, SDS-PAGE [23], x × 9000–10000,bovine, 3 subunits in equimolar amount, SDS-PAGE, gel filtration in presence of SDS [20], x × 38000, E. coli, SDS-PAGE [18], x × 36000, possibly tetramer, human erythrocyte membrane, SDS-PAGE [7], x × 34000, Acinetobacter calcoaceticus, SDS-PAGE [8], x × 17500, bovine, SDS-PAGE, (non-denatured enzyme has a MW of 480000) [15]) [7, 8, 15, 18, 20, 21, 23]

Glycoprotein/Lipoprotein

Lipoprotein (contains 0.0004 mmol phospholipid per mg protein, most of which is cardiolipin with traces of phosphatidylethanolamine and phosphatidylglycerol [8], contains no phospholipids [9], activity slightly increased by phospholipids [9]) [8, 9]

4 ISOLATION/PREPARATION

Source organism

Bovine [1, 15, 20]; Halophilic bacterium [2, 13]; Pig [3]; Rhodopseudomonas capsulata [4, 5]; Human [6, 7]; Acinetobacter calcoaceticus [8]; Alkalophilic bacillus [9]; Photobacterium phosphoreum [10]; Drosophila hydei [11]; E. coli [12, 18, 19]; Propionibacterium shermanii [14]; Arum maculatum [16]; Sulfolobus acidocaldarius [17]; Bacillus subtilis [21]; Agrobacterium tumefaciens [22]; Candida utilis [23]

Source tissue

Erythrocytes [1, 6, 7]; Cell [2, 4, 9, 22]; Heart [3, 20]; Neutrophil [15]; Larvae [11]; Embryo [11]

Localisation in source

Mitochondria [3, 11, 16, 20]; Soluble [4, 14]; Membrane-bound [5, 7–10, 15, 18, 21]; Membrane vesicles [21]

Purification

Bovine [1, 15, 20]; Halophilic bacterium [2]; Rhodopseudomonas capsulata [4, 5]; Human [7]; Acinetobacter calcoaceticus [8]; Alkalophilic bacillus [9]; Photobacterium phosphoreum (partial) [10]; Drosophila hydei [11]; E. coli [12]; Propionibacterium shermanii [14]; Arum maculatum [16]; Sulfolobus acidocaldarius [17]; Bacillus subtilis [21]; Agrobacterium tumefaciens [22]; Candida utilis [23]

Crystallization

[1]

Cloned

–

Renaturated

–

Enzyme Handbook © Springer-Verlag Berlin Heidelberg 1994
Duplication, reproduction and storage in data banks are only allowed with the prior permission of the publishers

5 STABILITY

pH

Temperature (°C)

25 (5 min, with FMN, riboflavin and NADH, inactivation [4], pH 4, 1 h, stable [17]) [4, 17]; 40 (40 min, 90% loss of activity) [14]; 100 (pH 7, 15 min, 30% loss of activity) [17]

Oxidation

Organic solvent

General stability information

Presence of salts increases stability of enzyme crystals in refrigerator [1]; Low ionic strength destabilizes [2]; Dialysis causes loss of activity [1]; 0.1% Triton X-100 and 20% glycerol stabilize [15]; Low concentration of enzyme causes loss of activity [1]; NADH stabilizes at low salt concentration [2]; FMN: stabilizes soluble enzyme [4], inactivates membrane-bound enzyme [5]; FAD stabilizes [4]; Repeated freezing and thawing inactivates [15]

Storage

Crystals kept in refrigerator for several months, stable [1]; 4°C, 1 month, less than 10% loss of activity [2]; –20 °C, 0.1% Triton X-100, 20% glycerol, several months [15]; –20°C, 50% glycerol, stable [17]

---

6 CROSSREFERENCES TO STRUCTURE DATABANKS

PIR/MIPS code

DEECR (Escherichia coli); JX0166 (Bacillus sp.); S09172 (chain 2 Chlamydomonas reinhardtii mitochondrion fragment); S04367 (chain 6 Chlamydomonas reinhardtii mitochondrion); B40358 (chain 1 Garden petunia mitochondrion); A25351 (Yeast Saccharomyces cerevisiae); A25096 (Neurospora crassa mitochondrion SGC3); B27311 (chain 3 Starfish Asterina pectinifera mitochondrion SGC8); C27311 (chain 4 Starfish Asterina pectinifera mitochondrion SGC8); D27311 (chain 5 Starfish Asterina pectinifera mitochondrion SGC8 fragment); A05099 (polypeptide II Bovine); S17191 (Neurospora crassa); S13801 (Neurospora crassa); A35693 (chain 7 Trypanosoma brucei mitochondrion SGC6); A34284 (polypeptide ND1 Sea urchin Paracentrotus lividus mitochondrion SGC8); B34284 (polypeptide ND2 Sea urchin Paracentrotus lividus mitochondrion SGC8); I34284 (polypeptide ND3 Sea urchin Paracentrotus lividus mitochondrion SGC8); A34285 (polypeptide ND4 Sea urchin Paracentrotus lividus mitochondrion SGC8); D34284 (polypeptide ND4L Sea urchin Paracentrotus lividus mitochondrion SGC8); B34285 (polypeptide ND5 Sea urchin Paracentrotus lividus mitochondrion SGC8); C34285 (polypeptide

ND6 Sea urchin Paracentrotus lividus mitochondrion SGC8); A26510 (polypeptide ND2 Sea urchin Paracentrotus lividus mitochondrion fragment); G26510 (polypeptide ND4 Sea urchin Paracentrotus lividus mitochondrion fragment); C26510 (polypeptide ND4L Sea urchin Paracentrotus lividus mitochondrion); S16967 (Bovine); E39746 (polypeptide ND1 Sea urchin Arbacia lixula fragment); A39746 (polypeptide ND3 Sea urchin Arbacia lixula); B39746 (polypeptide ND4 Sea urchin Arbacia lixula); C39746 (polypeptide ND5 Sea urchin Arbacia lixula fragment)

Brookhaven code

7 LITERATURE REFERENCES

[1] Adachi, K., Okuyama, T.: Biochim. Biophys. Acta,268,629–637 (1972)
[2] Hochstein, L.I., Dalton, B.P.: Biochim. Biophys. Acta,302,216–228 (1973)
[3] Kaniuga, Z.: Biochim. Biophys. Acta,73,550–564 (1963)
[4] Ohshima, T., Ohshima, M., Drews, G.: Z. Naturforsch.,39c,68–72 (1984)
[5] Ohshima, T., Drews, G.: Z. Naturforsch.,36c,400–406 (1981)
[6] Wang, C.-S.: Biochim. Biophys. Acta,616,22–29 (1980)
[7] Kitajima, S., Yasukochi, Y., Minakami, S.: Arch. Biochem. Biophys.,210,330–339 (1981)
[8] Borneleit, P., Kleber, H.-P.: Biochim. Biophys. Acta,722,94–101 (1983)
[9] Hisae, N., Aizawa, K., Koyama, N., Sekiguchi, T., Nosoh, Y.: Biochim. Biophys. Acta,743,232–238 (1983)
[10] Imagawa, T., Nakamura, T.: J. Biochem.,84,547–557 (1978)
[11] Hermans, L.: Biochim. Biophys. Acta,567,125–134 (1979)
[12] Dancey, G.F., Shapiro, B.M.: Biochim. Biophys. Acta,487,368–377 (1977)
[13] Hochstein, L.I.: Biochim. Biophys. Acta,403,58–66 (1975)
[14] Schwartz, A.C., Krause, A.E.: Z. Allg. Mikrobiol.,15,99–110 (1975)
[15] Nisimoto, Y., Wilson, E., Heyl, B.L., Lambeth, J.D.: J. Biol. Chem.,261,285–290 (1986)
[16] Cook, N.D., Cammack, R.: Eur. J. Biochem.,141,573–577 (1984)
[17] Wakao, H., Wakagi, T., Oshima, T.: J. Biochem.,102,255–262 (1987)
[18] Dancey, G.F., Levine, A.E., Shapiro, B.M.: J. Biol. Chem.,251,5911–5920 (1976)
[19] Dancey, G.F., Shapiro, B.M.: J. Biol. Chem.,251,5921–5928 (1976)
[20] Galante, Y.M., Hatefi, Y.: Arch. Biochem. Biophys.,192,559–568 (1979)
[21] Bergsma, J., Van Dongen, M.B.M., Konings, W.N.: Eur. J. Biochem.,128,151–157 (1982)
[22] Ramanarayanan, M., Rao, N.A., Vaidyanathan, C.S.: Indian J. Biochem. Biophys.,8,214–218 (1971)
[23] Mackler, B., Bevan, C., Person, R., Davis, K.A.: Biochem. Int.,3,9–17 (1981)

Enzyme Handbook © Springer-Verlag Berlin Heidelberg 1994
Duplication, reproduction and storage in data banks are only allowed with the prior permission of the publishers

1 NOMENCLATURE

EC number
1.6.99.5

Systematic name
NADH:(quinone-acceptor) oxidoreductase

Recommended name
NADH dehydrogenase (quinone)

Synonymes
Dehydrogenase, reduced nicotinamide adenine dinucleotide (quinone)
NADH-quinone oxidoreductase
DPNH-menadione reductase [1]
D-Diaphorase [1]

CAS Reg. No.
37256-36-3

2 REACTION AND SPECIFICITY

Catalysed reaction
NADH + acceptor →
→ NAD^+ + reduced acceptor

Reaction type
Redox reaction

Natural substrates

Substrate spectrum
1 NADH + acceptor (electron acceptors: menadione (poor) [1], benzoquinone [1], unsubstituted 1,2-naphthoquinone [1], unsubstituted 1,4-naphthoquinone [1], duroquinone [2], ubiquinone-1 [2])

Product spectrum
1 NAD^+ + reduced acceptor

Inhibitor(s)

Cofactor(s)/prostethic group(s)
No stimulation with FAD [1]

Metal compounds/salts

Enzyme Handbook © Springer-Verlag Berlin Heidelberg 1994
Duplication, reproduction and storage in data banks are only allowed with the prior permission of the publishers

Turnover number (min^{-1})

Specific activity (U/mg)
More [1]

K_m-value (mM)
0.012 (NADH) [1]; 0.1 (menadione) [1]

pH-optimum
5.5 (2 optima: 5.5 and 7.5) [1]; 7.5 (2 optima: 5.5 and 7. 5) [1]

pH-range

Temperature optimum (°C)

Temperature range (°C)

---

3 ENZYME STRUCTURE

Molecular weight

Subunits

Glycoprotein/Lipoprotein
–

---

4 ISOLATION/PREPARATION

Source organism
Pig (hog) [1]; Zea mays [2]; Zucchini [2]

Source tissue
Liver [1]; Etiolated plant tissues [2]; Hypocotyl [2]

Localisation in source
Soluble [1]; Extramitochondrial membranes [2]

Purification
Pig (hog) [1]

Crystallization
–

Cloned
–

Renaturated
–

5 STABILITY

pH

Temperature (°C)

Oxidation

Organic solvent

General stability information

Storage

6 CROSSREFERENCES TO STRUCTURE DATABANKS

PIR/MIPS code

Brookhaven code

7 LITERATURE REFERENCES

[1] Koli, A.K., Yearby, C., Scott, W., Donaldson, K.O.: J. Biol. Chem.,244,621–629 (1969)
[2] Pupillo, P., De Luca, L.: Dev. Plant Biol.,7,321–328 (1982)

Enzyme Handbook © Springer-Verlag Berlin Heidelberg 1994
Duplication, reproduction and storage in data banks are only allowed with the prior permission of the publishers

1 NOMENCLATURE

EC number
1.6.99.6

Systematic name
NADPH:(quinone-acceptor)oxidoreductase

Recommended name
NADPH dehydrogenase (quinone)

Synonymes
Dehydrogenase, reduced nicotinamide adenine dinucleotide phosphate (quinone)
NADPH oxidase

CAS Reg. No.
37256-37-4

---

2 REACTION AND SPECIFICITY

Catalysed reaction
NADPH + acceptor →
→ $NADP^+$ + reduced acceptor

Reaction type
Redox reaction

Natural substrates

Substrate spectrum
1 NADPH + acceptor (electron acceptors: menadione [1], 1,4-naphthoquinone [1], p-benzoquinone [1], methyl-p-benzoquinone [1], duroquinone [2], ubiquinone-1 [2])

Product spectrum
1 $NADP^+$ + reduced acceptor

Inhibitor(s)
Dicoumarol [1]; Folic acid derivatives [1]; More (not: 2,4-dinitrophenol) [1]

Cofactor(s)/prostethic group(s)
FAD (activity is dependent on added FAD) [1]

Metal compounds/salts

Enzyme Handbook © Springer-Verlag Berlin Heidelberg 1994
Duplication, reproduction and storage in data banks are only allowed with the prior permission of the publishers

Turnover number (min^{-1})

Specific activity (U/mg)
More [1]

K_m-value (mM)
0.011 (menadione) [1]; 0.046 (NADPH) [1]; More [2]

pH-optimum
6.5–7.5 [1]

pH-range

Temperature optimum (°C)

Temperature range (°C)

---

3 ENZYME STRUCTURE

Molecular weight

Subunits

Glycoprotein/Lipoprotein
–

---

4 ISOLATION/PREPARATION

Source organism
Pig (hog) [1]; Zea mays [2]; Zucchini [2]

Source tissue
Liver [1]; Etiolated plant tissue [2]

Localisation in source
Extramitochondrial membranes [2]

Purification
Pig [1]

Crystallization
–

Cloned
–

Renaturated
–

5 STABILITY

pH

Temperature (°C)

Oxidation

Organic solvent

General stability information

Storage

6 CROSSREFERENCES TO STRUCTURE DATABANKS

PIR/MIPS code

Brookhaven code

7 LITERATURE REFERENCES

[1] Koli, A.K., Yearby, C., Scott, W., Donaldson, K.O.: J. Biol. Chem.,244,621–629 (1969)
[2] Pupillo, P., De Luca, L.: Dev. Plant Biol.,7,321–328 (1982)

Enzyme Handbook © Springer-Verlag Berlin Heidelberg 1994
Duplication, reproduction and storage in data banks are only allowed with the prior permission of the publishers

1 NOMENCLATURE

EC number
1.6.99.7

Systematic name
NAD(P)H:6,7-dihydropteridine oxidoreductase

Recommended name
Dihydropteridine reductase

Synonymes
NADPH-specific dihydropteridine reductase [3, 42]
Reductase, dihydropteridine (reduced nicotinamide adenine dinucleotide)
Dihydropteridine reductase (NADH)
EC 1.6.99.10 (formerly)
DHPR [40]
NADH-dihydropteridine reductase [2]
NADPH-dihydropteridine reductase [37]
More (EC 1.6.99.10 and EC 1.6.99.7 are different enzymes: EC 1.6.99.10 (NADPH-specific dihydropteridine reductase) and NADH-dihydropteridine reductase (EC 1.6.99.7) have no common antigenic determinants, and show different physicochemical properties [42], enzyme is not identical with dihydrofolate reductase) [42]

CAS Reg. No.
9074-11-7; 70851-99-9

2 REACTION AND SPECIFICITY

Catalysed reaction
NAD(P)H + 6,7-dihydropteridine →
→ $NAD(P)^+$ + 5,6,7,8-tetrahydropteridine (ordered bi-bi mechanism [13])

Reaction type
Redox reaction

Enzyme Handbook © Springer-Verlag Berlin Heidelberg 1994
Duplication, reproduction and storage in data banks are only allowed with the prior permission of the publishers

Natural substrates

More (essential component of the hepatic phenylalanine hydroxylating system [1], essential component of catecholamine biosynthetic metabolism [32], may also have an important role in regulation of catecholamine synthesis [33], no marked specificity for the pteridine cofactor that occurs naturally in this organism (L-threo-neopterin) [6], essential enzyme component of the complex system catalyzing hydroxylation of phenylalanine, tyrosine and tryptophan [36]) [1, 6, 32, 33, 36]

Substrate spectrum

1 NAD(P)H + quinonoid 6,7-dihydropteridine (substrate is the quinonoid form of dihydropteridine)
2 Quinonoid 7,8-(6H)-dihydrobiopterin + NAD(P)H [1, 2, 20, 40]
3 Quinonoid 6,7-dimethyl-7,8-(6H)-dihydropterin + NAD(P)H [1, 20, 40]
4 Quinonoid 6-methyl-7,8-(6H)-dihydropterin + NAD(P)H [2, 20, 40]
5 Quinonoid dihydro-2-methylamino-4-hydroxy-6,7-dimethylpteridine + NAD(P)H [1]
6 2-Amino-4-hydroxy-6,7-dimethyl-quinonoid-dihydropteridine + NAD(P)H [1]
7 Quinonoid 6,6-Dimethyldihydropterin + NAD(P)H [20]
8 Quinonoid dihydroneopterin + NAD(P)H [40]
9 Quinonoid dihydromonapterin + NAD(P)H [40]
10 More (absolutely specific for quinonoid dihydro-isomer corresponding to 7,8-dihydropterins [1], no reaction with: 7,8-dihydropterin [35], dihydrofolate [35], enzyme also has pterin-independent NADH and NADPH oxidoreductase activity with potassium ferricyanide [40]) [1–3, 20, 35, 40]

Product spectrum

1 5,6,7,8-Tetrahydropteridine + $NAD(P)^+$
2 5,6,7,8-Tetrahydropterin + $NAD(P)^+$
3 6,7-Dimethyl-5,6,7,8-tetrahydropterin + $NAD(P)^+$
4 6-Methyl-5,6,7,8-tetrahydropterin + $NAD(P)^+$
5 Tetrahydro-2-methylamino-4-hydroxy-6,7-dimethylpteridine + $NAD(P)^+$
6 2-Amino-4-hydroxy-6,7-dimethyltetrahydropteridine + $NAD(P)^+$ [1]
7 6,6-Dimethyltetrahydropterin + $NAD(P)^+$
8 Tetrahydroneopterin + $NAD(P)^+$
9 Tetrahydromonapterin + $NAD(P)^+$
10 ?

Inhibitor(s)

Aminopterin (no inhibition up to 0.1 mM [33], poor inhibition [3], 1 mM [35]) [1–3, 6, 31, 35]; Methotrexate (no inhibition up to 0.1 mM [33]) [1]; Amethopterin (poor inhibition [3, 35]) [2, 3, 35]; $NADP^+$ [3, 43]; L-Thyrosine [3, 22, 27]; 2,6-Dichlorophenolindophenol [3, 43]; $HgCl_2$ [5, 6, 36]; p-Chloromercuribenzoate [5, 19, 36]; NAD^+ [5]; N-Ethylmaleimide [6, 36]; $MgCl_2$ (no effect [36]) [6]; $MnCl_2$ (no effect [36]) [6]; $CdCl_2$ [6]; Dopamine (no inhibition below 0.2 mM [21]) [16, 27, 31]; Serotonin [17]; Noradrenalin (no inhibition below 0.2 mM [21]) [17]; Tryptophan [17]; Phenylalanine [17]; Phenyllactate [17]; Phenylpyruvate [17, 31]; 5,5'-Dithiobis(2-nitrobenzoate) [19, 36]; N-Bromosuccinimide [19]; Aminochromes (oxidation products of dopamine, noradrenaline and adrenaline) [21]; Tyramine [22, 27]; p-Hydroxyphenylpyruvate [22]; L-beta-3,4-Dihydroxyphenylalanine (L-DOPA) [22]; p-Hydroxyphenyllactate [22]; p-Hydroxyphenylacetate [22]; Hydroxylated derivatives of 1-methyl-4-phenyl-1,2,3,6-tetrahydropyridine (1-methyl-4-phenyl-1,2,3,6-tetrahydropyridine and its chloro- and norderivatives: no inhibition at lower than millimolar concentration) [23]; Apomorphine (and its analogs) [24]; Quinoid 6,6,8-trimethyl-7,8-dihydro(6H)pterin (weak inhibition) [25]; Catecholamine [27]; 5-Hydroxydopamine [27]; 6-Hydroxydopamine [27]; N-Methyldopamine [27]; Octopamine [27]; Norepinephrine [27]; Epinephrine [27]; 3-Iodotyrosine (weak) [27]; 3-O-Methyldopamine [27]; 4-O-Methyldopamine [27]; 3-O-Methylepinephrine [27]; alpha-Methyltyrosine (weak) [27]; cis-Diaminodichloroplatinum [28]; trans-Diaminodichloroplatinum [28]; Dopamine-derived tetrahydroisoquinolines (e.g. 3', 4'-deoxynorlaudanosolinecarboxylic acid, higenamine-1-carboxylic acid, higenamine, saesolinol) [31]; Iodoacetamide [36]; Tetrahydrobiopterin (above 0.05 mM, substrate inhibition) [36]; 2,4-Diaminopteridine [39]; L-Thyroxine [43]; More (quinone products of oxidation of L-DOPA are responsible for the time-dependent inactivation of the enzyme) [30]

Cofactor(s)/prostethic group(s)

FAD (contains bound FAD) [40]; NADH (utilizes both NADH and NADPH [1, 2, 33–36, 38–40], NADH preferred [1, 2, 6, 15, 34–36, 38], bovine liver: 2 distinct dihydropteridine reductases which catalyze the conversion of the quinoid dihydropteridine to tetrahydropterin: 1. utilizes NADH as a better substrate than NADPH (NADH-dihydropteridine reductase, DPR) [2, 42, 43], 2. strict specificity for NADPH (NADPH-specific dihydropteridine reductase (TPR)) [3]) [1–3, 6, 15, 33–36, 38–40, 42, 43]; NADPH (utilizes both NADH and NADPH [1, 2, 33–36, 38–40], NADH preferred [1, 2, 6, 15, 34–36, 38], bovine liver: 2 distinct dihydropteridine reductases which catalyze the conversion of the quinoid dihydropteridine to tetrahydropterin: 1. utilizes NADH

Enzyme Handbook © Springer-Verlag Berlin Heidelberg 1994
Duplication, reproduction and storage in data banks are only allowed with the prior permission of the publishers

as a better substrate than NADPH (NADH-dihydropteridine reductase, DPR) [2, 42, 43], 2. strict specificity for NADPH (NADPH-specific dihydropteridine reductase (TPR)) [3], inactive with NADPH [6], specific for NADPH [6]) [1–3, 6, 15, 33–36, 38–40, 42, 43]

Metal compounds/salts

Turnover number (min^{-1})

2580–9540 (quinonoid 6,7-dihydropteridine, turnover number of rat liver enzyme and various E. coli-expressed mutants) [8]; 654 (20°C, quinonoid 7,8(6H)-dihydropterin) [13]; 1422 (25°C, quinonoid 7,8(6H)-dihydropterin) [13]; 1734 (30°C, quinonoid 7,8(6H)-dihydropterin) [13]; 6780 (37°C, quinonoid 7,8(6H)-dihydropterin) [13]; 1200 (quinonoid dihydrobiopterin) [6]

Specific activity (U/mg)

15.6 [1, 38]; 55.1 [16]; 20.7 [40]; 64.2 [32]; 422 [36]; More (immunoassay [11]) [2–5, 14, 37, 42]

K_m-value (mM)

0.001 (quinonoid dihydrobiopterin (+ NADH)) [1]; 0.009 (quinonoid dihydrobiopterin (+ NADPH)) [1]; 0.0152 (2-amino-4-hydroxy-6,7-dimethyldihydropteridine (+ NADH)) [1]; 0.006 (quinonoid 2-amino-4-hydroxy-6,7-dimethyldihydropteridine (+ NADPH)) [1]; 0.004 (NADH (+ quinonoid dihydrobiopterin)) [1]; 0.0709 (NADPH (+ quinonoid dihydrobiopterin)) [1]; 0.0057 (NADH (+ quinonoid 2-amino-4-hydroxy-6,7-dimetyldihydropteridine)) [1]; 0.080 (NADPH (+ quinonoid 2-amino-4-hydroxy-6,7-dimethyldihydropteridine)) [1]; 0.0056 (quinonoid 7,8-dihydro-6-methylpterin (+ NADH)) [2]; 0. 0058 (quinonoid 7,8-dihydro-6-methylpterin (+ NADPH)) [2]; 0.0016 (quinonoid 7,8-dihydrobiopterin (+ NADH)) [2]; 0.0011 (quinonoid 7,8-tetrahydrobiopterin (+ NADPH)) [2]; 0.00079 (NADH (+ quinonoid 7,8-dihydro-6-methylpterin)) [2]; 0.095 (NADPH (+ quinonoid 7,8-dihydrobiopterin)) [2]; 0.0015 (NADH (+ quinonoid 7,8-dihydrobiopterin)) [2]; 0.074 (NADPH (+ quinonoid 7,8-dihydrobiopterin)) [2]; 0.0014 (quinonoid 7,8-dihydro-6-methylpterin (+ NADPH), NADPH (+ quinonoid dihydro-6-methylpterin)) [3]; 0.16 (quinonoid dihydrobiopterin) [6]; 0.34 (quinonoid 2-amino-4-hydroxy-6,7-dimethyldihydropteridine) [6]; 1.45 (quinonoid L-threo-dihydroneopterin) [6]; More (K_m for quinoid 7,8(6H)-dihydrobiopterin increases with increasing temperature (from 20°C to 37°C) [13]) [4–6, 8, 13, 15, 20, 21, 26, 32, 34–36, 38–40]

pH-optimum

7 [5]; 7.2 [6, 36]; 7.6 [32]

pH-range

6.8–7.8 (6.8: about 45% of activity maximum, 7.8: about 15% of activity maximum) [1]

Temperature optimum (°C)

37 [36]

Temperature range (°C)

3 ENZYME STRUCTURE

Molecular weight

41000 (sheep, gradient PAGE) [1, 38]
43500–49000 (Pseudomonas sp., gel filtration) [6]
47000 (sheep, gel filtration) [5]
47000–54000 (human, sedimentation equilibrium analysis, gel filtration [36], human, sedimentation equilibrium studies, gel filtration, gradient PAGE [4]) [4, 36]
49000 (bovine, sedimentation equilibrium analysis [14], rat, gel filtration [32]) [14, 32]
49000–50000 (bovine, NADH-dihydropteridine reductase) [2]
51000 (rat, gel filtration) [41]
54000 (E. coli, gel filtration) [40]
55000 (Crithidia fasciculata, gel filtration) [35]
65000–70000 (bovine, NADPH-specific dihydropteridine reductase, gel filtration, polyacrylamide gel electrophoresis at various gel concentrations, equilibrium centrifugation analysis) [3]
68000 (bovine, gel filtration) [42]
69000 (human, gel filtration) [42]
70000 (bovine, gradient PAGE) [37]

Subunits

Dimer (2 × 21000, sheep liver, SDS-PAGE [1, 38], 2 × 25000, bovine, NADH-dihydropteridine reductase [2], 2 × 35000, bovine, NADPH-specific dihydropteridine reductase, SDS-PAGE [3], 2 × 27000, E. coli [40], 2 × 26000, human, SDS-PAGE [4], 2 × 27000, sheep brain, SDS-PAGE [5], 2 × 25000, rat, SDS-PAGE [32], 2 × 35000, human, bovine, SDS-PAGE [42], 2 × 27000, E. coli, SDS-PAGE [40], 2 × 25500, rat, SDS-PAGE [41], 2 distinct subunits alpha and beta identified, the forms I, II and III represent three different dimeric combinations: I (alpha,alpha), II (alpha,beta), III (beta,beta) [41]) [1–5, 32, 36, 38, 40–42]

Glycoprotein/Lipoprotein

–

Enzyme Handbook © Springer-Verlag Berlin Heidelberg 1994
Duplication, reproduction and storage in data banks are only allowed with the prior permission of the publishers

4 ISOLATION/PREPARATION

Source organism

Crithidia fasciculata [35]; Monkey [29]; Rabbit [34]; Cat [34]; Sheep [1, 5, 20, 22, 34, 38]; Bovine [2, 3, 14, 20, 29, 33, 34, 37, 42, 43]; Human [4, 7, 9, 18, 21–24, 26–31, 36, 38, 42]; Rat (multiple forms: I, II, II [41]) [8, 10, 12, 15–17, 19, 23, 24, 32, 34, 38, 39, 41]; E. coli [40]; Pseudomonas sp. (ATCC 11299a) [6]

Source tissue

Cultured fibroblasts [18]; Continuous lymphoid cells [18]; Striatal synaptosomes [23, 24]; Adrenal medulla [1, 33]; Liver [1–4, 8, 15–19, 22–24, 27, 29, 31, 32, 36–39, 41–43]; Kidney [1]; Brain [1, 5, 21, 26, 28, 30]; Heart (low activity) [1]; Skeletal muscle (low activity) [1]; Spleen (low activity) [1]

Localisation in source

Purification

Human [4, 21, 36, 42]; Rat [16, 32, 39]; Sheep [1, 5, 38]; Bovine [2, 3, 14, 37, 42]; Pseudomonas sp. (ATCC 11299a) [6]; E. coli [40]

Crystallization

[12, 14]

Cloned

(human [7], rat liver enzyme expressed in E. coli [8], high-level expression of human enzyme in E. coli [9], rat liver enzyme [10]) [7–10]

Renaturated

–

5 STABILITY

pH

6.5–8.0 (unstable below and above) [35]

Temperature (°C)

80 (1 min, inactivation) [33]

Oxidation

Organic solvent

General stability information

NADH protects against inactivation [6]; NADPH protects from inactivation [6]; Unstable to $(NH_4)_2SO_4$ precipitation [35]; Not stable during gel filtration [35]; Stable to freezing, –70°C, and thawing [41]

Storage

-80°C, stable for years [1]; -20°C, stable for at least a year (enzyme activity in ammonium sulfate precipitate) [6]; -70°C, stable for at least 2 years [14]; -80°C, 40 mM potassium phosphate buffer, pH 6.8, stable for at least a week [35]; -80°C, stable for at least 3 months [37]; 4°C, stable for up to 2 weeks [41]

6 CROSSREFERENCES TO STRUCTURE DATABANKS

PIR/MIPS code

Brookhaven code

7 LITERATURE REFERENCES

[1] Kaufman, S.: Methods Enzymol.,142,97–103 (1987)
[2] Hasegawa, H., Nakanishi, N.: Methods Enzymol.,142,103–110 (1987)
[3] Hasegawa, H., Nakanishi, N.: Methods Enzymol.,142,111–116 (1987)
[4] Firgaira, F.A., Cotton, R.G.H., Jennings, I., Danks, D.M.: Methods Enzymol.,142,116–126 (1987)
[5] Scrimgeour, K.G., Cheema-Dhadli, S.: Methods Enzymol.,142,127–132 (1987)
[6] Williams, C.D., Dickens, G., Letendre, C.H., Guroff, G., Haines, C., Shiota, T.: J. Bacteriol.,127,1197–1207 (1976)
[7] Lockyer, J., Cook, R.G., Milstein, S., Kaufman, S., Woo, S.L.C., Ledley, F.D.: Proc. Natl. Acad. Sci. USA,84 ,3329–3333 (1987)
[8] Matthews, D.A., Varughese, K.I., Skinner, M., Xuong, N.H., Hoch, J., Trach, K., Schneider, M., Bray, T., Whiteley, J.M.: Arch. Biochem. Biophys.,287,234–239 (1991)
[9] Armarego, W.L.F., Cotton, G.H., Dahl, H.-H.M., Dixon, N.E.: Biochem. J.,261,265–268 (1989)
[10] Shahbaz, M., Hoch, J.A., Trach, A., Hural, J.A., Webber, S., Whiteley, J.M.: J. Biol. Chem.,262,16412–16416 (1987)
[11] Kwan, S.-W., Shen, R.-S., Abell, C.W.: Anal. Biochem.,164,391–396 (1987)
[12] Matthews, D.A., Webber, S., Whiteley, J.M.: J. Biol. Chem.,261,3891–3893 (1986)
[13] Randles, D.: Eur. J. Biochem.,155,301–304 (1986)
[14] Hasegawa, H.: J. Biochem.,81,169–177 (1977)
[15] Webber, S., Whiteley, J.M.: J. Biol. Chem.,253,6724–6729 (1978)
[16] Purdy, S.E., Blair, J.A., Barford, P.A.: Biochem. J.,195,769–771 (1981)
[17] Gould, K.G., Engel, P.C.: Biochem. Soc. Trans.,8,565–566 (1980)
[18] Firgaira, F.A., Choo, K.H., Cotton, R.G.H., Danks, D. M.: Biochem. J.,197,45–53 (1981)
[19] Webber, S., Whiteley, J.M.: Arch. Biochem. Biophys.,206,145–152 (1981)
[20] Bailey, S.W., Ayling, J.E.: Biochemistry,22,1790–1798 (1983)
[21] Armarego, W.L.F., Waring, P.: Biochem. Biophys. Res. Commun.,113,895–899 (1983)
[22] Shen, R.-S.: Biochim. Biophys. Acta,785,181–185 (1984)
[23] Abell, C.W., Shen, R.-S., Gessner, W., Brossi, A.: Science,224,405–407 (1984)
[24] Shen, R.-S., Smith, R.V., Davis, P.J., Abell, C.W.: J. Biol. Chem.,259,8994–9000 (1984)

Enzyme Handbook © Springer-Verlag Berlin Heidelberg 1994
Duplication, reproduction and storage in data banks are only allowed with the prior permission of the publishers

[25] Randles, D., Armarego, L.F.: Eur. J. Biochem.,146,467–474 (1985)
[26] Armarego, W.L.F., Ohnishi, A., Taguchi, H.: Biochem. J.,234,335–342 (1986)
[27] Shen, R.-S.: Biochim. Biophys. Acta,743,129–135 (1983)
[28] Armarego, W.L.F., Ohnishi, A.: Eur. J. Biochem.,164,403–409 (1987)
[29] Nakanishi, N., Ozawa, K., Yamada, S.: J. Biochem.,99,1311–1315 (1986)
[30] Waring, P.: Eur. J.Biochem.,155,305–310 (1986)
[31] Shen, R.-S., Smith, R.V., Davis, P.J., Brubaker, A., Abell, C.W.: J. Biol. Chem.,257,7294–7297 (1982)
[32] Nakanishi, N., Hirayama, K., Yamada, S.: J. Biochem.,92,1033–1040 (1982)
[33] Musacchio, J.M.: Biochim. Biophys. Acta,191,485–487 (1969)
[34] Nielsen, K.H., Simonsen, V., Lind, K.E.: Eur. J. Biochem.,9,497–502 (1969)
[35] Hirayama, K., Nakanisi, N., Sueoka, T., Katoh, S., Yamada, S.: Biochim. Biophys. Acta,612,337–343 (1980)
[36] Firgaira, F.A., Cotton, G.H., Danks, D.M.: Biochem. J.,197,31–43 (1981)
[37] Nakanishi, N., Hasegawa, H., Watabe, S.: J. Biochem.,81,681–685 (1977)
[38] Craine, J.E., Hall, E.S., Kaufman, S.: J. Biol. Chem.,247,6082–6091 (1972)
[39] Lind, K.E.: Eur. J. Biochem.,25,560–562 (1972)
[40] Vasudevan, S.G., Shaw, D.C., Armarego, W.L.F.: Biochem. J.,255,581–588 (1988)
[41] Webber, S., Hural, J.A., Whiteley, J.M.: Arch. Biochem. Biophys.,248,358–367 (1986)
[42] Nakanishi, N., Hasegawa, H., Yamada, S., Akino, M.: J. Biochem.,99,635–644 (1986)
[43] Nakanishi, N., Hasegawa, H., Akino, M., Yamada, S.: J. Biochem.,99,645–652 (1986)

1 NOMENCLATURE

EC number
1.6.99.8

Systematic name
NADH:cob(III)alamin oxidoreductase

Recommended name
Aquacobalamin reductase

Synonymes
Reductase aquacobalamin
Aquocobalamin reductase
Vitamin B_{12a} reductase
Reductase, vitamin B_{12a}
NADH-linked aquacobalamin reductase
B_{12a} reductase [2]

CAS Reg. No.
37256-39-6

2 REACTION AND SPECIFICITY

Catalysed reaction
NADH + 2 aquacob(III)alamin →
→ NAD^+ + 2 cob(II)alamin

Reaction type
Redox reaction

Natural substrates
NADH + aquacob(III)alamin (conversion of vitamin B_{12} to its principal coenzyme form, adenosyl-B_{12}) [2]

Substrate spectrum
1 NADH + aquacob(III)alamin
2 More (dithioerythritol cannot replace NADH as reductant) [2]

Product spectrum
1 NAD^+ + cob(II)alamin
2 ?

Inhibitor(s)

Enzyme Handbook © Springer-Verlag Berlin Heidelberg 1994
Duplication, reproduction and storage in data banks are only allowed with the prior permission of the publishers

Cofactor(s)/prostethic group(s)
FAD (flavoprotein, utilizes FAD better than FMN) [2]; FMN (flavoprotein, utilizes FAD better than FMN) [2]; NADH

Metal compounds/salts

Turnover number (min^{-1})

Specific activity (U/mg)

K_m-value (mM)
0.015 (FAD or FMN) [2]

pH-optimum
7.0 (assay at) [1]

pH-range

Temperature optimum (°C)
40 (assay at) [1]

Temperature range (°C)

---

3 ENZYME STRUCTURE

Molecular weight

Subunits

Glycoprotein/Lipoprotein
–

---

4 ISOLATION/PREPARATION

Source organism
Clostridium tetanomorphum [2]; Rat [1, 3]; Monkey (Macaca) [1]; Bovine [1]; Pig [1]; Chicken [1]; Frog (Rana catesbeina) [1]; Girella punctate (sea fish) [1]; Cyprinus auratus (freshwater fish) [1]; Human [4]

Source tissue
Adrenal body [1]; Bone marrow [1, 4]; Liver [1, 3, 4]; Pancreas [1]; Kidney [1, 4]; Testis [1]; Jejunum [1, 4]; Ileum [1, 4]; Large intestine [1]; Spleen [1, 4]; Heart [1]; Lung [1, 4]; Stomach [1, 4]; Brain [1]; Cerebrum [4]; Cerebellum [4]; Adrenal glands [4]; Duodenum [4]; Colon [4]

Localisation in source
Mitochondria (inside of outer membrane) [3, 4]; Microsomes [3, 4]

Purification

Crystallization

–

Cloned

–

Renaturated

–

5 STABILITY

pH

Temperature (°C)

50 (5 min, about 80% loss of activity) [2]; 60 (5 min, about 90% loss of activity) [2]; 70 (5 min, complete loss of activity) [2]

Oxidation

Organic solvent

General stability information

Lyophilized preparations of cell-free extract: stable to storage at –20°C for several weeks [2]; Solutions: rapid loss of activity, especially if frozen and thawed [2]; Dithioerythritol stabilizes [2]

Storage

6 CROSSREFERENCES TO STRUCTURE DATABANKS

PIR/MIPS code

Brookhaven code

7 LITERATURE REFERENCES

[1] Watanabe, F., Nakano, Y., Tachikake, N., Tamura, Y., Yamanaka, H., Kitaoka, S.: J. Nutr. Sci. Vitaminol.,36,349–356 (1990)
[2] Walker, G.A., Murphy, S., Huennekens, F.M.: Arch. Biochem. Biophys.,134,95–102 (1969)
[3] Watanabe, F., Nakano, Y., Maruno, S., Tachikake, N., Tamura, Y., Kitaoka, S.: Biochem. Biophys. Res. Commun.,165,675–679 (1989)
[4] Watanabe, F., Nakano, Y., Tachikake, N., Kitaoka, S., Tamura, Y., Yamanaka, H., Haga, S., Imai, S., Saido, H.: Int. J. Biochem.,23,531–533 (1991)

Enzyme Handbook © Springer-Verlag Berlin Heidelberg 1994
Duplication, reproduction and storage in data banks are only allowed with the prior permission of the publishers

1 NOMENCLATURE

EC number
1.6.99.9

Systematic name
NADH:cob(II)alamin oxidoreductase

Recommended name
Cob(II)alamin reductase

Synonymes
Reductase, cob(II)alamin
Reductase, vitamin B_{12r}
Vitamin B_{12r} reductase
B_{12r} reductase [1]

CAS Reg. No.
37256-40-9

2 REACTION AND SPECIFICITY

Catalysed reaction
NADH + 2 cob(II)alamin →
→ NAD^+ + 2 cob(I)alamin

Reaction type
Redox reaction

Natural substrates
NADH + cob(II)alamin (conversion of vitamin B_{12} to its principal coenzyme form, adenosyl-B_{12}) [1]

Substrate spectrum
1 NADH + cob(II)alamin
2 More (dithioerythritol can replace NADH as reductant) [1]

Product spectrum
1 NAD^+ + cob(I)alamin
2 ?

Inhibitor(s)

Enzyme Handbook © Springer-Verlag Berlin Heidelberg 1994
Duplication, reproduction and storage in data banks are only allowed with the prior permission of the publishers

Cofactor(s)/prostethic group(s)
FAD (flavoprotein, stimulated equally well by FAD or FMN) [1]; FMN (flavoprotein, stimulated equally well by FAD or FMN) [1]; NADH

Metal compounds/salts

Turnover number (min^{-1})

Specific activity (U/mg)

K_m-value (mM)

pH-optimum

pH-range

Temperature optimum (°C)

Temperature range (°C)

---

3 ENZYME STRUCTURE

Molecular weight

Subunits

Glycoprotein/Lipoprotein
–

---

4 ISOLATION/PREPARATION

Source organism
Clostridium tetanomorphum [1]; Human [2]

Source tissue
Liver [2]; Kidney [2]; Pancreas [2]; Heart [2]; Spleen [2]; Lung [2]; Cerebrum [2]; Cerebellum [2]; Adrenal glands [2]; Stomach [2]; Duodenum [2]; Jejunum [2]; Ileum [2]; Colon [2]; Bone marrow [2]

Localisation in source
Mitochondria [2]; Microsomes [2]

Purification

Crystallization
–

Cloned
–

Renatured

–

5 STABILITY

pH

Temperature (°C)

50 (5 min, about 40% loss of activity) [1]; 60 (5 min, 75% loss of activity) [1]; 70 (5 min, about 90% loss of activity) [1]

Oxidation

Organic solvent

General stability information

Storage

6 CROSSREFERENCES TO STRUCTURE DATABANKS

PIR/MIPS code

Brookhaven code

7 LITERATURE REFERENCES

[1] Walker, G.A., Murphy, S., Huennekens, F.M.: Arch. Biochem. Biophys.,134,95–102 (1969)

[2] Watanabe, F., Nakano, Y., Tachikake, N., Kitaoka, S., Tamura, Y., Yamanaka, H., Haga, S., Imai, S., Saido, H.: Int. J. Biochem.,23,531–533 (1991)

Enzyme Handbook © Springer-Verlag Berlin Heidelberg 1994
Duplication, reproduction and storage in data banks are only allowed with the prior permission of the publishers

1 NOMENCLATURE

EC number
1.6.99.11

Systematic name
NADPH:aquacob(III)alamin oxidoreductase

Recommended name
Aquacobalamin reductase (NADPH)

Synonymes
Reductase, aquacobalamin (reduced nicotinamide adenine dinucleotide phosphate)
NADPH-linked aquacobalamin reductase

CAS Reg. No.
110777-32-7

2 REACTION AND SPECIFICITY

Catalysed reaction
NADPH + 2 aquacob(III)alamin →
→ $NADP^+$ + 2 cob(III)alamin

Reaction type
Redox reaction

Natural substrates
NADPH + aquacob(III)alamin

Substrate spectrum
1 NADPH + aquacob(III)alamin
2 NADPH + hydroxycobalamin [4]
3 More (specific for NADPH [4] not: cyanocobalamin)

Product spectrum
1 $NADP^+$ + cob(III)alamin
2 $NADP^+$ + 5'-deoxyadenosylcobalamin [4]
3 ?

Inhibitor(s)
5,5'-Dithiobis(2-nitrobenzoic acid) [4]; N-Ethylmaleimide [4]; Mersalyl [4]; Zn^{2+} [4]; Al^{3+} [4]

Enzyme Handbook © Springer-Verlag Berlin Heidelberg 1994
Duplication, reproduction and storage in data banks are only allowed with the prior permission of the publishers

Cofactor(s)/prostethic group(s)

NADPH (specific for NADPH) [4]; FAD (flavoprotein, contains 1 molecule of FAD or FMN as prosthetic group, FAD or FMN not required as cofactors) [4]; FMN (flavoprotein, contains 1 molecule of FMN or FAD as prosthetic group, FAD or FMN not required as cofactors) [4]; More (no activity with NADH) [3, 4]

Metal compounds/salts

Turnover number (min^{-1})

Specific activity (U/mg)

2.24 [4]

K_m-value (mM)

0.043 (NADH) [4]; 0.055 (hydroxycobalamin) [4]

pH-optimum

7.0 [4]

pH-range

Temperature optimum (°C)

40 [4]

Temperature range (°C)

---

3 ENZYME STRUCTURE

Molecular weight

65000–66000 (Euglena gracilis, gel filtration, SDS-PAGE) [4]

Subunits

Glycoprotein/Lipoprotein

–

---

4 ISOLATION/PREPARATION

Source organism

Rat [1, 2]; Monkey (Macaca) [1]; Bovine [1]; Pig [1]; Chicken [1]; Frog (Rana catesbeina) [1]; Girella punctate (sea fish) [1]; Cyprinus uratus (freshwater fish) [1]; Euglena gracilis [3, 4]

Source tissue

Liver [1, 2]; Pancreas [1]; Kidney [1]; Testis [1]; Spleen [1]; Heart [1]; Lung [1]; Stomach [1]; Brain [1]; Ileum [1]; Jejunum [1]; Large intestine [1]

Localisation in source
Mitochondria (inside of outer membrane [2]) [2–4]; Microsomes [2]

Purification
Euglena gracilis [4]

Crystallization
–

Cloned
–

Renaturated
–

5 STABILITY

pH
6.0–8.0 (10 min, 55°C, stable) [4]

Temperature (°C)
50 (10 min, stable [3], 10 min, pH 7.0, stable up to) [4]; 60 (10 min, pH 7.0, complete loss of activity) [4]

Oxidation

Organic solvent

General stability information

Storage
0°C, 2 weeks [3]

6 CROSSREFERENCES TO STRUCTURE DATABANKS

PIR/MIPS code

Brookhaven code

7 LITERATURE REFERENCES

[1] Watanabe, F., Nakano, Y., Tachikake, N., Tamura, Y., Yamanaka, H., Kitaoka, S.: J. Nutr. Sci. Vitaminol.,36,349–356 (1990)
[2] Watanabe, F., Nakano, Y., Maruno, S., Tachikake, N., Tamura, Y., Kitaoka, S.: Biochem. Biophys. Res. Commun.,165,675–679 (1989)
[3] Watanabe, F., Oki, Y., Nakano, Y., Kitaoka, S.: Agric. Biol. Chem.,51,273–274 (1987)
[4] Watanabe, F., Oki, Y., Nakano, Y., Kitaoka, S.: J. Biol. Chem.,262,11514–11518 (1987)

Enzyme Handbook © Springer-Verlag Berlin Heidelberg 1994
Duplication, reproduction and storage in data banks are only allowed with the prior permission of the publishers

1 NOMENCLATURE

EC number
1.6.99.12

Systematic name
NADPH:cyanocob(III)alamin oxidoreductase (cyanide-eliminating)

Recommended name
Cyanocobalamin reductase (NADPH, cyanide-eliminating)

Synonymes
Reductase, cyanocobalamin

CAS Reg. No.
131145-00-1

---

2 REACTION AND SPECIFICITY

Catalysed reaction
NADPH + cyanocob(III)alamin →
→ $NADP^+$ + cob(I)alamin + cyanide (mechanism [1])

Reaction type
Redox reaction
Decyanation [1]

Natural substrates
NADPH + cyanocob(III)alamin (enzyme in conversion of hydroxocobalamin to 5'-deoxyadenosyl cobalamin) [1]

Substrate spectrum
1 NADPH + cyanocob(III)alamin

Product spectrum
1 $NADP^+$ + cob(I)alamin + cyanide

Inhibitor(s)
SH-inhibitors [1]

Cofactor(s)/prostethic group(s)
FAD (required as cofactor) [1]; FMN (required as cofactor) [1]; NADPH [1]; More (NADH cannot replace NADPH) [1]

Metal compounds/salts

Enzyme Handbook © Springer-Verlag Berlin Heidelberg 1994
Duplication, reproduction and storage in data banks are only allowed with the prior permission of the publishers

Turnover number (min^{-1})

Specific activity (U/mg)

K_m-value (mM)
0.0071 (cyanocobalamin) [1]; 0.2 (NADPH) [1]

pH-optimum
8.0 [1]

pH-range

Temperature optimum (°C)
30 [1]

Temperature range (°C)

3 ENZYME STRUCTURE

Molecular weight

Subunits

Glycoprotein/Lipoprotein
–

4 ISOLATION/PREPARATION

Source organism
Euglena gracilis [1]

Source tissue
Cell [1]

Localisation in source
Mitochondria (outer side of inner or outer membrane) [1]

Purification

Crystallization
–

Cloned
–

Renaturated
–

5 STABILITY

pH

6.0–7.0 [1]

Temperature (°C)

30 (stable below) [1]

Oxidation

Organic solvent

General stability information

Storage

6 CROSSREFERENCES TO STRUCTURE DATABANKS

PIR/MIPS code

Brookhaven code

7 LITERATURE REFERENCES

[1] Watanabe, F., Oki, Y., Nakano, Y., Kitaoka, S.: J. Nutr. Sci. Vitaminol.,34,1–10 (1988)

Enzyme Handbook © Springer-Verlag Berlin Heidelberg 1994
Duplication, reproduction and storage in data banks are only allowed with the prior permission of the publishers

1 NOMENCLATURE

EC number
1.6.99.13

Systematic name
NADH:Fe^{3+} oxidoreductase

Recommended name
Ferric-chelate reductase

Synonymes
Reductase, iron chelate
Ferric chelate reductase
Iron chelate reductase

CAS Reg. No.
122097-10-3

2 REACTION AND SPECIFICITY

Catalysed reaction
NADH + 2 Fe^{3+} →
→ NAD^+ + 2 Fe^{2+}

Reaction type
Redox reaction

Natural substrates
NADH + Fe^{3+} (reaction as prerequisite for iron-uptake by plants [4], ferricyanide and Fe^{3+}-chelate reductase are possibly one enzyme [2], ferricyanide and Fe-chelate reductase are different enzymes [4], cytoplasmic reductant e.g. NADH [4]) [2, 4]

Substrate spectrum
1 NADH + Fe^{3+} (reduction of chelated Fe^{3+}: Fe^{3+}-EDTA-chelate [1, 4], Fe^{3+}-citrate chelate (best substrate from various Fe-citrate species: $Fe^{3+}(cit^{3-})_2$ [4]) [2–4], ferricyanide [2, 4, 5], Fe^{3+}-N-hydroxymethylethylene-diaminetriacetate [2, 4], Fe^{3+}-oxalate (best substrate, highest activity) [4], ethylene-diamine-di(o-hydroxyphenylacetate) [4], electron donor: not NADPH [1], NADPH-dependent Fe^{3+}-citrate reduction occurs at lower rate (3% of NADH) and is not influenced by Fe in growth medium [2]) [1–5]

Enzyme Handbook © Springer-Verlag Berlin Heidelberg 1994
Duplication, reproduction and storage in data banks are only allowed with the prior permission of the publishers

Product spectrum
1 NAD^{+} + Fe^{2+} [1–5]

Inhibitor(s)
Superoxide dismutase [1]; Oxygen [1]; Triazine dyes (e.g. Cibacron blue (competitive to NADH), Green HE-4BDA (strong), Red H-8B, Blue MX-2G, Red H-3 BN, Blue MX-R, Orange MX-2R (weak)) [4]

Cofactor(s)/prostethic group(s)
Citrate (activation, chelator) [2–4]; Oxalate (activation, best chelator) [4]; Ferricyanide (activation, chelator) [2, 4, 5]; EDTA (activation, synthetic chelator) [1, 4]; N-Hydroxymethyl-ethylenediaminetriacetate (activation) [2, 4]; Ethylenediamine-di(o-hydroxyphenylacetate) (activation, synthetic chelator) [4]; Detergents (activation [1], slight activation, e.g. lysophosphatidylcholine, Brij 58, Triton X-100, Nonidet P-40 [4]) [1, 4]; More (less activity with synthetic Fe-chelates) [4]

Metal compounds/salts
Mg^{2+} (activation at low NADH- and ferricyanide-concentration) [5]

Turnover number (min^{-1})

Specific activity (U/mg)
More [2]; 0.0042 (NADPH/Fe^{3+}-citrate, iron deficient grown tomato) [2]; 0.12 (NADH/Fe^{3+}-citrate, iron sufficient grown tomato) [2]; 0.25 (iron-rich grown tomato) [1]; 0.41 (iron deficient grown tomato) [1]; 0.7 (NADH/ferricyanide, iron sufficient grown tomato) [2]; 1.47 (NADH/ferricyanide, iron deficient grown tomato) [2]

K_m-value (mM)
0.008 (ferricyanide) [5]; 0.018 (NADH, iron sufficient grown tomato) [4]; 0.023 (NADH, iron sufficient grown tomato) [4]; 0.03 (NADH) [5]; 0.031 (Fe^{3+}, iron sufficient grown tomato) [4]; 0.035 (Fe^{3+}, iron deficient grown tomato) [4]; 0.20 (Fe^{3+}-EDTA) [1]; 0.23 (NADH) [1]

pH-optimum
More (3 isoforms with pI: 5.5, 5.8, 6.2) [4]; 6.5 [4]; 6.8 [1]; 7.3 (assay at) [5]

pH-range
5.5–8.0 (about half-maximal activity at pH 5.5 and 8.0) [4]

Temperature optimum (°C)
25 (assay at) [5]

Temperature range (°C)

3 ENZYME STRUCTURE

Molecular weight

Subunits

Glycoprotein/Lipoprotein

–

4 ISOLATION/PREPARATION

Source organism

Lycopersicum esculentum (tomato, cv Abunda [1], Mill. cv Rutgers [2, 4]) [1–4]; Beta vulgaris [5]

Source tissue

Root (primary and lateral root tips and hairs [2]) [1–4]

Localisation in source

Plasma membrane (donor and acceptor site on cytoplasmic surface [5]) [1–5]

Purification

Lycopersicum esculentum (partial, 3 isoforms by electrophoresis [3], 3 isoforms by isoelectric focusing [4]) [3, 4]

Crystallization

–

Cloned

–

Renaturated

–

5 STABILITY

pH

Temperature (°C)

Oxidation

Organic solvent

Enzyme Handbook © Springer-Verlag Berlin Heidelberg 1994
Duplication, reproduction and storage in data banks are only allowed with the prior permission of the publishers

General stability information

Triton X-100 stabilizes during solubilization [1]; Triton X-100 is required for optimal exposition of catalytic sites of plasma-membrane vesicles which are not exposed to external solution [2]; n-Octyl-beta-D-glucopyranoside solubilizes at a 22:1 w/w detergent/protein ratio [4]

Storage

–80°C, membranes stable in buffer containing proteinase inhibitors [4]

6 CROSSREFERENCES TO STRUCTURE DATABANKS

PIR/MIPS code

Brookhaven code

7 LITERATURE REFERENCES

[1] Brüggemann, W., Moog, P., Nakagawa, H., Janiesch, P., Kuiper, P.J.C.: Physiol. Plant.,79,339–346 (1990)
[2] Buckhout, T.J., Bell, P.F., Luster, D.G., Chaney, R.L.: Plant Physiol.,90,151–156 (1989)
[3] Holden, M.J., Luster, D.G., Chaney, R.L., Buckhout, T.J.: J. Plant Nutr.,15,1667–1678 (1992)
[4] Holden, M.J., Luster, D.G., Chaney, R.L., Buckhout, T.J, Robinson, C.: Plant Physiol.,97,537–544 (1991)
[5] Askerlund, P., Larsson, C., Widell, S.: FEBS Lett.,239,23–28 (1988)

1 NOMENCLATURE

EC number
1.7.2.1

Systematic name
Nitric-oxide:ferricytochrome-c oxidoreductase

Recommended name
Nitrite reductase (cytochrome)

Synonymes
Reductase, nitrite (cytochrome)

CAS Reg. No.
37256-41-0

2 REACTION AND SPECIFICITY

Catalysed reaction
Nitrite + 2 ferrocytochrome c →
→ nitric oxide + H_2O + 2 ferricytochrome c

Reaction type
Redox reaction

Natural substrates
Nitrite + ferrocytochrome c (role in respiration [4])

Substrate spectrum
1 Nitrite + ferrocytochrome c [1–5]
2 More (other electron donors: reduced benzyl viologen [1, 4], reduced methyl viologen [4], phenazine methosulfate [4], hydroquinone [4], highly purified enzyme has cytochrome c oxidase activity) [1, 4]

Product spectrum
1 Nitric oxide + H_2O + ferricytochrome c
2 ?

Inhibitor(s)
Urea [1]; Guanidine-HCl [1]: KCN [4]; NaN_3 [4]; NH_2OH [4]; More (CO, bathocuproine, diethyl dithiocarbamate, o-phenanthroline and alpha,alpha'-dipyridyl (at 1 mM): relatively ineffective) [4]

Enzyme Handbook © Springer-Verlag Berlin Heidelberg 1994
Duplication, reproduction and storage in data banks are only allowed with the prior permission of the publishers

Cofactor(s)/prostethic group(s)

Cytochrome (contains cytochrome, probably of the c and a_2 types [4], cytochrome c-552 or cytochrome c-553 from Pseudomonas denitrificans acts as acceptor [5]) [4, 5]

Metal compounds/salts

Copper (a copper protein [2, 4, 5], Alcaligenes sp.: contains 2 type 1 copper atoms per molecule, but no other types of copper [3], Micrococcus denitrificans: 5.5 ng/mg protein [4], Rhodopseudomonas sphaeroides: one subunit contains Cu^{2+} type 1, the second subunit Cu^{2+} type 2 [1]) [1–5]

Turnover number (min^{-1})

Specific activity (U/mg)

1.27 [4]; More [3]

K_m-value (mM)

0.0357 (nitrite) [1]; 0.046 (NO_2^-, cytochrome c) [4]; 0.027 (O_2, cytochrome c oxidase activity) [4]

pH-optimum

6.7 [4]

pH-range

Temperature optimum (°C)

30 (assay at) [4]

Temperature range (°C)

---

3 ENZYME STRUCTURE

Molecular weight

80000 (Rhodopseudomonas sphaeroides, gel filtration) [1]
130000 (Micrococcus denitrificans, gel filtration) [4]

Subunits

Dimer (2 × 39000–42000, Rhodopseudomonas sphaeroides, enzyme composed of 2 nonidentical subunits: one contains Cu^{2+} type 1, the second subunit contains Cu^{2+} type 2, SDS gel electrophoresis [1], 2 × 37000, Alcaligenes sp. [3], SDS-PAGE) [1, 3]

Glycoprotein/Lipoprotein

Contains no carbohydrate [3]

4 ISOLATION/PREPARATION

Source organism

Rhodopseudomonas sphaeroides (f. sp. denitrificans) [1]; Alcaligenes sp. (NCIB 11015) [2, 3]; Micrococcus denitrificans [4]; Pseudomonas denitrificans [5]

Source tissue

Cell [1–4]

Localisation in source

Soluble [5]

Purification

Rhodopseudomonas sphaeroides (f. sp. denitrificans) [1]; Alcaligenes sp. [3]; Micrococcus denitrificans [4]

Crystallization

–

Cloned

–

Renaturated

–

5 STABILITY

pH

Temperature (°C)

Oxidation

Organic solvent

General stability information

Storage

–15°C, 4 weeks [4]

Enzyme Handbook © Springer-Verlag Berlin Heidelberg 1994
Duplication, reproduction and storage in data banks are only allowed with the prior permission of the publishers

6 CROSSREFERENCES TO STRUCTURE DATABANKS

PIR/MIPS code

PIR2: A34255 (Pseudomonas aeruginosa (fragment)); PIR3: A39735 (Pseudomonas stutzeri (fragment))

Brookhaven code

7 LITERATURE REFERENCES

[1] Michalski, W.P., Nicholas, D.J.D.: Biochim. Biophys. Acta,828,130–137 (1985)
[2] Masuko, M., Iwasaki, H., Sakurai, T., Suzuki, S., Nakahara, A.: J. Biochem.,98,1285–1291 (1985)
[3] Masuko, M., Iwasaki, H., Sakurai, T., Suzuki, S., Nakahara, A.: J. Biochem.,96,447–454 (1984)
[4] Lam, Y., Nicholas, D.J.D.: Biochim. Biophys. Acta,180 ,459–472 (1969)
[5] Miyata, M., Mori, T.: J. Biochem.,66,463–471 (1969)

1 NOMENCLATURE

EC number
1.7.3.1

Systematic name
Nitroethane:oxygen oxidoreductase

Recommended name
Nitroethane oxidase

Synonymes
Oxidase, nitroethane

CAS Reg. No.
9029-36-1

2 REACTION AND SPECIFICITY

Catalysed reaction
Nitroethane + H_2O + O_2 →
→ acetaldehyde + nitrite + H_2O_2

Reaction type
Redox reaction

Natural substrates
Nitroethane + H_2O + O_2 (enzyme may participate in inorganic nitrogen nutrition of Neurospora) [1]

Substrate spectrum
1 Nitroethane + H_2O + O_2 [1]
2 beta-Nitropropionic acid + H_2O + O_2 [1]
3 1-Nitropropane + H_2O + O_2 [1]
4 2-Nitropropane + H_2O + O_2 [1]
5 More (slight activity (less than 5% of the activity towards nitroethane) with: nitromethane, nitroacetic acid, 1-chloro-1-nitropropane, methylene blue also can function as hydrogen acceptor (low reaction rate)) [1]

Product spectrum
1 Acetaldehyde + nitrite + H_2O_2
2 Malonic aldehyde + nitrite + H_2O_2
3 Propionaldehyde + nitrite + H_2O_2
4 Acetone + nitrite + H_2O_2
5 ?

Enzyme Handbook © Springer-Verlag Berlin Heidelberg 1994
Duplication, reproduction and storage in data banks are only allowed with the prior permission of the publishers

Inhibitor(s)
Maleic acid [1]; Tris [1]; Boric acid [1]; $CuSO_4$ [1]; KNO_3 [1]; NaCN (slight) [1]; NaCl (slight) [1]; NH_2OH (slight) [1]

Cofactor(s)/prostethic group(s)
More (probably not a hemoprotein) [1]

Metal compounds/salts

Turnover number (min^{-1})

Specific activity (U/mg)

K_m-value (mM)

pH-optimum
7.0 [1]

pH-range
6.2–7.6 (half-maximal activity at pH 6.2 and 7.6) [1]

Temperature optimum (°C)

Temperature range (°C)

---

3 ENZYME STRUCTURE

Molecular weight

Subunits

Glycoprotein/Lipoprotein
–

---

4 ISOLATION/PREPARATION

Source organism
Neurospora crassa [1]

Source tissue
Mycelium [1]

Localisation in source

Purification
Neurospora crassa (partial) [1]

Crystallization

–

Cloned

–

Renaturated

–

5 STABILITY

pH

11.0 (1 h, 0°C, full activity recovered after readjusting the solution to pH 7.0) [1]; 5.5 (1 h, 0°C, complete loss of activity) [1]

Temperature (°C)

50 (10 s, 70% loss of activity) [1]; 100 (2 min, complete loss of activity) [1]

Oxidation

Organic solvent

General stability information

Dialysis: stable to [1]

Storage

In refrigerator for several days [1]; –20°C, weeks, slow decrease of activity [1]

6 CROSSREFERENCES TO STRUCTURE DATABANKS

PIR/MIPS code

Brookhaven code

7 LITERATURE REFERENCES

[1] Little, H.N.: J. Biol. Chem.,193,347–358 (1951)

Enzyme Handbook © Springer-Verlag Berlin Heidelberg 1994
Duplication, reproduction and storage in data banks are only allowed with the prior permission of the publishers

1 NOMENCLATURE

EC number
1.7.3.2

Systematic name
N-Acetylindoxyl:oxygen oxidoreductase

Recommended name
Acetylindoxyl oxidase

Synonymes
Oxidase, acetylindoxyl

CAS Reg. No.
9029-37-2

2 REACTION AND SPECIFICITY

Catalysed reaction
N-Acetylindoxyl + O_2 →
→ N-acetylisatin + ?

Reaction type
Redox reaction

Natural substrates
N-Acetylindoxyl + O_2 [1]

Substrate spectrum
1 N-Acetylindoxyl + O_2 [1]

Product spectrum
1 N-Acetylisatin + ?

Inhibitor(s)
Cysteine [1]; Glutathione [1]; Indole (slight) [1]; Dioxindole (slight) [1]; Potassium indoxyl sulfate (slight) [1]; Indole-3-aldehyde (slight) [1]; Cyanide [1]; Thiourea [1]; Potassium ethyl xanthate [1]; Diethyl dithiocarbamate [1]; 8-Hydroxyquinoline [1]; Caffeic acid [1]; Catechol [1]; Dihydroxyphenylalanine [1]; o-Aminophenol [1]; m-Cresol [1]; m-Chlorophenol [1]; 1-Naphthol [1]; 2-Naphthol [1]

Enzyme Handbook © Springer-Verlag Berlin Heidelberg 1994
Duplication, reproduction and storage in data banks are only allowed with the prior permission of the publishers

Cofactor(s)/prostethic group(s)

Metal compounds/salts

Turnover number (min^{-1})

Specific activity (U/mg)

K_m-value (mM)

pH-optimum
9.0 (increase of activity from pH 7.0 to 9.0) [1]

pH-range

Temperature optimum (°C)

Temperature range (°C)

---

3 ENZYME STRUCTURE

Molecular weight

Subunits

Glycoprotein/Lipoprotein
–

---

4 ISOLATION/PREPARATION

Source organism
Corn [1]

Source tissue
Shoots [1]

Localisation in source

Purification

Crystallization
–

Cloned
–

Renaturated
–

5 STABILITY

pH

Temperature (°C)
100 (15 min, crude preparation, stable) [1]

Oxidation

Organic solvent

General stability information

Storage

6 CROSSREFERENCES TO STRUCTURE DATABANKS

PIR/MIPS code

Brookhaven code

7 LITERATURE REFERENCES

[1] Beevers, H., French, R.C.: Arch. Biochem. Biophys.,50 ,427–439 (1954)

Enzyme Handbook © Springer-Verlag Berlin Heidelberg 1994
Duplication, reproduction and storage in data banks are only allowed with the prior permission of the publishers

1 NOMENCLATURE

EC number
1.7.3.3

Systematic name
Urate:oxygen oxidoreductase

Recommended name
Urate oxidase

Synonymes
Oxidase, urate
Uric acid oxidase
Uricase
Uricase II [6]

CAS Reg. No.
9002-12-4

2 REACTION AND SPECIFICITY

Catalysed reaction
Urate + O_2 →
→ unidentified products (mechanism [20])

Reaction type
Redox reaction

Natural substrates
Urate + O_2
More (important general role in nitrogen metabolism [8], re-initiation of uricase activity in the mechanism by which Bacillus fastidiosus spores are triggered to emerge from the dormant state [11], purine catabolic enzyme [19], discussion of involvement in nitrogen assimilation processes in ureide plants [48]) [8, 11, 19, 48]

Enzyme Handbook © Springer-Verlag Berlin Heidelberg 1994
Duplication, reproduction and storage in data banks are only allowed with the prior permission of the publishers

Substrate spectrum

1 Urate + O_2 (catalase activity: Glycine max [12], cow pea [12], high specificity for uric acid [17, 21], 2,6-dichloroindophenol cannot substitute for O_2 as electron acceptor [41], can use only O_2 as electron acceptor [43]) [3, 11–17, 19–21, 27, 29, 35, 41–43, 45, 46, 49]

Product spectrum

1 Unidentified products (allantoin + H_2O_2 [3], urea and racemic allantoin formed in phosphate buffer, urea and other ninhydrin-positive materials formed in borate buffer [15], RS-(+)-allantoin formed by chemical decarboxylation and S-(+)-allantoin by enzymatic decarboxylation [20], forms H_2O_2 [43], the initial products decompose to form allantoin) [3, 15, 20, 43]

Inhibitor(s)

CN^- (mechanism of inhibition [18]) [11, 15–19, 21, 40, 42–44]; Diethyldithiocarbamic acid [11]; NaN_3 [11]; NH_2OH [11]; $Na_2S_2O_4$ [11]; Cu^{2+} [11, 13, 16, 17, 21, 32, 40, 42, 44, 45]; 5-Azaorotate [11, 19, 21, 43]; Glutamine (weak [12]) [12, 16]; NH_4^+ (weak [12]) [12, 16]; Adenine (weak [12], not [41]) [12, 16]; Allantoin (weak [12]) [12, 16]; Hg^{2+} (1 mM: not [21]) [13, 40, 49]; p-Chloromercuribenzoate (slight effect [15]) [13, 15, 17, 40]; 5,5'-Dithiobis(2-nitrobenzoic acid) [13]; o-Iodosobenzoate [13]; Xanthine [15, 16, 21, 41]; 2-Hydroxypurine [15]; Urate (substrate inhibition: above 0.120 mM [41], above 0.125 mM [43], not [15]) [41, 43]; Iodoacetate (low effect) [15]; Aspartic acid (slight) [10]; Amelide [32]; Cyanurate [32]; Phosphate (no inactivation by phosphate, in presence of borate or dithiothreitol) [34]; Mn^{2+} [16, 21]; Co^{2+} [16, 17, 21]; Mg^{2+} (slight [16], 1 mM: not [21]) [16]; 2,2'-Dipyridyl (weak [45]) [16, 45]; o-Phenanthroline (not [40], weak [45]) [16, 19, 45]; EDTA [16, 19]; Salicylhydroxamic acid [16]; Inosine 5'-monophosphate [16]; Guanine [16]; Hypoxanthine (weak [16, 41]) [16, 21, 41]; Urea (weak [16]) [16, 40]; Allantoic acid (weak) [16]; Glyoxylic acid (weak) [16]; Ag^+ [17]; Periodate [18]; Hydroxylamine [18]; Thiourea (slight) [19]; 9-Methyluric acid [21]; 8-Azaxanthine [21]; 3-Methyluric acid [21]; 7-Mcthyluric acid [21]; 3,7-Dimethylxanthine (i.e. theobromine, slight) [21]; 1,3,7-Trimethylxanthine (i.e. caffeine, slight) [21]; Zn^{2+} [21]; Ni^{2+} [21]; Cd^{2+} [21]; Cr^{3+} [21]; Pb^{2+} [21]; Fe^{2+} [21]; Guanidinium salts (inactivation is pH-dependent: slightly inhibitory below pH 10, rapid inactivation at high pH) [36]; Biguanidine salts (inactivation is pH-dependent: slightly inhibitory below pH 10, rapid inactivation at high pH) [36]; Dicyandiamide (inactivation is pH-dependent: small below pH 10, rapid increase at high pH) [36]; Arginine (weak) [36]; N-Ethylmaleimide [17, 36]; Sucrose [42]; Glycerol [42]; Oxopurines [43]; Trichloropurine [43, 46]; Hydroxypurines [43];

2,9-Dimethyl-1,10-phenanthroline (neo-cuproin) [45]; More (enzymes from bovine, kidney, Candida utilis and Streptomyces cyanogenus are sulfhydryl enzymes, enzymes from Bacillus fastidiosus and from Neurospora sp. are not [15], partially inhibited by various cations, inhibiting effect decreases in the order: Zn^{2+}, Ni^{2+}, Co^{2+}, Cu^{2+}, Cd^{2+}, Cr^{3+}, Mn^{2+}, Pb^{2+}, Fe^{2+} [21], overview: inhibition by substituted pyrimidines [28], bovine kidney enzyme: unaffected by a number of tertiary and quarternary ammonium bases [32], overview: effect of various metal ions on uricase activity in various buffers [49]) [15, 21, 28, 32, 49]

Cofactor(s)/prostethic group(s)

No cofactors [21]

Metal compounds/salts

Copper (neither copper nor iron detected: Candida utilis [13], enzyme contains 0.15 mol of copper (no function as cofactor) [15], copper: uricase ratio is 1:7 (no function as cofactor) [21], copper: uricase ratio is 1:1 [30], negligible amount of copper [40], contains 0.2 mol copper per mol enzyme protein [44]) [13, 15, 21, 30, 40, 44]; Iron (neither copper nor iron detected: Candida utilis [13], enzyme contains less than 0.05 mol of iron (no function as cofactor) [15], iron: uricase ratio is 1:50 (no function as cofactor) [21], contains nearly one mol per mol of enzyme [40], contains 0.1 mol iron per mol enzyme [44]) [15, 21, 40, 44]; Fe^{3+} (stimulation [16, 44], slight stimulation [45]) [14, 44, 45]; More (absence of Mg, Ni, Zn, Co, Cd and Mn) [13]

Turnover number (min^{-1})

Specific activity (U/mg)

6.85 [12]; 5.6 [16]; 12.3 [15]; 2.55 [25]; 6.6 [17]; 9.35 (male rats) [42]; 10.48 [49]; 15.3 (female rats) [42]; More [3, 13, 19, 21, 23, 45]

K_m-value (mM)

0.010 (uric acid) [12, 13]; 0.018 (uric acid) [16]; 0.031 (O_2) [12]; 0.029 (O_2) [16]; 0.099 (uric acid) [15]; 1.5 (uric acid) [17]; 0.042 (uric acid) [19]; 0.00588 (uric acid) [49]; More (K_m of poly(ethylene glycol):urate oxidase conjugates [14], of free and gel-entrapped enzyme [27], of immobilized enzyme [29], different K_m values of female and male rat liver enzyme [42]) [11, 14, 21, 27, 29, 35, 41–43, 45, 46]

pH-optimum

8–8.5 (borate and phosphate buffer) [15]; 8.5 [3, 46, 49]; 8.5–9.3 [42]; 8.6 (free enzyme) [27]; 8.9 [41]; 9–10 [16]; 9.2 [45]; 9.4 [43]; 9.5 (immobilized enzyme [27], glycine buffer [15]) [12, 15, 17, 21, 27]; More (increased optimal pH for poly(ethylene glycol):urate oxidase conjugate) [14]

Enzyme Handbook © Springer-Verlag Berlin Heidelberg 1994
Duplication, reproduction and storage in data banks are only allowed with the prior permission of the publishers

pH-range

7.4–9.6 (50% of activity maximum at pH 7.4 and 9.6, free enzyme) [27]; 7–11 (7: about 50% of activity maximum, 11: about 40% of activity maximum) [17]; 8–11 (50% of activity maximum at pH 8 and 11, immobilized enzyme) [27]; 8.0–10.2 (8.0: about 35% of activity maximum, 10.2: about 55% of activity maximum, Glycine max) [12]; 8.5–10.5 (50% of activity maximum at pH 8.5 and 10.5) [21]

Temperature optimum (°C)

30 [3]; 30–35 [21]; 35 (5 min incubation test [49]) [15, 49]; 40 [17]; More (decreased optimal temperature of poly(ethylene glycol):urate oxidase conjugates) [14]

Temperature range (°C)

20–50 (20°C: about 60% of activity maximum [17], about 70% of activity maximum [49], 50°C: about 50% of activity maximum [17], about 60% of activity maximum [49]) [17, 49]

3 ENZYME STRUCTURE

Molecular weight

32000–35000 (Glycine max, SDS-PAGE, gel filtration, native enzyme at pH 7.5 is a monomer with MW 32000–35000, at pH 8.8 the enzyme associates to form a dimer of about 68000) [12]

33274 (rat, determination of nucleotide sequence of cDNA and calculation of corresponding amino acid sequence) [4]

34000 (Phaseolus vulgaris, SDS-PAGE) [6]

50000 (Vigna unguiculata, gel filtration native enzyme at pH 7.5 is a monomer with MW 50000, at pH 8.8 the enzyme associates to form a dimer of about 100000) [16]

68000 (Glycine max, SDS-PAGE, gel filtration, native enzyme at pH 7.5 is a monomer with MW 32000–35000, at pH 8.8 the enzyme associates to form a dimer of about 68000) [12]

100000 (bovine, gel filtration [35], Vigna unguiculata, gel filtration native enzyme at pH 7.5 is a monomer with MW 50000, at pH 8.8 the enzyme associates to form a dimer of about 100000) [16]

102000 (Candida utilis, polyacrylamide disc electrophoresis) [3]

105000 (Enterobacter cloacae, gel filtration) [17]

109000 (Streptomyces cyanogenus, gel filtration) [15]

115000–123000 (Neurospora crassa, gel electrophoresis, gel filtration) [19]

120000 (Candida utilis, gel filtration, equilibrium sedimentation [40], Candida utilis, gel filtration [13]) [13, 40]

125000 (pig, short-column meniscus depletion sedimentation equilibrium) [30]
145000–150000 (Bacillus fastidiosus, gel filtration) [21]
230000 (rat, gel filtration) [25]

Subunits

? (x × 34000, Phaseolus vulgaris, SDS-PAGE [6], x × 33000, rat, SDS-PAGE [23]) [6, 23]
Monomer (Glycine max, (Vigna unguiculata), native enzyme at pH 7.5 is a monomer with MW 32000–35000 (50000), at pH 8.8 the enzyme associates to form a dimer of about 68000 (100000)) [12]
Dimer (Glycine max, (Vigna uniguiculata), native enzyme at pH 7.5 is a monomer with MW 32000–35000 (50000), at pH 8.8 the enzyme associates to form a dimer of about 68000 (10000)) [12]
Trimer (3 × 32000, Streptomyces cyanogenus, SDS-PAGE) [15]
Tetramer (4 × 30000, Candida utilis, SDS-PAGE [13], 4 × 33000, Neurospora crassa, SDS-PAGE [19], 2 × 36000 + 2 × 39000, Bacillus fastidiosus, SDS-PAGE [21], 4 × 32000, pig, SDS-PAGE [30]) [13, 19, 21, 30]
Hexamer (6 × 37000, rat, SDS-PAGE) [25]

Glycoprotein/Lipoprotein

–

4 ISOLATION/PREPARATION

Source organism

Human [1]; Rabbit [2, 38]; Candida utilis [3, 10, 13, 14, 40, 47, 49]; Rat [4, 23, 25, 27, 42]; Pig [5, 14, 27–30]; Phaseolus vulgaris [6]; Drosophila melanogaster [7]; Streptomyces aureofaciens [8]; Glycine max (inoculated with Bradyrhizobium japonicum [9]) [9, 12]; Bacillus fastidiosus [11, 21, 47]; Vigna unguiculata (cow pea) [16]; Streptomyces cyanogenus [15]; Enterobacter cloacae [17]; Aspergillus flavus [18, 47]; Neurospora crassa [19]; Candida tropicalis [26]; Bovine [31, 32, 34–37]; Tupaia glis (prosimian) [38]; Monkey (no activity in liver of New World monkeys, low activity in liver of Old World monkeys) [38]; Maize [39, 41]; Castor bean [41]; Peanut [41]; Joshua tree [41]; Sunflower [41]; Safflower (weak activity) [41]; One-leaved pinyon [41]; Acanthamoeba sp. [43]; Arthrobacter pascens [44, 45]; Dictyostelium discoideum [46]; Phaseolus coccineus [48]

Source tissue

Liver (no activity in liver of New World monkeys, low activity in liver of Old World monkeys [38]) [2, 4, 14, 23, 27–30, 38, 42]; Cell [3, 11, 15, 17, 19, 21]; Larval Malphigian tubules [7]; Spores [11]; Seeds [12]; Kidney [31, 32, 34, 35, 37]; Root tips [39]; Seedlings (fat-degrading tissues of) [41]; Endosperm [41]; Cotyledons [41]; Perisperm [41]; Scutella [41]; Megagametophyte [41]; Leaf [48]

Enzyme Handbook © Springer-Verlag Berlin Heidelberg 1994
Duplication, reproduction and storage in data banks are only allowed with the prior permission of the publishers

Localisation in source

Soluble [6]; Peroxisomes (limited to large peroxisomes in uninfected cells of root nodules of Glycine max inoculated with Bradyrhizobium japonicum [9]) [9, 25, 39, 43, 46, 48]; Intracellular [21]; Extracellular (soluble) [22]; Microbodies [26]; Mitochondria [35]; Glyoxysomes [41]

Purification

Candida utilis (overview: extraction methods [10]) [3, 10, 13, 49]; Glycine max [12]; Streptomyces cyanogenus [15]; Vigna unguiculata [16]; Enterobacter cloacae [17]; Neurospora crassa [19]; Bacillus fastidiosus [21]; Rat [23, 42]; Bovine (mitochondrial fraction of bovine kidney cortex contains 2 forms of urate oxidase) [35]; Arthrobacter pascens [45]; More (overview: effect of detergents on extraction [31], extraction methods [10], effect of treatment with dithiothreitol on extraction and purification [35]) [10, 31, 35]

Crystallization

[21]

Cloned

(Drosophila melanogaster [7], human [1], rat [4], rabbit [2], pig [5], Phaseolus vulgaris [6]) [1, 2, 4–7]

Renaturated

–

5 STABILITY

pH

6 (45°C, 30 min, 40% loss of activity) [17]; 6–11 (35°C, 1 h, stable) [15]; 6.5–10.5 (22°C, 30 min, immobilized enzyme, stable) [27]; 7.0–11.0 (10 min, stable) [49]; 8–9 (45°C, 30 min, stable) [17]; 8–11 (4°C, 60 days, 35% loss of activity) [3]; 10 (45°C, 30 min, 30% loss of activity) [17]; 12.0 (22°C, 30 min, 55% loss of activity) [27]

Temperature (°C)

18–20 (2 days, 70% loss of activity) [16]; 40 (30 min, stable below [17], below pH 9.5, 2 h, very slow loss of activity in absence or presence of Cu^{2+} [37], 10 min, stable below [49]) [17, 37, 49]; 50 (pH 7.8, 10 min, stable below [15], 10 min, stable [21]) [15, 21]; 60 (pH 7.8, 10 min, 50% loss of activity [15], 30 min, about 40% loss of activity [17], 10 min, 55% loss of activity [21]) [15, 17, 21]; More (thermal inactivation rises steeply as $CuSO_4$ concentration rises from 0.025 to 0.175 mM and as the pH of the medium exceedes 9.5) [37]

Oxidation

Sensitive to O_2 [41]

Organic solvent

General stability information

Dithiothreitol (prevents polymerization and stabilizes throughout purification [13], stabilization [33], overview: effect of treatment with dithiothreitol on extraction and purification [35]) [13, 33, 35]; Repeated freezing and thawing has no effect [42]; Frozen enzyme retains complete activity [16]; Unusually resistant to SDS [30]; Unusually resistant to guanidinium chloride [30]; Urea, 4 M, several h without loss of activity [42]; Stability of immobilized enzyme depends on the time of stirring during immobilization and on the quantity of enzyme used [24]; Borate stabilizes [33]; Cu^{2+} inactivates at low temperature, uric acid prevents inactivation [44]; EDTA stabilizes [33]; Proteolytic digestion by endopeptidases cause rapid loss of activity, exopeptidases have slight effect [47]; Urate oxidase from female rat livers is more stable than enzyme from male rat livers [42]; Fe^{3+} partially protects against inactivation at low pH or at low ionic strength, stimulates reactivation [44]; Ammonium sulfate protects against inactivation at low pH [44]; Lower stability in solutions of phosphate buffer than in borate buffer [49]; Little loss of activity by freeze-drying [49]

Storage

4°C, 20 mM phosphate buffer (pH 7.8), for at least 1 month [15]; 4°C, 7 days, 40% loss of activity [16]; 0–4°C, crystals in $(NH_4)_2SO_4$ solution, 3 months, stable [21]; 3°C, 0.15% sodium carbonate, several weeks, stable [42]; –15°C, 0.1 M borate buffer, pH 8.0, containing 0.1 M or more ammonium sulfate [45]

6 CROSSREFERENCES TO STRUCTURE DATABANKS

PIR/MIPS code

PIR3: S22867 (Aspergillus flavus); PIR3: A36227 (Baboon); PIR2: S08676 (Fruit fly (Drosophila melanogaster)); PIR3: A35892 (Human (fragment)); PIR3: B36227 (Mouse); PIR3: C36227 (Pig); PIR2: S04332 (Rabbit); PIR1: RDRTU (Rat); PIR3: A32267 (Rat); PIR2: A31261 (Rat (fragment))

Brookhaven code

Enzyme Handbook © Springer-Verlag Berlin Heidelberg 1994
Duplication, reproduction and storage in data banks are only allowed with the prior permission of the publishers

7 LITERATURE REFERENCES

[1] Yeldandi, A.V., Wang, X., Alvares, K., Kumar, S., Sambasiva Rao, M., Reddy, J.K.: Biochem. Biophys. Res. Commun.,171,641–646 (1990)
[2] Motojima, K., Goto, S.: Biochim. Biophys. Acta,1008,116–118 (1989)
[3] Adamek, V., Kralova, B., Suchova, M., Valentova, O., Demnerova, K.: J. Chromatogr.,497,268–275 (1989)
[4] Motojima, K., Kanaya, S., Goto, S.: J. Biol. Chem.,263,16677–16681 (1988)
[5] Lee, C.C., Wu, X., Gibbs, R.A., Cook, R.G., Muzny, D. M., Caskey, C.T.: Science,239,1288–1291 (1988)
[6] Sanchez, F., Campos, F., Padilla, J., Bonneville, J.-M., Enriquez, C., Caput, D.: Plant Physiol.,84,1143–1147 (1987)
[7] Kral, L.G., Johnson, D.H., Burnett, J.B., Friedman, T. B.: Gene,45,131–137 (1986)
[8] Demnerova, K., Kralova, B., Lehejckova, R., Adamek, V. : Biotechnol. Lett.,8,577–578 (1986)
[9] Van den Bosch, K.A., Newcomb, E.H.: Planta,167,425–436 (1986)
[10] Kralova, B., Lehejckova, R., Demnerova, K., Dobransky, T.: Biotechnol. Lett.,8,99–102 (1986)
[11] Salas, J.A., Johnstone, K., Ellar, D.J.: Biochem. J.,229,241–249 (1985)
[12] Lucas, K., Boland, M.J., Schubert, K.R.: Arch. Biochem. Biophys.,226,190–197 (1983)
[13] Nishimura, H., Yoshida, K., Yokota, Y., Matsushima, A., Inada, Y.: J. Biochem.,91,41–48 (1982)
[14] Chen, R.H.-L., Abuchowski, A., Van Es, T., Palczuk, N.C., Davis, F.F.: Biochim. Biophys. Acta,660,293–298 (1981)
[15] Ohe, T., Watanabe, Y.: J. Biochem.,89,1769–1776 (1981)
[16] Rainbird, R.M., Atkins, C.A.: Biochim. Biophys. Acta,659,132–140 (1981)
[17] Machida,, Y., Nakanishi, T.: Agric. Biol. Chem.,44,2811–2815 (1980)
[18] Conley, T.G., Priest, D.G.: Biochem. J.,187,733–738 (1980)
[19] Wang, L.-W. C., Marzluf, G.A.: Arch. Biochem. Biophys.,201,185–193 (1980)
[20] Bongaerts, G.P.A., Vogels, G.D.: Biochim. Biophys. Acta,567,295–308 (1979)
[21] Bongaerts, G.P.A., Uitzetter, J., Brouns, R., Vogels, G.D.: Biochim. Biophys. Acta,527,348–358 (1978)
[22] Mahler, J.L.: Anal. Biochem.,38,65–84 (1970)
[23] Watanabe, T., Suga, T.: Anal. Biochem.,89,343–347 (1978)
[24] Johnson, D.B., Coughlan, M.P.: Biotechnol. Bioeng.,20,1085–1095 (1978)
[25] Antonenkov, V.D., Pachenko, L.F.: FEBS Lett.,88,151–154 (1978)
[26] Tanaka, A., Yamamura, M., Kawamoto, S., Fukui, S.: Appl. Environ. Microbiol.,34,342–346 (1977)
[27] Hinberg, I., O'Driscoll, K.F.: Biotechnol. Bioeng.,17,1435–1441 (1975)
[28] Sedor, F.A., Sander, E.G.: Biochem. Biophys. Res. Commun.,75,406–413 (1977)
[29] Johnson, D.B., Coughlan, M.P.: Biochem. Soc. Trans.,2,1362–1363 (1974)
[30] Pitts, O.M., Priest, D.G., Fish, W.W.: Biochemistry,13,888–892 (1974)
[31] Truscoe, R.: Enzymologia,33,19–32 (1967)
[32] Truscoe, R.: Enzymologia,34,337–343 (1968)
[33] Truscoe, R.: Enzymologia,34,325–336 (1968)
[34] Truscoe, R.: Enzymologia,35,19–30 (1968)
[35] James, K.A.C., Tate, W.P., Truscoe, R.: Enzymologia,37,131–152 (1969)

[36] Bentley, K.W., Truscoe, R.: Enzymologia,37,285–313 (1969)
[37] Turner, J.C., Truscoe, R.: Enzymologia,43,57–70 (1972)
[38] Christen, P., Peacock, W.C., Christen, A.E., Wacker, W.E.C.: Eur. J. Biochem.,12,3–5 (1970) (Review)
[39] Parish, R.W.: Planta,104,247–251 (1972)
[40] Itaya, K., Fukumoto, J., Yamamoto, T.: Agric. Biol. Chem.,35,813–821 (1971)
[41] Theimer, R.R., Beevers, H.: Plant Physiol.,47,246–251 (1971)
[42] Townsend, D., Lata, G.F.: Arch. Biochem. Biophys.,135,166–172 (1969)
[43] Müller, M., Moller, K.M.: Eur. J. Biochem.,9,424–430 (1969)
[44] Nose, K., Arima, K.: Biochim. Biophys. Acta,151,63–69 (1968)
[45] Arima, K., Nose, K.: Biochim. Biophys. Acta,151,54–62 (1968)
[46] Parish, R.W.: Eur. J. Biochem.,58,523–531 (1975)
[47] Fitzpatrick, D.A., McGeeney, K.F.: Biochem. Soc. Trans.,3,1253–1255 (1975)
[48] Theimer, R.R., Heidinger, P.: Z. Pflanzenphysiol.,73,360–370 (1974)
[49] Itaya, K., Yamamoto, T., Fukumoto, J.: Agric. Biol. Chem.,31,1256–1264 (1967)

Enzyme Handbook © Springer-Verlag Berlin Heidelberg 1994
Duplication, reproduction and storage in data banks are only allowed with the prior permission of the publishers

1 NOMENCLATURE

EC number
1.7.3.4

Systematic name
Hydroxylamine:oxygen oxidoreductase

Recommended name
Hydroxylamine oxidase

Synonymes
Oxidase, hydroxylamine
HAO [2, 3]
Hydroxylamine oxidoreductase

CAS Reg. No.
9075-43-8

2 REACTION AND SPECIFICITY

Catalysed reaction
Hydroxylamine + O_2 →
→ nitrite + H_2O

Reaction type
Redox reaction

Natural substrates
Hydroxylamine + O_2 (energy production) [1]

Substrate spectrum
1 Hydroxylamine + O_2 (in presence of diethyldithiocarbamate, hydroxylamine is oxidized to nitrite, for each mol of nitrite produced approximately 1 mol of diethyldithiocarbamate is oxidized to bis(diethyldithiocarbamoyldisulfide) [12])
2 Pyrogallol + O_2 [11]

Product spectrum
1 Nitrite + H_2O
2 ?

Enzyme Handbook © Springer-Verlag Berlin Heidelberg 1994
Duplication, reproduction and storage in data banks are only allowed with the prior permission of the publishers

Inhibitor(s)

CO [5]; Mn^{2+} (inhibits one step of oxidation of hydroxylamine ([HNO]--> NO), 0.001 mM, inhibition of HNO_2 synthesis is the same at pH 6,7,8 and 9) [9]; H_2O_2 (inhibits one step of oxidation of hydroxylamine (NH_2OH--> [HNO]) [9], inactivation most rapid at pH 9–10, overview: substrates which protect against inactivation [14]) [9, 14]; $Ce(NH_4)_2(NO_3)_6$ (0.1 mM, stimulates the rate of oxidation of NH_2OH 2.6-fold and inhibits production of HNO_2 by 40%) [9]; $Co(NO_3)_2$ (0.1 mM, 70% inhibition of the rate of nitrite synthesis) [9]; KCN [10]

Cofactor(s)/prostethic group(s)

Heme (contains at least one b- and one c-type cytochrome [15], heme-like chromophore P-460 is part of substrate-binding site [3], resolution of the hemes by redox potentiometry and electron spin resonance spectroscopy [4], multi-heme enzyme containing at least 5 thermodynamically distinct c-type hemes and the heme-like moiety P-460 [4], contains hemes c-553, c-559 and P-460 in the ratio 5:2:1, P-460 is the site of electron entry [5], contains c hemes and the CO-binding heme P-460 in a 7:1 ratio [6], resolution of multiple heme centers by electron paramagnetic resonance spectroscopy [7], enzyme posseses heme c molecules in different states [10], each alpha,beta subunit of the $(alpha,beta)_3$ subunit structure contains 7–8 c-type hemes and one prosthetic group (P-460), P-460 is a MW 17000 protein fragment [13]) [3–7, 10, 13, 15]

Metal compounds/salts

Mn^{2+} (0.001 mM, stimulation of NH_2OH utilization, stimulation at pH 6 and 9 is approximately 40% greater compared with stimulation at pH 7 and 8) [9]; $Ce(NH_4)_2(NO_3)_6$ (0.1 mM, stimulates the rate of oxidation of NH_2OH 2.6-fold and inhibits production of HNO_2 by 40%) [9]; Iron (enzyme contains 20 mol of iron but no other metals) [11]

Turnover number (min^{-1})

23000 (hydroxylamine oxidized) [11]; 9500 (nitrite produced) [11]; More (overview: effect of pH on rate of reduction of heme c-553 of the enzyme by NH_2OH, rate constant for reduction of hemes of the enzyme by dithionite at 2°C and 19°C and by NH_2OH or NH_2NH_2 at 2°C) [5]

Specific activity (U/mg)

28 [11]; More [1]

K_m-value (mM)

0.040 (Chlorobium limicola forma sp. thiosulfatophilum, cytochrome c) [10]; 0.025 (Rhodospirillum rubrum, Saccharomyces oviformis and cow cytochrome c) [10]; 0.200 (Thiobacillus novellus cytochrome c) [10]; 0.017 (tuna cytochrome c) [10]; 0.036 (horse cytochrome c) [10]

pH-optimum
9.5 [10]

pH-range

Temperature optimum (°C)
25 (assay at) [11]

Temperature range (°C)
More (changes in enzyme conformation imposed by changes in solvent, temperature or pressure affect the rates of intramolecular electron transfer from the substrate site to c hemes) [3]

3 ENZYME STRUCTURE

Molecular weight
175000–180000 (Nitrosomonas europaea, gel filtration) [10]
180000–200000 (Nitrosomonas europaea, PAGE) [8]
200000 (Nitrosomonas europaea, sedimentation velocity experiments) [15]

Subunits
Hexamer (3 × 63000 + 3 × 11000, $(alpha,beta)_3$ subunit structure, Nitrosomonas europaea, SDS-PAGE, 200000 MW enzyme forms a 63000 MW monomer after heme removal, hydroxylamine oxidoreductase probably consists of 3 molecules of monoheme c-type cytochrome (MW 11000) and 3 tightly complexed molecules of a catalytically active MW 63000 protein containing 6 c-type hemes and one P-460 heme) [8]

Glycoprotein/Lipoprotein
–

4 ISOLATION/PREPARATION

Source organism
Nitrosomonas europaea (gene cloned and expressed in Pseudomonas putida [2]) [1–15]

Source tissue
Cell [1, 11]

Localisation in source
Soluble [11]

Purification
Nitrosomonas europaea [1]

Enzyme Handbook © Springer-Verlag Berlin Heidelberg 1994
Duplication, reproduction and storage in data banks are only allowed with the prior permission of the publishers

Crystallization

–

Cloned

(Nitrosomonas europaea gene cloned and expressed in Pseudomonas putida) [2]

Renaturated

–

5 STABILITY

pH

Temperature (°C)

More (changes in enzyme conformation imposed by changes in solvent, temperature or pressure affect the rates of intramolecular electron transfer from the substrate site to c hemes) [3]

Oxidation

Organic solvent

More (changes in enzyme conformation imposed by changes in solvent, temperature or pressure affect the rates of intramolecular electron transfer from the substrate site to c hemes) [3]

General stability information

Storage

6 CROSSREFERENCES TO STRUCTURE DATABANKS

PIR/MIPS code

Brookhaven code

7 LITERATURE REFERENCES

[1] Rees, M.K.: Biochemistry,7,353–366 (1968)
[2] Tatsuaki, T., Yusuke, T., Reiji, T.: Hakko Kogaku Kaishi,66,103–107 (1988)
[3] Balny, C., Hooper, A.B.: Eur. J. Biochem.,176,273–279 (1988)
[4] Prince, R.C., Hooper, A.B.: Biochemistry,26,970–974 (1987)
[5] Hooper, A.B., Tran, M., Balny, C.: Eur. J. Biochem.,141,565–571 (1984)
[6] Hooper, A.B., Debey, P., Andersson, K.K., Balny, C.: Eur. J. Biochem.,134,83–87 (1983)
[7] Lipscomb, J.D., Hooper, A.B.: Biochemistry,21,3965–3972 (1982)
[8] Terry, K.R., Hooper, A.B.: Biochemistry,20,7026–7032 (1981)
[9] Hooper, A.B., Terry, K.R.: Biochim. Biophys. Acta,571 ,12–20 (1979)
[10] Yamanaka, T., Shinra, M., Takahashi, K., Shibasaka, M.: J. Biochem.,86,1101–1108 (1979)
[11] Hooper, A.B., Maxwell, P.C., Terry, K.R.: Biochemistry,17,2984–2989 (1978)
[12] Hooper, A.B., Terry, K.R., Maxwell, P.C.: Biochim. Biophys. Acta,462,141–152 (1977)
[13] Andersson, K.K., Kent, T.A., Lipscomb, J.D., Hooper, A.B., Münck, E.: J. Biol. Chem.,259,6833–6840 (1984)
[14] Hooper, A.B., Terry, K.R.: Biochemistry,16,455–459 (1977)
[15] Rees, M.K.: Biochemistry,7,366–372 (1968)

Enzyme Handbook © Springer-Verlag Berlin Heidelberg 1994
Duplication, reproduction and storage in data banks are only allowed with the prior permission of the publishers

1 NOMENCLATURE

EC number
1.7.3.5

Systematic name
3-aci-Nitropropanoate:oxygen oxidoreductase

Recommended name
3-aci-Nitropropanoate oxidase

Synonymes
Propionate-3-nitronate oxidase [1]

CAS Reg. No.

---

2 REACTION AND SPECIFICITY

Catalysed reaction
3-aci-Nitropropanoate + O_2 →
→ 3-oxopropanoate + nitrite + H_2O_2 (primary products of the enzymatic reaction probably are the nitropropanoate free radical and superoxide)

Reaction type
Redox reaction

Natural substrates

Substrate spectrum
1 3-aci-Nitropropanoate + O_2 [1]
2 Butyrate-4-nitronate + O_2 [1]
3 More (neither cytochrome c nor $K_3Fe(CN)_6$ serve as oxidants, other substrates tested are oxidized less than 2% the rate of 3-nitropropanoate) [1]

Product spectrum
1 3-Oxopropanoate + nitrite + H_2O_2
2 Succinate + nitrite + H_2O_2
3 ?

Inhibitor(s)

Cofactor(s)/prostethic group(s)
FMN (flavoprotein) [1]

Metal compounds/salts

Enzyme Handbook © Springer-Verlag Berlin Heidelberg 1994
Duplication, reproduction and storage in data banks are only allowed with the prior permission of the publishers

Turnover number (min^{-1})
21000 (3-nitropropanoate) [1]

Specific activity (U/mg)
274 [1]

K_m-value (mM)

pH-optimum
8.1 (assay at) [1]

pH-range

Temperature optimum (°C)
25 (assay at) [1]

Temperature range (°C)

---

3 ENZYME STRUCTURE

Molecular weight
73000 (Penicillium atrovenetum, gel filtration) [1]

Subunits
Dimer (2 × 38000, Penicillium atrovenetum, SDS-PAGE in presence of mercaptoethanol) [1]

Glycoprotein/Lipoprotein
–

---

4 ISOLATION/PREPARATION

Source organism
Penicillium atrovenetum [1]

Source tissue
Mycelium [1]

Localisation in source
Soluble [1]

Purification
Penicillium atrovenetum [1]

Crystallization
–

Cloned

–

Renaturated

–

5 STABILITY

pH

Temperature (°C)

Oxidation

Organic solvent

General stability information

Storage

6 CROSSREFERENCES TO STRUCTURE DATABANKS

PIR/MIPS code

Brookhaven code

7 LITERATURE REFERENCES

[1] Porter, D.J.T., Bright, H.J.: J. Biol. Chem.,262,14428–14434 (1987)

Enzyme Handbook © Springer-Verlag Berlin Heidelberg 1994
Duplication, reproduction and storage in data banks are only allowed with the prior permission of the publishers

1 NOMENCLATURE

EC number
1.7.7.1

Systematic name
Ammonia:ferredoxin oxidoreductase

Recommended name
Ferredoxin-nitrite reductase

Synonymes
Reductase, ferredoxin-nitrite

CAS Reg. No.
37256-44-3

2 REACTION AND SPECIFICITY

Catalysed reaction
Nitrite + 3 reduced ferredoxin →
→ ammonia + 3 oxidized ferredoxin (mechanism [10, 11], sequential reaction scheme [4])

Reaction type
Redox reaction

Natural substrates
Nitrite + reduced ferredoxin (second step of nitrate assimilation [4], second enzyme of photosynthetic nitrate-reducing system [10]) [2, 4, 5, 10, 11]

Substrate spectrum
1 Nitrite + reduced ferredoxin (other electron donors: flavodoxin [2, 11], methyl viologen [2, 8, 11, 28], benzyl viologen [2, 8], not: flavin [2], NADH [2], NADPH [2], enzyme (from Chlorella fusca [27], from several sources [11]) can also catalyze the reduction of hydroxylamine to ammonia, at low rate [11, 27]) [1–28]

Product spectrum
1 Ammonia + oxidized ferredoxin

Enzyme Handbook © Springer-Verlag Berlin Heidelberg 1994
Duplication, reproduction and storage in data banks are only allowed with the prior permission of the publishers

Inhibitor(s)

Cyanide [2, 11, 25–28]; p-Hydroxymercuribenzoate (enzyme from higher plants and eukaryotic algae inhibited, cyanobacterial enzyme not [11]) [2, 11]; N-Ethylmaleimide [2]; NaN_3 (slight [1], not [18]) [3]; Bathophenanthroline (slight) [2]; Sulfite (10 mM) [2]; Hydroxylamine (10 mM) [2]; CO (enzyme from higher plants and algae [11]) [8, 11, 26, 27]; Mersalyl (reversible by glutathione [18]) [11, 18]; Phenyl mercury acetate [18]; p-Chloromercuribenzoate (not [28]) [18, 26, 27]; o-Phenanthroline (slight) [27]; EDTA [27]; More (anti-glutamate synthase antibodies recognize nitrite reductase, but inhibit only ferredoxin-linked activity and not methyl viologen-linked activity) [5]

Cofactor(s)/prostethic group(s)

Siroheme (prosthetic group [2, 10, 11, 13, 25, 26], 0.65 mol of siroheme per mol of enzyme [8], 1 mol siroheme per mol of enzyme [10, 11, 26], 0.93 mol siroheme per mol of enzyme [13], 2 mol of siroheme per mol of enzyme [25]) [2, 8, 10, 11, 13, 25, 26]

Metal compounds/salts

Iron (0.65 molecules of siroheme per enzyme molecule [8], enzyme contains 1 tetranuclear center (Fe4-S4) and 1 siroheme, minimum of 5 Fe atoms per active enzyme molecule [10, 11], 4.8 Fe atoms per molecule of MW 63000, 0.93 mol siroheme per mol of enzyme [13], Cucurbita: contains 2 mol of Fe per mol [18], 7 mol of iron per 86000 MW protein [24], spinach: contains about 2 mol of siroheme per mol of enzyme, 3 mol of iron per mol of enzyme (MW 61000), one in siroheme, one (Fe2-S2) [26], 2 atoms of iron probably bound in the chromophoric group [27]) [8, 10, 11, 13, 18, 24–27]

Turnover number (min^{-1})

5340 (nitrite, Cucurbita pepo) [11]; 3240 (nitrite, Chlorella fusca) [11]; 600 (nitrite, Porphyra yezoensis) [11]; 6600 (nitrite, Spinacia oleracea) [11]

Specific activity (U/mg)

625 [2]; 181 [3]; 89 [8]; 108 [10]; 207 [13]; 194 [16]; 196.8 [19]; 140 [20]; 53.4 [23]; 107.8 [26]; 51.7 [27]

K_m-value (mM)

0.040 (nitrite) [2]; 0.022 (ferredoxin) [2]; 0.215 (methyl viologen) [2]; 0.360 (nitrite) [3]; 0.063 (methyl viologen) [3]; 0.28 (nitrite) [6]; 0.025 (spinach ferredoxin) [6]; More [8, 11, 14, 16, 24, 28]

pH-optimum

7–8 [10]; 7.1–7.8 [28]; 7.3 (phosphate buffer) [2]; 7.5 [11]; 7.6 (Tris-HCl buffer [2]) [2, 3]; 7.7 [6]; 7.8 [16]

pH-range

6.8–8.3 (6.8: about 75% of activity maximum, 8.3: about 70% of activity maximum) [28]

Temperature optimum (°C)

50 (increase of activity up to 50°C) [2]

Temperature range (°C)

3 ENZYME STRUCTURE

Molecular weight

52000 (Spirulina platensis, gel filtration) [16]
54000 (Phormidium laminosum, gel filtration) [2]
59000 (Porphyra yezoensis, SDS-PAGE) [6]
60000–63000 (Spinacia oleracea, gel filtration, sedimentation equilibrium) [11, 13, 20, 26]
61000 (Oryza sativa, molecular sieve chromatography [3]) [3, 11]
62686 (Spinacia oleracea, amino acid composition) [26]
63000 (Chlorella fusca, Porphyra yezoensis, Cucurbita pepo [11], Spinacia oleracea, sedimentation equilibrium, SDS-PAGE [13], Chlorella fusca, gel filtration, SDS-PAGE [27]) [11, 13, 27]
68000 (Phaseolus angularis, gel filtration, SDS-PAGE) [15]
69000 (Phormidium laminosum, native PAGE) [1]
86000 (Spinacia oleracea, gel filtration [24], Chlamydomonas reinhardtii, gel filtration [8]) [8, 24]
100000 (Phaseolus angularis, gel filtration) [23]
More (spinach: the native enyzme (MW 85000) can be split into a modified form (MW 61000) that retains activity with the non-physiological electron-donor, methyl viologen but loses most of the ferredoxin-linked activity and an 24000 MW fragment (coupling protein) in which the ferredoxin-binding domain is located) [22, 25]

Subunits

Monomer (1 × 54000, Phormidium laminosum, SDS-PAGE [2], 1 × 60000, Oryza sativa, SDS-PAGE [3], 1 × 68000, Phaseolus angularis, SDS-PAGE [15], Spinacia oleracea, SDS-PAGE [13], 1 × 63000, Chlorella fusca, SDS-PAGE (+ mercaptoethanol) [27], Chlorella fusca [11], Spinacia oleracea [11]) [2, 3, 11, 13, 15, 27]
Dimer (1 × 63000 + 1 × 25000, Chlamydomonas reinhardtii, SDS-PAGE [8], 1 × 64000 + 1 × 35000, Phaseolus angularis, SDS-PAGE [23]) [8, 23]
More (spinach: the native enyzme (MW 85000) can be split into a modified form (MW 61000) that retains activity with the non-physiological electron-donor methyl viologen, but loses most of the ferredoxin-linked activity and an 24000 MW fragment (coupling protein) in which the ferredoxin-binding domain is located) [22, 25]

Enzyme Handbook © Springer-Verlag Berlin Heidelberg 1994
Duplication, reproduction and storage in data banks are only allowed with the prior permission of the publishers

Glycoprotein/Lipoprotein

–

4 ISOLATION/PREPARATION

Source organism

Phormidium laminosum (cyanobacterium) [1, 2]; Oryza sativa (rice) [3, 12]; Spinacia oleracea (spinach) [4, 6, 10, 11, 13, 14, 17, 19, 20, 22, 24–26, 28]; Chlamydomonas reinhardtii [5, 8]; Porphyra yezoensis [7, 11]; Anacystis nidulans [9, 10]; Cucurbita pepo [10, 11, 18]; Chlorella fusca [10, 11, 21, 27]; Dunaliella tertiolecta [10]; Anabaena cylindrica [10]; Phaseolus angularis [15, 17, 23]; Spirulina platensis (cyanobacterium) [16, 17]; Phaseolus vulgaris [17]; Allium tuberosum [17]; Symphytum officinale [17]; Zea mays [17]; Hordeum vulgare [17]; More (enzyme is characteristic for photosynthetic organisms: higher plants, green algae, red algae, cyanobacteria [11], immunological comparison of enzyme from different plant sources [17]) [11, 17]

Source tissue

Leaf [3, 10, 13, 19, 20, 28]; Thallus [7]; Shoot (green [15], etiolated [15, 23]) [15, 23]; Root [15]; Cell [16]

Localisation in source

Soluble (exception: Anacystis [11]) [2, 11]; Chloroplast [10, 19, 28]; Subcellular particles (Anacystis) [11]

Purification

Phormidium laminosum [2]; Oryza sativa (rice) [3]; Porphyra yezoensis [7]; Chlamydomonas reinhardtii [8]; Spinacia oleracea [10, 13, 19, 20, 26, 28]; Phaseolus angularis [15, 23]; Spirulina platensis [16]; Cucurbita pepo [18]; Chlorella fusca [27]

Crystallization

–

Cloned

[6, 12]

Renaturated

–

5 STABILITY

pH

7 (enzyme in crude extract is rapidly inactivated below pH 7) [27]

Temperature (°C)

23 (1 h, pH 8.0, 75 mM Tris-HCl, stable) [14]; 40 (stable) [10]; 60 (10 min, more than 95% loss of activity) [10]

Oxidation

Organic solvent

General stability information

Unstable in crude extracts [27]

Storage

–20°C, several months [2]; 4°C, 1 week [2]; –20°C, 50% glycerol, for at least 6 months [10]; 4°C, 0.1 M potassium phosphate buffer, pH 7.7, 1 week [10]; 4°C, 200 mM Tris-HCl buffer, pH 7.5, 200 mM NaCl, 10% glycerol, for at least 1 month [16]; –20°C, 10% glycerol, 10 mM beta-mercaptoethanol, stable for several weeks [23]; 4°C, 30% loss of activity within 3 weeks [27]; –24°C, 30% loss of activity within 4 months [27]

6 CROSSREFERENCES TO STRUCTURE DATABANKS

PIR/MIPS code

PIR2: PS0153 (Rice (fragment)); PIR2: S20495 (precursor European white birch); PIR2: JA0172 (precursor Maize (fragment)); PIR2: S16603 (precursor Spinach)

Brookhaven code

7 LITERATURE REFERENCES

[1] Arizmendi, J.M., Serra, J.L.: Biochem. Soc. Trans., 18, 637–638 (1990)
[2] Arizmendi, J.M., Serra, J.L.: Biochim. Biophys. Acta, 1040, 237–244 (1990)
[3] Ida, S., Iwagami, K., Minobe, S.: Agric. Biol. Chem., 53, 2777–2784 (1989)
[4] Mikami, B., Ida, S.: J. Biochem., 105, 47–50 (1989)
[5] Romero, L.C., Gotor, C., Marquez, A.J., Forde, B.G., Vega, J.M.: Biochim. Biophys. Acta, 957, 152–157 (1988)
[6] Back, E., Burkhart, W., Moyer, M., Privalle, L., Rothstein, S.: Mol. Gen. Genet., 212, 20–26 (1988)
[7] Ide, T., Tamura, G.: Agric. Biol. Chem., 51, 3391–3393 (1987)
[8] Romero, L.C., Galvan, F., Vega, J.M.: Biochim. Biophys. Acta, 914, 55–63 (1987)
[9] Herrero, A., Guerrero, M.G.: J. Gen. Microbiol., 132, 2463–2468 (1986)
[10] Vega, J.M., Cardenas, J., Losada, M.: Methods Enzymol., 69, 255–270 (1980) (Review)
[11] Guerrero, M.G., Vega, J.M., Losada, M.: Annu. Rev. Plant Physiol., 32, 169–204 (1981) (Review)

Enzyme Handbook © Springer-Verlag Berlin Heidelberg 1994
Duplication, reproduction and storage in data banks are only allowed with the prior permission of the publishers

[12] Matsui, J., Takeba, G., Ida, S.: Agric. Biol. Chem.,54,3069–3071 (1990)
[13] Ida, S., Mikami, B.: Biochim. Biophys. Acta,871,167–176 (1986)
[14] Stein Privalle, L., Privalle, C.T., Leonardy, N.J., Kamin, H.: J. Biol. Chem.,260,14344–14350 (1985)
[15] Ishiyama, Y., Shinoda, I., Fukushima, K., Tamura, G.: Plant Sci.,39,89–95 (1985)
[16] Yabuki, Y., Mori, E., Tamura, G.: Agric. Biol. Chem.,49,3061–3062 (1985)
[17] Ishiyama, Y., Shinoda, I., Fukushima, K., Tamura, G.: Agric. Biol. Chem.,49,2225–2226 (1985)
[18] Hucklesby, D.P., James, D.M., Banwell, M.J., Hewitt, E.J.: Phytochemistry,15,599–603 (1976)
[19] Hirasawa, M., Tamura, G.: Agric. Biol. Chem.,43,659–661 (1979)
[20] Ida, S.: J. Biochem.,82,915–918 (1977)
[21] Zumft, W.G., Paneque, A., Aparicio, P.J., Losada, M.: Biochem. Biophys. Res. Commun.,36,980–986 (1969)
[22] Hirasawa, M., Knaff, D.B.: Biochim. Biophys. Acta,830,173–180 (1985)
[23] Ishiyama, Y., Tamura, G.: Plant Sci. Lett.,37,251–256 (1985)
[24] Hirasawa-Soga, M., Tamura, G.: Agric. Biol. Chem.,45,1615–1620 (1981)
[25] Hirasawa-Soga, M., Horie, S., Tamura, G.: Agric. Biol. Chem.,46,1319–1328 (1982)
[26] Vega, J.M., Kamin, H.: J. Biol. Chem.,252,896–909 (1977)
[27] Zumft, W.G.: Biochim. Biophys. Acta,276,363–375 (1972)
[28] Ramirez, J.M., Del Campo, F.F., Paneque, A., Losada, M.: Biochim. Biophys. Acta,118,58–71 (1966)

1 NOMENCLATURE

EC number
1.7.7.2

Systematic name
Nitrite:ferredoxin oxidoreductase

Recommended name
Ferredoxin-nitrate reductase

Synonymes
Assimilatory nitrate reductase
Reductase, nitrate (ferredoxin)
Assimilatory ferredoxin-nitrate reductase [2]

CAS Reg. No.
60382-69-6

2 REACTION AND SPECIFICITY

Catalysed reaction
Nitrate + 2 reduced ferredoxin →
→ nitrite + H_2O + 2 oxidized ferredoxin

Reaction type
Redox reaction

Natural substrates
Nitrate + reduced ferredoxin (assimilatory nitrate reduction [2], nitrate reduction [4, 5]) [2, 4, 5]

Substrate spectrum
1 Nitrate + reduced ferredoxin (enzyme also has methyl viologen-linked activity [2, 9]) [1–9]

Product spectrum
1 Nitrite + H_2O + oxidized ferredoxin

Inhibitor(s)
CNO_3^- [1]; Cyanide [7, 9]; N_3^- [1, 9]; Dithionite [2]; Photoactivated FAD/EDTA [1]; p-Hydroxymercuribenzoate [7]; More (ammonia promotes conversion of ferredoxin-nitrate reductase into its reduced inactive form (metabolic interconversion)) [4]

Enzyme Handbook © Springer-Verlag Berlin Heidelberg 1994
Duplication, reproduction and storage in data banks are only allowed with the prior permission of the publishers

Cofactor(s)/prostethic group(s)

Heme (Rhodopseudomonas capsulata: enzyme contains 1.5 heme molecules of the c-type) [7]; Cyanate (Azotobacter chroococcum, Acinetobacter calcoaceticus: stimulation) [7]

Metal compounds/salts

Molybdenum (molybdoprotein (Azotobacter chroococcum, Acinetobacter calcoaceticus) [7], a molybdenum iron-sulfur protein [2, 7], contains 0.9 mol of molybdenum [2]) [2, 7]; Iron (a molybdenum iron-sulfur protein [2, 7], contains 4 iron atoms [2]) [2, 7]; Mn^{2+} (stimulation) [9]; Fe^{2+} (stimulation) [9]; Mg^{2+} (stimulation) [9]; Ca^{2+} (stimulation) [9]

Turnover number (min^{-1})

Specific activity (U/mg)

305 (ferredoxin-linked activity) [2]; 1020 (methyl viologen-linked activity) [2]; 52 [9]; More [7]

K_m-value (mM)

0.038 (reduced ferredoxin) [2]; 2.5 (reduced methyl viologen) [2]; 0.10 (nitrate) [9]

pH-optimum

8.5 (anaerobic assay, methyl viologen-linked activity, ferredoxin-linked activity) [2]; 9 [9]; 10.2 (aerobic assay, methyl viologen linked activity) [2]; 10.5 (Anacystis nidulans, electron donor: dithionite-reduced methyl viologen) [7]

pH-range

7.3–9.2 (at pH 7.3 and 9.2: 50% of activity maximum, aerobic assay, methyl viologen-linked activity and ferredoxin-linked activity) [2]; 8.3–9.8 (at pH 8.3 and 9.8 about 50% of activity maximum) [9]

Temperature optimum (°C)

30 (assay at) [2]

Temperature range (°C)

---

3 ENZYME STRUCTURE

Molecular weight

75000 (Anacystis nidulans) [7]
80000 (Clostridium perfringens, gel filtration) [9]
85000 (Plectonema boryanum, sedimentation equilibrium) [2]
100000 (about, Azotobacter chroococcum, Acinetobacter calcoaceticus) [7]
185000 (Rhodopseudomonas capsulata) [7]

Subunits

Monomer (1 × 75000, Anacystis nidulans [7], 1 × 80000 Clostridium perfringens, SDS-PAGE [9], 1 × 83000, Plectonema boryanum, SDS-PAGE [2]) [2, 7, 9]
Dimer (2 × 85000, Rhodopseudomonas capsulata) [7]

Glycoprotein/Lipoprotein

–

4 ISOLATION/PREPARATION

Source organism

Plectonema boryanum [1, 2]; Phormidium unicatum [3]; Nostoc muscorum [4, 7]; Anacystis nidulans [5–7]; Anabaena sp. (strain 7119) [6]; Nostoc sp. (strain 6719) [6]; Anabaena cylindrica [7]; Azotobacter chroococcum [7]; Acinetobacter calcoaceticus [7]; Ectothiorhodospira shaposhnikovii [8]; Clostridium perfringens [9]; Rhodopseudomonas capsulata [7]

Source tissue

Cell [2]

Localisation in source

Purification

Plectonema boryanum [2]; Anacystis nidulans [7]; Clostridium perfringens [9]

Crystallization

–

Cloned

–

Renaturated

–

5 STABILITY

pH

Temperature (°C)

Oxidation

More (reversible inactivation by exposure of dithionite solution of the enzyme to air) [1]

Organic solvent

Enzyme Handbook © Springer-Verlag Berlin Heidelberg 1994
Duplication, reproduction and storage in data banks are only allowed with the prior permission of the publishers

General stability information

Low ionic strength, unstable after 70 h, 0.002 mg protein/ml: unstable in 10 mM phosphate buffer or 50 mM Tris-HCl buffer, pH 7.5 [2]; In vivo stability [5]

Storage

-20°C, in 50 mM Tris-HCl buffer, pH 7.5, 0.5 M NaCl, 1 mM EDTA, no loss of activity after 1 month [2]

6 CROSSREFERENCES TO STRUCTURE DATABANKS

PIR/MIPS code

Brookhaven code

7 LITERATURE REFERENCES

[1] Mikami, B., Ida, S.: Plant Cell Physiol.,27,1013–1021 (1986)
[2] Mikami, B., Ida, S.: Biochim. Biophys. Acta,791,294–304 (1984)
[3] Nath Bagachi, S., Singh Chauhan, V., Palod, A.: Curr. Microbiol.,21,53–57 (1990)
[4] Ortega, T., Rivas, J., Cardenas, J., Losada, M.: Biochem. Biophys. Res. Commun.,78,185–193 (1977)
[5] Herrero, A., Flores, E., Guerrero, M.G.: Arch. Biochem. Biophys.,234,454–459 (1984)
[6] Herrero, A., Flores, E., Guerrero, M.G.: J. Bacteriol.,145,175–180 (1981)
[7] Guerrero, M.G., Vega, J.M., Losada, M.: Annu. Rev. Plant Physiol.,32,169–204 (1981) (Review)
[8] Malofeeva, I.V., Kondratieva, E.N., Rubin, A.B.: FEBS Lett.,53,188–189 (1975)
[9] Seki-Chiba, S., Ishimoto, M.: J. Biochem.,82,1663–1671 (1977)

1 NOMENCLATURE

EC number
1.7.99.1

Systematic name
Ammonia:(acceptor) oxidoreductase

Recommended name
Hydroxylamine reductase

Synonymes
Reductase, hydroxylamine (acceptor)

CAS Reg. No.
37256-42-1

2 REACTION AND SPECIFICITY

Catalysed reaction
Hydroxylamine + reduced acceptor →
→ ammonia + acceptor

Reaction type
Redox reaction

Natural substrates

Substrate spectrum
1 Hydroxylamine + reduced acceptor (e.g. pyocyanine [1], reduced methylene blue [1], reduced benzyl viologen [9], hydroxylamine reductase 1 (associated with nitrite reductase): reduced benzyl viologen [5, 7], ferredoxin [5, 7], $FMNH_2$ [5, 7], $FADH_2$, NADPH [5], hydroxylamine reductase 2 (not associated with nitrite reductase): only reduced benzyl viologen [5–7], or dithionite [6, 7]) [1, 5–7, 9]

Product spectrum
1 Ammonia + oxidized acceptor

Enzyme Handbook © Springer-Verlag Berlin Heidelberg 1994
Duplication, reproduction and storage in data banks are only allowed with the prior permission of the publishers

Inhibitor(s)
CO (with reduced benzyl viologen as electron donor, reductase 1: irreversible inhibition by light, reductase 2 : reversible inhibition by light) [4]; Bathophenanthroline (hydroxylamine reductase 2) [4]; Semicarbazide (hydroxylamine reductase 2, slight) [4]; Isonicotinic acid (hydroxylamine reductase 2, slight) [4]; Atabrine (hydroxylamine reductase 1) [4]; Guanidinium chloride (slight) [7]; Hydrazine (hydroxylamine reductase 2, slight [4]) [1, 4]; Cu^{2+} (inhibition of dialyzed enzyme) [1]; Zn^{2+} (inhibition of dialyzed enzyme) [1]; Nitrite [9]

Cofactor(s)/prostethic group(s)
FAD (flavoprotein [1], contains 0.9 nmol FAD per mg protein) [1]

Metal compounds/salts
Iron (iron-chlorin protein also catalyzes methyl viologen-linked reduction of hydroxylamine and that of nitrite at a slower rate [3], iron containing proteins: nitrite reductase-hydroxylamine reductase 1 and hydroxylamine reductase 2) [3, 4]; Mn^{2+} (required [1], enzyme contains manganese [6]) [1, 6]; Co^{2+} (can substitute for Mn^{2+}, only 10% as effective) [1]

Turnover number (min^{-1})

Specific activity (U/mg)
2.87 [1]; More [7]

K_m-value (mM)
0.0075 (benzyl viologen) [9]; 0.00001 ($FADH_2$) [1]; 0. 000012 (reduced pyocyanine) [1]; 0.000044 (reduced methylene blue) [1]; 0.05–0.9 (hydroxylamine) [9]

pH-optimum
More (no clear or consistent pH optima observed with hydroxylamine reductase 1 and hydroxylamine reductase 2) [7]

pH-range

Temperature optimum (°C)
37 [1]

Temperature range (°C)

---

3 ENZYME STRUCTURE

Molecular weight
35000 (Cucurbita pepo, hydroxylamine reductase 2, gel filtration) [5, 7]
60000 (Cucurbita pepo, hydroxylamine reductase 1, gel filtration) [5, 7]

Subunits

Glycoprotein/Lipoprotein

–

4 ISOLATION/PREPARATION

Source organism

Aspergillus niger (iron-chlorin protein also catalyzes methyl viologen-linked reduction of hydroxylamine and that of nitrite at a slower rate) [3]; Fungi [6]; Spinacia oleracea (hydroxylamine reductase 1, hydroxylamine reductase 2) [5]; Pseudomonas aeruginosa [1]; Pennisetum purpureum (low activity) [2]; Brachiaria mutica (low activity) [2]; Chloris gayana (low activity) [2]; Panicum maximum (low activity) [2]; Cucurbita pepo (hydroxylamine reductase 1) [4, 5, 7, 9]; Wheat (nitrite reductase and hydroxylamine reductase are distinct enzymes) [8]

Source tissue

Leaf [2, 5, 7, 8]; Cell [1]

Localisation in source

Chloroplast [8]

Purification

Pseudomonas aeruginosa [1]; Cucurbita pepo [7]

Crystallization

–

Cloned

–

Renaturated

–

5 STABILITY

pH

Temperature (°C)

55 (10 min, 25–40% loss of activity) [7]; 60 (10 min, 50–85% loss of activity) [7]

Oxidation

Organic solvent

Enzyme Handbook © Springer-Verlag Berlin Heidelberg 1994
Duplication, reproduction and storage in data banks are only allowed with the prior permission of the publishers

General stability information

Storage

−18°C, DEAE-cellulose fraction, often stable for several weeks, but unpredictable loss of activity [7]

6 CROSSREFERENCES TO STRUCTURE DATABANKS

PIR/MIPS code

Brookhaven code

7 LITERATURE REFERENCES

[1] Walker, G.C., Nicholas, D.J.D.: Biochim. Biophys. Acta,49,361–368 (1961)
[2] Venkataramana, S., Das, V.S.R.: Z. Pflanzenphysiol.,105,289–296 (1982)
[3] Horie, S., Watanabe, T., Nakamura, S.: J. Biochem.,80 ,579–593 (1976)
[4] Hucklesby, D.P., Hewitt, E.J., James, D.M.: Biochem. J.,117,30P (1970)
[5] Hewitt, E.J., Hucklesby, D.P.: Biochem. Biophys. Res. Commun.,25,689–693 (1966)
[6] Nicholas, D.J.D.: Symp. Soc. Exp. Biol.,13,1 (1959)
[7] Hucklesby, D.P., Hewitt, E.J.: Biochem. J.,119,615–627 (1970)
[8] Sawhney, S.K., Nicholas, D.J.D.: Phytochemistry,14,1499–1503 (1975)
[9] Cresswell, C.F., Hageman, R.H., Hewitt, E.J., Hucklesby, D.P.: Biochem. J.,94,40–53 (1965)

1 NOMENCLATURE

EC number
1.7.99.3

Systematic name
Nitric-oxide:(acceptor) oxidoreductase

Recommended name
Nitrite reductase

Synonymes
EC 1.6.6.5 (formerly)
Methyl viologen-nitrite reductase
Reductase, nitrite
Cd-cytochrome nitrite reductase
NADH-nitrite reductase

CAS Reg. No.
9080-03-9

2 REACTION AND SPECIFICITY

Catalysed reaction
2 Nitrite + reduced acceptor →
→ 2 nitric oxide + 2 H_2O + acceptor

Reaction type
Redox reaction

Natural substrates
Nitrite + reduced acceptor (enzyme involved in sequential process of nitrate reduction to gaseous products [1, 2], the physiological electron donor that transfers electrons of the copper nitrite reductase from Achromobacter cycloclastes, a small blue copper protein, does not transfer electrons to the copper nitrite reductase of Bacillus halodenitrificans [6]) [1, 2, 6]

Enzyme Handbook © Springer-Verlag Berlin Heidelberg 1994
Duplication, reproduction and storage in data banks are only allowed with the prior permission of the publishers

Substrate spectrum

1 Nitrite + reduced acceptor (electron donors for the reduction of nitrite: reduced pyocyanine [2], reduced flavins (riboflavin, $FMNH_2$, $FADH_2$) [2], reduced methyl viologen [1, 2], reduced methylene blue [1], reduced 1,4-naphthoquinone [1], ascorbate-phenazine methosulfate [3, 6], not: NADH [2], NADPH [2], reduced cytochrome c [2]) [1–7]
2 Nitrite + hydroxylamine [3, 6]

Product spectrum

1 Nitric oxide + H_2O + acceptor (nitric oxide is the main product [1])
2 Nitrous oxide + ? [6]

Inhibitor(s)

KCN [1–3, 6]; p-Chloromercuribenzoate (slight [1], reversed by glutathione [2]) [1, 2, 3, 6]; N-Ethylmaleimide (slight) [1]; Diethyldithiocarbamate [1–3, 6]; o-Phenanthroline [1, 2]; Monoiodoacetate (slight) [1]; 2-Mercaptoethanol [1]; 2,2'-Dipyridyl [2]; Neo-cuproine [2]; CO [3]; EDTA [6]; Hydroxyquinoline [6]; Thioglycolate [6]

Cofactor(s)/prostethic group(s)

FAD (Pseudomonas enzyme contains 1.5 nmol FAD per mg of protein [2], FAD is required for maximum enzyme activity when reduced pyocyanine is the donor, not with reduced methylene blue [2], Achromobacter cycloclastes: no flavin detected [3]) [2]

Metal compounds/salts

Copper (copper protein [1, 5, 6], iron and copper required for nitrite reduction [2], copper type: 1 and 2 [5, 6], 4 mol copper per mol of enzyme [1], 1.78–1.98 atoms of copper per molecule [3], function of copper [5]) [1–3, 5, 6]; Iron (iron and copper required for nitrite reduction) [2]; Phosphate (activates) [2]; More (activation by elevated salt concentrations) [6]

Turnover number (min^{-1})

Specific activity (U/mg)

0.38 [1]; 20.6 [6]; 4.8 [2]; More [3]

K_m-value (mM)

0.074 (nitrite) [1]; 0.0003 (reduced methyl viologen) [2]; 0.00008 (reduced pyocyanine) [2]; 0.031 (NO_2^-) [2]; 0.5 (NO_2^-) [3, 5]

pH-optimum

5.9 [1]; 6.2 [3, 5]; 6 [6]

pH-range

Temperature optimum (°C)
30 (assay at) [1, 3]

Temperature range (°C)

3 ENZYME STRUCTURE

Molecular weight
69000 (Achromobacter cycloclastes, gel filtration) [3]
82000 (Bacillus halodenitrificans, gel filtration) [6]
110000 (Alcaligenes faecalis, gel filtration) [1]

Subunits
Tetramer (4 × 30000, Alcaligenes faecalis, SDS-PAGE) [1]
Dimer (2 × 40000, Bacillus halodenitrificans, SDS-PAGE) [6]

Glycoprotein/Lipoprotein
–

4 ISOLATION/PREPARATION

Source organism
Alcaligenes faecalis (strain S-6) [1]; Pseudomonas aeruginosa [2]; Pseudomonas stutzeri [4]; Bacillus halodenitrificans [6]; Achromobacter cycloclastes [3, 5, 7]

Source tissue
Cell [1, 2]

Localisation in source
Cytoplasmic membrane (firmly bound) [6]

Purification
Alcaligenes faecalis (strain S-6) [1]; Pseudomonas aeruginosa [2]; Achromobacter cycloclastes [3]; Bacillus halodenitrificans [6]

Crystallization
[1, 7]

Cloned
–

Renaturated
–

Enzyme Handbook © Springer-Verlag Berlin Heidelberg 1994
Duplication, reproduction and storage in data banks are only allowed with the prior permission of the publishers

5 STABILITY

pH

Temperature (°C)
80 (5 min, 56% loss of activity) [3]; 100 (5 min, 100% loss of activity) [3]

Oxidation

Organic solvent

General stability information

Storage
–80°C, 50% glycerol, anaerobic conditions, 10 months [6]

6 CROSSREFERENCES TO STRUCTURE DATABANKS

PIR/MIPS code
PIR3: A37260 (Achromobacter cycloclastes); PIR2: A34141 (Pseudomonas aeruginosa (fragment))

Brookhaven code
1NRD (Achromobacter cycloclastes)

7 LITERATURE REFERENCES

[1] Kakutani, T., Watanabe, H., Arima, K., Beppu, T.: J. Biochem.,89,453–461 (1981)
[2] Walker, G.C., Nicholas, D.J.D.: Biochim. Biophys. Acta,49,350–360 (1961)
[3] Iwasaki, H., Matsubara, T.: J. Biochem.,71,645–652 (1972)
[4] Chung, C.W., Najjar, V.A.: J. Biol. Chem.,218,617–625 (1956)
[5] Iwasaki, H., Noji, S., Shidara, S.: J. Biochem.,78,355–361 (1975)
[6] Payne, W.J., LeGall, J.: Biochim. Biophys. Acta,1056,225–232 (1991)
[7] Turley, S., Adman, E.T., Sieker, L.C., Liu, M.-Y., Payne, W.J., LeGall, J.: J. Mol. Biol.,200,417–419 (1988)

1 NOMENCLATURE

EC number
1.7.99.4

Systematic name
Nitrite:(acceptor) oxidoreductase

Recommended name
Nitrate reductase

Synonymes
Respiratory nitrate reductase
Reductase, nitrate (acceptor)
Nitrate reductase (acceptor)
More (the enzymes EC 1.7.99.4 and EC 1.9.6.1 are probably identical, in vivo cytochrome serves as electron donor in the electron transport chain to NO_3^- [32, 33], in vitro artificial electron donors e.g. reduced viologen indicators, can replace cytochrome c)

CAS Reg. No.
37256-45-4

2 REACTION AND SPECIFICITY

Catalysed reaction
Nitrate + reduced acceptor →
→ nitrite + acceptor (mechanism [19], with reduced viologens: compulsory-order Theorell-chance [21], with quinols as electron donor: two-site, enzyme-substitution mechanism [21])

Reaction type
Redox reaction

Enzyme Handbook © Springer-Verlag Berlin Heidelberg 1994
Duplication, reproduction and storage in data banks are only allowed with the prior permission of the publishers

Natural substrates

Nitrate + reduced acceptor (the enzymes EC 1.7.99.4 and EC 1.9.6.1 are probably identical, in vivo cytochrome serves as electron donor in the electron transport chain to NO_3^- [32, 33], in vitro artificial electron donors e.g. reduced viologen indicators, can replace cytochrome c, reduction of cytochrome b_{556} by ubiquinol which releases two protons, electrons are passed to nitrate reductase and used to reduce NO_3^- to NO_2^- [31], physiological function: transfer of electrons from cytochrome b_{559} to nitrate [4], electron transport chain in vivo:--> cytochrome b_1--> Mo--> NO_3^- [32], Fe^{3+}--> specific cytochrome-nitrate reductase--> NO_3^- [33], metabolic role: nitrate respiration [32]) [4, 31–33]

Nitrite + ferricyanide (nitrification) [34]

Substrate spectrum

1 Nitrate + reduced acceptor (ir [19], r [34], reduced acceptor e.g.: reduced benzyl viologen [1, 7, 9, 15, 18, 32–34], reduced viologen indicators [2, 7, 16, 20, 21, 32], $FMNH_2$ [2, 9, 16], reduced methyl viologen [3, 4, 11, 22, 32, 34], reduced ferredoxin [11], $FADH_2$ [16, 32], duroquinol [20], quinols [20], NADPH [32], reduced cytochrome b_1 [32], phenosafranine [32], reduced methylene blue [32]) [1–34]

2 Nitrite + ferricyanide [34]

3 ClO_3^- + reduced acceptor [2, 4, 16, 21]

4 BrO_3^- + reduced acceptor [16, 21]

5 More (not: NADH [2, 4, 15, 16], NADPH [2, 4, 16], $FADH_2$ [4], $FMNH_2$ [4], menadione [4], the enzymes EC 1.7.99.4 and EC 1.9.6.1 are probably identical, in vivo cytochrome serves as electron donor in the electron transport chain to NO_3^- [32, 33], in vitro artificial electron donors (e.g. reduced viologen indicators) can replace cytochrome c) [2, 4, 9, 15, 16, 32, 33]

Product spectrum

1 Nitrite + acceptor (ir [19], r [34]) [1–34]

2 Nitrate + ferrocyanide [34]

3 ?

4 ?

5 ?

Inhibitor(s)

CN^- [2, 11, 15, 16, 24]; N_3^- [2, 11, 16, 20, 33]; NaCl (0.17 M NaCl + 0.5 mM MgCl activates, 4.27 M NaCl + 0.5 mM MgCl inhibits) [3]; p-Chloromercuribenzoate (above 0.1 mM [4], not [11]) [4, 15, 33]; ClO_3^- [4, 15]; Dithiol [15]; KCNS [15]; o-Phenanthroline [15]; Mepacrine [15]; ClO_4^- [15]; BrO_3^- [15]; IO_3^- [15]; 2-n-Heptyl-4-hydroxyquinoline N-oxide [20]; Ferricyanide [24]; Na_2S [33]; Hydrogensulfite [33]

Cofactor(s)/prostethic group(s)

Cytochrome b (Pseudomonas denitrificans: purified enzyme has a cytochrome b-type absorption spectrum) [15]; Cytochrome c (Nitrobacter hamburgensis, nitrite oxidoreductase enzyme complex) [34]; Bactopterin (Pseudomonas stutzeri, cofactor) [23]; More (contains no cytochrome: Aerobacter aerogenes [1], Klebsiella aerogenes [4], Rhodobacter sphaeroides [18], Bacillus stearothermophilus [22], no FAD: Paracoccus halodenitrificans [2], Paracoccus denitrificans [16], no FMN: Paracoccus halodenitrificans [2], Paracoccus denitrificans [16]) [1, 2, 4, 16, 18, 22]

Metal compounds/salts

Iron (Bacillus licheniformis: 6.9 atoms of non-heme iron per molecule of enzyme [27], Nitrobacter hamburgensis: contains iron [34], Paracoccus halodenitrificans: enzyme contains 2 Fe atoms, non-heme iron protein [2], Clostridium perfringens: molybdo-iron-sulfur protein [11], Paracoccus denitrificans: iron content about 8 mol per mol of enzyme, non-heme iron protein [16], Bacillus stearothermophilus: 6 gatom of non-heme iron in 1 mol of purified enzyme [22], E. coli: 16 gatom of iron per mol of enzyme [24], contains iron-sulfur centres [31], Klebsiella aerogenes, nitrate reductase I, MW 26000 (abc_2): 8 iron-sulfur groups and 4 tightly bound non-heme iron atoms per molecule of enzyme, nitrate reductase II: no tightly bound iron [6]) [2, 6, 11, 16, 18, 22, 24, 27, 31, 34]; Molybdenum (Nitrobacter hamburgensis: contains molybdenum [34], Paracoccus halodenitrificans: enzyme contains 1 molybdenum atom [2], Klebsiella aerogenes, nitrate reductase I, MW 260000 (abc_2): 0.24 atoms of molybdenum per molecule of enzyme [6], Clostridium perfringens: molybdo-iron-sulfur protein [11], Paracoccus denitrificans: molybdenum centre [17], Rhodobacter sphaeroides: 1 mol molybdenum per mol of enzyme [18], Pseudomonas stutzeri: molybdoenzyme [23], E. coli: 0.8 mol of molybdenum per mol of enzyme [24], molybdenoenzyme [31], Bacillus licheniformis: 0.93 atoms of molybdenum per molecule of enzyme, molybdenum seems to be a part of a low-molecular weight peptide to which it may be bound by interaction with thiol groups [27]) [2, 6, 11, 17, 18, 23, 24, 27, 31, 34]; Mn^{2+} (stimulates) [11]; Fe^{2+} (stimulates) [11]; Mg^{2+} (stimulates) [11]; Ca^{2+} (stimulates) [11]; Na^+ (0.17 M NaCl + 0.5 mM MgCl activates, 4.27 M NaCl + 0.5 mM MgCl inhibits) [3]

Turnover number (min^{-1})

Specific activity (U/mg)

More [2, 18, 33, 34]; 39.1 [4]; 52 [11]; 50 [20]; 62.53 [22]

K_m-value (mM)

More [15, 18, 20, 32–34]; 1.3 (NO_3^-) [2]; 5 (ClO_3^-) [2]; 1.6 (reduced benzyl viologen, NO_3^-) [4]; 0.10 (NO_3^-) [11]; 0.0002 (benzyl viologen) [18]; 3.6 (NO_2^-) [34]

Enzyme Handbook © Springer-Verlag Berlin Heidelberg 1994
Duplication, reproduction and storage in data banks are only allowed with the prior permission of the publishers

pH-optimum

5.6–5.8 (NO_3^-) [16]; 6.2–6.4 (ClO_3^-) [16]; 6.3 (NO_3^-) [2]; 6.4 (ClO_3^-) [2]; 6.8 [4]; 7.0 (nitrate + reduced acceptor) [34]; 7–8.5 [33]; 7.1 (NO_3^-, Enterobacter cloacae [25], E. coli [26]) [25, 26]; 8.0 (NO_3^- + reduced acceptor [15], NO_2^- + ferricyanide [34]) [15, 34]; 9.0 [11]

pH-range

7.0–9.8 (7.0: about 10% of activity maximum, 9.8: about 50% of activity maximum) [11]; 5.5–10.0 (sharp drop of activity below pH 5.5 and above 10.0) [33]

Temperature optimum (°C)

More (temperature optimum is a function of both: the concentration and the specific cation present) [3]; 56 (0.5 mM $MgCl_2$) [3]; 73 (2 M KCl) [3]; 75 (optimum of 5 min reaction) [25]; 85 (4.27 M NaCl) [3]

Temperature range (°C)

3 ENZYME STRUCTURE

Molecular weight

More [9]
90000 (Clostridium perfringens, gel filtration) [11]
165000 (Paracoccus halodenitrificans, polyacrylamide gel electrophoresis) [2]
160000 (Klebsiella aerogenes, nitrate reductase II, gel filtration [4], Paracoccus denitrificans, gel filtration, ultracentifugation [16]) [4, 16]
180000 (Rhodobacter sphaeroides, gel filtration of urea-treated enzyme) [18]
196000 (Bacillus licheniformis, polyacrylamide gel electrophoresis) [27]
200000 (Paracoccus denitrificans, gel filtration) [19]
210000 (Bacillus stearothermophilus, polyacrylamide gel electrophoresis) [22]
260000 (Klebsiella aerogenes, nitrate reductase I (abc_2), gel filtration) [4]
360000 (Nitrobacter hamburgensis, nitrite oxidoreductase enzyme complex, gel filtration) [34]
1060000 (Klebsiella aerogenes, nitrate reductase I, $(abc_2)_4$, gel filtration) [4]

Subunits

Monomer (Aerobacter aerogenes [1], 1 × 90000, Clostridium perfringens, SDS-PAGE [11]) [1, 11]
Dimer (1 × 150000 + 1 × 57000, Bacillus licheniformis, SDS-PAGE [27], 1 × 117000 + 1 × 57000, Klebsiella aerogenes, nitrate reductase II, SDS-PAGE [4], 1 × 120000 (alpha) + 1 × 60000 (beta), Rhodobacter sphaeroides, SDS-PAGE [18], 1 × 127000 (alpha) + 1 × 61000 (beta), Paracoccus denitrificans, SDS-PAGE [19], 1 × 150000 + 1 × 44000, Bacillus stearothermophilus, SDS-PAGE [22]) [4, 18, 19, 22, 27]
Trimer (1 × 127000 (alpha) + 1 × 61000 (beta) + 1 × 21000 (gamma), Paracoccus denitrificans, SDS-PAGE) [20]
Tetramer (1 × 117000 + 1 × 57000 + 2 × 52000, Klebsiella aerogenes, nitrate reductase II, SDS-PAGE) [4]
Pentamer ($alpha_2beta_2gamma_1$, 2 × 116000 (alpha) + 2 × 65000 (beta) + 1 × 32000 (gamma), Nitrobacter hamburgensis, nitrite reductase enzyme complex, SDS-PAGE) [34]
Hexadecamer (4 × 117000 + 4 × 57000 + 8 × 52000, Klebsiella aerogenes, nitrate reductase II, SDS-PAGE) [4]
? (x × 150000 (alpha) + x × 67000 ($beta_1$) + x × 65000 ($beta_2$), molar ratio alpha: $beta_1$ + $beta_2$ is 1: 1, E. coli, SDS-PAGE [7, 9], an additional subunit is present in haem-containing enzyme [9], x × 150000 (alpha) + x × 60000 (beta) + x × 20000 (gamma), E. coli [31]) [7, 9, 31]

Glycoprotein/Lipoprotein

No glycoprotein [27]

4 ISOLATION/PREPARATION

Source organism

Aerobacter aerogenes [1]; Paracoccus halodenitrificans [2]; Halobacterium sp. [3]; Klebsiella aerogenes (2 forms: nitrate reductase I (ab), nitrate reductase II (abc_2) and $(abc_2)_4$ [4, 6]) [4, 6, 28]; E. coli (5-aminolaevulinic acid auxotroph [5], E. coli K-12 [7, 9, 21]) [5, 7–9, 12–14, 21, 24, 26, 31, 32]; Clostridium perfringens [11]; Pseudomonas denitrificans [15]; Paracoccus denitrificans [16, 17, 19, 20]; Rhodobacter sphaeroides f.sp. denitrificans [18]; Bacillus stearothermophilus [22]; Pseudomonas stutzeri [23]; Bacillus licheniformis [27]; Thiosphaera pantotropha [29]; Neurospora crassa [30]; Achromobacter fischeri [33]; Nitrobacter hamburgensis [34]; Enterobacter cloacae [25]

Source tissue

Cell [2, 3]

Enzyme Handbook © Springer-Verlag Berlin Heidelberg 1994
Duplication, reproduction and storage in data banks are only allowed with the prior permission of the publishers

Localisation in source

Membrane (nitrate reductase of cytoplasmic membrane is vectorial, reducing nitrate on the outer aspect of the membrane with 2 H^+ and 2 electrons that have crossed from the inner aspect of the membrane [8], arrangement in cytoplasmic membrane [12], Klebsiella aerogenes: transmembrane protein [28], Bacillus licheniformis: the 2 subunits are localized on the cytoplasmic side of the membrane [28]) [8, 12, 28]

Purification

Aerobacter aerogenes [1]; Paracoccus halodenitrificans [2]; Halobacterium sp. (partial) [3]; Klebsiella aerogenes (2 forms: nitrate reductase I, nitrate reductase II) [4]; E. coli (K-12 [7, 9]) [7, 9, 24]; Clostridium perfringens [11]; Pseudomonas denitrificans [15]; Rhodobacter sphaeroides f.sp. denitrificans [18]; Paracoccus denitrificans (partial [19, 20]) [16, 19, 20]; Bacillus stearothermophilus [22]; Bacillus licheniformis [27]; Achromobacter fischeri [33]

Crystallization

–

Cloned

(large subunit gene of E. coli enzyme) [13]

Renaturated

–

5 STABILITY

pH

6.5–10 [25]

Temperature (°C)

6 (immobilized enzyme more stable at 6°C than at 23°C) [14]; 20 (24 h, complete loss of activity) [4]; 23 (immobilized enzyme more stable at 6°C than at 23°C) [14]; 50 (10 min, stable up to) [25, 33]; 60 (10 min, 15% loss of activity) [25, 33]; 70 (5 min, complete loss of activity) [33]; 80 (0.17 M NaCl: 90% loss of activity after 1 min, 0.85 M NaCl: 50% loss of activity after 7 min, 4.27–5.31 M NaCl: no loss of activity after 15 min) [3]

Oxidation

Organic solvent

General stability information

NaCl protects against heat inactivation 4.27–5.31 M NaCl [3]; Deoxycholate stabilizes [4]; Freezing (slow freezing inactivates [10], rapid freezing in liquid N_2 and thawing at room temperature can be repeated at least ten times without loss of activity [9]) [9, 10]; Immobilized enzyme more labile than free enzyme [14]; Sucrose stabilizes [25]; Stable to prolonged dialysis [33]

Storage

4°C, prolonged storage [4, 9]; 5°C, 0.1 M Tris-HCl, 10% sucrose, pH 8.8, 24 h, 10% loss of activity [25]; Frozen, several months [33]

6 CROSSREFERENCES TO STRUCTURE DATABANKS

PIR/MIPS code

PIR1: RDECNA (alpha chain Escherichia coli); PIR2: A31911 (alpha chain Escherichia coli (fragment)); PIR1: RDECNB (beta chain Escherichia coli); PIR2: B31911 (beta chain Escherichia coli (fragment)); PIR1: RDECNG (gamma chain Escherichia coli); PIR2: C31911 (gamma chain Escherichia coli (fragment))

Brookhaven code

7 LITERATURE REFERENCES

[1] Van 'T Riet, J., Planta, R.J.: FEBS Lett.,5,249–252 (1969)
[2] Rosso, J.-P., Forget, P., Pichinoty, F.: Biochim. Biophys. Acta,321,443–455 (1973)
[3] Marquez, E.D., Brodie, A.F.: Biochim. Biophys. Acta,321,84–89 (1973)
[4] Van 'T Riet, J., Planta, R.J.: Biochim. Biophys. Acta,379,81–94 (1975)
[5] Kemp, M.B., Haddock, B.A., Garland, P.B.: Biochem. J.,148,329–333 (1975)
[6] Van 'T Riet, J., Van Ee, J.H., Wever, R., Van Gelder, B.F., Planta, R.J.: Biochim. Biophys. Acta,405,306–317 (1975)
[7] Clegg, R.A.: Biochem. Soc. Trans.,3,691–694 (1975)
[8] Garland, P.B., Downie, J.A., Haddock, B.A.: Biochem. J.,152,547–559 (1975)
[9] Clegg, R.A.: Biochem. J.,153,533–541 (1976)
[10] MacGregor, C.H., Schnaitman, C.A., Normasell, D.E., Hodgins, M.G.: J. Biol. Chem.,249,5321–5327 (1974)
[11] Seki-Chiba, S., Ishimoto, M.: J. Biochem.,82,1663–1671 (1977)
[12] Graham, A., Boxer, D.H.: FEBS Lett.,113,15–20 (1980)
[13] McPherson, M.J., Baron, A.J., Pappin, D.J.C., Wootton, J.C.: FEBS Lett.,177,260–264 (1984)
[14] Schiller, J.G., Liu, C.C.: Biotechnol. Bioeng.,18,1643–1645 (1976)
[15] Radcliff, B.C., Nicholas, D.J.D.: Biochim. Biophys. Acta,205,273–287 (1970)
[16] Forget, P.: Eur. J. Biochem.,18,442–450 (1971)
[17] Turner, N., Ballard, A.L., Bray, R.C., Ferguson, S.: Biochem. J.,252,925–926 (1988)

Enzyme Handbook © Springer-Verlag Berlin Heidelberg 1994
Duplication, reproduction and storage in data banks are only allowed with the prior permission of the publishers

[18] Byrne, M.D., Nicholas, D.J.D.: Biochim. Biophys. Acta,915,120–124 (1987)
[19] Ballard, A.L., Ferguson, S.J.: Eur. J. Biochem.,174,207–212 (1988)
[20] Craske, A., Ferguson, S.J.: Eur. J. Biochem.,158,429–436 (1986)
[21] Morpeth, F.F., Boxer, D.H.: Biochemistry,24,40–46 (1985)
[22] Chikwem, J.O., Downey, R.J.: Anal. Biochem.,126,74–80 (1982)
[23] Krüger, B., Meyer, O., Nagel, M., Andreesen, J.R., Meincke, M., Bock, E., Blümle, S., Zumft, W.G.: FEMS Microbiol. Lett.,48,225–227 (1987)
[24] Adams, M.W.W., Mortenson, L.E.: J. Biol. Chem.,257,1791–1799 (1982)
[25] Aoki, K., Shinke, R., Nishira, H.: Agric. Biol. Chem.,45,817–822 (1981)
[26] Taniguchi, S., Itagaki, E.: Biochim. Biophys. Acta,44,263–279 (1960)
[27] Van 'T Riet, J., Wientjes, F.B., Van Doorn,J., Planta, R.J.: Biochim. Biophys. Acta,576,347–360 (1979)
[28] Wientjes, F.B., Kolk, A.H.J., Nanninga, N., Van 'T Riet, J.: Eur. J. Biochem.,95,61–67 (1979)
[29] Bell, L.C., Richardson, D.J., Ferguson, S.J.: FEBS Lett.,265,85–87 (1990)
[30] Subramanian, K.N., Sorger, G.J.: J. Bacteriol.,110,538–546 (1972)
[31] Boxer, D., Malcolm, A., Graham, A.: Biochem. Soc. Trans.,10,480–481 (1982) (Review)
[32] Nason, A. in "The Enzymes",2nd. Ed. (Boyer, P.D., Lardy, H., Myrbäck, K., eds.) 7,587–607 (1963) (Review)
[33] Sadana, J.C., McElroy, W.D.: Arch. Biochem. Biophys.,67,16–34 (1957)
[34] Sundermeyer-Klinger, H., Meyer, W., Warninghoff, B., Bock, E.: Arch. Microbiol.,140,153–158 (1984)

1 NOMENCLATURE

EC number
1.7.99.5

Systematic name
5-Methyltetrahydrofolate:(acceptor) oxidoreductase

Recommended name
5,10-Methylenetetrahydrofolate reductase ($FADH_2$)

Synonymes
EC 1.1.1.68 (formerly)
EC 1.1.99.15 (formerly)
5-Methyltetrahydrofolate:NAD oxidoreductase
5-Methyltetrahydrofolate:NAD^+ oxidoreductase [6, 7]
Reductase, methylenetetrahydrofolate (reduced riboflavin adenine dinucleotide)
5,10-Methylenetetrahydrofolate reductase
Methylenetetrahydrofolate reductase
N5,10-Methylenetetrahydrofolate reductase
5,10-Methylenetetrahydropteroylglutamate reductase
N^5,N^{10}–Methylenetetrahydrofolate reductase
Methylenetetrahydrofolic acid reductase

CAS Reg. No.
9028-69-7

2 REACTION AND SPECIFICITY

Catalysed reaction
5,10-Methylenetetrahydrofolate + reduced acceptor →
→ 5-methyltetrahydrofolate + acceptor (acceptor is free FAD, stereochemistry [1])

Reaction type
Redox reaction

Natural substrates
5,10-Methylenetetrahydrofolate + $FADH_2$ (bacterial enzyme system [2], enzyme in pathway of synthesis of acetate from CO_2 via formate and a series of reactions involving tetrahydrofolate and a corrinoid [4], biosynthesis of 5-methyltetrahydrofolate (a donor of methyl groups of methionine) [5]) [2, 4, 5]

Enzyme Handbook © Springer-Verlag Berlin Heidelberg 1994
Duplication, reproduction and storage in data banks are only allowed with the prior permission of the publishers

Substrate spectrum

1 5,10-Methylenetetrahydrofolate + reduced acceptor (r [3–5], equilibrium lies far in favor of 5-methyl-tetrahydrofolate formation [7], forward reaction (reduced acceptor): $FADH_2$ [1–4, 7], reduced ferredoxin [3, 4], NADH [6, 7], reverse reaction (oxidized acceptor): FAD [3, 4], rubredoxin [3, 4], menadione [3, 4, 6, 7], benzyl viologen [3, 4], methylene blue [3, 4], no activity with pyridine nucleotides [4])
2 5,10-Methylene-5,6,7,8-tetrahydropteroylmonoglutamate + reduced acceptor (e.g. menadione, not $NADP^+$) [7]
3 5,10-Methylene-5,6,7,8-tetrahydropteroylpentaglutamate + reduced acceptor (e.g. menadione, not $NADP^+$) [7]

Product spectrum

1 5-Methyltetrahydrofolate + oxidized acceptor
2 (+)-5-Methyl-5,6,7,8-tetrahydropteroylmonoglutamate + oxidized acceptor [7]
3 (+)-5-Methyl-5,6,7,8-tetrahydropteroylpentaglutamate + oxidized acceptor [7]

Inhibitor(s)

S-Adenosylmethionine [6]

Cofactor(s)/prostethic group(s)

FAD (flavoprotein, protein contains 1.7 molecules FAD, oxidized acceptor in reverse reaction) [1, 3, 4, 7]; NADH [6, 7]

Metal compounds/salts

Iron (enzyme contains 1.5–2 molecules iron [3], iron-sulphur clusters) [3, 4]; Zinc (enzyme contains 2.3 molecules zinc) [3, 4]

Turnover number (min^{-1})

Specific activity (U/mg)

139 [3, 4]; 15.7 [5]; More [7]

K_m-value (mM)

0.12 (DL-5-methyltetrahydrofolate) [3, 4]; 11.1 (benzyl viologen) [3, 4]; 0.045 ((+)-5-methyl-5,6,7,8-tetrahydropteroylmonoglutamate) [7]; 0.0064 ((+)-5-methyl-5,6,7,8-tetrahydropteroylpentaglutamate) [7]

pH-optimum

6.3–6.4 [5]; 6.7 (5-methyl-5,6,7,8-tetrahydropteroylmonoglutamate, 5-methyl-5,6,7,8-tetrahydropteroylpentaglutamate) [7]; 7.8–8.0 [4]

pH-range

6–8 (6: about 70% of activity maximum, 8: about 35% of activity maximum (5-methyl-5,6,7,8-tetrahydropteroylmonoglutamate), about 65% of activity maximum (5-methyl-5,6,7,8-tetrahydropteroylpentaglutamate)) [7]

Temperature optimum (°C)

37 (assay at) [1, 4, 5]

Temperature range (°C)

3 ENZYME STRUCTURE

Molecular weight

237000 (Clostridium formicoaceticum, gel filtration) [3, 4]

Subunits

Octamer (4 × 26000 (alpha) + 4 × 35000 (beta), Clostridium formicoaceticum, SDS-PAGE) [3, 4]

Glycoprotein/Lipoprotein

–

4 ISOLATION/PREPARATION

Source organism

Pig [1, 2]; E. coli [5]; Clostridium formicoaceticum [3, 4]; Rat [6, 7]

Source tissue

Liver [1, 6, 7]

Localisation in source

Purification

Rat [7]; Pig [2]; Clostridium formicoaceticum [3, 4]; E. coli [5]

Crystallization

–

Cloned

–

Renaturated

–

Enzyme Handbook © Springer-Verlag Berlin Heidelberg 1994
Duplication, reproduction and storage in data banks are only allowed with the prior permission of the publishers

5 STABILITY

pH

Temperature (°C)

Oxidation

Organic solvent

General stability information

Oxygen labile (half-life: less than 1 h in aerobic buffer, sodium dithionite prevents inactivation by oxygen) [3, 4]

Storage

–15°C, several months [1]; 10°C, pH 7.4, 20% glycerol, 2 mM dithionite, in an anaerobic chamber, a few weeks [4]

---

6 CROSSREFERENCES TO STRUCTURE DATABANKS

PIR/MIPS code

PIR1: RDECMH (Escherichia coli); PIR2: S03169 (Salmonella typhimurium)

Brookhaven code

---

7 LITERATURE REFERENCES

[1] Vanoni, M.A., Lee, S., Floss, H.G., Matthews, R.G.: J. Am. Chem. Soc.,112,3987–3992 (1990)
[2] Buchanan, J.M.: Methods Enzymol.,17B,371–378 (1971) (Review)
[3] Clark, J.E., Ljungdahl, L.G.: Methods Enzymol.,122,392–399 (1986) (Review)
[4] Clark, J.E., Ljungdahl, L.G.: J. Biol. Chem.,259,10845–10849 (1984)
[5] Katzen, H.M., Buchanan, J.M.: J. Biol. Chem.,240,825–835 (1965)
[6] Kutzbach, C., Stokstad, E.L.R.: Biochim. Biophys. Acta,139,217–220 (1967)
[7] Cheng, F.W., Shane, B., Stokstad, E.L.R.: Can. J. Biochem.,53,1020–1027 (1975)

1 NOMENCLATURE

EC number
1.7.99.6

Systematic name
Nitrogen:(acceptor) oxidoreductase (N_2O-forming)

Recommended name
Nitrous-oxide reductase

Synonymes
Reductase, nitrous oxide
Nitrous oxide reductase
N_2O reductase [1]

CAS Reg. No.
55576-44-8

2 REACTION AND SPECIFICITY

Catalysed reaction
Nitrous oxide + reduced acceptor →
→ nitrogen + acceptor

Reaction type
Redox reaction

Natural substrates
N_2O + reduced acceptor (final step in denitrification pathway [1], terminal step in the energy-conserving denitrification pathway of soil and marine denitrifying bacteria) [3]

Substrate spectrum
1 N_2O + reduced acceptor (e.g. benzyl viologen cation radical [3, 5–7, 12], methylene blue [6], methyl viologen cation radical [12]) [3, 5–7, 12, 17]
2 More (other classes of reductants including dithionite cannot substitute for viologen dyes [1, 2], enzyme exhibits a dithionite-N_2O oxidoreductase as well as a (benzyl viologen radical cation)-N_2O oxidoreductase activity [3]) [1–3]

Product spectrum
1 N_2 + acceptor
2 ?

Enzyme Handbook © Springer-Verlag Berlin Heidelberg 1994
Duplication, reproduction and storage in data banks are only allowed with the prior permission of the publishers

Inhibitor(s)

KCl (50 mM) [7]; NaCl (50 mM) [7]; Potassium phosphate (250 mM) [7]; Acetylene [3, 5, 12, 17]; CN^- [3, 5, 12]; N_3^- (turnover-dependent inactivation: $N_3^- > CNO^- > NO_2^- > F^-$ [4]) [3–5, 12]; F^- (turnover-dependent inactivation: $N_3^- >$ CNO- $> NO_2^- > F^-$) [4]; EDTA [5]; o-Phenanthroline [5]; CO [12]; CNO_3^- (turnover-dependent inactivation: $N_3^- > CNO^- > NO_2^- > F^-$) [4]; NO_2 (turnover-dependent inactivation: $N_3^- > CNO^- > NO_2^- > F^-$) [4]

Cofactor(s)/prostethic group(s)

Cytochrome c (Wolinella succinogenes: about 1 cytochrome c per subunit [3], Rhodopseudomonas sphaeroides: enzyme is devoid of heme [5], content and nature of the soluble c-type cytochromes depend strongly on the species and the particular growth conditions [16]) [3, 5, 16]

Metal compounds/salts

Copper (4 copper atoms per molecule: Achromobacter cycloclastes [1], Paracoccus denitrificans [4], Rhodopseudomonas sphaeroides f. sp. denitrificans [5], 8 copper atoms per molecule: Pseudomonas stutzeri [2], Pseudomonas perfectomarinus [6, 14], 3 copper atoms per subunit (i.e. 6 atoms per molecule): Wolinella succinogenes [3]) [1–6, 14, 15]; Iron (Wolinella succinogenes: contains 1 iron atom per subunit (i.e. 2 per molecule, as cytochrome c) [3], Pseudomonas perfectomarinus: iron not found in stoichiometric amounts [14]) [3, 14]; Calcium (Wolinella succinogenes: 1 calcium atom per subunit (i.e. 2 calcium per molecule)) [3]; Zinc (Rhodopseudomonas sphaeroides f.sp. denitrificans: purified enzyme contains 2.02 atoms of zinc per molecule [5], Pseudomons perfectomarinus: zinc not found in stoichiometric amounts [14]) [5, 14]; Nickel (Rhodopseudomonas sphaeroides f.sp. denitrificans: purified enzyme contains 0.76 atoms of nickel per molecule) [5]; More (the 120000 MW copper protein of N_2-producing denitrifying bacteria and nitrous oxide reductase are different entities) [15]

Turnover number (min^{-1})

Specific activity (U/mg)

86 [1]; 122 [4]; 30.0 [7]; 160 [3]; 50.37 [5]; More [6]

K_m-value (mM)

0.007 (N_2O) [3, 4]; 0.004 (benzyl viologen radical cation) [3]; 0.026 (N_2O) [5]; 0.0009 (reduced benzyl viologen cation) [12]; 0.005 (reduced methyl viologen) [12]

pH-optimum

8 (broad [7]) [7, 13]; 9 [5]

pH-range

6–9.2 (50% of activity maximum at pH 6.0 and 9.2) [5]

Temperature optimum (°C)

Temperature range (°C)

3 ENZYME STRUCTURE

Molecular weight

72000 (Achromobacter cycloclastes, gel filtration, SDS-PAGE) [1]
73000–95000 (Rhodopseudomonas sphaeroides f. sp. denitrificans, gel filtration, SDS-PAGE) [5]
85000 (Paracoccus denitrificans, gel filtration) [7]
118000–120000 (Pseudomonas perfectomarinus, gel filtration [6, 14, 16], the 120000 MW copper protein of N_2-producing denitrifying bacteria and nitrous oxide reductase are different entities [15]) [6, 14–16]
140000 (Paracoccus denitrificans, gel filtration) [4]
162000 (Wolinella succinogenes, gel filtration) [3]

Subunits

Monomer (1 × 72000, Achromobacter cycloclastes, SDS-PAGE [1], 1 × 73000–95000, Rhodopseudomonas sphaeroides, SDS-PAGE [5]) [1, 5]
Dimer (2 × 72000, Paracoccus denitrificans, SDS-PAGE [4], 2 × 74000, Pseudomonas perfectomarinus, SDS-PAGE [6], 2 × 62000, Pseudomonas perfectomarinus, SDS-PAGE [14], 2 × 88000, Wolinella succinogenes, SDS-PAGE [3]) [3, 4, 6, 14]

Glycoprotein/Lipoprotein

–

4 ISOLATION/PREPARATION

Source organism

Achromobacter cycloclastes [1]; Paracoccus denitrificans [4, 7, 12, 13]; Alcaligenes sp. (NCIB 11015) [1]; Pseudomonas perfectomarinus [6, 12, 14–16]; Pseudomonas stutzeri [2, 16]; Alcaligenes faecalis (IAM 1015) [1]; Rhizobium sp. (strain 8A55) [10]; Pseudomonas sp. G59 [11]; Rhodopseudomonas capsulata (anaerobically and phototrophically grown cultures) [8]; Rhodopseudomonas palustris (anaerobically and phototrophically grown cultures) [8]; Rhodospirillum rubrum (anaerobically and phototrophically grown cultures) [8]; Pseudomonas nautica 617 [9]; Pseudomonas fluorescens [16]; Rhodopseudomonas sphaeroides (f. sp. denitrificans [5], anaerobically and phototrophically grown cultures [8]) [5, 8]; Wolinella succinogenes [3]; Flexibacter canadensis [17]; Soil bacterium Is-11 [17]

Enzyme Handbook © Springer-Verlag Berlin Heidelberg 1994
Duplication, reproduction and storage in data banks are only allowed with the prior permission of the publishers

Source tissue
Cells [1, 3, 5, 7, 17]

Localisation in source
Soluble [12]; Periplasm [8]

Purification
Achromobacter cycloclastes [1]; Wolinella succinogenes [3]; Paracoccus denitrificans (partial [7]) [4, 7]; Rhodopseudomonas sphaeroides (f. sp. denitrificans) [5]; Pseudomonas perfectomarinus [6, 14]

Crystallization
–

Cloned
(Pseudomonas stutzeri gene) [2]

Renaturated
–

5 STABILITY

pH

Temperature (°C)

Oxidation
O_2 (rapid inactivation by O_2 [7], $t_{1/2}$ in presence of O_2 at room temperature: 65 min [4], $t_{1/2}$ in presence of O_2 at room temperature: 1 h [12]) [4, 7, 12]

Organic solvent

General stability information
Rapid inactivation by dithionite [7]; Rapid inactivation by mercaptoethanol [7]; Benzyl viologen causes rapid inactivation when incubated with benzyl viologen radical cation and either O_2 or N_2O, stable in presence of N_2O or benzyl viologen radical cation alone [4]

Storage
4°C, stable for at least 100 h under air and 3 h in presence of 5 mM EDTA [3]

6 CROSSREFERENCES TO STRUCTURE DATABANKS

PIR/MIPS code
PIR2: A31845 (precursor Pseudomonas stutzeri)

Brookhaven code

7 LITERATURE REFERENCES

[1] Hulse, C.L., Averill, B.A.: Biochem. Biophys. Res. Commun., 166, 729–735 (1990)
[2] Viebrock, A., Zumft, W.G.: J. Bacteriol., 170, 4658–4668 (1988)
[3] Teraguchi, S., Hollocher, T.C.: J. Biol. Chem., 264, 1972–1979 (1989)
[4] Snyder, S.W., Hollocher, T.G.: J. Biol. Chem., 262, 6515–6525 (1987)
[5] Michalski, W.P., Hein, D.H., Nicholas, D.J.D.: Biochim. Biophys. Acta, 872, 50–60 (1986)
[6] Coyle, C.L., Zumft, W.G., Kroneck, P.M.H., Körner, H., Jakob, W.: Eur. J. Biochem., 153, 459–467 (1985)
[7] Kristjansson, J.K., Hollocher, T.C.: Curr. Microbiol., 6, 247–251 (1981)
[8] McEwan, A.G., Greenfield, A.J., Wetzstein, H.G., Jackson, J.B., Ferguson, S.J.: J. Bacteriol., 164, 823–830 (1985)
[9] Bonin, P., Gilewicz, M., Bertrand, J.C.: Can. J. Microbiol., 35, 1061–1064 (1989)
[10] Coyne, M.S., Focht, D.D.: Appl. Environ. Microbiol., 53, 1168–1170 (1987)
[11] Aida, T., Hata, S., Kusunoki, H.: Can. J. Microbiol., 32, 543–547 (1986)
[12] Kristjansson, J.K., Hollocher, T.C.: J. Biol. Chem., 255, 704–707 (1980)
[13] Kucera, I., Boublikova, P., Dadak, V.: Collect. Czech. Chem. Commun., 49, 2709–2712 (1984)
[14] Zumft, W.G., Matsubara, T.: FEBS Lett., 148, 107–112 (1982)
[15] Snyder, S.W., Hollocher, T.C.: Biochem. Biophys. Res. Commun., 119, 588–592 (1984)
[16] Matsubara, T., Zumft, W.G.: Arch. Microbiol., 132, 322–328 (1982)
[17] Jones, A.M., Knowles, R.: Can. J. Microbiol., 36, 430–434 (1990)

Enzyme Handbook © Springer-Verlag Berlin Heidelberg 1994
Duplication, reproduction and storage in data banks are only allowed with the prior permission of the publishers

1 NOMENCLATURE

EC number
1.7.99.7

Systematic name
Nitrous-oxide:(acceptor) oxidoreductase (NO-forming)

Recommended name
Nitric-oxide reductase

Synonymes
Reductase, nitric oxide
Nitric oxide reductase
EC 1.7.99.2 (formerly)
Nitrogen oxide reductase

CAS Reg. No.
37256-43-2

2 REACTION AND SPECIFICITY

Catalysed reaction
Nitrous oxide + acceptor →
→ nitric oxide + reduced acceptor

Reaction type
Redox reaction

Natural substrates
Nitric oxide + acceptor (i.e. NO, major and critically important [3] enzyme of denitrification pathway [1, 4, 7], cytochrome c_{550} possible physiological electron-donor [5]) [1, 3–5, 7]

Substrate spectrum
1 Nitric oxide + reduced acceptor (i.e. NO, ir [1], one-electron reduction [8], very high affinity between enzyme and NO [9], in vitro acceptors: phenazine methosulfate [1], phenazine methosulfate/DTT [2], phenazine methosulfate/ascorbate [3], isoascorbate/2,3,5,6-tetramethylphenylene-1,4-diamine [5, 7], ascorbate/2,3,5,6-tetramethylphenylene-1,4-diamine [5, 6, 8]) [1–9]
2 More (no substrates are nitrate, nitrite, nitrous oxide) [1]

Enzyme Handbook © Springer-Verlag Berlin Heidelberg 1994
Duplication, reproduction and storage in data banks are only allowed with the prior permission of the publishers

Product spectrum

1 Nitrous oxide + acceptor (i.e. N_2O) [1–9]
2 ?

Inhibitor(s)

CHAPS (weak inhibition [1], no inhibition [8]) [1]; Cyanide [1]; NO (reversible) [2, 3, 7, 8]; Azide (no inhibitor, crude preparation [9]) [3, 5]; PMSF [5]; o-Phenanthroline (crude preparation [9], no inhibition [1]) [1]; Antimycin A (with NADH as electron-donor, not with ascorbate/2,3,5,6-tetra-methyl-phenylene-1,4-diamine, membrane vesicle preparation [6, 7], no inhibition [3]) [6, 7]; Myxothiazol (with NADH as electron-donor, not with isoascorbate/2,3,5,6-tetramethylphenylene-1,4-diamine, membrane vesicle preparation) [7]; Acetylene [7]; N-Bromosuccinimide [3]; p-Chloromercuribenzoate (crude preparation, partially restorable by cysteine) [9]; Diethyldithiocarbamate (crude preparation) [9]; Triton X-100 (restorable by the addition of CHAPS-solubilized extracts, Paracoccus denitrificans [8]) [3–5, 8]; Detergents (non-ionic: Lubol, Brij 35, Tergitol T15-S-9, anionic: Tergitol T-7, SDS, extensive loss of activity [3], sodium desoxycholate, cetyltrimethylammoniumbromide [4]) [3, 4]

Cofactor(s)/prostethic group(s)

Cytochrome b (requirement, the enzyme consists of a cytochrome bc complex) [3]; Cytochrome c (requirement, the enzyme consists of a cytochrome bc complex) [3]; Cytochromes (several cytochromes as judged from visible absorption spectrum) [4]; Cytochrome c_{550} (possible physiological electron donor) [5]; Phenazine methosulfate (requirement, in vitro acceptor, together with DTT [2] or ascorbate [3] it acts as reductant to NO, best electron-carrier between enzyme and ascorbate) [1–4]; N, N, N', N'-Tetramethylphenylene-1,4-diamine (activation, less effective than phenazine methosulfate [3], no cofactor [1]) [3]; 2,3,5,6-Tetramethylphenylene-1,4-diamine (activation, less effective than phenazine methosulfate [3], in vitro, together with isoascorbate and horse heart cytochrome c [5], no cofactor [1]) [3, 5]; NADH (slight activation, electron-donor, together with 2,3,5,6-tetramethylphenylene-1,4-diamine as mediator [6, 7], electrons from NADH pass down the electron-transfer chain as far as cytochrome c_1 or c_{550} [7]) [3, 6, 7]; Brilliant cresyl blue (activation, electron-donor) [9]; Methylene blue (activation, electron-donor) [9]; Leuco-forms of thionine (activation, electron-donor) [9]; Gramicidin D/ammonium acetate (stimulates NADH dependent activity) [7]; EDTA (stimulation of purified enzyme) [1]; Diethyldithiocarbamate (stimulation of purified enzyme) [1]; 2,2'-Bipyridyl (stimulation of purified enzyme) [1]; o-Phenanthroline (stimulation of purified enzyme) [1]; Soybean phospholipids (activation, not phosphatidyl choline or Triton X-100) [1]; n-Octyl-beta-D-glucopyranoside (activation, at its critical micelle concentration [1], solubilizes [2], best solubilizing agent [3]) [1–3]; Octyl-beta-D-thioglucopyranoside (activation, at its critical micelle concentration) [1]; More (reductase is linked to electron-transport-chain at cytochrome c-level [6], no cofactor: O_2, cytochrome o (co-type), tetramethylphenylene-1, 4-diamine [1], NADH [5, 9], cytochrome aa_3 [7], flavin, reduced dichlorophenol-indophenol, NADPH, NADH/FMN [9], no activation: CHAPSO [3]) [1, 3, 5–7, 9]

Metal compounds/salts

Iron (enzyme consists of heme b and c [1, 5], high-spin and low-spin ferric heme proteins [1], heme and non-heme iron, around 5.5 mol iron per MW 73000, 3 mol heme per MW 73000, 1 heme per subunit [3]) [1, 3, 5]; Ascorbate (activation, part of reducing system [2], together with phenazine methosulfate it acts as reductant [3], electron-donor, together with 2,3,5,6-tetramethylphenylene-1,4-diamine as mediator [6]) [1–4, 6]; Isoascorbate (activation, in vitro together with 2,3,5,6-tetramethylphenylene-1,4-diamine as electron-donor and horse heart cytochrome c as mediator [5], together with 2,3,5,6-tetramethylphenylene-1,4-diamine as mediator [7]) [5, 7]; Succinate (activation, electron-donor, together with 2,3,5,6-tetramethylphenylene-1,4-diamine as mediator [6, 7], no activation [9]) [6, 7]; Lactate (activation) [9]; More (no Cu^{2+} or Zn^{2+} detectable by atomic absorption spectroscopy [1], no Cu^{2+}, Co^{2+}, Mn^{2+}, Mo^{2+}, Ni^{2+}, Se^{2+} [2], no activation by malate, succinate, oxaloacetate [9]) [1, 2, 9]

Turnover number (min^{-1})

2160 (NO, pH 6.5) [3]; 2200 (NO) [1]

Specific activity (U/mg)

0.58 [7]; 6.0 [2]; 10.1 [3]; 11.0 [5]; 11.8 [1]; 22.0 (NO) [3]; 40.0 (with addition of stabilizers) [1]

K_m-value (mM)

More (far below 0.001 mM (NO), too low to measure reliably by amperometric method) [3]; 0.002 (phenazine methosulfate) [1]; 0.01 (or less [5], NO) [5, 9]; 0.06 (NO) [1]

pH-optimum

4.8 (purified) [1]; 5.0 (with phenazine methosulfate/ascorbate as reductant) [3]; 5.2 (membrane-bound) [1]; 5.3 [4]; 6.5 (assay at) [2]; 7.0–7.4 (crude preparation) [9]

Enzyme Handbook © Springer-Verlag Berlin Heidelberg 1994
Duplication, reproduction and storage in data banks are only allowed with the prior permission of the publishers

pH-range

5.5–6.0 (with phenazine methosulfate/ascorbate as reductant, about 85% of maximal activity at pH 5.5 and 50% at pH 6.0) [3]

Temperature optimum (°C)

25 (assay at) [2]; 30 (assay at) [1–4]

Temperature range (°C)

---

3 ENZYME STRUCTURE

Molecular weight

45000 (Pseudomonas stutzeri, PAGE) [1]

Subunits

Dimer (1 × 17000 + 1 × 38000, Pseudomonas stutzeri, SDS-PAGE [1], 1 × 17500 + 1 × 38000, Paracoccus denitrificans, SDS-PAGE [3], 1 × 18000 + 1 × 37000, Paracoccus denitrificans, lithium dodecylsulfate-PAGE [5]) [1, 3, 5]
More (the enzyme is composed of a tight cytochrome bc complex) [3]

Glycoprotein/Lipoprotein

–

---

4 ISOLATION/PREPARATION

Source organism

Pseudomonas stutzeri (strain ZoBell, syn. Pseudomonas perfectomarina [8]) [1, 8]; Paracoccus denitrificans (anaerobically grown, no activity in aerobically grown cells) [2, 3, 5–8]; Paracoccus halodenitrificans [4]; Achromobacter cycloclastes [8]; Pseudomonas aeruginosa (strain PAO1) [8]; Pseudomonas denitrificans [9]

Source tissue

Cell [1–9]

Localisation in source

Membranes [1–9]

Purification

Pseudomonas stutzeri (solubilized by Triton X-100, protein/detergent ratio 1:1, ion exchange chromatography) [1]; Paracoccus denitrificans (solubilized and fractionated by octyl-glucopyranoside, 0.05% w/v and 0.8% w/v, hydroxyapatite chromatography, amino acid analysis [3], solubilized by dodecylmaltoside [5]) [3, 5]; Paracoccus halodenitrificans (partial) [4]

Crystallization

–

Cloned

–

Renaturated

–

5 STABILITY

pH

Temperature (°C)

Oxidation

Purified enzyme gradually loses activity in air at room temperature or 0°C [1]

Organic solvent

Glycerol, 25% v/v, stable to [7]

General stability information

Triton X-100 stabilizes [1]; Detergents, such as Triton X-100, Lubol, Brij 35, Tergitol T15-S-9, Tergitol T-7 or SDS inactivate [3]; Triton X-100 inactivates [3, 5]; 2-Phenylethanol, 0.01% v/v, stabilizes during storage [5]; Benzamidine, 1 mM, stabilizes during storage [5]; Pepstatin, 0.0001 mM, stabilizes during storage [5]; EDTA does not stabilize [5]; Freezing in liquid nitrogen, protein-detergent complex (1:1) stable to [1]; Freezing in liquid nitrogen, stable to [3]; Freezing at –20°C and thawing, membrane-preparation stable to one cycle of, loss of activity during repeated cycles [7]; Glycerol, 25% v/v, stabilizes [7]

Storage

–24°C, stable in presence of 0.2% Triton X-100 [1]; –20°C, stable [3]; 4°C, $t_{1/2}$: 2 days [3, 5]

6 CROSSREFERENCES TO STRUCTURE DATABANKS

PIR/MIPS code

Brookhaven code

Enzyme Handbook © Springer-Verlag Berlin Heidelberg 1994
Duplication, reproduction and storage in data banks are only allowed with the prior permission of the publishers

7 LITERATURE REFERENCES

[1] Heiss, B., Frunzke, K., Zumft, W.G.: J. Bacteriol.,171,3288–3297 (1989)
[2] Turk, T., Hollocher, T.C.: Biochem. Biophys. Res. Commun.,183,983–988 (1992)
[3] Dermastia, M., Turk, T., Hollocher, T.C.: J. Biol. Chem.,266,10899–10905 (1991)
[4] Grant, M.A., Cronin, S.E., Hochstein, L.I.: Arch. Microbiol.,140,183–186 (1984)
[5] Carr, G.J., Ferguson, S.J.: Biochem. J.,269,423–429 (1990)
[6] Carr, G.J., Ferguson, S.J.: Biochem. Soc. Trans.,16,187–188 (1988)
[7] Carr, G.J., Page, M.D., Ferguson, S.J.: Eur. J. Biochem.,179,683–692 (1989)
[8] Shapleigh, J.P., Davies, K.J.P., Payne, W.J.: Biochim. Biophys. Acta,911,334–340 (1987)
[9] Miyata, M., Matsubara, T., Mori, T.: J. Biochem.,66,759–765 (1969)

1 NOMENCLATURE

EC number
1.8.1.2

Systematic name
Hydrogen-sulfide:$NADP^+$ oxidoreductase

Recommended name
Sulfite reductase (NADPH)

Synonymes
Reductase, sulfite (reduced nicotinamide adenine dinucleotide phosphate)
NADPH-sulfite reductase
NADPH-dependent sulfite reductase
H_2S-NADP oxidoreductase [19]

CAS Reg. No.
9029-35-0

2 REACTION AND SPECIFICITY

Catalysed reaction
Hydrogen sulfide + 3 $NADP^+$ + 3 H_2O →
→ hydrogen sulfite + 3 NADPH + 3 H^+

Reaction type
Redox reaction

Natural substrates
Sulfite + NADPH + H^+ (no physiological nitrite reductase [13]) [13, 18]

Substrate spectrum
1 Sulfite + NADPH (NADPH can be replaced by reduced methyl viologen [7, 13, 20], benzyl viologen [20], reduction of sulfite by reduced methyl viologen is catalyzed by hemoprotein subunit alone [8], sole compound reduced [9]) [1, 7–9, 13, 18, 20]
2 Nitrite + NADPH (NADPH can be replaced by reduced methyl viologen, the latter reaction is catalyzed by hemoprotein subunit alone [7]) [1, 7, 8, 13, 18, 20]
3 Hydroxylamine + NADPH (NADPH can be replaced by reduced methyl viologen) [13]

Enzyme Handbook © Springer-Verlag Berlin Heidelberg 1994
Duplication, reproduction and storage in data banks are only allowed with the prior permission of the publishers

4 $NADP^+$ + reduced methyl viologen [13]
5 More (reactions catalyzed by flavoprotein subunit alone: transfer of electrons from NADPH to cytochrome c, ferricyanide, dichlorphenolindophenol, menadione, FMN, FAD, O_2 [1, 8, 13, 17, 18, 20], transfer of hydrogen from NADPH to 3-acetylpyridineadenine dinucleotide phosphate [1, 8, 13], reduction of sulfite, nitrite, and hydroxylamine by NADPH requires catalytic activities of both subunits, reduction of sulfite, nitrite, and hydroxylamine by methyl viologen requires hemoprotein subunit [1, 7, 8], reactions of FNM depleted enzyme [8]) [1, 7, 8, 13, 17, 18, 20]

Product spectrum

1 Sulfide + $NADP^+$ + H_2O [13]
2 Ammonia + $NADP^+$ [13, 17]
3 Ammonia + $NADP^+$ [13]
4 NADPH + oxidized methyl viologen
5 ?

Inhibitor(s)

CN^- (severe inhibition of NADPH-sulfite,-nitrite,-hydroxylamine reductase, no inhibition of methyl viologen-sulfite reductase, small extent of inhibition of NADPH dependent reduction of cytochrome c, ferricyanide, quinones etc. [18, 20]) [7–9, 13, 16–18, 20]; CO (binding to siroheme [10]) [7–10, 13, 16, 17, 22]; p-Mercuriphenylsulfonate (NADPH-dependent reactions [8, 17]) [8, 13, 17]; Arsenite [8, 13, 17, 18, 20]; Sulfide (reduction of sulfite, nitrite, hydroxylamine by reduced methyl viologen) [8, 18]; $NADP^+$ (NADPH-dependent activities) [13, 18]; F^- [13]; Sulfate [13]; Cl^- [13]; Br^- [13]; NO_3^- [13]; SCN^- [13, 22]; p-Chloromercuribenzoate (NADPH-dependent activities) [18, 22]; 2'-AMP (NADPH-sulfite reductase and NADPH-cytochrome c reductase) [20]; Mepacrine [22]; 8-Hydroxyquinoline [22]; o-Phenanthroline [22]; Low ionic strength (NADPH-dependent activities) [4, 23]

Cofactor(s)/prostethic group(s)

FAD (4 mol per octamer of flavoprotein subunit, bound to flavoprotein subunit [1, 8, 16], 2 mol per mol holoenzyme [6], role in catalytic process [12]) [1, 6, 8, 11, 12, 18, 20, 22]; FMN (4 mol per octamer of flavoprotein subunit, bound to flavoprotein subunit [1, 8, 16], 2 mol per mol holoenzyme [6], role in catalytic process [12]) [1, 6, 8, 11, 12, 16, 18, 20, 22]; Siroheme (4 mol per tetramer of hemoprotein subunit, bound to hemoprotein subunit [3, 8, 15], 2 mol per mol of holoenzyme [6], structure [15, 27]) [1, 3, 6, 8, 15, 27]; Iron-sulfur cluster (Fe4-S4, 4 mol per tetramer of hemoprotein subunit, bound to hemoprotein subunit) [3]; NADPH (specific for as electron donor, no reaction with NADH) [16, 20]

Metal compounds/salts
EDTA (increase of activity) [9]; No divalent cation requirement [9]

Turnover number (min^{-1})
9500 (NADPH + 3-acetylpyridineadenine dinucleotide phosphate) [13]; 10300 (NADPH + FMN) [13]; 27700 (NADPH + dichlorphenolindophenol) [13]; 28100 (NADPH + ferricyanide) [13]; 19100 (NADPH + cytochrome c) [13]; 75 (NADPH + O_2) [13]; 36300 (reduced methyl viologen + $NADP^+$) [13]; 1800–2400 (sulfite + NADPH) [13, 17]; 2600–3100 (nitrite + NADPH) [13, 17]; 5900–13000 (hydroxylamine + NADPH) [13, 17]; 3900 (sulfite + reduced methyl viologen) [13]; 9000 (FAD + NADPH) [17]; More (values for reactions catalyzed by holoenzyme, hemoprotein subunit alone, and flavoprotein subunit alone with various electron acceptors) [1]

Specific activity (U/mg)
81 [1]; 2.73–2.87 [16, 17]; 1.85 [20]; More [22]

K_m-value (mM)
0.012–0.014 (sulfite) [7, 18, 20]; 0.0043–0.0074 (sulfite) [13, 17]; 0.6 (sulfite, Saccharomyces cerevisiae) [9]; 2.0 (sulfite, Saccharomyces bayanus) [9]; 0.8–1.5 (nitrite) [7, 13, 18, 20]; 4.5–10.5 (hydroxylamine) [7, 13, 18, 20]; 0.018 (NADPH, reduction of cytochrome c [18]) [17, 18]; 0.021 (NADPH, reduction of sulfite) [18]; 0.06 (cytochrome c) [18]; 0.063 (FMN) [22]; 0.083 (FAD) [22]

pH-optimum
7.0–7.5 (NADPH-sulfite reductase) [18, 20]; 7.2 (electron donor NADPH) [17]; 7.5–7.9 (NADPH-cytochrome c reductase) [18, 20]; 7.7–8.5 (methyl viologen-sulfite reductase) [18]; 7.9 (sulfite) [13]; 8.6 (nitrite) [13]; 9.5 (hydroxylamine) [13]

pH-range

Temperature optimum (°C)

Temperature range (°C)

Enzyme Handbook © Springer-Verlag Berlin Heidelberg 1994
Duplication, reproduction and storage in data banks are only allowed with the prior permission of the publishers

3 ENZYME STRUCTURE

Molecular weight

670000 (E. coli, sedimentation equilibrium centrifugation, sedimentation and diffusion coefficients) [16]
650000 (Saccharomyces cerevisiae, gel filtration, sedimentation analysis) [5]
604000 (Saccharomyces cerevisiae, gel filtration) [6]
488000 (Salmonella typhimurium, sedimentation under nondenaturing conditions) [1]
350000 (Saccharomyces cerevisiae, calculation from FAD and FMN content, sedimentation coefficient) [18, 20]
300000 (Saccharomyces cerevisiae, gel filtration, sedimentation analysis at low ionic strength, dissociated) [5]

Subunits

Dodecamer ($alpha_8$,$beta_4$, 8 × 58000–60000 + 4 × 54000–57000, E. coli, sedimentation under denaturing conditions [1, 8, 11], SDS-PAGE [11, 25]) [1, 8, 11, 25]
Tetramer ($alpha_2$,$beta_2$, 2 × 116000 + 2 × 167000, Saccharomyces cerevisiae, SDS-PAGE) [6]

Glycoprotein/Lipoprotein

–

---

4 ISOLATION/PREPARATION

Source organism

E. coli [1–3, 7, 8, 10–13, 15–17, 24–26]; Salmonella typhimurium [1, 8]; Saccharomyces cerevisiae (var. ellipsoideus [9]) [4–6, 8, 9, 18, 20–23]; Neurospora crassa [8]; Saccharomyces bayanus [9]; Desulfotomaculum nigrificans [14]; Salmonella pullorum [19]

Source tissue

Cell [1–26]

Localisation in source

Soluble part of cell [18, 20]

Purification

E. coli (preparation of FMN depleted enzyme [12]) [12, 16, 17]; Salmonella typhimurium [1]; Saccharomyces bayanus (partial) [9]; Saccharomyces cerevisiae (var. ellipsoideus (partial) [9], preparation of mutant form lacking FAD and hence with no NADPH-sulfite reductase activity but with methyl viologen-sulfite reductase activity [21]) [9, 18, 20, 21]

Crystallization
(hemoprotein subunit [26], x-ray structure of hemoprotein subunit [3]) [3, 26]

Cloned
[24, 25]

Renaturated
–

5 STABILITY

pH

Temperature (°C)
25 (stable at) [19]; 45 (instable at) [19]; 50 (inactivation above) [23]

Oxidation

Organic solvent
Acetone: inactivation [18]; Ethanol: inactivation [18]

General stability information
Low ionic strength such as 0.01 M phosphate buffer destroys NADPH-dependent activities [5, 18]

Storage
–20°C, 0.3 M phosphate buffer pH 7.3, 1 year, or 1 week at 4°C [18, 20]; –15°C, 0.05 M potassium phosphate buffer pH 7.7, 10 mM EDTA, at least 6 months [17]; –15°C 0.05 M potassium phosphate buffer pH 7.7, 0.1 mM EDTA, 2 years, or 1 month at 4°C [16]; –15°C, reduced hemoprotein subunit or reduced flavoprotein subunit [11]

6 CROSSREFERENCES TO STRUCTURE DATABANKS

PIR/MIPS code
PIR2: B34231 (Escherichia coli); PIR2: A34231 (Salmonella typhimurium); PIR2: S14220 (alpha chain Escherichia coli (strain K12) (fragment)); PIR3: B34354 (hemoprotein component Escherichia coli); PIR3: A34354 (hemoprotein component Salmonella typhimurium)

Brookhaven code

Enzyme Handbook © Springer-Verlag Berlin Heidelberg 1994
Duplication, reproduction and storage in data banks are only allowed with the prior permission of the publishers

7 LITERATURE REFERENCES

[1] Ostrowski, J., Barber, M.J., Rueger, D.C., Miller, B. E., Siegel, L.M., Kredich, N.M.: J. Biol. Chem.,264,15796–15808 (1989)
[2] Young, L.J., Siegel, L.M.: Biochemistry,27,4991–4999 (1988)
[3] McRee, D.E., Richardson, D.C., Richardson, J.S., Siegel, L.M.: J. Biol. Chem.,261,10277–10281 (1986)
[4] Kobayashi, K., Yoshimoto, A.: Biochim. Biophys. Acta,709,38–45 (1982)
[5] Kobayashi, K., Yoshimoto, A.: Biochim. Biophys. Acta,709,46–52 (1982)
[6] Kobayashi, K., Yoshimoto, A.: Biochim. Biophys. Acta,705,348–356 (1982)
[7] Siegel, L.M., Rueger, D.C., Barber, M.J., Krueger, R. J., Orme-Johnson, N.R., Orme-Johnson, W.H.: J. Biol. Chem.,257,6343–6350 (1982)
[8] Hatefi, Y., Stiggall, D.L. in "The Enzymes" (Boyer, P.D., ed.) 13,175–297 (1976) (Review)
[9] Dott, W., Trüper, H.G.: Arch. Microbiol.,108,99–104 (1976)
[10] Murphy, M.J., Siegel, L.M., Kamin, H.: J. Biol. Chem.,249,1610–1614 (1974)
[11] Siegel, L.M., Davis, P-S.: J. Biol. Chem.,249,1587–1598 (1974)
[12] Faeder, E.J., Davis, P.S., Siegel, L.M.: J. Biol. Chem.,249,1599–1609 (1974)
[13] Siegel, L.M., Davis, P.S., Kamin, H.: J. Biol. Chem.,249,1572–1586 (1974)
[14] Murphy, M.J., Siegel, L.M.: J. Biol. Chem.,248,6911–6919 (1973)
[15] Murphy, M.J., Siegel, L.M., Kamin, H.: J. Biol. Chem.,248,2801–2814 (1973)
[16] Siegel, L.M., Murphy, M.J., Kamin, H.: J. Biol. Chem.,248,251–264 (1973)
[17] Siegel, L.M., Kamin, H.: Methods Enzymol.,17B,539–545 (1971)
[18] Yoshimoto, A., Naiki, N., Sato, R.: Methods Enzymol.,17B,520–528 (1971)
[19] Hoeksema, W.D., Schoenhard, D.E.: J. Bacteriol.,108,154–158 (1971)
[20] Yoshimoto, A., Sato, R.: Biochim. Biophys. Acta,153,555–575 (1968)
[21] Yoshimoto, A., Sato, R.: Biochim. Biophys. Acta,153,576–588 (1968)
[22] Prabhakararao, K., Nicholas, D.J.D.: Biochim. Biophys. Acta,180,253–263 (1969)
[23] Yoshimoto, A., Sato, R.: Biochim. Biophys. Acta,220,190–205 (1970)
[24] Tei, H., Murata, K., Kimura, A.: Biotechnol. Appl. Biochem.,12,212–216 (1990)
[25] Li, C., Peck, H.D., Przybyla, A.E.: Gene,53,227–234 (1987)
[26] McRee, D.E., Richardson, D.C.: J. Mol. Biol.,154,179–180 (1982)
[27] Scott, A.I., Irwin, A.J., Siegel, L.M., Schoolery, J. N.: J. Am. Chem. Soc.,100,7987–7994 (1978)

1 NOMENCLATURE

EC number
1.8.1.3

Systematic name
Hypotaurine:NAD^+ oxidoreductase

Recommended name
Hypotaurine dehydrogenase

Synonymes

CAS Reg. No.
37256-46-5

---

2 REACTION AND SPECIFICITY

Catalysed reaction
Hypotaurine + H_2O + NAD^+ →
→ taurine + NADH

Reaction type
Redox reaction

Natural substrates
Hypotaurine + H_2O + NAD^+ [1]

Substrate spectrum
1 Hypotaurine + H_2O + NAD^+ [1]

Product spectrum
1 Taurine + NADH [1]

Inhibitor(s)
Phosphate [1]; KCN [1]; Ammonium sulfate [1]

Cofactor(s)/prostethic group(s)
NAD^+ [1]; Heme (hemoprotein) [1]

Metal compounds/salts
Molybdenum hemoprotein [1]

Turnover number (min^{-1})

Enzyme Handbook © Springer-Verlag Berlin Heidelberg 1994
Duplication, reproduction and storage in data banks are only allowed with the prior permission of the publishers

Specific activity (U/mg)

K_m-value (mM)

pH-optimum
7.5–7.8 [1]

pH-range

Temperature optimum (°C)

Temperature range (°C)

---

3 ENZYME STRUCTURE

Molecular weight

Subunits

Glycoprotein/Lipoprotein
–

---

4 ISOLATION/PREPARATION

Source organism
Rat [1]

Source tissue
Liver [1]

Localisation in source

Purification

Crystallization
–

Cloned
–

Renaturated
–

---

5 STABILITY

pH

Temperature (°C)

Oxidation

Organic solvent

General stability information

Storage

6 CROSSREFERENCES TO STRUCTURE DATABANKS

PIR/MIPS code

Brookhaven code

7 LITERATURE REFERENCES

[1] Sumizu, K.: Biochim. Biophys. Acta,63,210–212 (1962)

Enzyme Handbook © Springer-Verlag Berlin Heidelberg 1994
Duplication, reproduction and storage in data banks are only allowed with the prior permission of the publishers

1 NOMENCLATURE

EC number
1.8.1.4

Systematic name
Dihydrolipoamide:NAD^+ oxidoreductase

Recommended name
Dihydrolipoamide dehydrogenase

Synonymes
Dehydrogenase, lipoamide
Diaphorase
Lipoyl dehydrogenase
Lipoamide reductase
Dehydrolipoate dehydrogenase
Dihydrolipoic dehydrogenase
Dihydrolipoyl dehydrogenase
Dihydrothioctic dehydrogenase
Lipoate dehydrogenase
Lipoamide dehydrogenase (NADH)
Lipoic acid dehydrogenase
Lipoamide oxidoreductase (NADH)
LDP-Val [7]
LDP-Glc [7]
EC 1.6.4.3 (formerly)

CAS Reg. No.
9001-18-7

2 REACTION AND SPECIFICITY

Catalysed reaction
Dihydrolipoamide + NAD^+ →
→ lipoamide + NADH (mechanism [33, 46])

Reaction type
Redox reaction

Enzyme Handbook © Springer-Verlag Berlin Heidelberg 1994
Duplication, reproduction and storage in data banks are only allowed with the prior permission of the publishers

Natural substrates

Dihydrolipoamide + NAD^+ (physiological direction of reaction [33], dihydrolipoyl moiety covalently bound to epsilon amino group of a lysine residue in lipoyl reductase transacetylase [41, 44, 46], lipoic acid containing protein P2 which is part of glycine decarboxylase complex [9, 11], dihydrolipoamide reductase takes part in oxidative decarboxylation of pyruvate as E3 component of pyruvate dehydrogenase complex [10, 11, 33, 48, 49], in alpha-ketoglutarate dehydrogenase complex [10, 11], in branched-chain oxoacid dehydrogenase complex [10, 11], possible functions in membranes [13]) [9–11, 13, 33, 41, 44, 46, 48, 49]

Substrate spectrum

1 Lipoamide + NADH (r [2, 33, 41]) [1, 2, 4, 9, 12, 14, 22, 30, 33, 41, 44, 46]
2 Lipoic acid + NADH (lipoic acid can be replaced by lipoic acid containing protein P2 [9]) [9, 12, 41, 46]
3 NAD(P)H + thio-$NAD(P)^+$ [4]
4 More (diaphorase type reactions with NADH as electron donor: reduction of benzyl viologen [1, 4], dichlorophenolindophenol [1, 4, 24, 43, 44, 46], ferricyanide [1, 4, 22, 24, 43, 46], menadione [1, 4], methylene blue [1, 4, 44, 46], nitro blue tetrazolium [4], O_2 (week, under aerobic conditions) [4], iodonitrotetrazolium chloride [24], vitamin K_5 [43], ferric chelate of o-phenanthroline [46], vitamin K_3 [46], oxidized pyridine nucleotide analogs [46], transhydrogenase reactions between nicotinamide nucleotides [4, 11, 22, 31, 41], one electron transfer between NADPH and ferricyanide [22], nitroreductase [11]) [1, 4, 11, 22, 24, 31, 41, 43, 44, 46]

Product spectrum

1 Dihydrolipoamide + NAD^+ [41]
2 Dihydrolipoic acid + NAD^+
3 $NAD(P)^+$ + thio-NAD(P)H
4 ?

Inhibitor(s)

Cu^{2+} [4, 9, 31, 33, 41, 46]; Cd^{2+} [4, 31, 41, 45]; Zn^{2+} [4, 9]; Co^{2+} [4]; Pb^{2+} [4]; Ni^{2+} [4, 9]; Hg^{2+} [4]; Arsenite [4, 33, 41, 43, 45, 46]; Phenylarsinoxide [4]; 5,5'-Dithiobis-(2-nitrobenzoate) [4, 34, 41]; Ca^{2+} [9]; p-Aminophenyldichloroarsine (in presence of NADH) [14, 16, 18]; Iodoacetic acid (in presence of NADH or dihydrolipoamide) [21]; NADH [24, 31, 40, 46]; p-Aminophenylarsenoxide [27]; p-[(Bromoacetyl)-amino]phenylarsenoxide [27]; p-Chloromercuribenzoate [31, 41, 43]; Pteridines [42]

Cofactor(s)/prostethic group(s)

FAD (1 mol per mol of subunit [1, 4, 9, 10, 20, 24, 26, 30, 31], non-covalently bound [1, 30]) [1, 2, 4, 9, 10, 12, 20, 24, 26, 28, 30, 31, 44–46, 48, 49]; NADH (highly specific for) [1, 2, 4, 9, 12, 14, 21, 30, 40, 41, 43–46]; NADPH (Clostridium sporogenes active only with NADPH, Clostridium cylindrosporum active with NADPH and NADH) [4]

Metal compounds/salts

NaCl (required for activity) [18]

Turnover number (min^{-1})

More (comparison of values for enzyme from 2-oxoglutarate dehydrogenase complex and pyruvate dehydrogenase complex) [41]

Specific activity (U/mg)

297 [2]; 488 (Clostridium cylindrosporum) [4]; 36 (Clostridium sporogenes) [4]; 214 (Peptostreptococcus glycinophilus) [4]; 68 [7]; More (assay method [39, 47]) [1, 6, 9, 10, 14, 16, 17, 24, 26, 28, 30, 31, 39–42, 44, 47]

K_m-value (mM)

0.15 (NADH [1], NAD^+ [6, 24]) [1, 6, 24]; 3.7 (lipoamide) [1]; 0.62 (NAD^+) [1]; 0.13–0.83 (dihydrolipoamide) [1, 14, 16, 24, 28, 34]; 0.023 (NADH) [2]; 0.52 (NAD^+) [14]; 0.2–0.35 (NAD^+) [16, 33]; 0.017 (lipoamide) [21]; 1.1 (NAD^+, Natronobacterium gregoryi) [21]; 2 (lipoic acid) [26, 36, 43]; 1.83 (NAD^+, free enzyme) [28]; 0.4 (NAD^+, enzyme bound to pyruvate dehydrogenase complex) [28]; 0.28 (dihydrolipoamide, free enzyme) [28]; 0.08 (dihydrolipoamide, enzyme bound to pyruvate dehydrogenase complex) [28]; 0.01 (NADH) [31]; 0.037 (NAD^+) [31]; 5.0 (lipoamide, K_m increases with increasing hydrophobicity of side chain) [36, 42]

pH-optimum

5.7 (reduction of lipoic acid) [41]; 6.0 (reduction of lipoic acid) [43]; 6.1 (reduction of lipoyllysine) [41]; 6.2–6.5 (reduction of lipoamide) [31, 40, 42, 44]; 6.5 (oxidation of NADH) [33]; 7.2 [1]; 7.4 (reduction of dichlorophenolindophenol) [31]; 7.5 (reduction of thio-NAD^+) [31]; 7.5–8.1 (lipoylglycineamide) [43]; 8.0 (oxidation of dihydrolipoamide [44]) [28, 44]; 8.2 (oxidation of dihydrolipoamide) [30]; 8.5 [24]

pH-range

5.0–10.0 (less than 10% of maximal activity above and below) [30]; 7–8 [28]

Temperature optimum (°C)

Temperature range (°C)

Enzyme Handbook © Springer-Verlag Berlin Heidelberg 1994
Duplication, reproduction and storage in data banks are only allowed with the prior permission of the publishers

3 ENZYME STRUCTURE

Molecular weight

125000 (Pseudomonas putida, LPD-Glc, gel filtration) [20]
110000–118000 (Thermoplasma acidophilum, gel filtration [14], Saccharomyces cerevisiae, thin layer gel filtration [24], Pseudomonas putida, LPD-Glc, HPLC [20], E. coli, sucrose gradient centrifugation [28], sedimentation and diffusion coefficients [45], Halobacterium halobium, Halobacterium volcanii, Natronobacterium pharaonis, Natronobacterium gregoryi, Natronococcus occultus, gel filtration [21], Pelobacter carbinolicus, gradient gel electrophoresis [1]) [1, 14, 20, 21, 24, 28, 45]
98000–105000 (Peptostreptococcus glycinophilus, Clostridium cylindrosporum, Clostridium sporogenes, gel filtration [4], Ascaris lumbricoides, gel filtration [31], Pseudomonas putida, LPD-Val, gel filtration [20], Malbranchea pulchella, thin layer gel filtration [30], Trypanosoma cruzi, gel filtration [2], Azotobacter vinelandii, sedimentation equilibrium centrifugation [19], rat, gel filtration [26]) [2, 4, 19, 20, 26, 30, 31]
93000 (Pseudomonas putida, LPD-Val, HPLC) [20]
88000 (E. coli, enzyme from mutant lacking genes for pyruvate dehydrogenase and 2-oxoglutarate dehydrogenase complex, gel filtration) [6]
65000–70000 (Eubacterium acidaminophilum, native gel electrophoresis, sucrose gradient centrifugation, gel filtration) [9]

Subunits

Dimer (2 × 49193, Bacillus stearothermophilus, calculation from nucleotide sequence [3], 2 × 50554 (or 2 × 51274 if FAD is included), E. coli, calculation from nucleotide sequence [50], 2 × 49000–60000, Pelobacter carbinolicus, SDS-PAGE, N-terminal amino acid sequences identical [1], Trypanosoma cruzi, SDS-PAGE, N-terminal amino acid sequence [2], Clostridium sp., Peptostreptococcus glycinophilus, SDS-PAGE [4], Pseudomonas putida, SDS-PAGE [10, 20], Halobacterium halobium, SDS-PAGE [12, 18], rat, SDS-PAGE [17, 26], Bacillus subtilis, SDS-PAGE [23], Saccharomyces cerevisiae, SDS-PAGE [24], Salmonella typhimurium, SDS-PAGE [25], E. coli, SDS-PAGE [28], Malbranchea pulchella, SDS-PAGE, gel filtration in urea [30], Ascaris lumbricoides, SDS-PAGE [31], pig, amino acid analysis, sequences of tryptic and peptic peptides [34], 2 × 46000, E. coli, enzyme from mutant lacking genes for pyruvate dehydrogenase and 2-oxoglutarate dehydrogenase complexes [6], 2 × 34500, Eubacterium acidaminophilum, SDS-PAGE [9]) [1–4, 6, 9, 10, 12, 17, 18, 20, 23–26, 28, 30, 31, 34, 50]
Dodecamer (12 × 55000, mammalian pyruvate dehydrogenase complex, mammalian 2-oxoglutarate dehydrogenase complex) [51]

Glycoprotein/Lipoprotein

–

4 ISOLATION/PREPARATION

Source organism

Azotobacter vinelandii [19]; Pelobacter carbinolicus [1]; Trypanosoma cruzi [2]; Bacillus stearothermophilus [3, 29]; Clostridium cylindrosporum [4]; Clostridium sporogenes [4]; Peptostreptococcus glycinophilus [4]; Trypanosoma brucei [5, 16]; E. coli (additional form, no part of pyruvate dehydrogenase or 2-oxoglutarate dehydrogenase complex [6]) [6, 27, 28, 32, 36, 40, 45, 50, 56]; Pseudomonas aeruginosa [7]; Bacillus subtilis [8, 23, 40]; Eubacterium acidaminophilum [9]; Pseudomonas putida [7, 10, 20]; Halobacterium halobium [12, 18, 21]; Thermoplasma acidophilum [14]; Saccharomyces cerevisiae [15, 24, 36, 40]; Rat [17, 26]; Pig [22, 34, 36–39, 41, 42, 47]; Salmonella typhimurium [25]; Malbranchea pulchella [30]; Ascaris lumbricoides [31]; Saccharomyces carlsbergensis [35]; Pseudomonas fluorescens [40]; Azotobacter agilis [40]; Neurospora crassa [40]; Serratia marcescens [40]; Spinacia oleracea [43]; Candida krusei [44]; Halobacterium volcanii [21]; Natronobacterium pharaonis [21]; Natronobacterium gregoryi [21]; Natronococcus occultus [21]; Archaebacteria (including halophiles, thermophiles, methanogenes) [13]; Eubacteria (overview [11, 33]) [11, 13, 33]; Eukaryotes (overview [11, 33]) [11, 13, 33]; Mammals [51]; Rhodospirillum rubrum [52]; Broccoli [53]; Cauliflower [53]; Hansenula miso [54]; Pigeon [55]

Source tissue

Heart [22, 34, 36–39, 41, 47]; Cell [1, 4]; Epimastigotes [2]; Liver [17, 26]; Mycelium [30]; Muscle [31, 55]; Brain [42]; Leaves [43]; Blood stream form (of Trypanosoma) [16]; Floral buds [53]; More (overview mammalian tissues) [33]

Localisation in source

Mitochondria [17, 26, 31]; Soluble part of cell [1, 4]; Cytoplasm [6]; Plasma membrane [5, 13, 14, 16]; Chloroplasts [43]; Cytosol (small amount) [26]

Purification

Pelobacter carbinolicus [1]; Trypanosoma cruzi [2]; Clostridium cylindrosporum [4]; Clostridium sporogenes [4]; Peptostreptococcus glycinophilus [4]; E. coli [6, 28, 36, 45]; Pseudomonas aeruginosa [7]; Pseudomonas putida (LDP-Val, i.e. E3 component of branched-chain keto acid dehydrogenase complex [7], LDP-3, part of unknown complex [10]) [7, 10]; Halobacterium halobium [12]; Rat (from branched-chain keto acid dehydrogenase complex [17]) [17, 26]; Azotobacter vinelandii (from

Enzyme Handbook © Springer-Verlag Berlin Heidelberg 1994
Duplication, reproduction and storage in data banks are only allowed with the prior permission of the publishers

pyruvate dehydrogenase complex) [19]; Salmonella typhimurium (from pyruvate dehydrogenase complex) [25]; Bacillus stearothermophilus (from pyruvate dehydrogenase complex) [29]; Malbranchea pulchella [30]; Ascaris lumbricoides (two forms: L-I, free form, L-II, bound to pyruvate dehydrogenase complex) [31]; Saccharomyces carlsbergensis [35]; Pig (multiple charge isomers [37], preparation of FAD depleted enzyme [38, 39], from alpha-keto acid dehydrogenase complex [41]) [37–39, 41, 42, 47]; Spinacia oleracea [43]; Candida krusei [44]; Neurospora crassa [40]; Pseudomonas fluorescens [40]; Hansenula miso (pyruvate dehydrogenase complex composed of EC 1.2.4.1, EC 1.8.1.4, EC 2.3.1.12) [54]; Pigeon [55]; Rhodospirillum rubrum (pyruvate dehydrogenase complex composed of EC 1.2.4.1, EC 1.8.1.4, EC 2.3.1.12) [52]; Broccoli (pyruvate dehydrogenase complex composed of EC 1.2.4.1, EC 1.8.1.4, EC 2.3.1.12) [53]; Cauliflower (pyruvate dehydrogenase complex composed of EC 1.2.4.1, EC 1.8.1.4, EC 2.3.1.12) [53]

Crystallization

(complex of EC 1.8.1.4 and EC 2.3.1.12, i.e. components E3 and E2 of pyruvate dehydrogenase complex [56]) [44, 56]

Cloned

[3, 15]

Renaturated

(reconstitution of pyruvate dehydrogenase complex composed of EC 1.2.4.1, EC 2.3.1.12, EC 1.8.1.4) [19]

---

5 STABILITY

pH

Temperature (°C)

60 (stable at) [12]; 80 (stability depending on ionic strength of buffer [12], 6 min, 50% activity [30]) [12, 30]; 75 (half-life 1–20 min, depending on presence of NAD^+ (stabilization) and dihydrolipoamide (destabilization)) [30]; 95 (15 min in presence of 4 M NaCl, no loss of activity) [21]

Oxidation

Organic solvent

General stability information

FMN stabilizes FAD depleted oxidized form of enzyme [33]; Urea, 6 M, stable [33]; Guanidine-HCl denaturation is reversible [33]; Stability of reduced enzyme is species dependent [33]; Reduced form less stable than oxidized form [18]

Storage

–20°C, or 4°C, 50 mM potassium phosphate buffer pH 7.6, 50% v/v glycerol [24]; –20°C, 6 months 60–90% loss of activity depending on species [4]; –20°C, lyophilized or dissolved in 10 mM Tris-HCl pH 7.5 stable, but complete loss of activity at 4°C [30]; 4°C, ammonium sulfate precipitate, several months [2]; Room temperature, 1 year, no loss of activity [12]; Decline of transhydrogenase activity during storage [31]

6 CROSSREFERENCES TO STRUCTURE DATABANKS

PIR/MIPS code

PIR1: DEAVHL (Azotobacter vinelandii); PIR3: S13839 (Bacillus stearothermophilus); PIR2: A28514 (Bacillus stearothermophilus (fragment)); PIR3: A35156 (Clostridium cylindrosporum (fragment)); PIR3: B35156 (Clostridium sporogenes (fragment)); PIR1: DEECLP (Escherichia coli); PIR3: D35156 (Eubacterium acidaminophilum (fragment)); PIR3: S22384 (Garden pea); PIR3: A39138 (Pelobacter carbinolicus (fragment)); PIR3: C35156 (Peptostreptococcus glycinophilus (fragment)); PIR2: A21524 (Pig (fragments)); PIR1: DEPSLP (Pseudomonas putida); PIR3: S13883 (Rat); PIR3: S19723 (Staphylococcus aureus); PIR3: S19685 (3 Pseudomonas putida); PIR3: A39406 (LPD-glc Pseudomonas putida); PIR3: S18152 (precursor Garden pea); PIR1: DEHULP (precursor Human); PIR1: DEPGLP (precursor Pig); PIR2: A30151 (precursor Yeast (Saccharomyces cerevisiae))

Brookhaven code

3LAD (Azotobacter vinelandii)

7 LITERATURE REFERENCES

[1] Oppermann, F.B., Schmidt, B., Steinbüchel, A.: J. Bacteriol.,173,757–767 (1991)
[2] Lohrer, H., Krauth-Siegel, R.L.: Eur. J. Biochem.,194 ,863–869 (1990)
[3] Borges, A., Hawkins, C.F., Packman, L.C., Perham, R.N. : Eur. J. Biochem.,194,95–102 (1990)
[4] Dietrichs, D., Andreesen, J.R.: J. Bacteriol.,172,243–251 (1990)
[5] Jackman, S.A., Hough, D.W., Danson, M.J., Stevenson, K.J., Opperdoes, F.R.: Eur. J. Biochem.,193,91–95 (1990)
[6] Richarme, G.: J. Bacteriol.,171,6580–6585 (1989)
[7] Sokatch, J.R.: Methods Enzymol.,166,342–350 (1988)
[8] Perham, R.N., Lowe, P.N.: Methods Enzymol.,166,330–342 (1988)
[9] Freudenberg, W., Dietrichs, D., Lebertz, H., Andreesen, J.R.: J. Bacteriol.,171,1346–1354 (1989)
[10] Burns, G., Sykes, P.J., Hatter, K., Sokatch, J.R.: J. Bacteriol.,171,665–668 (1989)
[11] Carothers, D.J., Pons, G., Patel, M.S.: Arch. Biochem. Biophys.,268,409–425 (1989)

Enzyme Handbook © Springer-Verlag Berlin Heidelberg 1994
Duplication, reproduction and storage in data banks are only allowed with the prior permission of the publishers

[12] Sundquist, A.R., Fahey, R.C.: J. Bacteriol.,170,3459–3467 (1988)
[13] Danson, M.J.: Biochem. Soc. Trans.,16,87–89 (1988)
[14] Smith, L.D., Bungard, S.J., Danson, M.J., Hough, D.W. : Biochem. Soc. Trans.,15,1097 (1987)
[15] Roy, D.J., Dawes, I.W.: J. Gen. Microbiol.,133,925–933 (1987)
[16] Danson, M.J., Conroy, K., McQuattie, A., Stevenson, K.J.: Biochem. J.,243,661–665 (1987)
[17] Ono, K., Hakozaki, M., Kimura, A., Kochi, H.: J. Biochem.,101,19–27 (1987)
[18] Danson, M.J., McQuattie, A., Stevenson, K.J.: Biochemistry,25,3880–3884 (1986)
[19] Bosma, H.J., De Kok, A., Westphal, A.H., Veeger, C.: Eur. J. Biochem.,142,541–549 (1984)
[20] Delaney, R., Burns, G., Sokatch, J.R.: FEBS Lett.,168,265–270 (1984)
[21] Danson, M.J., Eisenthal, R., Hall, S., Kessell, S.R., Williams, D.L.: Biochem. J.,218,811–818 (1984)
[22] Tsai, C.S., Wand, A.J., Templeton, D.M., Weiss, P.M.: Arch. Biochem. Biophys.,225,554–561 (1983)
[23] Hodgson, J.A., Lowe, P.N., Perham, R.N.: Biochem. J.,211,463–472 (1983)
[24] Heinrich, P., Ronft, H., Schartau, W., Kresze, G.-B.: Hoppe-Seyler's Z. Physiol. Chem.,364,41–50 (1983)
[25] Seckler, R., Binder, R., Bisswanger, H.: Biochim. Biophys. Acta,705,210–217 (1982)
[26] Matuda, S., Saheki, T.: J. Biochem.,91,553–561 (1982)
[27] Adamson, S.R., Stevenson, K.J.: Biochemistry,20,3420–3424 (1981)
[28] Schmincke-Ott, E., Bisswanger, H.: Eur. J. Biochem.,114,413–420 (1981)
[29] Henderson, C.E., Perham, R.N.: Biochem. J.,189,161–172 (1980)
[30] McKay, D.J., Stevenson, K.J.: Biochemistry,18,4702–4707 (1979)
[31] Komuniecki, R., Saz, H.J.: Arch. Biochem. Biophys.,196,239–247 (1979)
[32] Perham, R.N., Harrison, R.A., Brown, J.P.: Biochem. Soc. Trans.,6,47–50 (1978)
[33] Williams, C.H. in "The Enzymes",3rd. Ed. (Boyer, P. D., ed.) 13,89–173 (1976) (Review)
[34] Matthews, R.G., Arscott, L.D., Williams, C.H.: Biochim. Biophys. Acta,370,26–38 (1974)
[35] Wais, U., Gillmann, U., Ulrich, J.: Hoppe-Seyler's Z. Physiol. Chem.,354,1378–1388 (1973)
[36] Scouten, W.H., Torok, F., Gitomer, W.: Biochim. Biophys. Acta,309,521–524 (1973)
[37] Cohn, M.L., McManus, I.R.: Biochim. Biophys. Acta,276,70–84 (1972)
[38] Kalse, J.F., Veeger, C.: Biochim. Biophys. Acta,159,244–256 (1968)
[39] Veeger, C., Visser, J.: Methods Enzymol.,18B,582–590 (1971)
[40] Scouten, W.H., McManus, I.R.: Biochim. Biophys. Acta,227,248–263 (1971)
[41] Koike, M., Hayakawa, T.: Methods Enzymol.,18A,298–307 (1970)
[42] Millard, S.A., Kubose, A., Gal, E.M.: J. Biol. Chem.,244,2511–2515 (1969)
[43] Jacobi, G., Öhlers, U.: Z. Pflanzenphysiol.,58,193–206 (1968)
[44] Kawahara, Y., Misaka, E., Nakanishi, K.: J. Biochem.,63,77–82 (1963)
[45] Reed, L.J., Willms, C.R.: Methods Enzymol.,9,247–265 (1966)
[46] Massey, V. in "The Enzymes",2nd. Ed. (Boyer, P.D., Lardy, H., Myrbäck, K., eds.) 7,275–306 (1963) (Review)

[47] Massey, V.: Biochim. Biophys. Acta,37,314–322 (1960)
[48] Perham, R.N., Pakman, L.C., Radford, S.E.: Biochem. Soc. Symp.,54,67–81 (1987) (Review)
[49] Perham, R.N., Packman, L.C.: Ann. N.Y. Acad. Sci.,573,1–20 (1989) (Review)
[50] Guest, J.R., Darlison, M.G., Spencer, M.E., Stephens, P.E.: Biochem. Soc. Trans.,12,220–223 (1984)
[51] Yeaman, S.J.: Trends Biochem. Sci.,11,293–296 (1986) (Review)
[52] Lüderitz, R., Klemme, J.-H.: Z. Naturforsch.,32c,351–361 (1977)
[53] Randall, D.D.: Methods Enzymol.,89,408–414 (1982)
[54] Harada, R., Hirabayashi, T.: Methods Enzymol.,89,420–423 (1982)
[55] Furuta, S., Hashimoto, T.: Methods Enzymol.,89,414–420 (1982)
[56] Fuller, C.C., Reed, L.J., Oliver, R.M., Hackert, M.L.: Biochem. Biophys. Res. Commun.,90,431–438 (1979)

Enzyme Handbook © Springer-Verlag Berlin Heidelberg 1994
Duplication, reproduction and storage in data banks are only allowed with the prior permission of the publishers

1 NOMENCLATURE

EC number
1.8.2.1

Systematic name
Sulfite:ferricytochrome-c oxidoreductase

Recommended name
Sulfite dehydrogenase

Synonymes
Dehydrogenase, sulfite
Sulfite cytochrome c reductase
Sulfite-cytochrome c oxidoreductase
Sulfite oxidase
More (enzyme nomenclature) [2]

CAS Reg. No.
37256-47-6

2 REACTION AND SPECIFICITY

Catalysed reaction
Sulfite + 2 ferricytochrome c + H_2O →
→ sulfate + 2 ferrocytochrome c + 4 H^+

Reaction type
Redox reaction

Natural substrates
Sulfite + ferricytochrome c (part of enzyme complex oxidizing thiosulfate to sulfate) [2–4]

Substrate spectrum
1 Sulfite + ferricytochrome c + H_2O (highly specific for sulfite, ferricytochrome c replaceable by $K_3[Fe(CN)]_6$ [7–10], cytochromes from different sources [2, 6, 7, 13], no O_2 or nitrate as acceptor [10, 13], no methylene blue as acceptor [13]) [1–15]

Product spectrum
1 Sulfate + ferrocytochrome c + H^+

Enzyme Handbook © Springer-Verlag Berlin Heidelberg 1994
Duplication, reproduction and storage in data banks are only allowed with the prior permission of the publishers

Inhibitor(s)

p-Hydroxymercuribenzoate [2, 7]; CN^- [2, 9]; Thiol-binding reagents (reactivation by dithiothreitol [10]) [2, 8, 10]; Monovalent anions [7, 13]; Hg^{2+} [7]; AsO_2^- [7]; N-Ethylmaleimide [7]; Ionic detergents [8]; Phosphate [2, 13, 14]; p-Chloromercuribenzoate [13]; 2,2'-Bipyridyl [13]; o-Phenanthroline [13]; EDTA [13]

Cofactor(s)/prostethic group(s)

Cytochrome c-551 [2, 15]; No cofactor (as contradicting cofactor requirements have been reported for the enzyme from Thiobacillus novellus an unequivocal classification according to EC 1.8.2.1 or EC 1.8.3.1 is impossible [7, 9]) [7]; Heme (as contradicting cofactor requirements have been reported for the enzyme from Thiobacillus novellus an unequivocal classification according to EC 1.8.2.1 or EC 1.8.3.1 is impossible [7, 9]) [9]; No heme [8]

Metal compounds/salts

Molybdenum (as contradicting cofactor requirements have been reported for the enzyme from Thiobacillus novellus an unequivocal classification according to EC 1.8.2.1 or EC 1.8.3.1 is impossible [7, 9]) [9]; Phosphate (activation) [8]; Non-heme iron [13]; More (no iron or molybdenum [8], no ferric or ferrous ions [14]) [8, 14]

Turnover number (min^{-1})

Specific activity (U/mg)

128 (cytochrome c) [2]; 21.1 [3]; 1.5 [4]; More [9, 10, 13, 14]

K_m-value (mM)

0.014 (sulfite) [2]; 0.04 (sulfite with cytochrome c) [7]; 0.02 (sulfite with $K_3[Fe(CN)]_6$, similar value [10]) [7]; 1 (sulfite with $K_3[Fe(CN)]_6$) [8]; 0.06 (sulfite with cytochrome c, similar value [13, 15]) [8]; 0.42 ($K_3[Fe(CN)]_6$) [8]; 0.58 (sulfite, with cytochrome c) [14]; 0.25 ($K_3[Fe(CN)]_6$) [14]; 0.0018 (yeast cytochrome c) [15]

pH-optimum

7.5 [15]; 7.8 (cytochrome c) [8]; 8.0 [2, 7]; 8.3 [10]; 8.5 [9]; More (no specific optimum with ferricyanide) [8, 14]

pH-range

5.4–9.0 [14]

Temperature optimum (°C)

30 [8]

Temperature range (°C)

3 ENZYME STRUCTURE

Molecular weight

54000 (Thiobacillus thioparus, sucrose density gradient centrifugation) [13]
40000–44000 (Thiobacillus sp., gel filtration [9], SDS-PAGE [2, 9], gel electrophoresis [14]) [2, 9, 14]

Subunits

Monomer (1 × 400000, Thiobacillus novellus, SDS-PAGE) [9]
? (x × 23000, Thiobacillus novellus, SDS-PAGE) [15]

Glycoprotein/Lipoprotein

–

4 ISOLATION/PREPARATION

Source organism

Thiobacillus versutus [1, 2]; Thiobacillus sp. A2 [3, 4]; Thiobacillus novellus [5–7, 9, 11, 15]; Thiobacillus denitrificans [10]; Rhodotorula sp. [8, 12]; Thiobacillus thioparus [13]; Thiobacillus ferrooxidans [14]

Source tissue

Cell [14]

Localisation in source

Periplasmic space [1]

Purification

Thiobacillus sp. (partial [3, 4, 7, 14]) [2–4, 9–11, 14, 15]; Rhodotorula sp. (non-separable complex with thiosulfate oxidase) [12]

Crystallization

–

Cloned

–

Renaturated

(reconstitution of sulfite oxidase system) [15]

5 STABILITY

pH

Temperature (°C)

3 (8 h) [14]; 55 (5 min, inactivation [7], 5 min, 40% activity [8]) [7, 8]; 58 (reduced enzyme stable) [9]; 60 (1 min, inactivation [7], 7 min, 30% activity [8]) [7, 8]

Enzyme Handbook © Springer-Verlag Berlin Heidelberg 1994
Duplication, reproduction and storage in data banks are only allowed with the prior permission of the publishers

Oxidation

Organic solvent

Ethanol (50% v/v, inactivation) [7]

General stability information

Stable above 10 microgram per ml protein concentration [8]; Freezing/thawing (purified enzyme unstable, crude enzyme stable [8], partially purified enzyme stable [7]) [7, 8]; Sulfite stabilizes [7]; Reduced enzyme has higher stability than oxidized form [9]

Storage

–20°C, more than 6 months [7]; –20°C [14]; –20°C, 20% glycerol, several days [8]; 4°C, 1 month, 80% activity, 20°C, 1 week, 65% activity, –20°C, several months [2]

---

6 CROSSREFERENCES TO STRUCTURE DATABANKS

PIR/MIPS code

Brookhaven code

---

7 LITERATURE REFERENCES

[1] Lu, W.-P.: FEMS Microbiol. Lett.,34,313–317 (1986)
[2] Lu, W.-P., Kelly, D.P.: J. Gen. Microbiol.,130,1683–1692 (1984)
[3] Lu, W.-P., Kelly, D.P.: J. Gen. Microbiol.,129,3549–3564 (1983)
[4] Lu, W.-P., Kelly, D.P.: J. Gen. Microbiol.,129,1673–1681 (1983)
[5] Oh, J.K., Suzuki, I.: J. Gen. Microbiol.,99,413–423 (1977)
[6] Yamanaka, T.: J. Biochem.,77,493–499 (1975)
[7] Charles, A.M., Suzuki, I.: Biochim. Biophys. Acta,128,522–534 (1966)
[8] Kurek, E.J.: Arch. Microbiol.,143,277–282 (1985)
[9] Toghrol, F., Southerland, W.M.: J. Biol. Chem.,258,6762–6766 (1983)
[10] Aminuddin, M., Nicholas, D.J.D.: J. Gen. Microbiol.,82,103–113 (1974)
[11] Charles, A.M., Suzuki, I.: Biochim. Biophys. Acta,128,510–521 (1966)
[12] Kurek, E.J.: Arch. Microbiol.,134,143–147 (1983)
[13] Lyric, R.M., Suzuki, I.: Can. J. Biochem.,48,334–343 (1970)
[14] Vestal, J.R., Lundgren, D.G.: Can. J. Biochem.,49,1125–1130 (1971)
[15] Yamanaka, T., Yoshioka, T., Kimura, K.: Plant Cell Physiol.,22,613–622 (1981)

1 NOMENCLATURE

EC number
1.8.2.2

Systematic name
Thiosulfate:ferricytochrome-c oxidoreductase

Recommended name
Thiosulfate dehydrogenase

Synonymes
Tetrathionate synthase
Oxidase, thiosulfate
Enzymes, thiosulfate-oxidizing
Thiosulfate oxidase
Thiosulfate-oxidizing enzyme
Thiosulfate-acceptor oxidoreductase [3]

CAS Reg. No.
9076-88-4

2 REACTION AND SPECIFICITY

Catalysed reaction
2 Thiosulfate + 2 ferricytochrome c →
→ tetrathionate + 2 ferrocytochrome c

Reaction type
Redox reaction

Natural substrates
Thiosulfate + cytochrome c [8]

Enzyme Handbook © Springer-Verlag Berlin Heidelberg 1994
Duplication, reproduction and storage in data banks are only allowed with the prior permission of the publishers

Substrate spectrum

1 Thiosulfate + ferricytochrome c (ir [9, 13], cytochrome c [1–12], cytochrome of yeast [5], of tuna [5], mammalian (not [5, 10]) [7], horse heart (poor [5]) [4, 5], only native cytochrome c [2, 3, 10], cytochrome $c_{553.5}$ [9], also reaction with ferricyanide [1–5, 7–10, 12, 13], Ferrobacillus ferrooxidans enzyme is active only with ferricyanide, not with native or mammalian cytochrome c [8, 13], strictly specific for thiosulfate [11]) [1–11]

2 More (no activity with: cytochrome c-552, cytochrome c', cytochrome c-553 [5], NAD^+, $NADP^+$, methylene blue, 2,6-dichlorophenol indophenol, O_2 [12]) [5, 12]

Product spectrum

1 Tetrathionate + ferrocytochrome c

2 ?

Inhibitor(s)

Diethyldithiocarbamate (slight) [7]; Deoxycholate [2]; $HgCl_2$ [2, 4, 11–13]; N-Ethylmaleimide [2, 4, 11, 12]; p-Chloromercuribenzoate [4, 11, 12]; Azide [7, 13]; Cupferron [7]; EDTA [11, 12]; SO_3^{2-} [2, 5, 7, 8, 11, 12]; Cyanide [5, 7]; 2,2'-Bipyridyl [12]

Cofactor(s)/prostethic group(s)

Metal compounds/salts

Iron (enzyme contains 2 mol of non-heme iron) [12]

Turnover number (min^{-1})

Specific activity (U/mg)

More [13]; 9.13 [3]; 11.38 [10]; 316 [1]; 146.1 [12]; 6010 [11]

K_m-value (mM)

More (temperature dependency of K_m [10]) [5, 6, 10, 12, 13]; 0.004 (thiosulfate (+ cytochrome c)) [1]; 0.110 (thiosulfate (+ ferricyanide)) [1]; 0.14 (Rhodotorula cytochrome c) [2]; 0.16 (thiosulfate (+ ferricyanide)) [2]; 0.34 (ferricyanide) [2]; 0.62 (thiosulfate) [3]; 0.67 (thiosulfate) [4]; 0.9 (ferricyanide) [3]; 1.1 (ferricyanide) [4]; 1.57 (thiosulfate, pH 6.1–6.3, 30°C) [10]

pH-optimum

4–6 [3]; 5.0 (below [12], thiosulfate + ferricyanide) [9, 11, 12, 13]; 5.2 (thiosulfate + ferricyanide) [7]; 5.5–6.5 [4]; 6.5 (broad, thiosulfate + cytochrome c) [12]; 7.2–7.6 (thiosulfate + cytochrome c) [11]; 7.8 [2]; More (no definite pH-optimum) [10]

pH-range

4.5–9.0 [4]; 5.0–9.0 [11]; 5.5–8.8 (little change in activity between) [12]

Temperature optimum (°C)

30 [2, 10]; 30–42 [4]; 40 [11]

Temperature range (°C)

22–48 (about 50% of activity maximum at 22°C and 48°C) [11]

3 ENZYME STRUCTURE

Molecular weight

90000 (Alcaligenes sp., gel filtration) [11]
115000 (Thiobacillus thioparus, sucrose density gradient method) [12]
138000 (Thiobacillus tepidarius, gel filtration) [1]
180000 (Rhodopseudomonas globiformis, gel filtration) [3]

Subunits

Trimer (3 × 45000, Thiobacillus tepidarius, SDS-PAGE) [1]

Glycoprotein/Lipoprotein

–

4 ISOLATION/PREPARATION

Source organism

Thiobacillus thioparus [8, 12]; Ferrobacillus ferrooxidans [8, 13]; Thiobacillus sp. X [9]; Thiobacillus tepidarius [1, 6]; Rhodotorula sp. [2]; Rhodopseudomonas globiformis [3]; Pseudomonas aeruginosa [4]; Chromatium vinosum [5]; Heterotrophic bacteria [7]; Pseudomonas sp. 16B (marine heterotroph) [10]; Alcaligenes sp. [11]

Source tissue

Localisation in source

Periplasm [1, 6]

Purification

Alcaligenes sp. [11]; Thiobacillus tepidarius (partial) [1]; Rhodopseudomonas globiformis (partial) [3]; Pseudomonas aeruginosa [4]; Chromatium vinosum (partial) [5]; Pseudomonas sp. 16B (partial) [10]; Thiobacillus thioparus [12]; Ferrobacillus ferrooxidans [13]

Crystallization

–

Enzyme Handbook © Springer-Verlag Berlin Heidelberg 1994
Duplication, reproduction and storage in data banks are only allowed with the prior permission of the publishers

Cloned

–

Renaturated

–

5 STABILITY

pH

4 (unstable below) [13]

Temperature (°C)

4 (6 h, 10% loss of activity, 50% loss of activity after 5 days) [12]; 25 (30 min, 7% loss of activity) [11]; 45 (30 min, complete loss of activity [11], 3 min, pH 7, complete loss of activity [9]) [9, 11]; 55 (5 min, stable) [12]; 60 (10 min, complete loss of activity, crude extract [7], 5 min, 50% loss of activity [12]) [7, 12]; 100 (1 min, complete loss of activity) [5, 12]

Oxidation

Organic solvent

General stability information

Lability of enzyme during purification and storage at –20°C [1]

Storage

Frozen, 0.05 M phosphate buffer, pH 6.2, stable for at least 90 days [11]; 4°C, 2 days, stable [10]; –20°C, 36% loss of activity after 9 days [10]; 4°C, 50% loss of activity after 3 days [4]; –20°C, 50% loss of activity after 2 months, crude extract [7]; –25°C, 0.02 M phosphate, 40% loss of activity after 2 months [9]

6 CROSSREFERENCES TO STRUCTURE DATABANKS

PIR/MIPS code

Brookhaven code

7 LITERATURE REFERENCES

[1] Lu, W.-P., Kelly, D.P.: J. Gen. Microbiol.,134,877–885 (1988)
[2] Kurek, E.J.: Arch. Microbiol.,143,277–282 (1985)
[3] Then, J., Trüper, H.G.: Arch. Microbiol.,130,143–146 (1981)
[4] Schook, L.B., Berk, R.S.: J. Bacteriol.,140,306–308 (1979)
[5] Fukumori, Y., Yamanaka, T.: Curr. Microbiol.,3,117–120 (1979)
[6] Lu, W.-P., Kelly, D.P.: J. Gen. Microbiol.,134,865–876 (1988)
[7] Trudinger, P.A.: J. Bacteriol.,93,550–559 (1967)
[8] Suzuki, I.: Annu. Rev. Microbiol.,28,85–101 (1974) (Review)
[9] Trudinger, P.A.: Biochem. J.,78,680–686 (1961)
[10] Tuttle, J.H., Schwartz, J.H., Whited, G.M.: Appl. Environ. Microbiol.,46,438–445 (1983)
[11] Hall, M.R., Berk, R.S.: Can. J. Microbiol.,18,235–245 (1972)
[12] Lyric, R.M., Suzuki, I.: Can. J. Biochem.,48,355–363 (1970)
[13] Silver, M., Lundgren, D.G.: Can. J. Biochem.,46,1215–1220 (1968)

Enzyme Handbook © Springer-Verlag Berlin Heidelberg 1994
Duplication, reproduction and storage in data banks are only allowed with the prior permission of the publishers

1 NOMENCLATURE

EC number
1.8.3.1

Systematic name
Sulfite:oxygen oxidoreductase

Recommended name
Sulfite oxidase

Synonymes
Oxidase, sulfite
Sulphite oxidase

CAS Reg. No.
9029-38-3

2 REACTION AND SPECIFICITY

Catalysed reaction
Sulfite + O_2 + H_2O →
→ sulfate + H_2O_2 (this direct reduction of O_2 is prevented completely in presence of cytochrome c [19]); Sulfite + H_2O + 2 ferricytochrome c →
→ sulfate + 2 ferrocytochrome c + 2 H^+

Reaction type
Redox reaction

Natural substrates
Sulfite + cytochrome c (natural acceptor [24], catalytic cycle [30], detoxification [6, 11]) [6, 11, 24, 30]

Substrate spectrum
1 SO_3^{2-} + H_2O + A (A: electron acceptor, i.e. O_2, cytochrome c, $K_3[Fe(CN)]_6$, 2,6-dichloroindophenol, methylene blue, highly specific for sulfite as electron donor) [12, 19, 24–26, 28]

Product spectrum
1 SO_4^{2-} + AH_2 (reduced acceptor, i.e. H_2O_2, ferrocytochrome c, $K_4[Fe(CN)_6]$, reduced 2,6-dichloroindophenol, reduced methylene blue)

Enzyme Handbook © Springer-Verlag Berlin Heidelberg 1994
Duplication, reproduction and storage in data banks are only allowed with the prior permission of the publishers

Inhibitor(s)

CN^- (more than 10 mM [11], not: [17, 27]) [2, 11, 20, 26]; N-Cyclohexyl-N'-[2-(N-methylmorpholino)-ethyl]carbodiimide p-toluene sulfonate (CMC) [3]; N-Ethyl-5-phenylisoxazolium-3'-sulfonate (Woodward's reagent K) [4]; 1-Ethyl-3-(3-dimethylaminopropyl)carbodiimide hydrochloride (EDC) [4]; Heavy metal ions [11, 28]; Anions (e.g. sulfate [11], arsenite [20, 21, 26], only with cytochrome c or ferricyanide as electron acceptor [20, 22]) [11, 20–22, 26]; Mannitol (only with O_2 as electron acceptor) [26]; Cytochrome c (inhibition of O_2 consumption) [26]; Ferricyanide (inhibition of O_2 consumption) [26]; 2,6-Dichloroindophenol (inhibition of O_2 consumption) [26]; p-Chloromercuribenzoate [28]

Cofactor(s)/prostethic group(s)

Di-(carboxamidomethyl)molybdopterin [5, 13, 14]; Heme (cytochrome b_5, 1 per subunit, molybdohemoprotein) [2, 11, 20, 25, 27]

Metal compounds/salts

Mo (molybdohemoprotein) [11, 20, 25, 29]

Turnover number (min^{-1})

5100 (cytochrome c) [3]

Specific activity (U/mg)

35.4 [2]; 75 [17]; 90 [7]; More [6, 11, 23, 25, 26, 28]

K_m-value (mM)

0.00092 (cytochrome c) [3]; 0.39 (sulfite, similar value [17], bovine liver [20]) [12]; 0.015 (O_2) [12]; 0.002 (cytochrome c, similar value [7, 16]) [12, 20, 25]; 0.03 (sulfite, similar value chicken liver [20, 25]) [7]; More (kinetic studies [30]) [28, 30]

pH-optimum

7.5 [12]; 8.6 (with O_2 as acceptor [20]) [20, 26]; 8.7 [11]; 9.0 [7]

pH-range

7.0 (less than 50% of maximal activity below) [11]

Temperature optimum (°C)

Temperature range (°C)

---

3 ENZYME STRUCTURE

Molecular weight

115000–120000 (Merluccius productus, gel filtration [11], rat, sedimentation equilibrium centrifugation [16], bovine, sedimentation equilibrium centrifugation, gel filtration [27]) [11, 16, 27]

Subunits

Dimer (2 × 50545, chicken, amino acid sequence [1], 2 × 55000–60000, Merluccius productus, SDS-PAGE [11], rat, sedimentation equilibrium centrigugation with 6 M guanidium-HCl [16], SDS-PAGE [23], chicken, SDS-PAGE [25], bovine, SDS-PAGE [27]) [1, 11, 16, 23, 25, 27]
More (after tryptic cleavage: 9500, rat, heme-containing fragment, gel filtration with guanidine-HCl, 47400, rat, molybdenum-containing fragment, SDS-PAGE) [18]

Glycoprotein/Lipoprotein

–

4 ISOLATION/PREPARATION

Source organism

Vertebrates (immunochemical comparison [7]) [6, 7]; Chicken [1–5, 20–23, 25, 29, 30]; Rat [7–9, 12, 15–21, 23, 24, 28, 31, 32]; Mouse [12]; Merluccius productus (pacific hake) [11]; Bovine [20, 26–28]; Rabbit [20, 23]; Pig [20, 23]; Human [23]; Dog [28]

Source tissue

Liver [1–28]; Kidney [28]; Heart [28]; Brain (low activity) [28]; Testis (low activity) [28]; Lung (low activity) [28]; Spleen (low activity) [28]; More (activity in tissues and cells [6], distribution in fish organs [11], not: brain, muscle, blood, adipose tissue, thymus [23]) [6, 11, 23]

Localisation in source

Mitochondria (intermembrane space) [3, 8, 20, 23, 24]; Cytosol (precursor enzyme [8–10]) [8–10, 17]; Microsomes (inside vesicles) [17]; More (subcellular localisation) [17, 24]

Purification

Chicken (70–77% pure [2]) [2, 25]; Rat (isolation of heme- and molybdenum-containing fragments [18]) [7, 17, 18, 23, 28]; Merluccius productus (partial) [11]; Bovine (75% pure) [26]

Crystallization

(isolated molybdenum- and heme-fragments) [18]

Cloned

–

Renaturated

[31, 32]

Enzyme Handbook © Springer-Verlag Berlin Heidelberg 1994
Duplication, reproduction and storage in data banks are only allowed with the prior permission of the publishers

5 STABILITY

pH

Temperature (°C)

25 (10 min stable) [11]; 50 (rapid inactivation) [11]; 52 (inactivation, protection by sulfate) [26]; 54 (reduced form has higher stability than oxidized form) [15]

Oxidation

Organic solvent

General stability information

Trypsin inactivates [28]

Storage

–80°C, several weeks [4]

6 CROSSREFERENCES TO STRUCTURE DATABANKS

PIR/MIPS code

PIR2: A34180 (hepatic Chicken); PIR2: S10446 (hepatic Chicken (fragment)); PIR3: A38328 (hepatic Rat (fragments))

Brookhaven code

7 LITERATURE REFERENCES

[1] Neame, P.J., Barber, M.J.: J. Biol. Chem.,264,20894–20901 (1989)
[2] Kipke, C.A., Enemark, J.H., Sunde, R.A.: Arch. Biochem. Biophys.,270,383–390 (1989)
[3] Ritzman, M., Bosshard, H.R.: Eur. J. Biochem.,172,377–381 (1988)
[4] Ritzmann, M., Bosshard, H.R.: Eur. J. Biochem.,159,493–497 (1986)
[5] Kramer, S.P., Johnson, J.L., Ribeiro, A.A., Millington, D.S., Rajagopalan, K.V.: J. Biol. Chem.,262,16357–16363 (1987)
[6] Beck-Speier, I., Hinze, H., Hozer, H.: Biochim. Biophys. Acta,841,81–89 (1985)
[7] Kuwahara, T., Yoshimoto, I., Ito, A.: J. Biochem.,92,1925–1931 (1982)
[8] Ono, H., Ito, A.: Biochem. Biophys. Res. Commun.,107,258–264 (1982)
[9] Mihara, K., Omura, T., Harano, T., Brenner, S., Fleischer, S., Rajagopalan, K.V., Blobel, G.: J. Biol. Chem.,257,3355–3358 (1982)
[10] Ono, H., Ito, A.: J. Biochem.,91,117–123 (1982)
[11] Onoue, Y.: Biochim. Biophys. Acta,615,48–58 (1980)
[12] Shibuya, A., Horie, S.: J. Biochem.,87,1773–1784 (1980)
[13] Johnson, J.L., Hainline, B.E., Rajagopalan, K.V.: J. Biol. Chem.,255,1783–1786 (1980)
[14] Berg, J.M., Hodgson, K.O., Cramer, S.P., Corbin, J.L. , Elsberry, A., Pariyadath, N., Stiefel, E.I.: J. Am. Chem. Soc.,101,2774–2776 (1979)

[15] Southerland, W.M., Rajagopalan, K.V.: J. Biol. Chem.,253,8753–8758 (1978)
[16] Southerland, W.M., Winge, D.R., Rajagopalan, K.V.: J. Biol. Chem.,253,8747–8752 (1978)
[17] Ito, A., Kuwahara, T., Mitsunari, Y., Omura, T.: J. Biochem.,81,1531–1541 (1977)
[18] Johnson, J.L., Rajagopalan, K.V.: J. Biol. Chem.,252 ,2017–2025 (1977)
[19] Oshino, N., Chance, B.: Arch. Biochem. Biophys.,170,514–528 (1975)
[20] Bray, R.C. in "The Enzymes",3rd. Ed. (Boyer, P.D., ed.) 12,299–419 (1975) (Review)
[21] Kessler, D.L., Rajagopalan, K.V.: Biochim. Biophys. Acta,370,399–409 (1974)
[22] Kessler, D.L., Rajagopalan, K.V.: Biochim. Biophys. Acta,370,389–398 (1974)
[23] Kessler, D.L., Johnson, J.L., Cohen, H.J., Rajagopalan, K.V.: Biochim. Biophys. Acta,334,86–96 (1974)
[24] Cohen, H.J.: J. Biol. Chem.,247,7759–7766 (1972)
[25] Kessler, D.L., Rajagopalan, K.V.: J. Biol. Chem.,247 ,6566–6573 (1972)
[26] Cohen, H.J., Fridovich, I.: J. Biol. Chem.,246,359–366 (1971)
[27] Cohen, H.J., Fridovich, I.: J. Biol. Chem.,246,367–373 (1971)
[28] MacLeod, R.M., Farkas, W., Fridovich, I., Handler, P. : J. Biol. Chem.,236,1841–1846 (1961)
[29] George, G.N., Kipke, C.A., Prince, R.C., Sunde, R.A., Enemark, J.H., Cramer, S.P.: Biochemistry,28,5075–5080 (1989)
[30] Kipke, C.A., Cusanovich, M.A., Tolin, G., Sunde, R.A. , Enemark, J.H.: Biochemistry,27,2918–2926 (1988)
[31] Jones, H.P., Johnson, J.L., Rajagopalan, K.V.: J. Biol. Chem.,252,4988–4993 (1977)
[32] Johnson, J.L., Jones, H.P., Rajagopalan, K.V.: J. Biol. Chem.,252,4994–5003 (1977)

Enzyme Handbook © Springer-Verlag Berlin Heidelberg 1994
Duplication, reproduction and storage in data banks are only allowed with the prior permission of the publishers

1 NOMENCLATURE

EC number
1.8.3.2

Systematic name
Thiol:oxygen oxidoreductase

Recommended name
Thiol oxidase

Synonymes
Oxidase, thiol
Sulfhydryl oxidase

CAS Reg. No.
9029-39-4

2 REACTION AND SPECIFICITY

Catalysed reaction
4 R'C(R)SH + O_2 →
→ 2 R'C(R)S-S(R)CR' + 2 H_2O

Reaction type
Redox reaction

Natural substrates
R'C(R)SH + O_2 (R may be S or O or a variety of other groups, the enzyme is not specific for R', formation of disulfide bonds in epidermis [2], oxidation of thiols to disulfides [18]) [2, 18]

Substrate spectrum
1 R'C(R)SH + O_2 (e.g.: glutathione + O_2 (low activity [2, 13]) [1–6, 13], N-acetyl-L-cysteine + O_2 (low activity [2]) [1, 2, 5, 6], glycyl-glycyl-L-cysteine + O_2 [1], 5,5'-dithiobis(2-nitrobenzoate) + O_2 [2], reduced form of ribonuclease + O_2 (not [5]) [2–4, 13, 15], dithiothreitol + O_2 (not [6]) [2, 5, 13], D-penicillamine + O_2 [2, 13], cysteine + O_2 [2–6, 13], cysteamine + O_2 [3, 5, 6], reduced form of xanthine oxidase + O_2 [3], reduced form of chymotrypsinogen + O_2 (not [5]) [3], reduced cysteine containing peptides + O_2 [6], 2-mercaptoethanol + O_2 (slight [13]) [13, 14], thioglycolate + O_2 [14, 16], dithioglycolate + O_2 [16], alkylxanthanes + O_2 [16], thioacetate + O_2 [16], dithioacetate + O_2 [16], dithiooxalate + O_2 [16], thiophenol + O_2 [16, 17], thiohistidine +

Enzyme Handbook © Springer-Verlag Berlin Heidelberg 1994
Duplication, reproduction and storage in data banks are only allowed with the prior permission of the publishers

O_2 [16], ergothioneine + O_2 [16], catechol + O_2 [17], diethyldithiocarbamate + O_2 [17], methylmercaptoimidazol + O_2 [17], resorcinol + O_2 (not [14]) [17], R may be S or O or a variety of other groups, the enzyme is not specific for R') [1–6, 13–17]

2 More (not: butyl mercaptan, thioglycerol, cysteine, glutathione, mercaptosuccinic acid [16], thiouracil [17]) [16, 17]

Product spectrum

1 R'C(R)S-S(R)CR' + H_2O

2 ?

Inhibitor(s)

Iodoacetamide [2, 12]; Iodoacetate [2, 12]; Reduced ribonuclease (above 0.040 mM, substrate inhibition) [4]; Reduced glutathione (substrate inhibition [4, 15] above 0.8 mM [15]) [4, 15]; H_2O_2 [5, 9]; L-(alphaS, 5S)-alpha-Amino-3-chloro-4,5-dihydro-5-isoxazoleacetic acid (acividin) [5]; N-Ethymaleimide [5, 9]; EDTA [9]; Bathocuproine disulfonate [9]; o-Phenylenediamine [12]; o-Dianisidine [12]; Guanine [12]; $HgCl_2$ [12]; Azide (2 mM) [14, 16]; Cyanide [16, 17]; Cysteine [16]; Mercaptoethanol [17]; 2,2'-Dipyridyl [17]; 8-Hydroxyquinoline [17]

Cofactor(s)/prostethic group(s)

More (does not require loosely bound cofactor for activity) [14]

Metal compounds/salts

Copper (rat: activity increases by incubation with copper [2], metalloenzyme: contains 1 mol copper per subunit, copper is in a binuclear complex [5], involved in catalysis [9], bovine: about 0.15 gatom of copper per subunit [15]) [2, 5, 9, 15]; Iron (bovine: metalloprotein containing iron [10], enzyme-bound iron essential for catalytic activity [12], 0.5 atom of iron per subunit [15]) [10, 15]

Turnover number (min^{-1})

Specific activity (U/mg)

More [1–3, 5, 13, 17]

K_m-value (mM)

2.35 (N-acetyl-L-cysteine) [1]; 1.13 (N-acetyl-L-cysteine) [6]; 3.85 (N-acetyl-L-cysteine) [10, 12]; 6.31 (Gly-Gly-L-Cys) [1]; 0.805 (L-cysteine) [6]; 0.966 (D-cysteine) [6]; 1.25 (cysteamine) [6]; 0.80 (L-cysteine) [10, 12]; 1.33 (D-cysteine) [10, 12]; 10.0 (cysteamine) [10, 12]; 8.16 (glutathione) [1]; 0.73 (glutathione) [5]; 0.34 (glutathione) [10]; 0.0174 (reduced ribonuclease [4], corresponds to a sulfhydryl concentration of 0.14 mM) [4]; 0.77 (O_2) [6]; 100 (2-nitro-5-thiobenzoic acid) [12]; 0.41 (thiophenol) [17]; 0.92 (catechol) [17]

pH-optimum
6.0 [16]; 7.0 [15]; 8.0–8.2 [2]

pH-range
6.0–8.0 (activity considerably less at pH 7.0 than at pH 6.0, too low to be measured at pH 8.0) [16]

Temperature optimum (°C)
35 [15]

Temperature range (°C)

3 ENZYME STRUCTURE

Molecular weight
66000 (rat skin, gel filtration, SDS-PAGE [13]) [13]
89000 (bovine milk, subunit weight, SDS-PAGE) [12, 15]
120000 (pig kidney, gel filtration) [5]
More (exists in aggregated molecular form) [1]

Subunits
Dimer (2 × 70000, pig kidney, SDS-PAGE, enzyme also aggregates to larger multimers) [5]
? (x × 300000, bovine pancreas, SDS-PAGE, perhaps aggregated form) [1]

Glycoprotein/Lipoprotein
Glycoprotein (11–15% carbohydrate [10]) [10, 12, 15]

4 ISOLATION/PREPARATION

Source organism
Bovine [1, 3, 4, 6–8, 10, 12, 15]; Rat [2, 7, 9, 11, 13]; Pig [5, 7]; Goat [7]; Human [7]; Piricularia oryzae [16, 17]; Polyporus vesicolor [16]; Hamster [14]

Source tissue
Caudal epididymis [14]; Pancreas [1, 7]; Milk [3, 4, 6–8, 10, 12, 15]; Kidney [5, 7, 11]; Skin [2, 13] Small intestine (epithelium) [9]; Culture filtrate [16]; More (no activity: rat thymus, brain, heart, liver, spleen, lung or small intestine) [7]

Enzyme Handbook © Springer-Verlag Berlin Heidelberg 1994
Duplication, reproduction and storage in data banks are only allowed with the prior permission of the publishers

Localisation in source

Membrane (enzyme is closely associated with the plasma membrane of lactating cow and rat mammary tissues and the basal-lateral membrane of rat kidney cortex [7], membrane-associated [9], basal-lateral region of plasma membrane [11]) [7, 9, 11]; More (associated with endothelial cells lining the capillaries of rat kidney, heart and small intestine) [7]

Purification

Bovine [1, 10, 12, 15]; Rat [2, 13]; Pig [5]; Piricularia oryzae [17]

Crystallization

–

Cloned

–

Renaturated

–

---

5 STABILITY

pH

Temperature (°C)

60 (15 min, 85% loss of activity [2, 13], 13 min, complete loss of activity [2, 13], 5 min, stable [17]) [2, 13, 17]; 100 (destroyed by boiling) [17]; More (dithiothreitol increases heat stability) [2]

Oxidation

Organic solvent

General stability information

Immobilized enzyme, operational stability [12]; Dithiothreitol increases heat stability [12]

Storage

4°C, 0.5 M sucrose [3]; 4°C, 2 weeks [13]

---

6 CROSSREFERENCES TO STRUCTURE DATABANKS

PIR/MIPS code

Brookhaven code

7 LITERATURE REFERENCES

[1] Clare, D.A., Pinnix, I.B., Lecce, J.G., Horton, H.R.: Arch. Biochem. Biophys.,265,351–361 (1988)
[2] Goldsmith, L.A.: Methods Enzymol.,143,510–515 (1987)
[3] Swaisgood, H.E., Horton, H.R.: Methods Enzymol.,143,504–510 (1987)
[4] Janolino, V.G., Swaisgood, H.E.: Arch. Biochem. Biophys.,258,265–271 (1987)
[5] Lash, L.H., Jones, D.P.: Arch. Biochem. Biophys.,247,120–130 (1986)
[6] Sliwkowski, M.X., Swaisgood, H.E., Clare, D.A., Horton, H.R.: Biochem. J.,220,51–55 (1984)
[7] Clare, D.A., Horton, H.R., Stabel, T.J., Swaisgood, H. E., Lecce, J.G.: Arch. Biochem. Biophys.,230,138–145 (1984)
[8] Schmelzer, C.H., Phillips, C., Swaisgood, H.E., Horton, H.R.: Arch. Biochem. Biophys.,228,681–685 (1984)
[9] Lash, L.H., Jones, D.P.: Arch. Biochem. Biophys.,225,344–352 (1983)
[10] Swaisgood, H.-E., Sliwkowski, M.-X., Skudder, P.-J., Janolino, V.-G. in "Util. Enzymes Technol. Aliment., Symp. Int." (Dupuy, P., ed.) ,229–235 (1982)
[11] Lash, L.H., Jones, D.P.: Biochem. J.,203,371–376 (1982)
[12] Swaisgood, H.E.: Enzyme Microb. Technol.,2,265–272 (1986) (Review)
[13] Takamori, K., Thorpe, J.M., Goldsmith, L.A.: Biochim. Biophys. Acta,615,309–323 (1980)
[14] Chang, T.S.K., Morton, B.: Biochem. Biophys. Res. Commun.,66,309–315 (1975)
[15] Janolino, V.G., Swaisgood, H.E.: J. Biol. Chem.,250,2532–2538 (1975)
[16] Neufeld, H.A., Green, L.F., Latterell, F.M., Weintraub, R.L.: J. Biol. Chem.,232,1093–1099 (1958)
[17] Aurbach, G.D., Jakoby, W.B.: J. Biol. Chem.,237,565–568 (1962)
[18] Lash, L.H., Jones, D.P., Orrenius, S.: Biochim. Biophys. Acta,779,191–200 (1984) (Review)

Enzyme Handbook © Springer-Verlag Berlin Heidelberg 1994
Duplication, reproduction and storage in data banks are only allowed with the prior permission of the publishers

1 NOMENCLATURE

EC number
1.8.3.3

Systematic name
Glutathione:oxygen oxidoreductase

Recommended name
Glutathione oxidase

Synonymes
Oxidase, glutathione

CAS Reg. No.
55467-56-6

2 REACTION AND SPECIFICITY

Catalysed reaction
2 Glutathione + O_2 →
→ oxidized glutathione + H_2O_2

Reaction type
Redox reaction

Natural substrates
Glutathione + O_2

Substrate spectrum
1 Glutathione + O_2 [1–3]
2 N-Acetyl-L-cysteine + O_2 (slowly) [3]
3 L-Cysteine methyl ester + O_2 (slowly) [3]
4 Cysteamine + O_2 (slowly) [3]
5 Thiophenol + O_2 (slowly) [3]
6 2-Mercaptoethanol + O_2 (slowly [3]) [1, 3]
7 Cysteine + O_2 (slowly [3]) [1, 3]
8 Dithiothreitol + O_2 (low activity [3]) [1, 3]
9 Reduced RNase A + O_2 (no activity [1]) [3]
10 More (not: D, L-homocysteine [1]) [1, 3]

Enzyme Handbook © Springer-Verlag Berlin Heidelberg 1994
Duplication, reproduction and storage in data banks are only allowed with the prior permission of the publishers

Product spectrum

1 Oxidized glutathione + H_2O_2
2 ?
3 ?
4 ?
5 ?
6 ?
7 ?
8 ?
9 ?
10 ?

Inhibitor(s)

$ZnSO_4$ [3]

Cofactor(s)/prostethic group(s)

FAD (flavoprotein [1–3], contains 2 mol of FAD per mol of enzyme [3], 1 mol FAD per mol of enzyme [1]) [1–3]

Metal compounds/salts

Turnover number (min^{-1})

Specific activity (U/mg)

272 [3]

K_m-value (mM)

0.69 (glutathione) [3]; 3.6 (L-cysteine) [3]; 6.7 (dithiothreitol) [3]; 4.4 (glutathione) [1]; 9.1 (cysteine) [1]; 32 (2-mercaptoethanol) [1]; 0.7 (dithiothreitol) [1]

pH-optimum

7.0–7.8 [3]; 7.0 (substrate: dithioerythritol) [1]

pH-range

6.2–8.8 (sharp decline in activity below pH 6.2 and above pH 8.8) [3]; 5–10 (minimal activity below pH 5, active at least up to pH 10) [1]

Temperature optimum (°C)

30 (assay at) [3]

Temperature range (°C)

3 ENZYME STRUCTURE

Molecular weight
95000 (Penicillium sp. K-6–5, gel filtration) [3]
66000 (rat seminal vesicles, gel filtration, SDS-PAGE) [1]

Subunits
Monomer (1 × 66000, rat seminal vesicles, gel filtration, SDS-PAGE) [1]
Dimer (2 × 47000, Penicillium sp. K-6–5, SDS-PAGE) [3]

Glycoprotein/Lipoprotein
–

4 ISOLATION/PREPARATION

Source organism
Rat [1]; Penicillium sp. (K-6–5 [3]) [2, 3]; Fusarium sp. [2]; Aspergillus sp. [2]; More (not: molds belonging to Rhizopus and Mucor) [2]

Source tissue
Seminal vesicles [1]

Localisation in source
Extracellular [2, 3]; Intracellular [3]

Purification
Penicillium sp. K-6–5 [3]

Crystallization
–

Cloned
–

Renaturated
–

5 STABILITY

pH
5.2–8.6 (45°C, 30 min) [3]

Temperature (°C)
55 (30 min, no loss of activity) [3]; 60 (3.5 min, 50% loss of activity) [1]

Oxidation

Enzyme Handbook © Springer-Verlag Berlin Heidelberg 1994
Duplication, reproduction and storage in data banks are only allowed with the prior permission of the publishers

Organic solvent

General stability information

Storage

–20°C, 0.1 M, potassium phosphate buffer, pH 7.4, 6 months [3]

6 CROSSREFERENCES TO STRUCTURE DATABANKS

PIR/MIPS code

Brookhaven code

7 LITERATURE REFERENCES

[1] Ostrowski, M.C., Kistler, W.S.: Biochemistry,19,2639–2645 (1980)
[2] Kusakabe, H., Midorikawa, Y., Kuninaka, A., Yoshino, H.: Agric. Biol. Chem.,47,1385–1387 (1983)
[3] Kusakaba, H., Kuninaka, A., Yoshino, H.: Agric. Biol. Chem.,46,2057–2067 (1982)

1 NOMENCLATURE

EC number
1.8.3.4

Systematic name
Methanethiol:oxygen oxidoreductase

Recommended name
Methanethiol oxidase

Synonymes
Methylmercaptan oxidase
Oxidase, methyl mercaptan
Methyl mercaptan oxidase
(MM)-oxidase [1]
MT-oxidase [3]

CAS Reg. No.
112821-28-0

2 REACTION AND SPECIFICITY

Catalysed reaction
Methanethiol + O_2 + H_2O →
→ formaldehyde + H_2S + H_2O_2

Reaction type
Redox reaction

Natural substrates
Methanethiol + O_2 + H_2O (enzyme in oxidation pathway of dimethyl disulphide) [3]

Substrate spectrum
1 Methanethiol + O_2 + H_2O [1–3]
2 Ethyl mercaptan + O_2 + H_2O [1, 2]
3 Sulphide + O_2 (not [2]) [1]

Product spectrum
1 Formaldehyde + H_2S + H_2O_2
2 Acetaldehyde + H_2S + H_2O_2
3 S + H_2O_2 (expected products) [1]

Enzyme Handbook © Springer-Verlag Berlin Heidelberg 1994
Duplication, reproduction and storage in data banks are only allowed with the prior permission of the publishers

Inhibitor(s)
Ethyl mercaptan [1, 2]; N-Ethylmaleimide [2]; Cu^{2+} [2]; Methylamine [1]; Cyanide [1, 2]; Methanethiol (above 0.014 mM [1]) [1–3]; Sulphide (non-competitive inhibition of methyl mercaptan oxidation [1], no inhibition [2]) [1]

Cofactor(s)/prostethic group(s)

Metal compounds/salts

Turnover number (min^{-1})

Specific activity (U/mg)
1.27 [2]; 2.13 [1]

K_m**-value** (mM)
0.005–0.010 (methanethiol) [1]; 0.0097 (methyl mercaptan) [1]; 0.018 (ethyl mercaptan) [1]; 0.0313 (methyl mercaptan) [2]; 0.060 (sulphide) [1]

pH-optimum
8.2 [1]

pH-range

Temperature optimum (°C)
25 (assay at) [2]; 30 (assay at) [1]

Temperature range (°C)

---

3 ENZYME STRUCTURE

Molecular weight
29000–40000 (Thiobacillus thioparus, gel filtration, SDS-PAGE) [2]
40000–50000 (Hyphomicrobium sp. EG, gel filtration) [1]

Subunits
Monomer (1 × 49000, Hyphomicrobium sp. EG, SDS-PAGE [1], 1 × 40000, Thiobacillus thioparus [2]) [1, 2]

Glycoprotein/Lipoprotein
–

---

4 ISOLATION/PREPARATION

Source organism
Hyphomicrobium sp. EG [1]; Thiobacillus thioparus (TK-m [2], E6 [3]) [2, 3]

Source tissue

Cell [2]

Localisation in source

Purification

Hyphomicrobium sp. EG [1]; Thiobacillus thioparus (TK-m) [2]

Crystallization

–

Cloned

–

Renaturated

–

5 STABILITY

pH

Temperature (°C)

50 (10 min, stable) [1]; 65 (10 min, complete loss of activity) [1]

Oxidation

Organic solvent

General stability information

Storage

–20°C, stable for months [1]

6 CROSSREFERENCES TO STRUCTURE DATABANKS

PIR/MIPS code

Brookhaven code

7 LITERATURE REFERENCES

[1] Suylen, G.M.H., Large, P.J., van Dijken, J.P., Kuenen, J.G.: J. Gen. Microbiol.,133,2989–2997 (1987)
[2] Gould, W.D., Kanagawa, T.: J. Gen. Microbiol.,138,217–221 (1992)
[3] Smith, N.A., Kelly, D.P.: J. Gen. Microbiol.,134,3031–3039 (1988)

Enzyme Handbook © Springer-Verlag Berlin Heidelberg 1994
Duplication, reproduction and storage in data banks are only allowed with the prior permission of the publishers

1 NOMENCLATURE

EC number
1.8.4.1

Systematic name
Glutathione:homocystine oxidoreductase

Recommended name
Glutathione-homocystine transhydrogenase

Synonymes
Transhydrogenase, glutathione-homocystine

CAS Reg. No.
9029-40-7

2 REACTION AND SPECIFICITY

Catalysed reaction
2 Glutathione + homocystine →
→ oxidized glutathione + 2 homocysteine (the reactions catalyzed by this enzyme and by other in this subclass may be similar to those catalyzed by EC 2.5.1.18)

Reaction type
Redox reaction

Natural substrates
Glutathione + homocystine [1]

Substrate spectrum
1 Glutathione + homocystine (r) [1]

Product spectrum
1 Oxidized glutathione + homocysteine [1]

Inhibitor(s)

Cofactor(s)/prostethic group(s)

Metal compounds/salts

Enzyme Handbook © Springer-Verlag Berlin Heidelberg 1994
Duplication, reproduction and storage in data banks are only allowed with the prior permission of the publishers

Turnover number (min^{-1})

Specific activity (U/mg)

K_m**-value** (mM)

pH-optimum

pH-range

Temperature optimum (°C)
37 (assay at) [1]

Temperature range (°C)

---

3 ENZYME STRUCTURE

Molecular weight

Subunits

Glycoprotein/Lipoprotein
–

---

4 ISOLATION/PREPARATION

Source organism
Bovine [1]

Source tissue
Liver [1]

Localisation in source

Purification

Crystallization
–

Cloned
–

Renaturated
–

5 STABILITY

pH

Temperature (°C)

Oxidation

Organic solvent

General stability information

Storage

6 CROSSREFERENCES TO STRUCTURE DATABANKS

PIR/MIPS code

Brookhaven code

7 LITERATURE REFERENCES

[1] Racker, E.: J. Biol. Chem.,217,867–874 (1955)

Enzyme Handbook © Springer-Verlag Berlin Heidelberg 1994
Duplication, reproduction and storage in data banks are only allowed with the prior permission of the publishers

1 NOMENCLATURE

EC number
1.8.4.2

Systematic name
Glutathione:protein-disulfide oxidoreductase

Recommended name
Protein-disulfide reductase (glutathione)

Synonymes
Glutathione-insulin transhydrogenase
Insulin reductase
Reductase, protein disulfide (glutathione)
Protein disulfide transhydrogenase
Glutathione-protein disulfide oxidoreductase
Protein disulfide reductase (glutathione)
GSH-insulin transhydrogenase
Protein-disulfide interchange enzyme [1, 2, 19]
Protein-disulfide isomerase/oxidoreductase [3]
Thiol:protein-disulfide oxidoreductase [6, 19, 20, 25, 26]
Thiol-protein disulphide oxidoreductase [36]
More (thiol:protein-disulfide oxidoreductase (EC 1.8.4.2) and thiol:protein-disulfide isomerase (EC 5.3.4.1) are immunological identical [1, 2, 18], the two activities (cleavage and formation of protein-disulfide bonds) present alternate activities of the same enzyme [1, 2, 18, 36], glutathione-insulin transhydrogenase (EC 1.8.4.2) and protein disulfide-isomerase (EC 5.3.4.1) activities are not both catalyzed by a single enzyme species [13, 30, 41, 43]) [1, 2, 13, 18, 30, 36, 41, 43]

CAS Reg. No.
9082-53-5

---

2 REACTION AND SPECIFICITY

Catalysed reaction
2 Glutathione + protein-disulfide →
→ oxidized glutathione + protein-dithiol (random mechanism [2, 42])

Reaction type
Redox reaction

Enzyme Handbook © Springer-Verlag Berlin Heidelberg 1994
Duplication, reproduction and storage in data banks are only allowed with the prior permission of the publishers

Natural substrates

Glutathione + protein disulfide (modulation of enzymatic activity [1], reductive degradation and assembly of proteins [1], enzyme plays a role in formation of intramonomer bonds common to all immunoglobulin molecules [35], synthesis of protein disulfide bond [35], enzyme not directly involved in the subcellular processing of receptor-bound internalized insulin [4]) [1, 4, 35]

Substrate spectrum

1 Glutathione + protein disulfide (or other thiol substrates, e.g. cysteamine [1, 16], dithiothreitol dihydrolipoic acid [1], 2,3-dimercaptopropanol [1], 2-mercaptoethanol [6, 16, 39, 42], cysteine [6, 16, 39], dihydrolipoate [6], dihydrolipoamide [6], dithiothreitol [6, 39], enzyme can also utilize thiol groups of proteins: e.g. of alcohol dehydrogenase [10], hexokinase [10], fructose-1,6-diphosphatase [10], malate dehydrogenase [10], glyceraldehyde phosphate dehydrogenase [10], glycerol phosphate dehydrogenase [10], reduced ribonuclease [10, 42], reduced insulin A and B chain [10, 42], protein-disulfide e.g.: insulin [1–42], proinsulin [1, 2], oxytoxin [1, 2, 39], vasopressin [1, 2, 39], prolactin (poor substrate) [31]) [1–42]

2 More (enzyme also catalyzes reactivation and folding of protein containing incorrectly paired disulfide bond, e.g.: scrambled ribonuclease [1, 2, 39], scrambled lysozyme [1, 2], scrambled trypsin inhibitor [2], scrambled proinsulin [2], immunoglobulin (IgM [1], IgG [1]) [1, 2], reduction of ricin and other plant thiols [23], reduction of choleragen [25], thiol: protein-disulfide oxidoreductase (EC 1.8.4.2) and thiol: protein-disulfide isomerase (EC 5.3.4.1) are immunological identical [1, 2, 18], the two activities (cleavage and formation of protein-disulfide bonds) present alternate activities of the same enzyme [1, 2, 18, 36], glutathione-insulin transhydrogenase (EC 1.8.4.2) and protein disulfide-isomerase (EC 5.3.4.1) activities are not both catalyzed by a single enzyme species [13, 30, 41, 43]) [1, 2, 13, 18, 23, 25, 30, 36, 39, 41, 43]

Product spectrum

1 Oxidized glutathione + protein-dithiol [1–42]

2 ?

Inhibitor(s)

Lysolecithin [1, 2]; Phosphatidic acid [1, 2, 38]; Se^{2+} [1, 40]; Hg^{2+} [1, 31, 40]; Ca^{2+} (slight [1]) [1, 31, 39, 40]; Zn^{2+} [1, 39]; Cu^{2+} [1, 39]; Ni^{2+} (slight) [1]; Aprotinin [1]; Bacitracin [1, 21]; Cysteine (at high concentration) [6]; EDTA (5 mM) [6]; Iodoacetate (treatment with thiol prior to incubation with glutathione and substrate: no inhibition, preincubation with glutathione and thiol reagent: inhibition) [22]; Iodoacetamide (treatment with thiol prior to incubation with glutathione and substrate: no inhibition, preincubation with glutathione and thiol reagent: inhibition) [22]; N-Ethylmaleimide (treatment with thiol prior to incubation with glutathione and substrate: no inhibition, preincubation with glutathione and thiol reagent: inhibition) [22]; Insulin analogs (inhibition of insulin degradation) [28]; Scrambled forms of ribonuclease and lysozyme (inhibition of insulin degradation) [28]; Vasopressin [28]; Oxytoxin [28]; Glucagon [28]; Acetyl-L-tyrosine ethyl ester [31]; Ribonuclease [31]; Deoxycholate [39]; Lysophosphatidylcholine [38]; Mg^{2+} [39]; Oxidized glutathione (product inhibition) [42]; S-Sulfonated A-chain or B-chain of insulin (product inhibition) [42]

Cofactor(s)/prostethic group(s)

Phosphatidylethanolamine (activates at low concentration) [1, 3, 8]; EDTA (and other chelating agents activate [1, 40], activates [35], activation negated by metal ions [40], 5 mM: inhibition [6]) [1, 35, 40]; Phospholipids (slight activation by all phospholipids tested except for phosphatidic acid and phosphatidylserine (inhibition), highest activation by phosphatidylethanolamine) [38]; Histidine (activates, activation negated by metal ions) [40]; Chelating agents (e.g. EDTA, EGTA, 8-hydroxyquinoline, 1,10-phenanthroline, activation) [1, 40]

Metal compounds/salts

Mn^{2+} (1 mM: activates) [6]; Cs^{+} (0.3 M: activates) [6]; Phosphate buffer (enhances activity) [36]

Turnover number (min^{-1})

Specific activity (U/mg)

More (assay methods [1, 13, 24], specific activity in bovine tissues [19]) [1, 7, 13, 16, 19, 20, 24, 30, 31, 36]

Enzyme Handbook © Springer-Verlag Berlin Heidelberg 1994
Duplication, reproduction and storage in data banks are only allowed with the prior permission of the publishers

K_m-value (mM)

0.31 (reduced glutathione) [6]; 0.8–1.7 (reduced glutathione, enzyme from different rat tissues) [8]; 1.43 (reduced glutathione) [11]; 1.28 (reduced glutathione) [15]; 0.010 (insulin) [1]; 0.003–0.012 (insulin, enzymes from different rat tissues) [8]; 0.031 (insulin) [11]; 0.014 (insulin) [15]; 0.27 (cysteine) [6]; 0.20 (2-mercaptoethanol, dihydrolipoate) [6]; 0.14 (dihydrolipoamide) [6]; 0.13 (reduced dithiothreitol) [6]; More (comparison of K_m values in different assay methods [24], K_m values for insulin at various fixed thiol concentrations and for various thiol compounds at fixed insulin concentrations [42]) [1, 7, 31, 24, 36, 42]

pH-optimum

7.0 [11]; 7.2 [15]; 7.5 (reduced glutathione [6]) [6, 39]; 7.5–8.5 [31]; 7.8 [36]; 8.0 [35]; 9.0 (DL-dihydrolipoate) [6]

pH-range

6.2–10.2 (6.2: about 10% of activity maximum, 10.2: about 20% of activity maximum) [39]

Temperature optimum (°C)

25 (assay at) [11]; 37 (assay at) [8, 9]

Temperature range (°C)

---

3 ENZYME STRUCTURE

Molecular weight

50000–55000 (rat, gel filtration) [11]
58000 (bovine, gel filtration) [22]
60000 (rat, SDS-PAGE, a 120000 MW protein also found [11], bovine, SDS-PAGE, polyacrylamide disc-gel electrophoresis [26], bovine, sedimentation equilibrium centrifugation [29]) [11, 26, 29]
92000 (bovine, gel filtration) [29]
120000 (rat, SDS-PAGE, MW 60000 determined for the main component) [11]
More (3 molecular forms: MW 56000 (major form), MW 51000 (only in spleen, possibly a proteolytic product of 56000 MW protein, active), MW 67000 (possibly a precursor of 56000 MW protein)) [1]

Subunits

Monomer (1 × 60000, bovine, SDS-PAGE [27], 1 × 62500, bovine, SDS-PAGE [29]) [27, 29]
Dimer (2 × 60000, bovine, SDS-PAGE, dimerization after prolonged storage at –20°C, freeze-thawing or heating at 60°C, monomers held together by an intermolecular disulfide bond) [27]
Trimer (bovine, enzyme contains 3 amino-terminal residues, therefore might be composed of 3 polypeptide chains or subunits) [17]

Glycoprotein/Lipoprotein

Glycoprotein (bovine pancreas: 1.6% carbohydrate [17], bovine liver: 12% carbohydrate [29]) [17, 29]

---

4 ISOLATION/PREPARATION

Source organism

Bovine [1, 16–20, 22, 23, 25–27, 29–31, 36, 38, 40–43]; Rat (Wistar rat [15], sand rat (Psammomys obesus) [15]) [1, 4–9, 11–15, 21, 23, 34, 38]; Mouse [1, 15, 32, 35]; Human [1, 3, 32, 37]

Source tissue

Liver [1, 3–9, 11–14, 16, 18–21, 23, 25, 27, 29, 31, 38, 41, 43]; Pancreas (pancreatic islets [15, 34]) [1, 8, 15, 17, 19, 22, 34, 38, 40, 42]; Kidney [1, 8, 19]; Heart [1, 8, 19]; Intestine [1, 8, 19]; Spleen [1, 8, 19]; Testis [1, 8, 19]; Thymus [1, 8]; Fat tissue [1, 8, 19]; Lung [1, 8, 19]; Brain [1, 8, 19]; Diaphragm [1, 8]; Skeletal muscle [1, 8, 19]; Lymph node [19]; Parotid gland [19]; Placenta [1, 5]; Retina [1]; Eye lens [1]; Erythrocytes [1]; Leukocytes [1]; Spinal cord [19]; Paunch [19]; Aorta [19]; Skin fibroblast [32]; Mammary gland [33]; Lymphoid tissue [35]; Islet cell tumors (insulinomas, glutathione-insulin transhydrogenase is present in an inactive state as a divalent metal ion complex that can be activated by EDTA and/or glutathione) [37]; More (higher activity in mononuclear cells than in polymorphonuclear cells [1], relative activities of enzyme in various tissues [8, 19]) [1, 8, 19]

Localisation in source

Membrane (bound to [1], plasma membrane [12], weakly bound to membrane phospholipid components [14]) [1, 9, 12–14]; Microsomes (majority of the enzyme located at [1], cisternal surface of microsomes [1, 14]) [1, 9, 11, 13, 14]; Zymogen granules (of acinar cells) [1]; Secretory granules (of alpha and beta cells of islets of Langerhans) [1]; Endoplasmic reticulum [1, 4]

Purification

Rat [7, 11]; Bovine [16, 20, 29, 31]

Enzyme Handbook © Springer-Verlag Berlin Heidelberg 1994
Duplication, reproduction and storage in data banks are only allowed with the prior permission of the publishers

Crystallization

–

Cloned

(characterization of cDNA for human glutathione-insulin transhydrogenase) [3]

Renaturated

–

5 STABILITY

pH

Temperature (°C)

60 (30 min, 30% loss of activity [31], 30 min, 50% loss of activity [39]) [31, 39]; 65 (30 min, 90% loss of activity) [39]; 80 (30 min, about 60% loss of activity) [31]

Oxidation

Organic solvent

General stability information

EDTA stabilizes [6]; EDTA stabilizes during purification [39]; Metal ions required for activity and maintenance of the proper conformation of the enzyme [6]; Highly susceptible to proteolytic attack [35]; Glutathione, 10 mM, enhances heat stability [39]; Insulin, 0.087 mM, decreases heat stability [39]; Dimerization after prolonged storage at –20°C, freeze-thawing or heating at 60°C [27]

Storage

–25°C, pH 7.5, phosphate buffer [16]

6 CROSSREFERENCES TO STRUCTURE DATABANKS

PIR/MIPS code

PIR3: A41077 (Q-5 Rat (fragment))

Brookhaven code

7 LITERATURE REFERENCES

[1] Varandani, P.T. in "Coenzymes Cofactors",3Pt. A,753–765 (1989) (Review)
[2] Varandani, P.T.: Dev. Biochem.,1,29–42 (1978) (Review)

[3] Morris, J.I., Varandani, P.T.: Biochim. Biophys. Acta,949,169–180 (1988)
[4] Chowdhary, B.K., Smith, G.D., Mahler, R., Peters, T.J. : Biosci. Rep.,3,323–329 (1983)
[5] Hern, E.P., Varandani, P.T.: Biochem. Biophys. Res. Commun.,116,909–915 (1983)
[6] Spolter, P.D., Vogel, J.M.: Biochim. Biophys. Acta,167,525–537 (1968)
[7] Varandani, P.T.: Biochim. Biophys. Acta,286,126–135 (1972)
[8] Chandler, M.L., Varandani, P.T.: Biochim. Biophys. Acta,286,136–145 (1972)
[9] Varandani, P.T.: Biochim. Biophys. Acta,304,642–659 (1973)
[10] Chandler, M.L., Varandani, P.T.: Biochim. Biophys. Acta,320,258–266 (1973)
[11] Ansorge, S., Bohley, P., Kirschke, H., Langner, J., Wiederanders, B., Hanson, H.: Eur. J. Biochem.,32,27–35 (1973)
[12] Varandani, P.T.: Biochem. Biophys. Res. Commun.,55,689–696 (1973)
[13] Ibbetson, A.L., Freedman, R.B.: Biochem. J.,159,377–384 (1976)
[14] Hern, E.P., Varandani, P.T.: Biochim. Biophys. Acta,732,170–178 (1983)
[15] Kohnert, K.-D., Hahn, H.-J., Zühlke, H., Schmidt, S., Fiedler, H.: Biochim. Biophys. Acta,338,68–77 (1974)
[16] Schneider, F., Schauer, R., Martini, O., Hahn, J.: Hoppe-Seyler's Z. Physiol. Chem.,348,391–400 (1967)
[17] Varandani, P.T.: Biochim. Biophys. Acta,371,577–581 (1974)
[18] Bjelland, S., Wallevik, K., Kroll, J., Dixon, J.E., Morin, J.E., Freedman, R.B., Lambert, N., Varandani, P.T., Nafz, M.A.: Biochim. Biophys. Acta,747,197–199 (1983)
[19] Bjelland, S.: Comp. Biochem. Physiol.,87B,907–914 (1987)
[20] Bjelland, S., Foltmann, B., Wallevik, K.: Anal. Biochem.,142,463–466 (1984)
[21] Roth, R.A.: Biochem. Biophys. Res. Commun.,98,431–438 (1981)
[22] Varandani, P.T., Plumley, H.: Biochim. Biophys. Acta,151,273–275 (1968)
[23] Barbieri, L., Batelli, M.G., Stirpe, F.: Arch. Biochem. Biophys.,216,380–383 (1982)
[24] Chandler, M.L., Varandani, P.T.: Biochim. Biophys. Acta,397,307–317 (1975)
[25] Moss, J., Stanley, S.J., Morin, J.E., Dixon, J.E.: J. Biol. Chem.,255,11085–11087 (1980)
[26] Pace, M., Pietta, P.G., Fiorino, A., Pocaterra, E., Dixon, J.E.: Experientia,41,1332–1335 (1985)
[27] Pace, M., Dixon, J.E.: Int. J. Pept. Protein Res.,14,409–413 (1979)
[28] Varandani, P.T., Nafz, M.A., Chanmdler, M.L.: Biochemistry,14,2115–2120 (1975)
[29] Carmichael, D.F., Morin, J.E., Dixon, J.E.: J. Biol. Chem.,252,7163–7167 (1977)
[30] Hillson, D.A., Freedman, R.B.: Biochem. J.,191,389–393 (1980)
[31] Tomizawa, H.H.: Methods Enzymol.,17B,515–519 (1971)
[32] Morin, J.E., Dixon, J.E., Chang, P.P., Moss, J.: Biochem. Biophys. Res. Commun.,111,872–877 (1983)
[33] Ferrier, B.M., Hendrie, J.M., Cardy, C.A.: Can. J. Biochem.,55,340–345 (1977)
[34] Kohnert, K.-D., Jahr, H., Schmidt, S., Hahn, H.-J., Zühlke, H.: Biochim. Biophys. Acta,422,254–259 (1976)
[35] Roth, R.A., Koshland, M.E.: Biochemistry,20,6594–6599 (1981)
[36] Lambert, N., Freedman, R.B.: Biochem. J.,213,235–243 (1983)
[37] Varandani, P.T.: Biochem. Biophys. Res. Commun.,60,1119–1126 (1974)
[38] Varandani, P.T., Nafz, M.A.: Biochim. Biophys. Acta,438,358–369 (1976)
[39] Morin, J.E., Carmichael, C.F., Dixon, J.E.: Arch. Biochem. Biophys.,189,354–363 (1978)
[40] Varandani, P.T., Nafz, M.A.: Biochim. Biophys. Acta,832,7–13 (1985)
[41] Hawkins, H.C., Freedman, R.B.: Biochem. J.,159,385–393 (1976)
[42] Chandler, M.L., Varandani, P.T.: Biochemistry,14,2107–2115 (1975)
[43] Hillson, D.A., Freedman, R.B.: Biochem. J.,191,373–388 (1980)

Enzyme Handbook © Springer-Verlag Berlin Heidelberg 1994
Duplication, reproduction and storage in data banks are only allowed with the prior permission of the publishers

1 NOMENCLATURE

EC number
1.8.4.3

Systematic name
Coenzyme A:oxidized-glutathione oxidoreductase

Recommended name
Glutathione-CoA-glutathione transhydrogenase

Synonymes
Transhydrogenase, glutathione-coenzyme A glutathione disulfide
Glutathione-coenzyme A glutathione disulfide transhydrogenase
Glutathione coenzyme A-glutathione transhydrogenase
Glutathione:coenzyme A-glutathione transhydrogenase [2]

CAS Reg. No.
37256-48-7

2 REACTION AND SPECIFICITY

Catalysed reaction
CoA + oxidized glutathione →
→ CoA-glutathione + glutathione

Reaction type
Redox reaction

Natural substrates
CoA-glutathione + reduced glutathione (possible functions: 1. salvage of the coenzyme form of pantothenic acid which might be lost to the cell as CoA-glutathione, 2. maintenance of the free CoA-SH cellular level) [1]

Substrate spectrum
1 CoA-glutathione + reduced glutathione (r) [1]
2 Unsymmetrical disulfide + reduced glutathione (e.g. pantetheine-glutathione [1], thiolethanolamine-glutathione [1], cysteine-glutathione) [1]
3 More (catalyzes GSH-GSSG exchange reaction) [1]

Enzyme Handbook © Springer-Verlag Berlin Heidelberg 1994
Duplication, reproduction and storage in data banks are only allowed with the prior permission of the publishers

Product spectrum
1 CoA + oxidized glutathione
2 Sulfides + oxidized glutathione
3 ?

Inhibitor(s)

Cofactor(s)/prostethic group(s)

Metal compounds/salts

Turnover number (min^{-1})

Specific activity (U/mg)
0.513 [1]

K_m-value (mM)
0.33 (reduced glutathione) [1]; 0.045 (CoA-glutathione) [1]

pH-optimum
8.2 [1]

pH-range
7–9 (7: about 30% of activity maximum, 9: about 75% of activity maximum) [1]

Temperature optimum (°C)
25 (assay at) [1]

Temperature range (°C)

3 ENZYME STRUCTURE

Molecular weight
120000 (bovine, gel filtration) [1]

Subunits

Glycoprotein/Lipoprotein
–

4 ISOLATION/PREPARATION

Source organism
Bovine [1]; Rat [1, 2]

Source tissue
Kidney [1]; Pancreas [1]; Brain [1]; Liver [1]; Lung [1]; Muscle [1]; Heart (low activity) [1]

Localisation in source

Purification
Bovine [1]

Crystallization
–

Cloned
–

Renaturated
–

5 STABILITY

pH

Temperature (°C)

Oxidation

Organic solvent

General stability information
Inactivation during storage is largely reversed by glutathione [1]

Storage
–12°C, 6 weeks, stable [1]; –12°C, 7 months, 70% loss of activity [1]

6 CROSSREFERENCES TO STRUCTURE DATABANKS

PIR/MIPS code

Brookhaven code

7 LITERATURE REFERENCES

[1] Chang, S.H., Wilken, D.R.: J. Biol. Chem.,241,4251–4260 (1966)
[2] Dyar, R.E., Wilken, D.R.: Arch. Biochem. Biophys.,153 ,619–626 (1972)

Enzyme Handbook © Springer-Verlag Berlin Heidelberg 1994
Duplication, reproduction and storage in data banks are only allowed with the prior permission of the publishers

1 NOMENCLATURE

EC number
1.8.4.4

Systematic name
Glutathione:cystine oxidoreductase

Recommended name
Glutathione-cystine transhydrogenase

Synonymes
Transhydrogenase, glutathione-cystine
GSH-cystine transhydrogenase
NADPH-dependent GSH-cystine transhydrogenase

CAS Reg. No.
37256-49-8

2 REACTION AND SPECIFICITY

Catalysed reaction
2 Glutathione + cystine →
→ oxidized glutathione + 2 cysteine

Reaction type
Redox reaction

Natural substrates
Cystine + glutathione [1]

Substrate spectrum
1 Cystine + glutathione [1–6]
2 Disulfide + glutathione (disulfide of low molecular weight, e.g.: D-cystine [1], L-cystine [1], cystine [2, 3], L-cystinyldiglycine [1], beta-hydroxyethyl disulfide [1], diacetyl-L-cystine [1], L-cystine diamide [1], homocystine [1, 3], oxidized glutathione [3], dimethyl disulfide [3], dithiodiglycolic acid [3], lipoic acid oxidized [3], at low substrate concentration the reaction rate is greater with L-cystine than with any of the other substances [1]) [1–3]
3 More (inactive with: insulin and other proteins [2], albumin [3], keratin [3]) [2, 3]

Enzyme Handbook © Springer-Verlag Berlin Heidelberg 1994
Duplication, reproduction and storage in data banks are only allowed with the prior permission of the publishers

Product spectrum

1 Cysteine + oxidized glutathione [1–3]
2 Sulfide + oxidized glutathione
3 ?

Inhibitor(s)

Cu^{2+} [3]; Zn^{2+} [3]; Cd^{2+} [3]; Mn^{2+} [3]; Co^{2+} [3]; Hg^{2+} [3]; Iodoacetate [3]; MoO_3 [3]; Na_2HAsO_4 [3]

Cofactor(s)/prostethic group(s)

EDTA (activates) [3]

Metal compounds/salts

Turnover number (min^{-1})

Specific activity (U/mg)

61.6 (beta-hydroxyethyldisulfide) [1, 2]; More [3]

K_m-value (mM)

0.23 (cystine) [6]; 1.67 (glutathione) [6]

pH-optimum

8.6 (lower value (pH 7.8) chosen for the assay to achieve smaller non-enzymatic rate) [1, 2]

pH-range

6.0–8.5 (6.0: 10% of activity maximum, 8.5: activity maximum) [1]

Temperature optimum (°C)

23 (assay at) [1]

Temperature range (°C)

3 ENZYME STRUCTURE

Molecular weight

150000 (Saccharomyces cerevisiae, gel filtration) [1]

Subunits

Glycoprotein/Lipoprotein

–

4 ISOLATION/PREPARATION

Source organism
Saccharomyces cerevisiae [1–3]; Rat [4–6]

Source tissue
Cells [1, 2]; Intestinal mucosa [5]; Liver [4]; Kidney [4]; Small intestine [4, 6]; More (no activity in brush border) [5]

Localisation in source
Soluble (supernatant fraction) [5, 6]

Purification
Saccharomyces cerevisiae [1–3]

Crystallization
–

Cloned
–

Renaturated
–

5 STABILITY

pH

Temperature (°C)
4 (24 h, complete inactivation) [6]; 54 (crude state, quite labile to heat above 54°C) [2]; 60 (denaturation, crude state) [6]

Oxidation

Organic solvent

General stability information
Ammonium sulfate, 1 M, complete stabilization, 6 days at 20°C [1]; Glutathione stabilizes, stabilizing effect increased by cystine [1, 2]; Ethylene glycol, 10%, stabilizes [1, 2]; Glycerol, no stabilization [1]; EDTA stabilizes [1]; Cystine, alone: no stabilization, in addition to glutathione: increases stabilizing effect [1]

Storage
–20°C, pH 5.8, 1 mM glutathione, several months [1, 2]; 4°C, 24 h, complete inactivation, crude state [6]

Enzyme Handbook © Springer-Verlag Berlin Heidelberg 1994
Duplication, reproduction and storage in data banks are only allowed with the prior permission of the publishers

6 CROSSREFERENCES TO STRUCTURE DATABANKS

PIR/MIPS code

Brookhaven code

7 LITERATURE REFERENCES

[1] Nagai, S., Black, S.: J. Biol. Chem.,243,1942–1947 (1968)
[2] Nagai, S.: Methods Enzymol.,17B,510–515 (1971)
[3] Minoda, Y., Kurane, R., Yamada, K.: Agric. Biol. Chem.,37,2511–2516 (1973)
[4] Wendell, P.L.: Biochim. Biophys. Acta,159,179–181 (1968)
[5] States, B., Segal, S.: Biochem. J.,113,443–444 (1969)
[6] States, B., Segal, S.: Biochem. J.,132,623–631 (1973)

1 NOMENCLATURE

EC number
1.8.4.5

Systematic name
L-Methionine:oxidized-thioredoxin S-oxidoreductase

Recommended name
Methionine-S-oxide reductase

Synonymes
Reductase, methionine sulfoxide
Acetylmethionine sulfoxide reductase
Methionine sulfoxide reductase
Methionine S-oxide reductase

CAS Reg. No.
70248-65-6

2 REACTION AND SPECIFICITY

Catalysed reaction
L-Methionine S-oxide + reduced thioredoxin →
→ L-methionine + oxidized thioredoxin

Reaction type
Redox reaction

Natural substrates
L-Methionine S-oxide + reduced thioredoxin (methionine sulfoxide can support the growth of a methionine auxotroph of E. coli) [4]

Substrate spectrum
1 L-Methionine S-oxide + reduced thioredoxin (or dithiothreitol) (r) [1, 2]
2 More (E. coli: reductase catalyzes reduction of methionine sulfoxide to methionine, no reduction of protein bound sulfoxide [1], yeast: enzyme system consisting of 3 separable proteins which catalyze the specific reduction of L(-)-methionine sulfoxide to methionine by NADPH [3]) [1, 3]

Product spectrum
1 L-Methionine + oxidized thioredoxin (r)
2 ?

Enzyme Handbook © Springer-Verlag Berlin Heidelberg 1994
Duplication, reproduction and storage in data banks are only allowed with the prior permission of the publishers

Inhibitor(s)

Dimethylsulfoxide [2]; Arsenite [2]; Methyl ethyl sulfoxide [2]; Iodoacetamide [2]; Formaldehyde [2]

Cofactor(s)/prostethic group(s)

Metal compounds/salts

Turnover number (min^{-1})

Specific activity (U/mg)

More [1]

K_m-value (mM)

0.0003 (methionine sulfoxide) [1]

pH-optimum

pH-range

Temperature optimum (°C)

Temperature range (°C)

---

3 ENZYME STRUCTURE

Molecular weight

Subunits

Glycoprotein/Lipoprotein

–

---

4 ISOLATION/PREPARATION

Source organism

E. coli (reductase catalyzes reduction of methionine sulfoxide to methionine, no reduction of protein bound sulfoxide [1]) [1, 4]; Saccharomyces cerevisiae [2]; Yeast (enzyme system consisting of 3 separable proteins which catalyze the specific reduction of L(-)-methionine sulfoxide to methionine by NADPH) [3]

Source tissue

Localisation in source

Purification

Crystallization
–

Cloned
–

Renaturated
–

---

5 STABILITY

pH

Temperature (°C)
90 (2 min, 50% loss of activity) [1]

Oxidation

Organic solvent

General stability information

Storage

---

6 CROSSREFERENCES TO STRUCTURE DATABANKS

PIR/MIPS code

Brookhaven code

---

7 LITERATURE REFERENCES

[1] Ejiri, S., Weissbach, H., Brot, N.: Anal. Biochem.,102,393–398 (1980)
[2] Gibson, R.M., Large, P.J.: FEMS Microbiol. Lett.,26,95–99 (1985)
[3] Black, S., Harte, E.M., Hudson, B., Wartolofsky, L.: J. Biol. Chem.,235,2910–2916 (1960)
[4] Ejiri, S., Weissbach, H., Brot, N.: J. Bacteriol.,139,161–164 (1979)

Enzyme Handbook © Springer-Verlag Berlin Heidelberg 1994
Duplication, reproduction and storage in data banks are only allowed with the prior permission of the publishers

1 NOMENCLATURE

EC number
1.8.4.6

Systematic name
Protein-L-methionine:oxidized-thioredoxin S-oxidoreductase

Recommended name
Protein-methionine-S-oxide reductase

Synonymes
Reductase, methionine sulfoxide (protein)
Methionine sulfoxide peptide reductase
Protein (methionine sulfoxide) reductase
Met(O)-peptide reductase [1]
Peptide methionine sulfoxide reductase [3]

CAS Reg. No.
78206-57-2

2 REACTION AND SPECIFICITY

Catalysed reaction
Protein L-methionine S-oxide + reduced thioredoxin →
→ protein L-methionine + oxidized thioredoxin

Reaction type
Redox reaction

Natural substrates
Protein L-methionine S-oxide + reduced thioredoxin (biological reducing system consists of NADPH, thioredoxin and thioredoxin reductase, enzymatic reduction can restore biological activity of an inactive oxidized protein or peptide [1, 5, 6]: e.g.
N-formyl-L-methionyl-sulfoxide-L-leucyl-L-phenylalanine [5], alpha-1-proteinase inhibitor [6]) [1, 5, 6]

Enzyme Handbook © Springer-Verlag Berlin Heidelberg 1994
Duplication, reproduction and storage in data banks are only allowed with the prior permission of the publishers

Substrate spectrum

1 Protein L-methionine S-oxide + reduced thioredoxin (dithiothreitol can replace reduced thioredoxin [1, 4], peptide L-methionine S-oxide can replace protein L-methionine, e.g. E. coli ribosomal protein [1, 2, 4], alpha-1-proteinase inhibitor [1, 6], N-formylmethionine-leucine-phenylalanine (product is a chemotactic peptide) [1, 5], methionine sulfoxide in oxidized [Met]enkephalin [4]) [1, 2, 4–6]
2 N-Acetyl-L-methionine sulfoxide + reduced acceptor [3]
3 More (does not act on free methionine)

Product spectrum

1 Protein L-methionine + oxidized thioredoxin (or peptide L-methionine, or oxidized dithiothreitol)
2 N-Acetyl-L-methionine + oxidized acceptor [3]
3 ?

Inhibitor(s)

Methionine sulfoxide (0.1 mM, less than 20% inhibition) [4]

Cofactor(s)/prostethic group(s)

Metal compounds/salts

Turnover number (min^{-1})

Specific activity (U/mg)

More [1]

K_m-value (mM)

pH-optimum

7.4 (assay at) [1, 4]

pH-range

Temperature optimum (°C)

37 (assay at) [1, 4]

Temperature range (°C)

---

3 ENZYME STRUCTURE

Molecular weight

23000–24000 (E. coli, SDS-PAGE) [1]
18000–20000 (E. coli, gel filtration) [4]
18000 (human, gel filtration) [1]

Subunits

Glycoprotein/Lipoprotein

–

4 ISOLATION/PREPARATION

Source organism

E. coli [1, 4, 5, 6]; Human (HeLa cells [4]) [1, 2, 4]; Bovine [2]; Rat [4]; Euglena gracilis [4]; Tetrahymena pyriformis [4]; Spinach [4]

Source tissue

Leukocytes [1]; Lenses (high activity in outer epithelial layer with decreasing activity in the inner layer) [2]; Lung [4]; Liver [4]; Kidney [4]; Brain [4]

Localisation in source

Soluble [2]

Purification

E. coli (partial [4]) [1, 4]; Human (partial) [1]

Crystallization

–

Cloned

–

Renaturated

–

5 STABILITY

pH

Temperature (°C)

Oxidation

Organic solvent

General stability information

Storage

Enzyme Handbook © Springer-Verlag Berlin Heidelberg 1994
Duplication, reproduction and storage in data banks are only allowed with the prior permission of the publishers

6 CROSSREFERENCES TO STRUCTURE DATABANKS

PIR/MIPS code

Brookhaven code

7 LITERATURE REFERENCES

[1] Brot, N., Fliss, H., Coleman, T., Weissbach, H.: Methods Enzymol.,107,352–360 (1984)
[2] Spector, A., Scotto, R., Weissbach, H., Brot, N.: Biochem. Biophys. Res. Commun.,108,429–436 (1982)
[3] Brot, N., Werth, J., Koster, D., Weissbach, H.: Anal. Biochem.,122,291–294 (1982)
[4] Brot, N., Weissbach, L., Werth, J., Weissbach, H.: Proc. Natl. Acad. Sci. USA,78,2155–2158 (1981)
[5] Fliss, H., Vasanthakumar, G., Schiffmann, E., Weissbach, H., Brot, N.: Biochem. Biophys. Res. Commun.,109,194–201 (1982)
[6] Abrams, W.R., Weinbaum, G., Weissbach, L., Weissbach, H., Brot, N.: Proc. Natl. Acad. Sci. USA,78,7483–7486 (1981)

1 NOMENCLATURE

EC number
1.8.4.7

Systematic name
[Xanthine-dehydrogenase]:oxidized-glutathione S-oxidoreductase

Recommended name
Enzyme-thiol transhydrogenase (oxidized-glutathione)

Synonymes
Glutathione-dependent thiol:disulfide oxidoreductase
Thiol:disulphide oxidoreductase [1]

CAS Reg. No.

2 REACTION AND SPECIFICITY

Catalysed reaction
[Xanthine dehydrogenase] + oxidized glutathione →
→ [xanthine oxidase] + glutathione (converts EC 1.1.1.204 into EC 1.1.3.22 in presence of oxidized glutathione, also reduces the disulfide bond of ricin)

Reaction type
Redox reaction

Natural substrates
[Xanthine dehydrogenase] + oxidized glutathione [1]

Substrate spectrum
1 [Xanthine dehydrogenase] + oxidized glutathione [1, 2]
2 More (other disulfide compounds are either inactive or far less active than oxidized glutathione, reduction of disulfide bonds in ricin, thioltransferase activity, GSH: insulin transhydrogenase activity) [1]

Product spectrum
1 [Xanthine oxidase] + glutathione [1, 2]
2 ?

Inhibitor(s)

Cofactor(s)/prosthethic group(s)

Enzyme Handbook © Springer-Verlag Berlin Heidelberg 1994
Duplication, reproduction and storage in data banks are only allowed with the prior permission of the publishers

Metal compounds/salts

Turnover number (min^{-1})

Specific activity (U/mg)
More [1, 2]

K_m-value (mM)

pH-optimum
7.0–8.5 [2]

pH-range
6–9.5 [2]

Temperature optimum (°C)
37 (assay at) [1]

Temperature range (°C)

---

3 ENZYME STRUCTURE

Molecular weight
40000–43000 (rat, gel filtration, SDS-PAGE) [1]

Subunits

Glycoprotein/Lipoprotein
–

---

4 ISOLATION/PREPARATION

Source organism
Rat [1, 2]

Source tissue
Liver [1, 2]; Lung [1]; Spleen [1]; Skeletal muscle [1]; Brain [1]; Skin [1]; Kidney [1]

Localisation in source

Purification
Rat [1, 2]

Crystallization
–

Cloned

–

Renaturated

–

5 STABILITY

pH

Temperature (°C)

Oxidation

Organic solvent

General stability information

Storage

–25°C [2]

6 CROSSREFERENCES TO STRUCTURE DATABANKS

PIR/MIPS code

Brookhaven code

7 LITERATURE REFERENCES

[1] Battelli, M.G., Lorenzoni, E.: Biochem. J.,207,133–138 (1982)
[2] Battelli, M.G.: FEBS Lett.,113,47–51 (1980)

Enzyme Handbook © Springer-Verlag Berlin Heidelberg 1994
Duplication, reproduction and storage in data banks are only allowed with the prior permission of the publishers

1 NOMENCLATURE

EC number
1.8.5.1

Systematic name
Glutathione:dehydroascorbate oxidoreductase

Recommended name
Glutathione dehydrogenase (ascorbate)

Synonymes
Dehydrogenase, glutathione (ascorbate)
Dehydroascorbic reductase
Dehydroascorbic acid reductase
Glutathione dehydroascorbate reductase
DHA reductase [1]
Dehydroascorbate reductase [4]
GDOR [2]
Glutathione:dehydroascorbic acid oxidoreductase [10]

CAS Reg. No.
9026-38-4

2 REACTION AND SPECIFICITY

Catalysed reaction
2 Glutathione + dehydroascorbate →
→ oxidized glutathione + ascorbate (ordered or random mechanism involving the formation of a ternary complex [5], zero-ordered kinetics only observed for hydrogen acceptor, not for glutathione [10])

Reaction type
Redox reaction

Natural substrates
Dehydroascorbic acid + glutathione (vitamin C-conserving mechanism [2], regeneration of ascorbate from monodehydroascorbate and dehydroascorbate produced by ascorbate peroxidase, EC 1.11.1.11, for scavenging hydrogen peroxide [5]) [2, 5]

Enzyme Handbook © Springer-Verlag Berlin Heidelberg 1994
Duplication, reproduction and storage in data banks are only allowed with the prior permission of the publishers

Substrate spectrum

1 Dehydroascorbic acid + glutathione (specific for glutathione as hydrogen donor [4, 5, 10], specific for dehydroascorbate [5], L-threo-dehydroascorbate: best substrate [4, 10], D-threo-dehydroascorbate: 20% of the activity with L-threo-dehydroascorbate, activity with both erythro-dehydroascorbates lies between the threo-stereoisomers [4, 10]) [1–10]

2 More (L-cysteinyl-L-glycine not active as hydrogen donor [4], NADH, NADPH and cysteine cannot substitute for glutathione [5, 10], cysteine cannot replace glutathione [6]) [5, 6, 10]

Product spectrum

1 L-Ascorbate + oxidized glutathione [1–10]

2 ?

Inhibitor(s)

Zn^{2+} [5]; Fe^{3+} [5]; Cu^{2+} [5, 6]; Co^{2+} [5]; N-Ethylmaleimide [5, 6, 10]; p-Chloromercuribenzoate [5, 13]; Dehydroascorbate (high concentration) [6]; p-Hydroxymercuribenzoate [6, 13]; Iodoacetic acid [6, 10]; $HgCl_2$ [13]; Mersalyl [13]

Cofactor(s)/prostethic group(s)

2-Mercaptoethanol (1 mM, stimulation) [5]; Dithiothreitol (1 mM, stimulation) [5]

Metal compounds/salts

Turnover number (min^{-1})

Specific activity (U/mg)

0.264 [5]; More (assay method [3], spectrometric assay [12]) [3, 6, 10, 12]

K_m-value (mM)

1.3 (dehydroascorbate) [7]; 3.8 (glutathione) [7]; 0.85 (glutathione) [5]; 0.26 (dehydroascorbate) [5]; 0.34 (dehydroascorbate) [6]; 4.45 (glutathione) [6]; 0.39 (dehydroascorbate) [13]; 4.35 (glutathione) [13]; More [10]

pH-optimum

7.0 (maximal activity above [10]) [5, 10]; 7.5 [6, 8]; 8 [13]; More (non-enzymatic reduction occurs rapidly at pH 7.5–8.0, therefore the assay should be performed between pH 6.3 and 6.8) [8]

pH-range

Temperature optimum (°C)

38 [5]

Temperature range (°C)

3 ENZYME STRUCTURE

Molecular weight

23000 (potato, gel filtration) [13]
24000 (wheat, SDS-PAGE) [10]
25000 (spinach, gel filtration) [6]
28000 (Euglena gracilis Z, gel filtration) [5]

Subunits

Glycoprotein/Lipoprotein

–

4 ISOLATION/PREPARATION

Source organism

Euglena gracilis Z [5]; Wheat [1, 3, 10]; Guinea pig [2]; Spinach [6, 8, 12]; Human [7, 11]; Rat [9]; Potato [13]

Source tissue

Blood neutrophile [7]; Lymphocytes [7]; Grains (germinating) [1]; Brain [2]; Adrenal [2]; Stomach [2]; Liver [2, 9]; Leaves [6, 8, 12]; Wheat flour [10]; Tubers [13]; Erythrocytes [11]

Localisation in source

Cytoplasm [5, 7, 13]

Purification

Euglena gracilis Z (partial) [5]; Spinach (partial [6]) [6, 8]; Wheat [10]; Potato [13]

Crystallization

–

Cloned

–

Renaturated

–

Enzyme Handbook © Springer-Verlag Berlin Heidelberg 1994
Duplication, reproduction and storage in data banks are only allowed with the prior permission of the publishers

5 STABILITY

pH

7.0–8.0 (stable up to 42°C) [5]

Temperature (°C)

42 (stable up to, pH 7.0–8.0) [5]; 50 (10 min, 35% loss of activity [6], 7 min, stable up to [13]) [6, 13]; 60 (10 min, complete inactivation) [6]

Oxidation

Organic solvent

General stability information

Purified dehydroascorbate reductase is extremely unstable [6]

Storage

4°C, in presence of 2-mercaptoethanol stable for several days [13]

6 CROSSREFERENCES TO STRUCTURE DATABANKS

PIR/MIPS code

Brookhaven code

7 LITERATURE REFERENCES

[1] Redman, D.G.: Chem. Ind. (London) ,10,414–415 (1974)
[2] Grimble, R.F., Hughes, R.E.: Life Sci.,7,383–386 (1968)
[3] Kahnt, W.-D., Mundy, V., Grosch, W.: Z. Lebensm. Unters. Forsch.,158,77–82 (1975)
[4] Walther, C., Grosch, W.: J. Cereal Sci.,5,299–305 (1987)
[5] Shigeoka,, S., Yasumoto, R., Onishi, T., Nakano, Y., Kitaoka, S.: J. Gen. Microbiol.,133,227–232 (1987)
[6] Foyer, C.H., Halliwell, B.: Phytochemistry,16,1347–1350 (1977)
[7] Bigley, R., Riddle, M., Layman, D., Stankova, L.: Biochim. Biophys. Acta,659,15–22 (1981)
[8] Stahl, R.L., Liebes, L.F., Silber, R.: Methods Enzymol.,122,10–12 (1986) (Review)
[9] Hsu, J.M., Geller, S.: Arch. Biochem. Biophys.,118,85–89 (1967)
[10] Boeck, D., Grosch, W.: Z. Lebensm. Unters. Forsch.,162,243–251 (1976)
[11] Basu, S., Som, S., Deb, S., Mukherjee, D., Chatterjee, I.B.: Biochem. Biophys. Res. Commun.,90,1335–1340 (1979)
[12] Stahl, R.L., Liebes, L.F., Farber, C.M., Silber, R.: Anal. Biochem.,131,341–344 (1983)
[13] Dipierro, S., Borraccino, G.: Phytochemistry,30,427–429 (1991)

1 NOMENCLATURE

EC number
1.8.7.1

Systematic name
Hydrogen-sulfide:ferredoxin oxidoreductase

Recommended name
Sulfite reductase (ferredoxin)

Synonymes
Reductase, sulfite (ferredoxin)
Ferredoxin-sulfite reductase

CAS Reg. No.
37256-50-1

2 REACTION AND SPECIFICITY

Catalysed reaction
Sulfite + 3 reduced ferredoxin →
→ hydrogen sulfide + 3 oxidized ferredoxin + 3 H_2O

Reaction type
Redox reaction

Natural substrates
Sulfite + reduced ferredoxin (enzyme of assimilatory sulfate reduction [5], enzyme of dissimilatory sulfite reduction [10]) [5, 10]

Substrate spectrum
1 Sulfite + reduced ferredoxin [1–14]
2 S-Sulfoglutathione + reduced ferredoxin [6]
3 NO_2^- + reduced ferredoxin (ferredoxin replaceable by methyl viologen) [9]
4 NH_2OH + reduced ferredoxin (ferredoxin replaceable by methyl viologen) [9, 13]
5 More (electron acceptors: ferredoxin [1], methyl viologen (low activity [1]) [1, 2, 9, 11, 14], benzyl viologen (low activity) [14]) [1, 2, 9, 11, 14]

Enzyme Handbook © Springer-Verlag Berlin Heidelberg 1994
Duplication, reproduction and storage in data banks are only allowed with the prior permission of the publishers

Product spectrum

1 Hydrogen sulfide + oxidized ferredoxin + H_2O [1–14]
2 Glutathione persulfide + oxidized ferredoxin
3 NH_3 + oxidized ferredoxin (or methyl viologen)
4 ?
5 ?

Inhibitor(s)

CN^- [9]; CO [9]; NaCl [13]; Tris buffer (inhibition from pH 7.0 to 8.5) [14]

Cofactor(s)/prostethic group(s)

Siroheme (prosthetic group) [1–4, 7–10, 12]

Metal compounds/salts

Iron (contains one siroheme and one 4Fe-4S center [8, 9], iron is in high spin Fe^{3+} state [9], siroheme-containing iron-sulfur protein [10]) [8–10]

Turnover number (min^{-1})

Specific activity (U/mg)

61.1 [2]; 49.0 [3]; 100 [1]; 44.6 [4]; 5.09 [12]; More [14]

K_m-value (mM)

0.06 (S-sulfoglutathione) [6]; 0.011 (SO_3^{2-} (with reduced methyl viologen as electron donor), 63K sulfite reductase) [9]; 3.5 (NO_2^- (with reduced methyl viologen as electron donor), 63K sulfite reductase) [9]; 14.3 (NH_2OH (with reduced methyl viologen as electron donor), 63K sulfite reductase) [9]; 0.021 (SO_3^{2-} (with reduced ferredoxin as electron donor), reduced ferredoxin (with SO_3^{2-}as electron acceptor) 63K sulfite reductase) [9]; 3.8 (NO_2^- (with reduced ferredoxin as electron donor), 63K sulfite reductase) [9]; 0.025 (reduced ferredoxin (with NO_2^- as electron acceptor), SO_3^{2-} (with reduced ferredoxin as electron donor) 69K sulfite reductase) [9]; 0.012 (SO_3^{2-} (with reduced methyl viologen as electron donor), 69K sulfite reductase) [9]; 3.1 (NO_2^- (with reduced methyl viologen as electron donor), 69K sulfite reductase) [9]; 14.5 (NH_2OH (with reduced methyl viologen as electron donor), 69K sulfite reductase) [9]; 4.1 (NO_2^- (with reduced ferredoxin as electron donor), 69K sulfite reductase) [9]; 0.019 (reduced ferredoxin (with SO_3^{2-}as electron acceptor), 69K sulfite reductase) [9]; 0.028 (reduced ferredoxin (with NO_2^- as electron acceptor), 69K sulfite reductase [9], sulfite [3]) [3, 9]

pH-optimum

7.0 [14]; 8.0–8.5 [11]

pH-range

Temperature optimum (°C)

Temperature range (°C)

3 ENZYME STRUCTURE

Molecular weight

64000 (Brassica chinensis, SDS-PAGE) [2]
65000–70000 (Porphyra yezoensis, SDS-PAGE, gel filtration) [1]
112000 (Spinacia oleracea, gel filtration [13], sedimentation equilibrium analysis under nondenaturing conditions, 63K sulfite reductase [9]) [9, 13]
120000 (Spirulina platensis, gel filtration) [3]
134000 (Spinacia oleracea, sedimentation equilibrium analysis under non-denaturing conditions, 69K sulfite reductase) [9]
270000 (Spinacia oleracea, gel-electrophoresis under nondenaturing conditions) [4]

Subunits

Dimer (2 × 63000, Spirulina platensis [3], Spinacia oleracea, 63K sulfite reductase [9], SDS-PAGE [3, 9], 2 × 69000, Spinacia oleracea, SDS-PAGE, 69K sulfite reductase [9]) [3, 9]
Tetramer (4 × 71000, Spinacia oleracea, SDS-PAGE) [4]

Glycoprotein/Lipoprotein

–

4 ISOLATION/PREPARATION

Source organism

Desulfovibrio gigas [10]; Hordeum vulgare (barley) [11]; Clostridium pasteurianum [14]; Porphyra yezoensis [1]; Brassica chinensis (rape) [2]; Spirulina platensis (Cyanobacterium) [3, 6]; Spinacia oleracea [4, 7–9, 12, 13]; Pisum sativum [5]

Source tissue

Leaves [2, 4, 9, 13]; Seedlings [5]; Roots [5, 11]

Localisation in source

Proplastids (of pea root) [5]

Purification

Spinacia oleracea (2 forms: 63K sulfite reductase, 69K sulfite reductase [9]) [4, 9]; Porphyra yezoensis [1]; Brassica chinensis [2]; Spirulina platensis [3]

Crystallization

–

Enzyme Handbook © Springer-Verlag Berlin Heidelberg 1994
Duplication, reproduction and storage in data banks are only allowed with the prior permission of the publishers

Cloned

–

Renaturated

–

5 STABILITY

pH

Temperature (°C)

Oxidation

Organic solvent

General stability information

Stable at low ionic strength [13]

Storage

–70°C [9]; –20°C, 3 months [13]

6 CROSSREFERENCES TO STRUCTURE DATABANKS

PIR/MIPS code

PIR1: RDYCS7 (Synechococcus sp. (PCC 7942))

Brookhaven code

7 LITERATURE REFERENCES

[1] Koguchi, O., Tamura, G.: Agric. Biol. Chem.,53,1653–1662 (1989)
[2] Koguchi, O., Takahashi, H., Tamura, G.: Agric. Biol. Chem.,52,1867–1868 (1988)
[3] Koguchi, O., Tamura, G.: Agric. Biol. Chem.,52,373–380 (1988)
[4] Aketagawa, J., Tamura, G.: Agric. Biol. Chem.,44,2371–2378 (1980)
[5] Brunold, C., Suter, M.: Planta,179,228–234 (1989)
[6] Koguchi, O., Tamura, G.: Agric. Biol. Chem.,53,783–788 (1989)
[7] Hirasawa, M., Boyer, J.M., Gray, K.A., Davis, D.J., Knaff, D.B.: FEBS Lett.,221,343–348 (1987)
[8] Krueger, R.J., Siegel, L.M.: Biochemistry,21,2905–2909 (1982)
[9] Krueger, R.J., Siegel, L.M.: Biochemistry,21,2892–2904 (1982)
[10] Hall, M.H., Prince, R.H., Cammack, R.: Biochim. Biophys. Acta,581,27–33 (1979)
[11] Tamura, G., Hosoi, T.: Agric. Biol. Chem.,43,1601–1602 (1979)
[12] Tamura, G., Hosoi, T., Aketagawa, J.: Agric. Biol. Chem.,42,2165–2167 (1978)
[13] Asada, K., Tamura, G., Bandurski, R.S.: Methods Enzymol.,17B,528–539 (1971) (Review)
[14] Laishley, L.E.J., Lin, P.-M., Peck, H.D.: Can. J. Microbiol.,17,889–895 (1971)

1 NOMENCLATURE

EC number
1.8.99.1

Systematic name
Hydrogen-sulfite:(acceptor) oxidoreductase

Recommended name
Sulfite reductase

Synonymes
Reductase, sulfite
Assimilatory sulfite reductase
Assimilatory-type sulfite reductase

CAS Reg. No.
37256-51-2

2 REACTION AND SPECIFICITY

Catalysed reaction
Hydrogen sulfide + acceptor + 3 H_2O →
→ sulfite + reduced acceptor

Reaction type
Redox reaction

Natural substrates
Sulfite + reduced acceptor (no information about physiological electron acceptor)

Substrate spectrum
1 Sulfite + reduced acceptor (e.g. methyl viologen (sole artificial electron donor for Desulfovibrio vulgaris enzyme [3]) [1, 3, 5], benzyl viologen [1, 5], phenazine methosulfate [1], janus green [1], methylene blue [1], nile blue [1], toluidine blue [1]) [1–3, 5]
2 Hydroxylamine + reduced methyl viologen [5]
3 More (no substrates: neutral red, phenosafranine, janus green, nile blue, methylene blue, toluidine blue, thionine, phenazine methosulfate, FMN, FAD, riboflavin, cytochrome c_3, ferredoxin, NAD(P)H) [5]

Enzyme Handbook © Springer-Verlag Berlin Heidelberg 1994
Duplication, reproduction and storage in data banks are only allowed with the prior permission of the publishers

Product spectrum

1 Sulfide + oxidized acceptor (no products: trithionate, thiosulfate)
2 ?
3 ?

Inhibitor(s)

KCN [1, 5]; Arsenite [1, 5]; p-Chloromercuribenzoate (not [5]) [1]; EDTA (not [5]) [1]; More (not inhibitory: thiourea, N-ethylmaleimide, diethyldithiocarbamate, NH_2NH_2, Atabrine, 8-hydroxyquinoline, $CuSO_4$, NaN_3) [5]

Cofactor(s)/prostethic group(s)

4Fe-4S center (1 per mol of enzyme) [4]; More (not NAD(P)H as electron donor) [1, 2, 5]

Metal compounds/salts

Fe (contains Fe [1], 1 4Fe-4S center per mol of enzyme [4]) [1, 4]; Cu (contains Cu) [1]; More (no Co, Mn, Mo, Zn) [1]

Turnover number (min^{-1})

Specific activity (U/mg)

4.98 [5]; 1.48 [1]; 0.9 [3]

K_m**-value** (mM)

0.65 (sulfite) [1]; 0.025 (sulfite) [5]; 20 (hydroxylamine) [5]

pH-optimum

7.2–7.8 (Tris-HCl buffer) [5]; 7.5–8.0 [1]; 8.0 (above, phosphate buffer) [5]

pH-range

Temperature optimum (°C)

Temperature range (°C)

3 ENZYME STRUCTURE

Molecular weight

84000–87000 (Porphyra yezoensis, gel filtration) [1]
26800–27200 (Desulfovibrio vulgaris, sedimentation equilibrium [3], amino acid composition [4]) [3, 4]

Subunits

Glycoprotein/Lipoprotein

–

4 ISOLATION/PREPARATION

Source organism

Porphyra yezoensis [1]; Chlorobium sp. (strain PM and T) [2]; Rhodopseudomonas viridis [2]; Rhodomicrobium vannielii [2]; Rhodospirillum rubrum [2]; Rhodopseudomonas gelatinosa [2]; Rhodopseudomonas palustris [2]; Chromatium sp. strain D [2]; Desulfovibrio vulgaris [3, 4]; Aspergillus nidulans [5]

Source tissue

Thallus [1]; Mycelium [5]

Localisation in source

Soluble part of cell [2]

Purification

Porphyra yezoensis [1]; Desulfovibrio vulgaris [3, 4]; Aspergillus nidulans [5]

Crystallization

–

Cloned

–

Renaturated

–

5 STABILITY

pH

Temperature (°C)

50 (5 min, 50% loss of activity) [5]

Oxidation

Organic solvent

General stability information

Storage

–80°C, 0.05 M Tris-HCl buffer, pH 7.6, several months [4]; –20°C, several months [5]; –13.5°C, 6 months, no loss of activity [1]

Enzyme Handbook © Springer-Verlag Berlin Heidelberg 1994
Duplication, reproduction and storage in data banks are only allowed with the prior permission of the publishers

6 CROSSREFERENCES TO STRUCTURE DATABANKS

PIR/MIPS code

Brookhaven code

7 LITERATURE REFERENCES

[1] Saito, E., Tamura, G.: Agric. Biol. Chem.,35,491–500 (1971)
[2] Peck, H.D., Tedro, S., Kamen, M.D.: Proc. Natl. Acad. Sci. USA,71,2404–2406 (1974)
[3] Lee, J.-P., LeGall, J., Peck, H.D.: J. Bacteriol.,115 ,529–542 (1973)
[4] Huynh, B.H., Kang, L., DerVatanian, D.V., Peck, H.D., LeGall, J.: J. Biol. Chem.,259,15373–15376 (1984)
[5] Yoshimoto, A., Nakamura, T., Sato, R.: J. Biochem.,62 ,756–766 (1967)

1 NOMENCLATURE

EC number

1.8.99.2

Systematic name

AMP,sulfite:(acceptor) oxidoreductase

Recommended name

Adenylylsulfate reductase

Synonymes

Reductase, adenylylsulfate
Adenosine phosphosulfate reductase
Adenosine 5'-phosphosulfate reductase
APS-reductase
APS reductase [1]

CAS Reg. No.

9027-75-2

2 REACTION AND SPECIFICITY

Catalysed reaction

AMP + sulfite + acceptor →
→ adenylylsulfate + reduced acceptor (mechanism [13, 18])

Reaction type

Redox reaction

Natural substrates

Adenosine 5'-phosphosulfate + reduced acceptor (acts as terminal reductase in the respiratory sequence from sulfate to sulfide) [20]

Enzyme Handbook © Springer-Verlag Berlin Heidelberg 1994
Duplication, reproduction and storage in data banks are only allowed with the prior permission of the publishers

Substrate spectrum

1 AMP + sulfite + oxidized acceptor (acceptor: methyl viologen [1], ferricyanide (reverse reaction) [3, 7, 12, 15], cytochrome c from Candida krusei [15, 19], cytochrome c from horse heart [7, 15], no cytochrome c as electron acceptor [12, 17] [1, 3, 6–8, 12, 15, 16, 19])
2 IMP + sulfite + oxidized acceptor [15]
3 CMP + sulfite + oxidized acceptor [15]
4 GMP + sulfite + oxidized acceptor [6, 8, 12, 15]
5 UMP + sulfite + oxidized acceptor [12, 15]
6 Deoxy-AMP + sulfite + oxidized acceptor [7]
7 More (no reaction with ATP [3, 8], ADP [8], UMP [8], CMP [8], 3', 5'-cyclic AMP [8], CMP [12], IMP [12]) [3, 8, 12]

Product spectrum

1 Adenosine 5'-phosphosulfate + reduced acceptor
2 Inosine 5'-phosphosulfate + reduced acceptor [15]
3 Cytosine 5'-phoshosulfate + reduced acceptor
4 Guanosine 5'-phosphosulfate + reduced acceptor [6–8, 12]
5 Uracil 5'-phosphosulfate + reduced acceptor [12]
6 Deoxyadenosine 5'-phosphosulfate + reduced acceptor [7]
7 ?

Inhibitor(s)

AMP (substrate inhibition) [8, 19]; p-Chloromercuribenzoate [1, 18, 19]; N-Ethylmaleimide [19]; Iodoacetate [19]

Cofactor(s)/prostethic group(s)

FAD (1 mol per mol of enzyme [1, 4, 6, 15, 19]) [1–4, 6–8, 15, 16, 19]; 4Fe-4S center (2 per mol) [3]; Cytochrome type heme groups (2 per molecule, Thiocapsa roseopersicina) [1]

Metal compounds/salts

Nonheme iron [1–4, 6–8, 15, 19]; Acid-labile sulfide [1, 3, 6–8, 15, 19]

Turnover number (min^{-1})

24000 (AMP, Desulfovibrio desulfuricans) [2]; 22800 (AMP, Desulfovibrio vulgaris) [2]

Specific activity (U/mg)

4 [1]; 8.93 [7]; 13.5 [8]; 5.75 [15, 16]; More (immunoassay [2]) [2, 4, 6, 11, 19]

K_m-value (mM)

0.4 (AMP, Desulfovibrio desulfuricans [2], ferricyanide [6]) [2, 6]; 0.3 (AMP, Desulfovibrio vulgaris) [2]; 1–1.5 (sulfite with ferricyanide as acceptor [4, 19], AMP [6], GMP [15], IMP [15]) [4, 6, 15, 19]; 0.041 (AMP) [4]; 1 (AMP) [6]; 0.09–0.093 (AMP [7], ferricyanide [16], sulfite with cytochrome c as acceptor [19]) [7, 16, 19]; 0.071 (sulfite with cytochrome c as acceptor) [7]; 0.021 (cytochrome c) [7]; 0.13 (AMP [8], sulfite [8], ferricyanide [19]) [8, 19]; 0.16–0.2 (ferricyanide [12], AMP [15, 16]) [12, 15, 16]; 4.2–4.5 (ferricyanide [8], UMP [15]) [8, 15]; 3.1 (CMP) [15]; 0.019 (adenylylsulfate) [15]; 0.25 (guanylylsulfate) [15]; 0.91 (sulfite) [16]; 0.05 (AMP with cytochrome c as acceptor) [19]; 0.03 (cytochrome c) [19]; 0.073 (AMP with ferricyanide as acceptor) [19]; More (overview) [1]

pH-optimum

7.0–7.5 (methyl viologen) [15]; 7.4 [3]; 7.5 (UMP, CMP) [15]; 7.7 (ferricyanide) [7]; 8.0 (ferricyanide [19]) [4, 6, 8, 19]; 8.7 [16]; 8.8 (horse heart cytochrome c) [7]; 9.0 (cytochrome c [19], IMP, GMP [15]) [15, 19]; 9.5 (AMP) [15]; More (overview) [1]

pH-range

Temperature optimum (°C)

55 [7]; 75 [4]; 85 [6]

Temperature range (°C)

25–80 (less than 20% of maximal activity above and below) [4]; 55–85 [6]

3 ENZYME STRUCTURE

Molecular weight

439500 (Desulfovibrio vulgaris, sedimentation equilibrium in phosphate buffer, dimer) [15]
400000 (Desulfovibrio gigas, HPLC gel filtration) [3]
218500 (Desulfovibrio vulgaris, sedimentation equilibrium in Tris-maleate buffer, monomeric form) [15]
210000 (Chlorobium limicola, gel filtration) [16]
200000 (Thiobacillus denitrificans, estimation from sedimentation coefficient) [4]
190000 (Desulfovibrio desulfuricans, gel filtration) [2]
180000 (Thiocapsa roseopersicina, low speed sedimentation equilibrium [19], Chlorobium vibrioforme, gel filtration [8]) [8, 19]
175000 (Desulfobulbus propionicus, gel filtration) [7]
170000 (Thiobacillus thioparus) [1]
160000 (Archaeoglobus fulgidus VC-16, FPLC) [6]

Enzyme Handbook © Springer-Verlag Berlin Heidelberg 1994
Duplication, reproduction and storage in data banks are only allowed with the prior permission of the publishers

Subunits

Dimer (2 × 80000, Archaeoglobus fulgidus VC-16, SDS-PAGE) [6]
Tetramer (3 × 72000 + 1 × 20000, Desulfovibrio vulgaris, SDS-PAGE) [15]
Multimer (x × 70000 + x × 23000, Desulfovibrio gigas, SDS-PAGE [3], x × 70000 + x × 26000, Desulfovibrio desulfuricans, SDS-PAGE [2]) [2, 3]
More (electrophoretic pattern of Desulfovibrio and Desulfotomaculum species) [17]

Glycoprotein/Lipoprotein

–

4 ISOLATION/PREPARATION

Source organism

Desulfovibrio desulfuricans [2]; Desulfovibrio vulgaris [1, 2, 5, 11, 14, 15, 18]; Desulfovibrio gigas [2, 3, 10]; Desulfovibrio salexigens [2]; Desulfovibrio multispirans [2]; Desulfotomaculum orientis [2, 9]; Desulfotomaculum ruminis [2]; Desulfotomaculum nigrificans [2]; Desulfobulbus propionicus [2, 7]; Desulfosarcina variabilis [2, 7]; Thiobacillus denitrificans [2, 4, 21]; Chromatium vinosum [2, 12]; Archaeoglobus fulgidus (strain VC-16) [6]; Chlorobium vibrioforme (f. sp. thiosulfatophilum) [8]; Thiobacillus thioparus [1, 13, 22]; Chlorobium limicola [16]; Thiocapsa roseopersicina [1, 19]; Chlorella pyrenoidosa [1]; Desulfovibrio sp. (9 species) [17]; Desulfotomaculum sp. (4 species) [17]; Desulfobacter postgatei [7]; Desulfococcus multivorans [7]; Chromatiaceae [23]

Source tissue

Cell

Localisation in source

Cytoplasm (Desulfovibrio gigas, Desulfovibrio vulgaris [5]) [5, 10, 12]; Cytoplasmic membrane (Desulfovibrio thermophilus) [5]; Chromatophores [12, 23]

Purification

Desulfovibrio gigas [3]; Archaeoglobus fulgidus (strain VC-16) [6]; Chlorobium vibrioforme [8]; Desulfovibrio vulgaris [15]; Chlorobium limicola [16]; Thiocapsa roseopersicina [19]; Thiobacillus thioparus [22]; Thiobacillus denitrificans [21]

Crystallization

–

Cloned

–

Renatured

–

5 STABILITY

pH

Temperature (°C)

4 (2 days, 30% loss of activity) [16]; 22 (2 h, 30% loss of activity) [16]; 60 (precipitation, followed by loss of activity [16], 1 h stable [4]) [4, 16]; 65 (denaturation above [7], 15 min, no loss of activity [19]) [7, 19]; 70 (inactivation) [19]; 80 (inactivation) [4]

Oxidation

Organic solvent

General stability information

Storage

–20°C, inactivation in 30 days [3]; –20°C, several weeks [16]; –20°C, 3 months, or at 4°C under Ar, 10 days, 8% loss of activity [19]; –18°C, 68% loss in 16 days but only 5% loss with 50% glycerol in storage medium [7]; 4°C, 8 h, 20% loss of activity, 1 week 90% loss of activity [7]

6 CROSSREFERENCES TO STRUCTURE DATABANKS

PIR/MIPS code

PIR3: S18928 (Archaeoglobus fulgidus)

Brookhaven code

Enzyme Handbook © Springer-Verlag Berlin Heidelberg 1994
Duplication, reproduction and storage in data banks are only allowed with the prior permission of the publishers

7 LITERATURE REFERENCES

[1] Hatefi, Y., Stiggall, D.L. in "The Enzymes",3rd. Ed. (Boyer, P.D., ed.) 13,175–297 (1976) (Review)
[2] Odom, J.M., Jessie, K., Knodel, E., Emptage, M.: Appl. Environ. Microbiol.,57,727–733 (1991)
[3] Lampreia, J., Moura, I., Teixeira, M., Peck, H.D., LeGall, J., Huynh, B.H., Moura, J.J.G.: Eur. J. Biochem.,188,653–664 (1990)
[4] Taylor, B.F.: FEMS Microbiol. Lett.,59,351–354 (1989)
[5] Kremer, D.R., Veenhuis, M., Fauque, G., Peck, H.D., LeGall, J., Lampreia, J., Moura, J.J.G., Hansen, T.A.: Arch. Microbiol.,150,296–301 (1988)
[6] Speich, N., Trüper, H.G.: J. Gen. Microbiol.,134,1419–1425 (1988)
[7] Stille, W., Trüper, H.G.: Arch. Microbiol.,137,145–150 (1984)
[8] Khanna, S., Nicholas, D.J.D.: J. Gen. Microbiol.,129,1365–1370 (1983)
[9] Liu, C.-L., Hart, N., Peck, H.D.: Science,217,363–364 (1982)
[10] Odom, J.M., Peck, H.D.: J. Bacteriol.,147,161–169 (1981)
[11] Badziong, W., Thauer, R.K.: Arch. Microbiol.,125,167–174 (1980)
[12] Schwenn, J.D., Biere, M.: FEMS Microbiol. Lett.,6,19–22 (1979)
[13] Adachi, K., Suzuki, I.: Can. J. Microbiol.,55,91–98 (1977)
[14] Kobayashi, K., Morisawa, Y., Ishituka, T., Ishimoto, M.: J. Biochem.,78,1079–1085 (1975)
[15] Bramlett, R.N., Peck, H.D.: J. Biol. Chem.,250,2979–2986 (1975)
[16] Kirchhoff, J., Trüper, H.G.: Arch. Microbiol.,100,115–120 (1974)
[17] Skyring, G.W., Trudinger, P.A.: Can. J. Microbiol.,19,375–380 (1973)
[18] Peck, H.D., Bramlett, R., DerVartanian, D.V.: Z. Naturforsch.,27b,1084–1086 (1972)
[19] Trüper, H.G., Rogers, L.A.: J. Bacteriol.,108,1112–1121 (1971)
[20] Lampreia, J., Moura, I., Xavier, A.V., LeGall, J., Peck, H.D., Moura, J.J.G. in "Chemistry and Biochemistry of Flavoenzymes" (Moller, F., ed.) Vol.3, CRC Press Inc. Boca Raton, Fla. (1990)
[21] Bowen, T.J., Happold, F.C., Taylor, B.F.: Biochim. Biophys. Acta,118,566–576 (1966)
[22] Lyric, R.M., Suzuki, I.: Can. J. Biochem.,48,344–354 (1970)
[23] Trüper, H.G., Fisher, V.: Philos. Trans. R. Soc. Lond. B Biol. Sci.,298,529–542 (1982)

1 NOMENCLATURE

EC number
1.8.99.3

Systematic name
Trithionate:(acceptor) oxidoreductase

Recommended name
Hydrogensulfite reductase

Synonymes
Bisulfite reductase
Dissimilatory sulfite reductase
Desulfoviridin
Desulforubidin
Desulfofuscidin
Reductase, bisulfite
Dissimilatory-type sulfite reductase [7]

CAS Reg. No.
9059-42-1; 9045-15-2; 42612-25-9; 85876-01-3

2 REACTION AND SPECIFICITY

Catalysed reaction
$(O_3S.S.SO_3)^{2-}$ + acceptor + 2 H_2O + OH^- →
→ 3 HSO_3^- + reduced acceptor

Reaction type
Redox reaction

Natural substrates
Sulfite + electron donor (physiological donor: cytochrome c_3 [27], dissimilatory pathways of reduction of bisulfite to sulfite [17]) [17, 27]

Substrate spectrum
1 Sulfite + electron donor (electron donor: methyl viologen [10, 12, 14, 15, 20, 22, 26], benzyl viologen [10, 12, 14], cytochrome c_3 no direct electron donor [26], no reduction of thiosulfate, tri- or tetrathionate [12, 14, 20]) [7, 8, 10–16, 18, 20–22, 24, 26, 28]
2 Nitrite + electron donor (no reaction [21, 22]) [10]
3 Hydroxylamine + electron donor [10, 21]
4 More (reaction with dithionite [15], trimethylamine N-oxide [22], no substrates: chloride, cyanide, azide, adenosine N-oxide, taurine [22]) [15, 22]

Enzyme Handbook © Springer-Verlag Berlin Heidelberg 1994
Duplication, reproduction and storage in data banks are only allowed with the prior permission of the publishers

Product spectrum

1 Trithionate + thiosulfate + sulfide + polythionate + oxidized electron donor (trithionate main product [7, 8, 11, 18], trithionate sole product [28], no trithionate [16], sulfide main product [12], type and amount of products depending on experimental conditions, e.g. concentration of methyl viologen and hydrogenase [20, 22]) [7, 8, 10–12, 16, 18, 20–22, 24, 26–28]
2 NH_3 + H_2O + oxidized electron donor [10]
3 NH_3 + H_2O + oxidized electron donor [10]
4 ?

Inhibitor(s)

$NaAsO_2$ (not: [16]) [22]; More (not inhibitory: EDTA [16, 22], alpha,alpha'-dipyridyl [16, 22], KCN [16], NaN_3 [16, 22]) [16, 22]

Cofactor(s)/prostethic group(s)

Sirohydrochlorin (in desulfoviridin, demetalized siroheme) [2, 29]; Siroheme (i.e. iron chelates of sirohydrochlorin a class of tetrahydroporphyrin compounds [29], 2 siroheme groups per molecule [2], 4 siroheme groups per molecule [10], Mössbauer studies [2], EPR studies [2, 15]) [2, 10, 12, 14, 15, 23, 29]; 4Fe-4S center (4 per molecule) [10]

Metal compounds/salts

Fe (24 gatom per mol [12], 51 gatom per mol [14]) [12, 14]; Sulfur (20 acid-labile sulfur atoms per molecule [12], 47 acid-labile sulfur atoms per molecule [14]) [12, 14]

Turnover number (min^{-1})

53 (sulfite) [12]

Specific activity (U/mg)

0.632 [28]; 0.41 [25]; 0.331 [12]; More (dependency on growth conditions [19]) [7, 8, 10, 13, 14, 19, 27]

K_m-value (mM)

3.6 (sulfite) [22]

pH-optimum

6.0 [12, 14, 21, 28]; 5.5–6.0 (sulfite reduction) [22]; 7.5 (hydroxylamine reduction) [22]

pH-range

6–7.5 [16]

Temperature optimum (°C)

65–75 [10]

Temperature range (°C)

35 (very low activity below) [10]

3 ENZYME STRUCTURE

Molecular weight

280000 (Chromatium vinosum, gel filtration) [14]
226000 (Desulfovibrio vulgaris, sedimentation equilibrium centrifugation) [26]
175000–190000 (Desulfovibrio thermophilus, gel filtration, sedimentation equilibrium centrifugation [1], Desulfovibrio vulgaris, gel filtration [8], Chromatium vinosum, gel electrophoresis [16]) [1, 8, 16]
160000–167000 (Thermodesulfobacterium commune, sedimentation equilibrium centrifugation [10], Thiobacillus denitrificans, gel filtration [12]) [10, 12]

Subunits

Tetramer (alpha$_2$-beta$_2$, 2 × 44000 + 2 × 48000, Desulfovibrio thermophilus, SDS-PAGE [1], 2 × 50000–55000 + 2 × 39000–45000, Desulfovibrio vulgaris, SDS-PAGE [8, 26, 27], 4 × 48000, Thermodesulfobacterium commune, two different protein chains, SDS-PAGE, N-terminal amino acid sequence [10], 2 × 38000 + 2 × 43000, Thiobacillus denitrificans, SDS-PAGE [12]) [1, 8, 10, 12, 26, 27]
Multimer (x × 37000 + x × 42000, Chromatium vinosum, SDS-PAGE) [14]

Glycoprotein/Lipoprotein

–

4 ISOLATION/PREPARATION

Source organism

Desulfovibrio thermophilus (desulfofuscidin) [1, 4]; Desulfovibrio gigas (desulfoviridin) [2, 4, 15, 20, 23, 28, 29]; Desulfovibrio baculatus (desulforubidin) [2]; Desulfobacterium anilini (P-582) [3]; Desulfovibrio vulgaris (desulfoviridin) [4, 8, 13, 17–19, 22, 26, 27]; Sulfate reducing bacterial strain (DCB-1, desulfoviridin) [5]; Desulfovibrio sulfodismutans (desulfoviridin) [6]; Desulfovibrio africanus (desulfoviridin) [7]; Desulfonema limicola (desulfoviridin) [9]; Thermodesulfobacterium commune (desulfofuscidin) [10]; Desulfotomaculum nigrificans (P-582) [11, 21, 24, 29]; Thiobacillus denitrificans (desulfoviridin) [12]; Chromatium vinosum [14, 16]; Desulfovibrio desulfuricans (desulforubidin) [15, 23, 25]; Desulfotomaculum ruminis (P-582) [15]; More (overview) [1, 2]

Enzyme Handbook © Springer-Verlag Berlin Heidelberg 1994
Duplication, reproduction and storage in data banks are only allowed with the prior permission of the publishers

Source tissue
Cell [1–29]

Localisation in source
Cytoplasm [4, 9, 16]

Purification
Desulfovibrio thermophilus [1]; Desulfovibrio gigas [2, 28]; Desulfovibrio baculatus [2]; Desulfovibrio africanus [7]; Desulfovibrio vulgaris (separation of two forms [13]) [8, 13, 18, 26, 27]; Thermodesulfobacterium commune [10]; Thiobacillus denitrificans [12]; Chromatium vinosum (partial) [16]; Desulfotomaculum nigrificans (preparation contaminated with assimilatory sulfite reductase [21]) [21, 24]; Desulfovibrio desulfuricans [25]

Crystallization
–

Cloned
–

Renaturated
–

---

5 STABILITY

pH

Temperature (°C)

Oxidation

Organic solvent

General stability information
Slow loss of activity by repeated freezing/thawing [18]; Degradation by freezing [20]

Storage
–20°C, 50 mM Tris buffer pH 7.6, several months [26, 28]; –20°C, 50 mM potassium phosphate buffer pH 7.0 [12, 14]; –20°C, 1 month no loss of activity, 4°C, 1 month, 60% loss of activity [16]; –20°C, several months [18]; 4°C, at least 2 weeks [21]

---

6 CROSSREFERENCES TO STRUCTURE DATABANKS

PIR/MIPS code
PIR3: S11556 (Desulfovibrio thermophilus)

Brookhaven code

7 LITERATURE REFERENCES

[1] Fauque, G., Lino, A.R., Czechowski, M., Kang, L., DerVartanian, D.V., Moura, J.J.G., LeGall, J., Moura, I.: Biochim. Biophys. Acta,1040,112–118 (1990)
[2] Moura, I., LeGall, J., Lino, A.R., Peck, H.D., Fauque, G., Xavier, A.V., DerVartanian, D.V., Moura, J.J.G., Huynh, B.H.: J. Am. Chem. Soc.,110,1075–1082 (1988)
[3] Schnell, S., Bak, F., Pfennig, N.: Arch. Microbiol.,152,556–563 (1989)
[4] Kremer, D.R., Veenhuis, M., Fauque, G., Peck, H.D., LeGall, J., Lampreia, J., Moura, J.J.G., Hansen, T.A.: Arch. Microbiol.,150,296–301 (1988)
[5] Stevens, T.O., Linkfield, T.G., Tiedje, J.M.: Appl. Environ. Microbiol.,54,2938–2943 (1988)
[6] Bak, F., Pfennig, N.: Arch. Microbiol.,147,184–189 (1987)
[7] Seki, Y., Nagai, Y., Ishimoto, M.: J. Biochem.,98,1535–1543 (1985)
[8] Aketagawa, J., Kojo, K., Ishimoto, M.: Agric. Biol. Chem.,49,2359–2365 (1985)
[9] Widdel, F., Kohring, G.-W., Mayer, F.: Arch. Microbiol.,134,286–294 (1983)
[10] Hatchikian, E.C., Zeikus, J.G.: J. Bacteriol.,153,1211–1220 (1983)
[11] Akagi, J.M.: Biochem. Biophys. Res. Commun.,117,530–535 (1983)
[12] Schedel, M., Trüper, H.G.: Biochim. Biophys. Acta,568 ,454–467 (1979)
[13] Seki, Y., Kobayashi, K., Ishimoto, M.: J. Biochem.,85,705–711 (1979)
[14] Schedel, M., Vanselow, M., Trüper, H.G.: Arch. Microbiol.,121,29–36 (1979)
[15] Liu, C.L., DerVartanian, D.V., Peck, H.D.: Biochem. Biophys. Res. Commun.,91,962–970 (1979)
[16] Kobayashi, K., Katsura, E., Kondo, T., Ishimoto, M.: J. Biochem.,84,1209–1215 (1978)
[17] Drake, H.L., Akagi, J.M.: J. Bacteriol.,136,916–923 (1978)
[18] Drake, H.L., Akagi, J.M.: J. Bacteriol.,126,733–738 (1976)
[19] Kobayashi, K., Morisawa, Y., Ishituka, T., Ishimoto, M.: J. Biochem.,78,1079–1085 (1975)
[20] Jones, H.E., Skyring, G.W.: Biochim. Biophys. Acta,377,52–60 (1975)
[21] Akagi, J.M., Chan, M., Adams, V.: J. Bacteriol.,120,240–244 (1974)
[22] Kobayashi, K., Seki, Y., Ishimoto, M.: J. Biochem.,75,519–529 (1974)
[23] Murphy, M.J., Siegel, L.M., Kamin, H., DerVartanian, D.V., Lee, J.-P., LeGall, J., Peck, H.D.: Biochem. Biophys. Res. Commun.,54,82–88 (1973)
[24] Akagi, J.M., Adams, V.: J. Bacteriol.,116,392–396 (1973)
[25] Lee, J.-P., Yi, C.-S., LeGall, J., Peck, H.D.: J. Bacteriol.,115,453–455 (1973)
[26] Lee, J.-P., LeGall, J., Peck, H.D.: J. Bacteriol.,115,529–542 (1973)
[27] Kobayashi, K., Takahashi, E., Ishimoto, M.: J. Biochem.,72,879–887 (1972)
[28] Lee, J.P., Peck, H.D.: Biochem. Biophys. Res. Commun.,45,583–589 (1971)
[29] Murphys, M.J., Siegel, L.M.: J. Biol. Chem.,248,6911–6919 (1973)

Enzyme Handbook © Springer-Verlag Berlin Heidelberg 1994
Duplication, reproduction and storage in data banks are only allowed with the prior permission of the publishers

1 NOMENCLATURE

EC number

1.9.3.1

Systematic name

Ferrocytochrome-c:oxygen oxidoreductase

Recommended name

Cytochrome-c oxidase

Synonymes

Cytochrome oxidase
Cytochrome a_3
Cytochrome aa_3
Oxidase, cytochrome
Warburg's respiratory enzyme
Cytochrome c oxidase
Indophenol oxidase
Indophenolase
Complex IV (mitochondrial electron transport)
Ferrocytochrome c oxidase
NADH cytochrome c oxidase

CAS Reg. No.

9001-16-5

2 REACTION AND SPECIFICITY

Catalysed reaction

4 Ferrocytochrome c + O_2 →
→ 4 ferricytochrome c + 2 H_2O (mechanism of O_2 reduction [1, 7, 14, 16, 22], mechanism of proton pumping [1, 4, 11, 22, 120])

Reaction type

Redox reaction

Natural substrates

Ferrocytochrome c + O_2

Enzyme Handbook © Springer-Verlag Berlin Heidelberg 1994
Duplication, reproduction and storage in data banks are only allowed with the prior permission of the publishers

Substrate spectrum

1 Ferrocytochrome c + O_2 (ferrocytochromes of various sources e.g. horse [46, 55, 58, 91, 92], Candida krusei [55, 58, 91], Nitrosomonas europaea [55], Nitrobacter agilis [91], other electron donors: N, N, N', N'-tetramethyl-p-phenylendiamin [6, 20, 25, 32, 39, 69, 72, 86, 88], rusticyanin [28], phenazine methosulfate [39, 41, 61, 71, 87], ascorbate/hexaaminruthenium [69], ascorbate/diaminodurene [61]) [6, 20, 25, 28, 32, 39, 41, 46, 55, 58, 61, 69, 71, 72, 86–88, 91, 92]

2 More (proton translocation across mitochondrial (eukaryotes) and cytoplasmic (prokaryotes) membrane, overview proposed mechanims [1, 120], reaction intermediates [24], Rhodopseudomonas sphaeroides: no proton translocation [72, 89], overview additional activities i.e. catalase activity, peroxidase activity, superoxide dismutase activity, carbomonoxygenase activity [7]) [1, 4–5, 7, 11, 14–16, 22, 24, 25, 39, 46, 72, 77, 85, 87, 89, 120]

Product spectrum

1 Ferricytochrome c + H_2O [27, 92]

2 ?

Inhibitor(s)

Dicyclohexylcarbodiimide (inhibition of redox-linked proton translocation) [5]; CO (competitive to O_2) [17, 20, 40, 41, 60, 98]; NO (competitive to O_2) [17]; CN^- [17, 18, 20, 22, 25, 27, 35, 39, 40, 58, 61, 69, 72, 79, 81, 86, 89, 91, 92, 96, 98, 101]; Azide [17, 20, 39, 40, 69, 72, 79, 86, 89, 96]; Sulfide [17, 20]; Phosphate (more than 15 mM [56], not with yeast cytochrome c [92], more than 10 mM [96, 101], more than 70 mM [98]) [17, 56, 85, 87, 91, 92, 96, 98, 101]; Alkaline pH [17]; F^- [17]; Sulfate [20]; Cl^- [20]; Salicyl aldoxime [40]; Nonionic detergents [55]; Poly-L-lysine (oxidation of horse and Candida krusei cytochrome c) [55]; KCl [56, 86]; Triton X-100 [56]; High ionic strength (above 200 mM KCl) [72, 89]; Sodium deoxycholate [101]; N, N-Dimethyllauryl amine oxide [101]; Lithium diiodosalicylate [101]

Cofactor(s)/prostethic group(s)

Heme a_3 (located in subunit I [1, 52], closely associated with CuB [11]) [1, 2, 4, 6, 8, 11, 13, 19, 20, 22, 25, 31, 33, 39, 52, 58, 60]; Heme a (located in subunit I [1], spectral studies [16, 18], chemical structure [18]) [1, 2, 4, 6, 8, 10–13, 16, 18–20, 22, 25, 28, 31–34, 37, 39, 44–46, 58, 60, 81, 83–87]; Heme c (firmly bound [39]) [39, 87]; Cardiolipin (Nitrobacter: stimulation with non-physiological electron donors [6]) [6, 55, 79]; Cholate (reaction enhancement) [25]; n-Octylglucoside (stimulation) [40]; Phospholipids (activation) [85–87, 100]; More (location of hemes in subunits, eukaryotic and prokaryotic enzyme) [60]

Metal compounds/salts

Cu (models for metal binding [119], discussion of Cu content [2], location in subunits [3, 4, 60], CuA located in subunit II, binding of O_2 and reduction [1, 44], CuB located in subunit I, electron flow from ferrocytochrome c to binuclear center [1, 44], 3 mol Cu per monomer [26, 45], 5 mol Cu per dimer [47], Cu-binding sites [44], XAS-studies [29], EXAFS studies [29], CuA-protein coordination model [51], CuA-ligand structure [66]) [1–4, 6, 10–13, 16, 18, 20, 22, 26, 29, 30, 34, 36, 37, 39, 44–48, 51, 59, 60, 66, 79, 81, 83–85, 87, 91, 93, 95, 98, 102, 119]; Fe (bound to heme a, heme a_3, 2 mol per functional unit) [1–4, 6, 10, 16, 18, 26, 29, 39, 45, 47, 48, 59, 60, 102]; Zn (tightly bound to subunit VI, 1 mol per functional unit, EXAFS studies [43], PIXE studies [45], 1 mol per 2 mol of heme a [47], no Zn: Nitrobacter winogradskyi [34], Erythrobacter longus [58]) [2, 3, 34, 43, 45, 47, 58–60]; Mg (functional role of Mg and ATP binding site in association with subunit IV [59], 1 mol per monomeric functional unit [34, 48]) [2, 34, 48, 59]; Anionic detergents (increase of activity, eukaryotic enzyme) [55]; More (variation of metal content with purification procedure) [59]

Turnover number (min^{-1})

1800–36000 (cytochrome c) [1]; 3000–6000 (cytochrome c, soluble enzyme, higher values for particulate enzyme) [60]; 1460 (ascorbate/phenazine methosulfate) [61]; 918 (ascorbate/diaminodurene) [61]; 236 (ascorbate/N, N, N', N'-tetramethyl-p-phenylendiamine) [61]; More (dependency on ionic strength [62, 63], presence of detergents and phospholipids [37, 69, 88], dependency on osmotic pressure [121]) [13, 17, 21, 37, 46, 61–63, 69, 80, 88, 92, 93, 99, 121]

Specific activity (U/mg)

17.9 (cytochrome c) [25]; 24 (cytochrome c) [81]; 147 (cytochrome c) [56]; More [27, 32, 35, 41, 49, 55, 58, 84, 92, 95, 98, 102, 103]

K_m-value (mM)

0.067 (cytochrome c-551 from thermophilic bacterium PS3, bacterial enzyme) [6]; 0.011 (cytochrome c-552 from Thermus thermophilus, bacterial enzyme) [6]; 0.13 (cytochrome c from horse heart, bacterial enzyme) [6]; 0.006–0.077 (cytochrome c, dependency on pH, phosphate and K^+ concentration) [17]; 0.0017 (horse cytochrome c, presence of cardiolipin) [37]; 0.012 (horse cytochrome c, absence of cardiolipin) [37]; 0.00005–0.11 (cytochrome c, values depending on source of cytochrome c [34, 46, 58, 91], on ionic strength [23, 62, 63, 79]) [23, 34, 46, 58, 62, 63, 79, 91]; More (kinetic studies [1, 22, 23, 60, 120]) [1, 20–23, 28, 34, 37, 46, 52, 55, 58, 60, 63, 65, 72, 79, 80, 85, 87, 89, 93, 96, 99, 101, 109–112, 120]

Enzyme Handbook © Springer-Verlag Berlin Heidelberg 1994
Duplication, reproduction and storage in data banks are only allowed with the prior permission of the publishers

pH-optimum

3.5 [28]; 5.0 (MES buffer [92]) [23, 92]; 5.5 (phosphate buffer) [92]; 5.5–6.0 (glycylglycine/NaOH buffer) [101]; 5.6 (electron donor: cytochrome c from Nitrosomonas europaea) [79]; 5.6–5.9 (depending on type of cytochrome c) [91]; 6.0 (electron donor: cytochrome c from Candida krusei [81], purified enzyme [84], phosphate buffer [101]) [17, 81, 84, 98, 101]; 6.0–6.3 [99]; 6.1 (electron donor: cytochrome c from horse heart and Candida krusei [79]) [79, 87]; 6.4 (electron donor: cytochrome c from Pseudomonas sp. AM1) [81]; 6.5 (proton translocation [39]) [20, 39, 96]; 7.0–7.5 [100]; 7.5–8.0 (presence of reducing agents) [101]; 7.5–8.5 [85]

pH-range

5.0–7.1 (less than 75% of maximal activity above and below) [87]; 5.5–8.0 (proton translocation) [39]; 6–8 [100]; More (redox potential of cytochrome a is pH dependent) [1]

Temperature optimum (°C)

25 (assay) [21]; 40 [17]; 50 [25]; 77 [96]

Temperature range (°C)

60 (up to, Thermus thermophilus) [46]

---

3 ENZYME STRUCTURE

Molecular weight

204000 (vertebrate, amino acid sequence of 12 subunits + 6000 for subunit VIIb) [75]
350000 (bovine heart, dimeric form, Triton X-100 solubilized, hydrodynamic measurements) [60]
326000 (bovine, dimeric enzyme, sedimentation equilibrium centrifugation) [49]
290000–315000 (Bacillus subtilis, gel filtration, value depending on ionic strength [63], enzyme associated with detergents [69]) [63, 69]
250000 (thermophilic bacterium PS3, gel filtration, presence of N-lauryl sarcosinate) [87]
180000–280000 (eukaryotes, minimal MW, calculated from subunit composition, heme content) [15, 17]
210000 (vertebrate, theoretical value of monomer composed of 12–13 different subunits) [9]
200000 (bovine heart, monomeric enzyme, deoxycholate solubilized, hydrodynamic measurements) [60]
162000 (Ipomoea batatas, calculation from heme content) [100]
140000–158000 (elasmobranchs, sedimentation equilibrium analysis, monomeric form [49], calculation from heme content [93]) [49, 93]

150000 (Sulfolobus acidocaldarius, gel filtration) [25]
120000 (Sulfolobus acidocaldarius, HPLC) [25]
100000–104000 (Sulfolobus acidocaldarius, HPLC) [20]
92000 (Erythrobacter longus, SDS-PAGE, no mercaptoethanol) [58]
83000 (Nitrosomonas europaea, calculation from heme content) [79]
76000 (Pseudomonas sp. AM1, SDS-PAGE, no mercaptoethanol) [81]
67000 (Nitrosomonas europaea, calculation from heme content) [37]
More (comparison of values from hydrodynamic measurements [119]) [16, 60, 104, 105, 119]

Subunits

Dimer of multimer (bovine heart, 13 subunits, nomenclature system of subunits according to [106], I: 56993, II: 26049, III: 29918, IV: 17153, Va: 12436, Vb: 10670, VIa: 9419, VIb: 10068, VIc: 8480, VIIc: 5441, VIIa: 6244, VIIb: 6350, VIII: 4962, amino acid sequences [3, 60, 68] and literature cited therein, other nomenclature systems [3, 12, 60, 68, 107, 108, 119], separation of subunits [19], sequence alignment various organisms [119], differences in small subunit composition depending on source of enzyme [62], definitions of functional unit [60]) [3, 12, 19, 60, 68, 107, 108, 119]
Monomer or dimer of multimer (Saccharomyces cerevisiae, 9 subunits I: 56000, II: 26678, III: 30340, IV: 14858, Va: 12627, Vb: 14570, VIIa: 6603, VIIc: 5364, VIII: 6303, amino acid sequences [3, 4, 68], overview nomenclature systems [4, 68], sequence alignment with bovine heart [68]) [3, 4, 68]
Oligomer (Paracoccus denitrificans: 1 × 45000 + 1 × 28000 [6], third subunit of 23000 [54], thermophilic bacterium PS3: 1 × 56000 + 1 × 38000 + 1 × 22000 [6, 8, 46, 87], Thermus thermophilus HB8: 1 × 55000 or 71000 + 1 × 33000 [6, 46, 85, 95, 96], Bacillus subtilis and Bacillus stearothermophilus: 1 × 57000 + 1 × 37000 + 1 × 21000 [6, 39, 69, 86], Nitrobacter agilis: 1 × 51000 + 1 × 31000 [6, 91], Thiobacillus novellus: x × 32000 + x × 23000, [6, 92], Rhodopseudomonas sphaeroides: x × 45000 + x × 37000 + x × 35000 [6, 72, 89], Pseudomonas sp. AM1: x × 50000 + x × 30000 [6, 81], Sulfolobus acidocaldarius: x × 37000 + x × 23000 + x × 14000 [25], only one type of subunit, 38000–40000 [20, 32], Anacystis nidulans: x × 55000 + x × 32000 [36], Nitrosomonas europaea: x × 39000 + x × 28000 [37], Micrococcus luteus: x × 47000 + x × 31000 + x × 19000 [61], Ipomoea batatas: I: 39000, II: 33500, III: 26000, IV: 20000, V: 5700, probably 2 more small subunits [56], Dictyostelium discoideum: I: 55000, II: 29500, IV: 19000, V: 13000, VI: 11000, VII: 5700, subunit III lost during purification [80], sequence alignment various organisms [119], analytical method for all values: SDS-PAGE) [6, 8, 20, 25, 32, 36, 37, 39, 46, 54, 56, 61, 69, 72, 80, 81, 85–87, 89, 91, 92, 95, 96, 119]
Monomer (shark heart, at pH 7.4, bovine heart at pH 8.5, after treatment with detergent) [9, 49, 62]

Enzyme Handbook © Springer-Verlag Berlin Heidelberg 1994
Duplication, reproduction and storage in data banks are only allowed with the prior permission of the publishers

More (amino acid sequences of isolated subunits [50, 82, 94], anormalous migration of subunit I on SDS-gels, correction of values by calculation from relationship of retardation coefficient versus MW of marker proteins i.e. Ferguson plot [85, 87, 91], additional values for eukaryotic enzyme from SDS-PAGE [21, 90, 98, 99, 102–104], comparison of subunit composition in eukaryotes and prokaryotes [2, 12, 16], comparison of human and bovine isozymes [65]) [2, 12, 16, 21, 50, 65, 82, 85, 87, 90, 91, 94, 98, 99, 102–104]

Glycoprotein/Lipoprotein

Phospholipoprotein [12, 14–16, 26, 46, 56, 60, 88, 98, 100]

---

4 ISOLATION/PREPARATION

Source organism

Eukaryotes [1, 12, 15, 17, 60]; Prokaryotes (not: E. coli [6]) [1, 13, 60]; Bacillus subtilis [8, 63, 69, 86]; Paracoccus denitrificans [8, 13, 52, 54, 70]; Rhodopseudomonas sphaeroides [8, 72, 89]; Bacillus cereus [8]; Thiobacillus novellus [13, 92]; Thermus thermophilus [13, 33, 46, 85, 95, 96]; Nitrobacter agilis [13, 55, 91]; Thermophilic bacterium PS3 [8, 13, 71, 83, 87, 97]; Sulfolobus acidocaldarius [20, 25, 32]; Bacillus stearothermophilus [39]; Phormidium foveolarum [27]; Anacystis nidulans [35, 36]; Plectonema boryanum [35]; Anabaena variabilis [35, 40, 41]; Synechocystis sp. [35]; Nitrobacter winogradskyi [34]; Nitrosomonas europaea [37, 55, 79]; Erythrobacter longus [58]; Micrococcus luteus (lysodeikticus) [61]; Thiobacillus ferrooxidans [28]; Bradyrhizobium japonicum [64]; Saccharomyces cerevisiae [68, 82, 103–105]; Neurospora crassa [74, 78]; Dictyostelium discoideum [80]; Pseudomonas sp. AM1 [81]; Bacillus firmus [84]; Rhodopseudomonas palustris [101]; Bovine [17, 26, 29, 38, 42, 43, 45, 47, 49–51, 53, 59, 60, 62, 68, 73, 74, 90, 94, 102, 114]; Rat [53, 57, 98, 99]; Triticum aestivus (var. Titan Red, wheat) [21]; Streptomyces erythraeus [31]; Pig [45]; Sphyrna lewini (scallop hammerhead shark) [49, 93]; Elasmobranchs (e.g. Squalus acanthias: sping dogfish, Carcharinus limbatus: black tip shark, Carcharinus obscurus: dusty shark, Galleocerdo cuvieri: tiger shark) [93]; Ipomoea batatas (sweet potato) [56, 100]; Human [65]; Halobacterium halobium [113]; Pea [115]

Source tissue

Heart [9, 26, 29, 38, 43, 45, 47, 49–51, 53, 59, 62, 65, 68, 73–75, 77, 87, 90, 93, 94, 102, 114]; More (mammalian tissue-specific isoforms) [42]; Liver [45, 62, 74, 75, 88, 98, 99]; Diaphragm [45, 75]; Root [56, 100]; Skeletal muscle [62]; Kidney [62, 75]; Germ [21]; Filaments from ammonia-grown cultures [27]; Shoots [115]

Localisation in source

Mitochondrial membrane (possible arrangement of subunits in membrane [67]); Cytoplasmic membrane [6]; Thylakoid membrane (Cyanobacteria) [35]

Purification

Mammalian tissues [12]; Bacteria (literature overview) [13]; Sulfolobus acidocaldarius [20, 25, 32]; Triticum aestivus [21]; Phormidium foveolarum [27]; Thiobacillus ferrooxidans [28]; Thermus thermophilus [33, 85, 96]; Nitrosomonas europaea (normal and CuA-deficient forms of enzyme [37]) [37, 79]; Bacillus stearothermophilus (large scale) [39]; Anabaena variabilis (partial) [41]; Paracoccus denitrificans (one-subunit enzyme, fully active [52]) [52, 54, 70]; Ipomoea batatas [56, 100]; Erythrobacter longus [58]; Micrococcus luteus [61]; Human [65]; Bacillus subtilis [69]; Rhodopseudomonas sphaeroides [72, 89]; Bovine (affinity chromatography [74]) [74, 90, 102, 114]; Vertebrates (tissue-specific isozymes) [75]; Dictyostelium discoideum [80]; Pseudomonas sp. AM1 [81]; Thermophilic bacterium PS3 [83, 87, 97]; Bacillus firmus [84]; Nitrobacter agilis [91]; Thiobacillus novellus [92]; Rhodopseudomonas palustris [101]; Saccharomyces cerevisiae [103–105]; Halobacterium halobium (Cu-deficient enzyme) [113]; Pea (inactivation during purification) [115]; Neurospora crassa [74]; More (method for subunit III depleted enzyme) [5]

Crystallization

(proposed folding patterns of subunits [2], two-dimensional crystals [76], dimeric in crystal lattice [116]) [2, 9, 76, 116]

Cloned

(subunits I, II, III of eukaryotes encoded by mitochondrial DNA, other subunits nuclear encoded) [1, 4, 57, 78, 82]

Renaturated

(reconstitution into phospholipid vesicles [38], overview methods for reconstitution [60], reconstitution [39, 69, 71], reconstitution of proton pumping activity [73]) [38, 39, 60, 69, 71, 73]

5 STABILITY

pH

5 (irreversible denaturation below, protection by incorporation into proteoliposomes) [23]

Enzyme Handbook © Springer-Verlag Berlin Heidelberg 1994
Duplication, reproduction and storage in data banks are only allowed with the prior permission of the publishers

Temperature (°C)

–70 (stable) [56]; 4 (12 h: inactivation [56], instable [83]) [56, 83]; 50 (Bacillus sp. PS3: loss of activity above) [46]; 60 (10 min, 100% activity) [96]; 63 (10 min, 100% activity) [97]; 70 (10 min, 50% activity) [97]; 80 (10 min, 76% activity) [96]; 81 (inactivation, no protection by phospholipids) [39]; More (Bacillus sp. PS3: cold labile, Thermus thermophilus: not cold labile [46], maltoside: increase in thermal stability [118], thermal denaturation in lipid phase consisting of 4 sequential melting steps, beginning with denaturation of subunit III [117]) [46, 117, 118]

Oxidation

Organic solvent

Ethanol (40% v/v, stable [97], 60%, no inactivation [121]) [97, 121]; Methanol, 60%, no inactivation [121]; Propanol, 60%, no inactivation [121]; Ethylene glycol, 60%, no inactivation [121]; Propanediol, no inactivation [121]; Dimethylformamide, no inactivation [121]; Dimethylsulfoxide, no inactivation [121]

General stability information

Urea, 1 M, stable [97]; LiCl, 5 M, stable [97]; Inactivation by repeated freezing/thawing [101]; Unstable during purification [21]; Inactivation by centrifugation of pure complex [56]; Depletion of lipids causes inactivation [60]; Ultrafiltration causes inactivation [69]

Storage

–80°C, 5 mM Tris-cacodylate buffer, pH 7.6 [80]; –60°C, 100 mM phosphate buffer, pH 7.4, 1% Tween 80 [26]; 4°C, solubilized enzyme, several h, liquid N_2, several months [27]; –70°C, concentrated solution [47]; –70°C, 10 mM Tris-HCl buffer, pH 7.3, 0.1% cholate [70]; –70°C [59]; Liquid N_2, 10 mM Tris-HCl buffer, pH 8.0, 0.5% Tween 20 [55]; –20°C, 10 mM Tris-HCl buffer, pH 8.0, 0.5% Tween 20 [79]; 4°C, 0.2 M phosphate buffer, 2% cholate, decrease of solubility during long term storage [102]; Glycerol: no stabilization during storage [101]; On ice, up to 7 days [91, 99]

6 CROSSREFERENCES TO STRUCTURE DATABANKS

PIR/MIPS code

PIR3: S24569 (Crab-eating macaque); PIR3: A23711 (c993 chain IIc precursor Thermus aquaticus); PIR2: PS0025 (chain 1 Podospora anserina mitochondrion (SGC3) (fragment)); PIR3: C34284 (chain 1 Sea urchin (Paracentrotus lividus) mitochondrion (SGC8)); PIR2: I29968 (chain AED hepatic Bovine (fragment)); PIR1: ODXL1 (chain I African clawed frog mitochondrion (SGC1)); PIR3: S14397 (chain I Bacillus subtilis); PIR3:

S09684 (chain I Bovine mitochondrion); PIR1: ODBO1 (chain I Bovine mitochondrion (SGC1)); PIR3: S13076 (chain I Bradyrhizobium japonicum); PIR2: S01218 (chain I Brine shrimp mitochondrion (SGC4) (fragment)); PIR3: S26034 (chain I Caenorhabditis elegans mitochondrion (SGC4)); PIR2: S10189 (chain I Chicken mitochondrion (SGC1)); PIR2: A22684 (chain I Chlamydomonas reinhardtii (fragment)); PIR2: A24707 (chain I Chlamydomonas reinhardtii mitochondrion); PIR2: A22735 (chain I Emericella nidulans mitochondrion (SGC3)); PIR1: ODAS1 (chain I Emericella nidulans mitochondrion (SGC3) (fragment)); PIR1: ODOB1M (chain I Evening primrose mitochondrion); PIR1: ODFF1 (chain I Fruit fly (Drosophila melanogaster) mitochondrion (SGC4)); PIR1: ODFF1Y (chain I Fruit fly (Drosophila yakuba) mitochondrion (SGC4)); PIR2: S05290 (chain I Garden pea mitochondrion); PIR3: S19887 (chain I Giant arborvitae mitochondrion (fragment)); PIR2: JT0974 (chain I Halobacterium halobium); PIR3: A32431 (chain I Honeybee mitochondrion (SGC4)); PIR1: ODHU1 (chain I Human mitochondrion (SGC1)); PIR1: ODZM1 (chain I Maize mitochondrion); PIR1: ODMS1 (chain I Mouse mitochondrion (SGC1)); PIR1: ODNC1 (chain I Neurospora crassa mitochondrion (SGC3)); PIR3: S07649 (chain I Neurospora crassa mitochondrion (SGC3)); PIR2: S03809 (chain I Paracoccus denitrificans); PIR1: ODPP1 (chain I Paramecium sp. mitochondrion (SGC6)); PIR2: S07751 (chain I Paramecium tetraurelia mitochondrion (SGC6)); PIR3: S26022 (chain I Pig roundworm mitochondrion (SGC4)); PIR3: A36357 (chain I Plasmodium berghei yoelli (fragment)); PIR2: S03420 (chain I Podospora anserina mitochondrion (SGC3) (fragments)); PIR2: S14139 (chain I Radish mitochondrion); PIR2: S04749 (chain I Rat mitochondrion (SGC1)); PIR3: A41521 (chain I Rat mitochondrion (SGC1) (fragment)); PIR2: S20534 (chain I Rhodobacter sphaeroides); PIR1: ODRZ1 (chain I Rice mitochondrion); PIR2: D30010 (chain I Sauroleishmania tarentolae mitochondrion (SGC6)); PIR3: B26510 (chain I Sea urchin (Paracentrotus lividus) mitochondrion (fragment)); PIR2: S01501 (chain I Sea urchin (Strongylocentrotus purpuratus) mitochondrion (SGC8)); PIR1: OBSY1 (chain I Soybean mitochondrion); PIR3: S08470 (chain I Starfish (Asterina pectinifera) mitochondrion (SGC8)); PIR3: S14205 (chain I Starfish (Pisaster ochraceus) mitochondrion); PIR3: S08471 (chain I Starfish (Pisaster ochraceus) mitochondrion (SGC8)); PIR2: S14138 (chain I Sugar beet mitochondrion); PIR2: S21042 (chain I Sulfolobus acidocaldarius); PIR2: JT0965 (chain I Synechococcus sp. (fragment)); PIR2: S00742 (chain I Tetrahymena pyriformis mitochondrion (SGC6)); PIR3: A41457 (chain I Thermophilic bacterium PS-3); PIR3: S23631 (chain I Thiobacillus versutus (fragment)); PIR1: ODUTMB (chain I Trypanosoma brucei mitochondrion (SGC6)); PIR3: S16256 (chain I Wheat mitochondrion); PIR3: S17993 (chain I Yeast (Kluyveromyces marxianus var. lactis) mitochondrion); PIR1: ODBY1 (chain I Yeast (Saccharomyces cerevisiae) mitochondrion (SGC2)); PIR2: A22736 (chain I Yeast (Schizosac-

Enzyme Handbook © Springer-Verlag Berlin Heidelberg 1994
Duplication, reproduction and storage in data banks are only allowed with the prior permission of the publishers

charomyces pombe)); PIR2: A25568 (chain I Yeast (Schizosaccharomyces pombe) mitochondrion (SGC2) (fragment)); PIR3: C35121 (chain I homolog Paracoccus denitrificans (fragment)); PIR3: A42378 (chain I homolog Rhodobacter sphaeroides (fragment)); PIR2: JX0140 (chain I precursor Thermophilic bacterium PS-3); PIR3: A26804 (chain I mitochondrial Wheat); PIR2: S08270 (chain I-beta Paracoccus denitrificans); PIR2: G29968 (chain IHQ hepatic Bovine (fragment)); PIR1: OBXL2 (chain II African clawed frog mitochondrion (SGC1)); PIR3: S14396 (chain II Bacillus subtilis); PIR3: S17300 (chain II Beet mitochondrion); PIR1: OBRT2B (chain II Black rat mitochondrion (SGC1)); PIR1: OBBO2 (chain II Bovine mitochondrion (SGC1)); PIR3: S26035 (chain II Caenorhabditis elegans mitochondrion (SGC4)); PIR1: OBPZ2M (chain II Carrot mitochondrion); PIR2: S10190 (chain II Chicken mitochondrion (SGC1)); PIR2: A27420 (chain II Crab-eating macaque mitochondrion (SGC1)); PIR3: S05629 (chain II Emericella nidulans mitochondrion (SGC3)); PIR1: OBOB2M (chain II Evening primrose mitochondrion); PIR2: A40076 (chain II Evening primrose mitochondrion (fragment)); PIR1: OBFF2 (chain II Fruit fly (Drosophila melanogaster) mitochondrion (SGC4)); PIR1: OBFF2Y (chain II Fruit fly (Drosophila yakuba) mitochondrion (SGC4)); PIR1: OBPM2 (chain II Garden pea mitochondrion); PIR3: S14455 (chain II Garden petunia mitochondrion); PIR3: B32431 (chain II Honeybee mitochondrion (SGC4)); PIR1: OBHU2 (chain II Human mitochondrion (SGC1)); PIR1: OBZM2 (chain II Maize mitochondrion); PIR1: OBMS2 (chain II Mouse mitochondrion (SGC1)); PIR1: OBNC2 (chain II Neurospora crassa mitochondrion (SGC3)); PIR2: A19032 (chain II Neurospora crassa mitochondrion (SGC3) (fragment)); PIR2: B26530 (chain II Paramecium primaurelia mitochondrion (SGC6)); PIR2: A26530 (chain II Paramecium tetraurelia mitochondrion (SGC6)); PIR3: S26023 (chain II Pig roundworm mitochondrion (SGC4)); PIR1: OBUNMP (chain II Pneumocystis carinii mitochondrion); PIR1: OBJJMP (chain II Podospora anserina mitochondrion (SGC3)); PIR1: OBRT2 (chain II Rat mitochondrion (SGC1)); PIR1: OBRZ2 (chain II Rice mitochondrion); PIR2: I22848 (chain II Sauroleishmania tarentolae mitochondrion (SGC6)); PIR3: D26510 (chain II Sea urchin (Paracentrotus lividus) mitochondrion (fragment)); PIR3: E34284 (chain II Sea urchin (Paracentrotus lividus) mitochondrion (SGC8)); PIR2: S01503 (chain II Sea urchin (Strongylocentrotus purpuratus) mitochondrion (SGC8)); PIR2: S07169 (chain II Soybean mitochondrion); PIR3: S14207 (chain II Starfish (Pisaster ochraceus) mitochondrion); PIR2: S14157 (chain II Sugar beet mitochondrion); PIR2: S21041 (chain II Sulfolobus acidocaldarius); PIR1: OBUTMB (chain II Trypanosoma brucei mitochondrion (SGC6)); PIR1: OBWT2 (chain II Wheat mitochondrion); PIR1: OBHQMS (chain II Yeast (Hansenula saturnus) mitochondrion (SGC2)); PIR1: OBBY2 (chain II Yeast (Saccharomyces cerevisiae) mitochondrion (SGC2)); PIR2: A30010 (chain II homolog Sauroleishmania tarentolae mitochondrion (SGC6)); PIR3: A39653 (chain II precursor Cowpea); PIR1: OBPC2N (chain II precursor Paracoccus denitrificans); PIR2: JQ1013 (chain II precursor Rhodobacter sphaeroides); PIR2: JT0964 (chain II precursor Synechococcus sp.); PIR2: JX0141 (chain II precursor Thermophilic bacterium PS-3); PIR1: OBVK2M (chain II precursor Yeast (Kluyveromyces marxianus var. lactis) mitochondrion (SGC2)); PIR3: S19533 (chain II.1 Garden petunia mitochondrion); PIR1: OTXL3 (chain III African clawed frog mitochondrion (SGC1)); PIR3: S14398 (chain III Bacillus subtilis); PIR1: OTBO3 (chain III Bovine mitochondrion (SGC1)); PIR2: S07557 (chain III Bracket fungus (Schizophyllum commune) mitochondrion); PIR2: S01216 (chain III Brine shrimp mitochondrion (SGC4) (fragment)); PIR3: S26032 (chain III Caenorhabditis elegans mitochondrion (SGC4)); PIR1: OTCA3 (chain III Carp mitochondrion (SGC1)); PIR2: S10193 (chain III Chicken mitochondrion (SGC1)); PIR3: C30396 (chain III Chinook salmon mitochondrion); PIR3: B30396 (chain III Coho salmon mitochondrion); PIR2: S14123 (chain III Common sunflower mitochondrion); PIR3: A25877 (chain III Crithidia fasciculata mitochondrion (fragment)); PIR3: D30396 (chain III Cutthroat trout mitochondrion (SGC1)); PIR1: OTAS3 (chain III Emericella nidulans mitochondrion (SGC3)); PIR1: OTOB3M (chain III Evening primrose mitochondrion); PIR2: B40076 (chain III Evening primrose mitochondrion (fragment)); PIR1: OTFF3 (chain III Fruit fly (Drosophila melanogaster) mitochondrion (SGC4)); PIR2: S02250 (chain III Fruit fly (Drosophila melanogaster) mitochondrion (SGC4) (fragment)); PIR1: OTFF3Y (chain III Fruit fly (Drosophila yakuba) mitochondrion (SGC4)); PIR1: OTHU3 (chain III Human mitochondrion (SGC1)); PIR3: S20801 (chain III Maize); PIR3: S17911 (chain III Moss (Physcomitrella patens)); PIR1: OTMS3 (chain III Mouse mitochondrion (SGC1)); PIR1: OTNC3 (chain III Neurospora crassa mitochondrion (SGC3)); PIR2: S03807 (chain III Paracoccus denitrificans); PIR2: A24371 (chain III Paracoccus denitrificans (fragment)); PIR3: S26020 (chain III Pig roundworm mitochondrion (SGC4)); PIR3: H34012 (chain III Pink salmon mitochondrion); PIR2: S06029 (chain III Podospora anserina mitochondrion (SGC3)); PIR3: S22198 (chain III Potato buckeye rot agent); PIR3: G34012 (chain III Rainbow trout mitochondrion (SGC1)); PIR2: S04753 (chain III Rat mitochondrion (SGC1)); PIR1: OTRZ3M (chain III Rice mitochondrion); PIR2: G22848 (chain III Sauroleishmania tarentolae mitochondrion (SGC6)); PIR2: PH0855 (chain III Sea urchin (Arbacia lixula) mitochondrion (SGC8) (fragment)); PIR3: F26510 (chain III Sea urchin (Paracentrotus lividus) mitochondrion (fragment)); PIR3: H34284 (chain III Sea urchin (Paracentrotus lividus) mitochondrion (SGC8)); PIR2: S01506 (chain III Sea urchin (Strongylocentrotus purpuratus) mitochondrion (SGC8)); PIR3: A30396 (chain III Sockeye salmon mitochondrion); PIR1: OTSY3M (chain III Soybean mitochondrion); PIR2: A27311 (chain III Starfish (Asterina pectinifera) mitochondrion (SGC8) (fragment)); PIR3: S14210 (chain III Starfish

Enzyme Handbook © Springer-Verlag Berlin Heidelberg 1994
Duplication, reproduction and storage in data banks are only allowed with the prior permission of the publishers

(Pisaster ochraceus) mitochondrion); PIR2: JX0142 (chain III Thermophilic bacterium PS-3); PIR2: A28782 (chain III Trypanosoma brucei mitochondrion (SGC6) (fragment)); PIR1: OTWT3M (chain III Wheat mitochondrion); PIR1: OTBY3 (chain III Yeast (Saccharomyces cerevisiae) mitochondrion (SGC2)); PIR3: S10080 (chain III Yeast (Schizosaccharomyces pombe) mitochondrion (SGC2)); PIR1: OLBO4 (chain IV Bovine); PIR3: S21077 (chain IV Human); PIR3: S16114 (chain IV Mouse); PIR2: A25629 (chain IV Neurospora crassa (fragment)); PIR3: S12724 (chain IV Rat); PIR1: OLBY4 (chain IV Yeast (Saccharomyces cerevisiae) (fragment)); PIR3: S12508 (chain IV precursor Mouse); PIR2: S04070 (chain IV precursor Rat); PIR3: S17660 (chain IV precursor Slime mold (Dictyostelium discoideum)); PIR2: JX0143 (chain IV precursor Thermophilic bacterium PS-3); PIR2: A22786 (chain IV precursor Yeast (Saccharomyces cerevisiae)); PIR2: A29968 (chain IV hepatic Bovine (fragments)); PIR3: S14399 (chain IVb Bacillus subtilis); PIR3: A35537 (chain IX precursor heart Bovine); PIR3: B35537 (chain IX precursor hepatic Bovine); PIR2: A28969 (chain IX hepatic Bovine (fragments)); PIR2: B29968 (chain SSG hepatic Bovine (fragment)); PIR2: F29968 (chain STA hepatic Bovine (fragment)); PIR1: CABO (chain V Bovine); PIR2: B25629 (chain V Neurospora crassa (fragment)); PIR3: S19718 (chain V Slime mold (Dictyostelium discoideum)); PIR1: OTNCV (chain V precursor Neurospora crassa); PIR2: D29968 (chain V hepatic Bovine (fragment)); PIR1: OTBY5A (chain V.A precursor Yeast (Saccharomyces cerevisiae)); PIR1: OTBY5B (chain V.B precursor Yeast (Saccharomyces cerevisiae)); PIR1: OTHU5A (chain Va precursor Human); PIR2: S05495 (chain Va precursor Mouse); PIR2: S04592 (chain Va precursor Rat); PIR3: A39063 (chain Vb Human); PIR3: A39425 (chain Vb precursor Mouse); PIR2: S09297 (chain Vc Sweet potato (fragment)); PIR3: S11029 (chain Vc precursor Sweet potato); PIR2: C25629 (chain VI Neurospora crassa (clone 9) (fragment)); PIR1: OGDO6 (chain VI Slime mold (Dictyostelium discoideum)); PIR1: OTBY6 (chain VI precursor Yeast (Saccharomyces cerevisiae)); PIR2: E29968 (chain VI hepatic Bovine (fragment)); PIR1: OGBO6A (chain VIa Bovine); PIR2: S05304 (chain VIa Human); PIR2: S05318 (chain VIa Rat); PIR2: S16242 (chain VIa precursor cardiac Bovine); PIR2: S01157 (chain VIa cardiac Rat); PIR2: S01156 (chain VIa hepatic Rat); PIR3: S18314 (chain VIa liver Bovine (fragment)); PIR2: A24659 (chain VIb Bovine); PIR2: JS0677 (chain VIb Human); PIR3: JQ0765 (chain VIb Human); PIR2: S16245 (chain VIb Human (fragment)); PIR2: S05432 (chain VIb cardiac Bovine); PIR1: OGBO6C (chain VIc Bovine); PIR2: S01960 (chain VIc Human); PIR3: S16083 (chain VIc Mouse); PIR2: S00114 (chain VIc Rat); PIR2: A32741 (chain VIc hepatic Rat (fragment)); PIR2: B29907 (chain VIc-1 Rat); PIR2: A29907 (chain VIc-2 Rat); PIR1: OGBO7 (chain VII Bovine); PIR2: A60325 (chain VII Pig (fragments)); PIR1: OBBY7 (chain VII Yeast (Saccharomyces cerevisiae)); PIR2: C29968 (chain VII hepatic Bovine (fragment)); PIR3: S13002 (chain VIIa Bovine); PIR3: S13099 (chain VIIa Rat); PIR1: OBBY7A (chain VIIa Yeast (Saccharomyces cerevisiae)); PIR3: S16384 (chain VIIa precursor Mouse); PIR2: A25352 (chain VIIa precursor Yeast (Saccharomyces cerevisiae)); PIR2: S18187 (chain VIIa precursor cardiac Bovine); PIR2: B41034 (chain VIIa precursor hepatic Bovine); PIR2: S06897 (chain VIIa precursor hepatic Human); PIR2: S09724 (chain VIIa.1 skeletal muscle Human (fragment)); PIR2: S09725 (chain VIIa.2 skeletal muscle Human (fragment)); PIR3: S20290 (chain VIIa.H cardiac Human (fragment)); PIR3: S20374 (chain VIIa.L cardiac Human (fragment)); PIR2: A34410 (chain VIIb precursor cardiac Bovine); PIR3: S20289 (chain VIIb cardiac Human (fragment)); PIR3: S10303 (chain VIIc Mouse); PIR1: OSBO8A (chain VIIc precursor Bovine); PIR1: OSHU7C (chain VIIc precursor Human); PIR3: S20288 (chain VIIc cardiac Human (fragment)); PIR3: S13105 (chain VIIe Slime mold (Dictyostelium discoideum)); PIR2: A34254 (chain VIIe Slime mold (Dictyostelium discoideum) (fragment)); PIR2: S01785 (chain VIII Human); PIR2: S00115 (chain VIII Rat (fragment)); PIR2: A05020 (chain VIII Yeast (Saccharomyces cerevisiae)); PIR2: A25353 (chain VIII precursor Yeast (Saccharomyces cerevisiae)); PIR3: S13003 (chain VIII.1 Bovine); PIR3: S13004 (chain VIII.2 Bovine); PIR2: A29180 (chain VIIIb Bovine); PIR2: A25879 (chain VIIIc Bovine); PIR3: B34254 (chain VIIs Slime mold (Dictyostelium discoideum) (fragment)); PIR3: A34103 (precursor chain VIII Human); PIR2: JT0325 (Vb chain Bovine (fragment)); PIR2: JT0324 (Vb chain precursor Human); PIR3: JC1304 (VIa-M chain precursor Human); PIR3: JC1303 (VIIa-M chain precursor Human); PIR3: S07650 (Neurospora crassa mitochondrion (SGC3))

Brookhaven code

7 LITERATURE REFERENCES

[1] Chan, S.I., Li, P.M.: Biochemistry,29,1–12 (1990) (Review)
[2] Capaldi, R.A.: Arch. Biochem. Biophys.,280,252–262 (1990) (Review)
[3] Azzi, A., Müller, M.: Arch. Biochem. Biophys.,280,242–251 (1990) (Review)
[4] Capaldi, R.A.: Annu. Rev. Biochem.,59,569–596 (1990) (Review)
[5] Brunori, M., Antonini, G., Malatesta, F., Sarti, P., Wilson, M.T.: Eur. J. Biochem.,169,1–8 (1987) (Review)
[6] Ludwig, B.: FEMS Microbiol. Rev.,46,41–56 (1987) (Review)
[7] Naqui, A., Chance, B.: Annu. Rev. Biochem.,55,137–166 (1986) (Review)
[8] Poole, R.K.: Biochim. Biophys. Acta,726,205–243 (1983) (Review)
[9] Capaldi, R.A., Malatesta, F., Darley-Usmar, V.M.: Biochim. Biophys. Acta,726,135–148 (1983) (Review)
[10] Kadenbach, B., Merle, P.: FEBS Lett.,135,1–11 (1981) (Review)
[11] Wikström, M., Krab, K., Saraste, M.: Annu. Rev. Biochem.,50,623–655 (1981) (Review)
[12] Azzi, A.: Biochim. Biophys. Acta,594,231–252 (1980) (Review)
[13] Ludwig, B.: Biochim. Biophys. Acta,594,177–189 (1980) (Review)

Enzyme Handbook © Springer-Verlag Berlin Heidelberg 1994
Duplication, reproduction and storage in data banks are only allowed with the prior permission of the publishers

[14] Malmström, B.G.: Biochim. Biophys. Acta,549,281–303 (1979) (Review)
[15] Wikström, M., Krab, K.: Biochim. Biophys. Acta,549,177–222 (1979) (Review)
[16] Caughey, W.S., Wallace, W.J., Volpe, J.A., Yoshikawa, S. in "The Enzymes",3rd. Ed. (Boyer, P.D., ed.) 13,299–344 (1976) (Review)
[17] Nicholls, P., Chance, B. in "Mol. Mech. Oxygen Activ." (Hayashi, O., ed.) 479–534 (1974) (Review)
[18] Wharton, D.C. in "Inorg. Biochem." (Eichhorn, G.L., ed.) 2,955–987 (1973) (Review)
[19] Robinson, N.C., Dale, M.P., Talbert, L.H.: Arch. Biochem. Biophys.,281,239–244 (1990)
[20] Anemüller, S., Schäfer, G.: Eur. J. Biochem.,191,297–305 (1990)
[21] Peiffer, W.E., Ingle, R.T., Ferguson-Miller, S.: Biochemistry,29,8696–8701 (1990)
[22] Malmström, B.G.: Arch. Biochem. Biophys.,280,233–241 (1990)
[23] Cooper, C.E.: Biochim. Biophys. Acta,1017,187–203 (1990) (Review)
[24] Han, S., Ching, Y.-C., Rousseau, D.L.: Proc. Natl. Acad. Sci. USA,87,2491–2495 (1990)
[25] Wakagi, T., Yamauchi, T., Oshima, T., Müller, M., Azzi, A., Sone, N.: Biochem. Biophys. Res. Commun.,165,1110–1114 (1989)
[26] Öblad, M., Selin, E., Malmström, B., Strid, L., Aasa, R., Malmström, B.G.: Biochim. Biophys. Acta,975,267–270 (1989)
[27] Häfele, U., Scherer, S., Böger, P.: Z. Naturforsch.,44c,378–383 (1989)
[28] Kai, M., Yano, T., Fukumori, Y., Yamanaka, T.: Biochem. Biophys. Res. Commun.,160,839–843 (1989)
[29] Scott, R.A.: Annu. Rev. Biophys. Biophys. Chem.,18,137–158 (1989)
[30] Li, P.M., Malmström, B.G., Chan, S.I.: FEBS Lett.,248,210–211 (1989)
[31] Scott, R.I., Poole, R.K., Williams, H.: Curr. Microbiol.,18,297–302 (1989)
[32] Anemüller, S., Schäfer, G.: FEBS Lett.,244,451–455 (1989)
[33] Buse, G., Hensel, S., Fee, J.A.: Eur. J. Biochem.,181,261–268 (1989)
[34] Yamanaka, T., Fukumori, Y.: FEMS Microbiol. Rev.,54,259–270 (1988)
[35] Peschek, G.A., Molitor, V., Trnka, M., Wastyn, M., Erber, W.: Methods Enzymol.,167,437–449 (1988)
[36] Peschek, G.A., Wastyn, M., Trnka, M., Molitor, V., Fry, I.V., Packer, L.: Biochemistry,28,3057–3063 (1989)
[37] Numata, M., Yamazaki, T., Fukumori, Y., Yamanaka, T.: J. Biochem.,105,245–248 (1989)
[38] Sarti, P., Antonini, G., Malatesta, F., Vallone, B., Villaschi, S., Brunori, M., Hider, R.C., Hamed, K.: Biochem. J.,257,783–787 (1989)
[39] De Vrij, W., Heyne, R.I., Konings, W.N.: Eur. J. Biochem.,178,763–770 (1989)
[40] Wastyn, M., Achatz, A., Molitor, V., Peschek, G.A.: Biochim. Biophys. Acta,935,217–224 (1988)
[41] Häfele, U., Scherer, S., Böger, P.: Biochim. Biophys. Acta,934,186–190 (1988)
[42] Lomax, M.I., Grossman, L.I.: Trends Biochem. Sci.,14 ,501–503 (1989)
[43] Naqui, A., Powers, L., Lundeen, M., Constantinescu, A., Chance, B.: J. Biol. Chem.,263,12342–12345 (1988)
[44] Holm, L., Saraste, M., Wikström, M.: EMBO J.,6,2819–2823 (1987)
[45] Bombelka, E., Richter, F.-W., Stroh, A., Kadenbach, B.: Biochem. Biophys. Res. Commun.,140,1007–1014 (1986)
[46] Fee, J.A., Kuila, D., Mather, M.W., Yoshida, T.: Biochim. Biophys. Acta,853,153–185 (1986) (Review)

[47] Einarsdottir, O., Caughey, W.S.: Biochem. Biophys. Res. Commun.,124,836–842 (1984)
[48] Einarsdottir, O., Caughey, W.S.: Biochem. Biophys. Res. Commun.,129,840–847 (1985)
[49] Georgevich, G., Darley-Usmar, V.M., Capaldi, R.A.: Biochemistry,22,1317–1322 (1983)
[50] Steffens, G.J., Buse, G.: Hoppe-Seyler's Z. Physiol. Chem.,360,613–619 (1979)
[51] Martin, C.T., Scholes, C.P., Chan, S.I.: J. Biol. Chem.,263,8420–8429 (1988)
[52] Müller, M., Schläpfer, B., Azzi, A.: Biochemistry,27,7546–7551 (1988)
[53] Rich, P.R., West, I.C., Mitchell, P.: FEBS Lett.,233 ,25–30 (1988)
[54] Haltia, T., Puustinen, A., Finel, M.: Eur. J. Biochem.,172,543–546 (1988)
[55] Yamazaki, T., Fukumori, Y., Yamanaka, T.: J. Biochem.,103,499–503 (1988)
[56] Maeshima, M., Hattori, T., Asahi, T.: Methods Enzymol.,148,491–501 (1987)
[57] Suske, G., Mengel, T., Cordingley, M., Kadenbach, B.: Eur. J. Biochem.,168,233–237 (1987)
[58] Fukumori, Y., Watanabe, K., Yamanaka, T.: J. Biochem.,102,777–784 (1987)
[59] Yewey, G.L., Caughey, W.S.: Ann. N.Y. Acad. Sci.,550 ,22–32 (1988)
[60] Brunori, M., Antonini, G., Malatesta, F., Sarti, P., Wilson, M.T.: Adv. Inorg. Biochem.,7,93–154 (1987) (Review)
[61] Artzabatov, V., Müller, M., Azzi, A.: Arch. Biochem. Biophys.,257,476–480 (1987)
[62] Sinjorgo, K.M.C., Durak, I., Dekker, H.L., Edel, C.M. , Hakvoort, T.B.M., van Gelder, B.F., Muijsers, A.O.: Biochim. Biophys. Acta,893,251–258 (1987)
[63] de Vrij, W., Konings, W.N.: Eur. J. Biochem.,166,581–587 (1987)
[64] O'Brian, M.R., Maier, R.J.: Proc. Natl. Acad. Sci. USA,84,3219–3223 (1987)
[65] Sinjorgo, K.M.C., Hakvoort, T.B.M., Durak, I., Draijer, J.W., Post, J.K.P., Muijsers, A.O.: Biochim. Biophys. Acta,850,144–150 (1987)
[66] Li, M.P., Gelles, J., Chan, S.I., Sullivan, R.J., Scott, R.A.: Biochemistry,26,2091–2095 (1987)
[67] Hill, B.C., Greenwood, C., Nicholls, P.: Biochim. Biophys. Acta,853,91–113 (1986) (Review)
[68] Capaldi, R.A., Gonzales-Halphen, D., Takamiya, S.: FEBS Lett.,207,11–17 (1986)
[69] de Vrij, W., Poolman, B., Konings, W.N., Azzi, A.: Methods Enzymol.,126,159–173 (1986)
[70] Ludwig, B.: Methods Enzymol.,126,153–159 (1986)
[71] Hamamoto, T., Montal, M.: Methods Enzymol.,126,123–138 (1986)
[72] Azzi, A., Gennis, R.B.: Methods Enzymol.,126,138–145 (1986)
[73] Müller, M., Thelen, M., O'Shea, P., Azzi, A.: Methods Enzymol.,126,78–87 (1986)
[74] Brogler, C., Bill, K., Azzi, A.: Methods Enzymol.,126,64–72 (1986)
[75] Kadenbach, B., Stroh, A., Ungibauer, M., Kuhn-Nentwig, L., Büge, U., Jarausch, J.: Methods Enzymol.,126 ,32–45 (1986)
[76] Capaldi, R., Zhang, Y.-Z.: Methods Enzymol.,126,22–31 (1986)
[77] Casey, R.P.: Methods Enzymol.,126,13–21 (1986)
[78] Sachs, M.S., David, M., Werner, S., RajBhandary, U.L. : J. Biol. Chem.,261,869–873 (1986)
[79] Yamazaki, T., Fukumori, Y., Yamanaka, T.: Biochim. Biophys. Acta,810,174–183 (1985)
[80] Bisson, R., Schiavo, G., Papini, E.: Biochemistry,24 ,7845–7852 (1985)
[81] Fukumori, Y., Nakayama, K., Yamanaka, T.: J. Biochem.,98,493–499 (1985)
[82] Koerner, T.J., Hill, J., Tzagoloff, A.: J. Biol. Chem.,260,9513–9515 (1985)

Enzyme Handbook © Springer-Verlag Berlin Heidelberg 1994
Duplication, reproduction and storage in data banks are only allowed with the prior permission of the publishers

[83] Baines, B., Hubbard, J.A.M., Poole, R.K.: Biochim. Biophys. Acta,766,438–445 (1984)
[84] Kitada, M., Krulwich, T.A.: J. Bacteriol.,158,963–966 (1984)
[85] Hon-nami, K., Oshima, T.: Biochemistry,23,454–460 (1984)
[86] de Vrij, W., Azzi, A., Konings, W.N.: Eur. J. Biochem.,131,97–103 (1983)
[87] Sone, N., Yanagita, Y.: Biochim. Biophys. Acta,682,216–226 (1982)
[88] Merle, P., Kadenbach, B.: Eur. J. Biochem.,125,239–244 (1982)
[89] Gennis, R.B., Casey, R.P., Azzi, A., Ludwig, B.: Eur. J. Biochem.,125,189–195 (1982)
[90] Wei, Y.-H., King, T.E.: J. Biol. Chem.,256,10999–11003 (1981)
[91] Yamanaka, T., Kamita, Y., Fukumori, Y.: J. Biochem.,89,265–273 (1981)
[92] Yamanaka, T., Fujii, K.: Biochim. Biophys. Acta,591,53–62 (1980)
[93] Wilson, M.T., Lalla-Maharajh, W., Darley-Usmar, V., Bonaventura, J., Bonaventura, C., Brunori, M.: J. Biol. Chem.,255,2722–2728 (1980)
[94] Sacher, R., Steffens, G.J., Buse, G.: Hoppe-Seyler's Z. Physiol. Chem.,360,1385–1392 (1979)
[95] Fee, J.A., Choc, M.G., Findling, K.L., Lorence, R., Yoshida, T.: Proc. Natl. Acad. Sci. USA,77,147–151 (1980)
[96] Hon-nami, K., Oshima, T.: Biochem. Biophys. Res. Commun.,92,1023–1029 (1980)
[97] Sone, N., Ohyana, T., Kagawa, Y.: FEBS Lett.,106,39–42 (1979)
[98] Rascati, R.J., Parsons, P.: J. Biol. Chem.,254,1586–1593 (1979)
[99] Höchli, L., Hackenbrock, C.R.: Biochemistry,17,3712–3719 (1978)
[100] Maeshima, M., Asahi, T.: Arch. Biochem. Biophys.,187 ,423–430 (1978)
[101] King, M.-T., Drews, G.: Eur. J. Biochem.,68,5–12 (1976)
[102] Steffens, G., Buse, G.: Hoppe-Seyler's Z. Physiol. Chem.,357,1125–1137 (1976)
[103] Mason, T.L., Poyton, R.O., Wharton, D.C., Schatz, G. : J. Biol. Chem.,248,1346–1354 (1973)
[104] Rubin, M.S., Tzagoloff, A.: J. Biol. Chem.,248,4269–4274 (1973)
[105] Shakespeare, P.C., Mahler, H.R.: J. Biol. Chem.,246 ,7649–7655 (1971)
[106] Kadenbach, B., Jarausch, J., Hartmann, R., Merle, P. : Anal. Biochem.,129,517–521 (1983)
[107] Capaldi, R.A., Malatesta, F., Darley-Usmar, V.M.: Biochim. Biophys. Acta,726,135–148 (1983)
[108] Buse, G., Steffens, G.C.M., Meinecke, L. in " Structure and Function of Membrane Proteins" (Quagliarello, E., Palmieri, F., eds.) 6,131ff., Elsevier, Amsterdam (1983)
[109] Brunori, M., Antonini, G., Wilson, M.T.: Met. Ions Biol. Syst.,13,187–228 (1981)
[110] Wikström, M., Krab, K., Saraste, M. in "Cytochrome Oxidase- A Synthesis", Academic Press, N.Y. (1981) (Review)
[111] Hill, B.C., Greenwood, C.: Biochem. J.,218,913–921 (1984)
[112] Hill, B.C., Greenwood, C.: FEBS Lett.,166,362–366 (1984)
[113] Fujiwara, T., Fukumori, Y., Yamanaka, T.: J. Biochem.,105,287–292 (1989)
[114] van Buuren, K.J.H.: PH.D. Thesis, University of Amsterdam (1972)
[115] Matsuoka, M., Maeshima, M. Asahi, I.: J. Biochem.,90,649–655 (1981)
[116] Yoshikawa, S.T., Tera, Y., Takahashi, T., Tsukihara, T., Caughey, W.S.: Proc. Natl. Acad. Sci. USA,85,1354–1358 (1988)
[117] Rigell, C.W., de Saussare, C., Freire, E.: Biochemistry,24,5638–5646 (1985)
[118] Robinson, N.C., Neumann, J., Wiginton, D.: Biochemistry,24,6298–6304 (1985)
[119] Saraste, M.: Quart. Rev. Biophys.,23,331–366 (1990) (Review)
[120] Malmström, G.: Chem. Rev.,90,1247–1260 (1990) (Review)
[121] Kornblatt, J.A., Hui Bon Hoa, G.: Biochemistry,29,9370–9376 (1990)

1 NOMENCLATURE

EC number
1.9.3.2

Systematic name
Ferrocytochrome-c:oxygen oxidoreductase

Recommended name
Pseudomonas cytochrome oxidase

Synonymes
Cytochrome c-551:O_2, NO_2^- oxidoreductase
Cytochrome cd
Cytochrome cd_1
Oxidase, Pseudomonas cytochrome

CAS Reg. No.
9027-00-3

2 REACTION AND SPECIFICITY

Catalysed reaction
4 Ferrocytochrome c_2 + O_2 →
→ 4 ferricytochrome c_2 + 2 H_2O
Pseudomonas ferrocytochrome c + NO_2^- + H^+ →
→ Pseudomonas ferricytochrome c + NO + OH^- [29]

Reaction type
Redox reaction

Natural substrates
Azurin + O_2 [3, 21, 28]
Ferrocytochrome c-551 + O_2 [3, 21, 28]
Nitrite + Pseudomonas cytochrome c [29]

Enzyme Handbook © Springer-Verlag Berlin Heidelberg 1994
Duplication, reproduction and storage in data banks are only allowed with the prior permission of the publishers

Substrate spectrum

1 Pseudomonas ferrocytochrome c-551 + O_2 (O_2 can be replaced by NO_2^-) [21]
2 NH_2OH + pyocyanine (reduced, pyocyanine can be replaced by reduced methylene blue, no other electron donor) [26]
3 O_2 + Pseudomonas copper protein (reduced, O_2 can be replaced by NO_2^-, mechanism [27]) [1, 11, 13, 21, 27]
4 More (electron donor: sodium ascorbate, hydroquinone [21], cytochrome c of various origins [11], metabisulphite, dithionite [25], no donors: reduced pyridine nucleotides [26]) [11, 21, 25, 26]

Product spectrum

1 H_2O + Pseudomonas ferricytochrome c-551
2 NH_4OH + pyocyanine [26]
3 H_2O + Pseudomonas copper protein (oxidized, NO_2^- reduced to NO, N_2O [11, 13]) [6, 11, 13]
4 ?

Inhibitor(s)

CN^- (nitrite reduction, O_2 and NO_2^- reduction) [1, 2, 10, 21, 23, 26, 28]; CO (with O_2 as electron acceptor only) [1, 21, 23, 26, 28]; NO (depending on pH) [16]; NO_2^- (hydroxylamine reduction [26]) [2, 26]; Oxidized azurin (product inhibition) [3]; Oxidized cytochrome c (product inhibition) [3, 21]; SO_2^- [21]; Triton X-100 (enzyme incorporated into liposomes, slight inhibition of free enzyme) [18]

Cofactor(s)/prostethic group(s)

Heme c (2 mol per mol of dimeric enzyme, location in protein [14]) [1, 2, 7, 14, 28]; Heme d_1 (structure [4], location in protein [14]) [1, 2, 4, 14, 28]; Liposomes (activation when incorporated into) [18]; Subtilisin (activation by limited proteolysis) [9]

Metal compounds/salts

Fe (hemoprotein) [1, 7, 16, 24, 28]

Turnover number (min^{-1})

240 (cytochrome c) [11]

Specific activity (U/mg)

0.179 (O_2, component II) [2]; 0.814 (O_2, component I) [2]; 1.9–7.0 (O_2) [8]; 3.78 (copper protein) [29]; 20.0 (nitrite) [17]; More (assay method) [17, 22, 26, 28]

K_m-value (mM)

0.004 (reduced methylene blue) [26]; 0.0056 (ferrocytochrome c-551) [21]; 0.01 (pyocyanine, cytochrome c-551 with O_2 as electron acceptor) [26, 28]; 0.017 (reduced azurin) [26, 28]; 0.049 (reduced azurin) [21]; 0.05–0.08 (O_2, Erythrobacter sp. component I) [2, 11]; 0.054 (NO_2^-, Pseudomonas cytochrome c as electron donor) [26]; 0.1 (NO_2^-, pyocyanine as electron donor) [26]; 0.148 (O_2, Erythrobacter sp. component II) [2]; 4.0 (ascorbate, reduction of O_2) [21]; 30.0 (quinol, reduction of O_2) [21]

pH-optimum

5.3 [29]; 5.5–5.8 [11]; 6.5 (nitrite reduction, pyocyanine as electron donor) [26]; 6.8 (component II) [2]; 7.2 (reduction of hydroxylamine, pyocyanine as electron donor) [26]; 7.5 (component I) [2]

pH-range

Temperature optimum (°C)

30 (assay at) [21, 26, 28]; More (Arrhenius plots) [17]

Temperature range (°C)

3 ENZYME STRUCTURE

Molecular weight

119000–135000 (Erythrobacter sp. OCh 114, HPLC [2], Paracoccus denitrificans [11], Pseudomonas aeruginosa, SDS-PAGE after cross-linkage with dimethylsuberimidate [1, 16], sedimentation and diffusion data [28, 29]) [1, 2, 11, 16, 28, 29]

Subunits

Dimer (2 × 58000–67000, Erythrobacter sp. OCh 114, SDS-PAGE [2], Pseudomonas aeruginosa, SDS-PAGE [16, 28, 29], analysis of iron content [29], sedimentation velocity of succinylated enzyme [29], Pseudomonas stutzeri, SDS-PAGE [7]) [2, 7, 16, 28, 29]

More (dimeric when incorporated into liposomes [18], quarternary structure [20], proteolytic digestion products [10]) [10, 18, 20]

Glycoprotein/Lipoprotein

–

Enzyme Handbook © Springer-Verlag Berlin Heidelberg 1994
Duplication, reproduction and storage in data banks are only allowed with the prior permission of the publishers

4 ISOLATION/PREPARATION

Source organism

Alcaligenes faecalis [31]; Bacteria (grown anaerobically in presence of nitrate) [34]; Erythrobacter sp. OCh 114 [2]; Paracoccus denitrificans [11, 30]; Pseudomonas aeruginosa [1, 3, 5, 6, 8–10, 12–29]; Pseudomonas perfectomarinus [32]; Pseudomonas stutzeri [7]; Spirillum itersonii [34]; Thiobacillus denitrificans [33]

Source tissue

Localisation in source

Inner surface of cytoplasmic membrane (associated with) [1, 17]

Purification

Erythrobacter sp. OCh 114 (two types of complex cd_1) [2]; Paracoccus denitrificans [11]; Pseudomonas aeruginosa (large scale [22]) [8, 20, 22, 28, 29]; Pseudomonas stutzeri [7]

Crystallization

(method, X-ray crystallography, electron microscopy [15]) [1, 15, 29]

Cloned

–

Renaturated

(reconstitution of heme d depleted enzyme) [26]

---

5 STABILITY

pH

4.0–11.0 (dissociation of subunits above and below) [24]

Temperature (°C)

80–100 (denaturation temperature depending on redox state of hemes) [12]

Oxidation

Organic solvent

General stability information

Guanidine chloride, 2 M, 95% denaturation [24]; Unstable in reduced form [5]; Urea, 4 M, 91% denaturation [24]

Storage

–18°C [2]; Liquid N_2, rapid freezing, undefinitely [11]

6 CROSSREFERENCES TO STRUCTURE DATABANKS

PIR/MIPS code

PIR2: A32975 (precursor Pseudomonas aeruginosa); PIR1: OSPSZ (precursor Pseudomonas stutzeri)

Brookhaven code

7 LITERATURE REFERENCES

[1] Poole, R.K.: Biochim. Biophys. Acta,726,205–243 (1983) (Review)
[2] Doi, M., Shioi, Y., Morita, M., Takamiya, K.: Eur. J. Biochem.,184,521–527 (1989)
[3] Blatt, Y., Pecht, I.: Eur. J. Biochem.,160,149–153 (1986)
[4] Chang, C.K., Wu, W.: J. Biol. Chem.,261,8593–8596 (1986)
[5] Carson, S.D., Ching, Y.C., Wells, C.A., Wharton, D.C., Ondrias, M.R.: Biochemistry,25,787–790 (1986)
[6] Tordi, M.G., Silvestrini, M.C., Colosimo, A., Tuttobello, L., Brunori, M.: Biochem. J.,230,797–805 (1985)
[7] Liu, M.-Y., Liu, M.-C., Payne, W.J., Peck, H.D., LeGall, J.: Curr. Microbiol.,9,87–92 (1983)
[8] Silvestrini, M.C., Citro, G., Colosimo, A., Chersi, A., Zito, R., Brunori, M.: Anal. Biochem.,129,318–325 (1983)
[9] Horowitz, P.M., Falksen, K., Muhoberac, B.B., Wharton, D.C.: J. Biol. Chem.,257,9258–9260 (1982)
[10] Horowitz, P.M., Muhoberac, B.B., Falksen, K., Wharton, D.C.: J. Biol. Chem.,257,2140–2143 (1982)
[11] Timkovich, R., Dhesi, R., Martinkus, K.J., Robinson, M.K., Rea, T.M.: Arch. Biochem. Biophys.,215,47–58 (1982)
[12] Mitra, S., Donovan, J.W., Bersohn, R.: Biochem. Biophys. Res. Commun.,98,140–146 (1981)
[13] Wharton, D.C., Weintraub, S.T.: Biochem. Biophys. Res. Commun.,97,236–242 (1980)
[14] Mitra, S., Bersohn, R.: Biochemistry,19,3200–3203 (1980)
[15] Akey, C.W., Moffat, K., Wharton, D.C., Edelstein, S.J.: J. Mol. Biol.,136,19–43 (1980)
[16] Silvestrini, M.C., Colosimo, A., Brunori, M., Walsh, R.A., Barber, D., Greenwood, C.: Biochem. J.,183,701–709 (1979)
[17] Saraste, M., Kuronen, T.: Biochim. Biophys. Acta,513,117–131 (1978)
[18] Saraste, M.: Biochim. Biophys. Acta,507,17–25 (1978)
[19] Hill, K.E., Wharton, D.C.: J. Biol. Chem.,253,489–495 (1978)
[20] Saraste, M., Virtanen, I., Kuronen, T.: Biochim. Biophys. Acta,492,156–162 (1977)
[21] Barber, D., Parr, S.R., Greenwood, C.: Biochem. J.,157,431–438 (1976)
[22] Parr, S.R., Barber, D., Greenwood, C., Phillips, B.W., Melling, J.: Biochem. J.,157,423–430 (1976)
[23] Wharton, D.C., Gibson, Q.H.: Biochim. Biophys. Acta,430,445–453 (1976)
[24] Kuronen, T., Saraste, M., Ellfolk, N.: Biochim. Biophys. Acta,393,48–54 (1975)
[25] Parr, S.R., Wilson, M.T., Greenwood, C.: Biochem. J.,139,273–276 (1974)
[26] Singh, J.: Biochim. Biophys. Acta,333,28–36 (1973)

Enzyme Handbook © Springer-Verlag Berlin Heidelberg 1994
Duplication, reproduction and storage in data banks are only allowed with the prior permission of the publishers

[27] Wharton, D.C., Gudat, J.C., Gibson, Q.H.: Biochim. Biophys. Acta,292,611–620 (1973)
[28] Gudat, J.C., Singh, J., Wharton, D.C.: Biochim. Biophys. Acta,292,376–390 (1973)
[29] Kuronen, T., Ellfolk, N.: Biochim. Biophys. Acta,275,308–318 (1972)
[30] Newton, N.: Biochim. Biophys. Acta,185,316–331 (1969)
[31] Iwasaki, H., Marsubara, T.: J. Biochem.,69,847–857 (1971)
[32] Cox, C.D., Payne, W.J.: Can. J. Microbiol.,19,861–872 (1973)
[33] LeGall, J., Payne, W., Morgan, T.V., Der Vatanian, D.: Biochem. Biophys. Res. Commun.,87,355–362 (1979)
[34] Lemberg, R., Barrett, J. in "Cytochromes", Academic Press, London (1970) (Review)

1 NOMENCLATURE

EC number

1.9.6.1

Systematic name

Ferrocytochrome:nitrate oxidoreductase

Recommended name

Nitrate reductase (cytochrome)

Synonymes

Respiratory nitrate reductase
Reductase, nitrate (cytochrome)
Benzyl viologen-nitrate reductase
More (the enzymes EC 1.7.99.4 and EC 1.9.6.1 are probably identical, in vivo cytochrome serves as electron donor in the electron transport chain to NO_3^- [32, 33], in vitro artificial electron donors (e.g. reduced viologen indicators) can replace cytochrome c)

CAS Reg. No.

9029-42-9

2 REACTION AND SPECIFICITY

Catalysed reaction

Ferrocytochrome + nitrate →
→ ferricytochrome + nitrite

Reaction type

Redox reaction

Natural substrates

Nitrate + ferrocytochrome (in vivo cytochrome serves as electron donor in the electron transport chain to NO_3^- [32, 33], in vitro artificial electron donors (e.g. reduced viologen indicators) can replace cytochrome c, reduction of cytochrome b_{556} by ubiquinol which releases two protons, electrons are passed to nitrate reductase and used to reduce NO_3^- to NO_2^- [31], physiological function: transfer of electrons from cytochrome b_{559} to nitrate

Enzyme Handbook © Springer-Verlag Berlin Heidelberg 1994
Duplication, reproduction and storage in data banks are only allowed with the prior permission of the publishers

[4], electron transport chain in vivo:--> cytochrome b_1--> Mo--> NO_3^- [32], Fe^{3+}--> specific cytochrome-nitrate reductase--> NO_3^- [33], metabolic role: nitrate respiration [32]) [4, 31, 32, 33]
Nitrite + ferricyanide (nitrification) [34]

Substrate spectrum

1 Ferrocytochrome + nitrate [32, 33]
2 Nitrate + reduced acceptor (ir [19], r [34], acceptor e.g.: reduced benzyl viologen [1, 7, 9, 15, 18, 32–34], reduced viologen indicators [2, 7, 16, 20, 21, 32], $FMNH_2$ [2, 9, 16], reduced methyl viologen [3, 4, 11, 22, 32, 34], reduced ferredoxin [11], $FADH_2$ [16, 32], duroquinol [20], quinols [20], NADPH [32], reduced cytochrome b_1 [32], phenosafranine [32], reduced methylene blue [32]) [1–34]
3 Nitrite + ferricyanide [34]
4 ClO_3^- + reduced acceptor [2, 4, 16, 21]
5 BrO_3^- + reduced acceptor [16, 21]
6 More (not: NADH [2, 4, 15, 16], NADPH [2, 4, 16], $FADH_2$ [4], $FMNH_2$ [4], menadione [4], the enzymes EC 1.7.99.4 and EC 1.9.6.1 are probably identical, in vivo cytochrome serves as electron donor in the electron transport chain to NO_3^- [32, 33], in vitro artificial electron donors (e.g. reduced viologen indicators) can replace cytochrome c) [2, 4, 9, 15, 16, 32, 33]

Product spectrum

1 Ferricytochrome + nitrite [32, 33]
2 Nitrite + acceptor [1–34]
3 Nitrate + ferrocyanide [34]
4 ?
5 ?
6 ?

Inhibitor(s)

CN^- [2, 11, 15, 16, 24]; N_3^- [2, 11, 16, 20, 33]; NaCl (0.17 M NaCl + 0.5 mM MgCl activates, 4.27 M NaCl + 0.5 mM MgCl inhibits) [3]; p-Chloromercuribenzoate (above 0.1 mM [4], not [11]) [4, 15, 33]; ClO_3^- [4, 15]; Dithiol [15]; KCNS [15]; o-Phenanthroline [15]; Mepacrine [15]; ClO_4^- [15]; BrO_3^- [15]; IO_3^- [15]; 2-n-Heptyl-4-hydroxyquinoline N-oxide [20]; Ferricyanide [24]; Na_2S [33]; Hydrogensulfite [33]

Cofactor(s)/prostethic group(s)

Cytochrome b (Pseudomonas denitrificans: purified enzyme has a cytochrome b-type absorption spectrum) [15]; Cytochrome c (Nitrobacter hamburgensis, nitrite oxidoreductase enzyme complex) [34]; Bactopterin (Pseudomonas stutzeri, cofactor) [23]; More (contains no cytochrome: Aerobacter aerogenes [1], Klebsiella aerogenes [4], Rhodobacter sphaeroides [18], Bacillus stearothermophilus [22], no FAD: Micrococcus halodenitrificans [2], Micrococcus denitrificans [16], no FMN: Micrococcus halodenitrificans [2], Micrococcus denitrificans [16]) [1, 2, 4, 16, 18, 22]

Metal compounds/salts

Iron (Bacillus licheniformis: 6.9 atoms of non-heme iron per molecule of enzyme [27], Nitrobacter hamburgensis: contains iron [34], Micrococcus halodenitrificans: enzyme contains 2 Fe atoms, non-heme iron protein [2], Clostridium perfringens: molybdo-iron-sulfur protein [11], Micrococcus denitrificans: iron content about 8 mol per mol of enzyme, non-heme iron protein [16], Bacillus stearothermophilus: 6 mol of nonheme iron in 1 mol of purified enzyme [22], E. coli: 16 gatom of iron per mol of enzyme [24], contains iron-sulfur centres [31], Klebsiella aerogenes, nitrate reductase I, MW 26000 (abc_2): 8 iron-sulfur groups and 4 tightly bound non-heme iron atoms per molecule of enzyme, nitrate reductase II: no tightly bound iron [6]) [2, 6, 11, 16, 18, 22, 24, 27, 31, 34]; Molybdenum (Nitrobacter hamburgensis: contains molybdenum [34], Micrococcus halodenitrificans: enzyme contains 1 molybdenum atom [2], Klebsiella aerogenes, nitrate reductase I, MW 260000 (abc_2): 0.24 atoms of molybdenum per molecule of enzyme [6], Clostridium perfringens: molybdo-iron-sulfur protein [11], Paracoccus denitrificans: molybdenum centre [17], Rhodobacter sphaeroides: 1 mol molybdenum per mol of enzyme [18], Pseudomonas stutzeri: molybdoenzyme [23], E. coli: 0.8 gatom of molybdenum per mol of enzyme [24], molybdenoenzyme [31], Bacillus licheniformis: 0.93 atoms of molybdenum per molecule of enzyme, molybdenum seems to be a part of a low-molecular weight peptide to which it may be bound by interaction with thiol groups [27]) [2, 6, 11, 17, 18, 23, 24, 27, 31, 34]; Mn^{2+} (stimulates) [11]; Fe^{2+} (stimulates) [11]; Mg^{2+} (stimulates) [11]; Ca^{2+} (stimulates) [11]; Na^+ (0.17 M NaCl + 0.5 mM MgCl activates, 4.27 M NaCl + 0.5 mM MgCl inhibits) [3]

Turnover number (min^{-1})

Specific activity (U/mg)

More [2, 18, 33, 34]; 39.1 [4]; 52 [11]; 50 [20]; 62.53 [22]

Enzyme Handbook © Springer-Verlag Berlin Heidelberg 1994
Duplication, reproduction and storage in data banks are only allowed with the prior permission of the publishers

K_m-value (mM)

More [15, 18, 20, 32–34]; 1.3 (NO_3^-) [2]; 5 (ClO_3^-) [2]; 1.6 (reduced benzyl viologen, NO_3^-) [4]; 0.10 (NO_3^-) [11]; 0.0002 (benzyl viologen) [18]; 3.6 (NO_2^-) [34]

pH-optimum

5.6–5.8 (NO_3^-) [16]; 6.2–6.4 (ClO_3^-) [16]; 6.3 (NO_3^-) [2]; 6.4 (ClO_3^-) [2]; 6.8 [4]; 7.0 (nitrate (+ reduced acceptor)) [34]; 7–8.5 [33]; 7.1 (NO_3^-, Enterobacter cloacae [25], E. coli [26]) [25, 26]; 8.0 (NO_3^- (+ reduced acceptor) [15], NO_2^- (+ ferricyanide) [34]) [15, 34]; 9.0 [11]

pH-range

5.5–10.0 (sharp drop of activity below pH 5.5 and above 10.0) [33]; 7.0–9.8 (7.0: about 10% of activity maximum, 9.8: about 50% of activity maximum) [11]

Temperature optimum (°C)

More (temperature optimum is a function of both: the concentration and the specific cation present) [3]; 56 (0.5 mM $MgCl_2$) [3]; 73 (2 M KCl) [3]; 75 (optimum of 5 min reaction) [25]; 85 (4.27 M NaCl) [3]

Temperature range (°C)

3 ENZYME STRUCTURE

Molecular weight

90000 (Clostridium perfringens, gel filtration) [11]
165000 (Micrococcus halodenitrificans, electrophoresis) [2]
160000 (Klebsiella aerogenes, nitrate reductase II, gel filtration [4], Micrococcus denitrificans, gel filtration, ultracentifugation [16]) [4, 16]
180000 (Rhodobacter sphaeroides, gel filtration of urea-treated enzyme) [18]
196000 (Bacillus licheniformis, polyacrylamide gel electrophoresis) [27]
200000 (Paracoccus denitrificans, gel filtration) [19]
210000 (Bacillus stearothermophilus, polyacrylamide gel electrophoresis) [22]
260000 (Klebsiella aerogenes, nitrate reductase I (abc_2), gel filtration) [4]
360000 (Nitrobacter hamburgensis, gel filtration, nitrite oxidoreductase enzyme complex) [34]
1060000 (Klebsiella aerogenes, nitrate reductase I, $(abc_2)_4$, gel filtration) [4]

Subunits

Monomer (1 × 90000, Clostridium perfringens, SDS-PAGE) [11]
Dimer (1 × 150000 + 1 × 57000, Bacillus licheniformis, SDS-PAGE [27], 1 × 117000 + 1 × 57000, Klebsiella aerogenes, nitrate reductase II, SDS-PAGE [4], 1 × 120000 (alpha) + 1 × 60000 (beta), Rhodobacter sphaeroides, SDS-PAGE [18], 1 × 127000 (alpha) + 1 × 61000 (beta), Paracoccus denitrificans [19], 1 × 150000 + 1 × 44000, Bacillus stearothermophilus, SDS-PAGE [22]) [4, 18, 19, 22, 27]
Trimer (1 × 127000 (alpha) + 1 × 61000 (beta) + 1 × 21000 (gamma), Paracoccus denitrificans, SDS-PAGE) [20]
Tetramer (1 × 117000 + 1 × 57000 + 2 × 52000, Klebsiella aerogenes, nitrate reductase II, SDS-PAGE) [4]
Pentamer ($alpha_2beta_2gamma_1$, 2 × 116000 (alpha) + 2 × 65000 (beta) + 1 × 32000 (gamma), Nitrobacter hamburgensis, nitrite reductase enzyme complex, SDS-PAGE) [34]
Hexadecamer (4 × 117000 + 4 × 57000 + 8 × 52000, Klebsiella aerogenes, nitrate reductase II, SDS-PAGE) [4]
? (x × 150000 (alpha) + x × 67000 ($beta_1$) + x × 65000 ($beta_2$), molar ratio alpha: $beta_1$ + $beta_2$ is 1:1, E. coli, SDS-PAGE [7, 9], an additional subunit is present in haem-containing enzyme [9], x × 150000 (alpha) + x × 60000 (beta) + x × 20000 (gamma), E. coli [31]) [7, 9, 31]

Glycoprotein/Lipoprotein

More (does not contain carbohydrate) [27]

4 ISOLATION/PREPARATION

Source organism

Aerobacter aerogenes [1]; Micrococcus halodenitrificans [2]; Halobacterium sp. [3]; Klebsiella aerogenes (2 forms: nitrate reductase I (ab), nitrate reductase II (abc_2) and $(abc_2)_4$ [4, 6]) [4, 6, 25, 28]; E. coli (5-aminolaevulinic acid auxotroph [5], K-12 [7, 9, 21]) [5, 7–9, 12, 14, 21, 24, 25, 31, 32]; Clostridium perfringens [11]; Pseudomonas denitrificans [15]; Micrococcus denitrificans [16, 25]; Paracoccus denitrificans [17, 19, 20]; Rhodobacter sphaeroides (f.sp. denitrificans) [18]; Bacillus stearothermophilus [22]; Pseudomonas stutzeri [23]; Bacillus licheniformis [27]; Thiosphaera pantotropha [29]; Neurospora crassa [30]; Achromobacter fischeri [33]; Nitrobacter hamburgensis [34]

Source tissue

Cell [2, 3]

Enzyme Handbook © Springer-Verlag Berlin Heidelberg 1994
Duplication, reproduction and storage in data banks are only allowed with the prior permission of the publishers

Localisation in source

Membrane (nitrate reductase of cytoplasmic membrane is vectorial, reducing nitrate on the outer aspect of the membrane with 2 H^+ and 2 electrons that have crossed from the inner aspect of the membrane [8], arrangement in cytoplasmic membrane [12], Klebsiella aerogenes: transmembrane protein [28], Bacillus licheniformis: the 2 subunits are localized on the cytoplasmic side of the membrane [27]) [8, 12, 27]

Purification

Aerobacter aerogenes [1]; Micrococcus halodenitrificans [2]; Halobacterium sp. (partial) [3]; Klebsiella aerogenes (2 forms: nitrate reductase I, nitrate reductase II [4, 6]) [4]; E. coli (K-12 [7, 9]) [7, 9, 24]; Clostridium perfringens [11]; Pseudomonas denitrificans [15]; Micrococcus denitrificans [16]; Rhodobacter sphaeroides [18]; Paracoccus denitrificans (partial) [19, 20]; Bacillus stearothermophilus [22]; Bacillus licheniformis [27]; Achromobacter fischeri [33]

Crystallization

–

Cloned

(large subunit gene of E. coli enzyme) [13]

Renaturated

–

5 STABILITY

pH

6.5–10 [25]

Temperature (°C)

6 (immobilized enzyme more stable at 6°C than at 23°C) [14]; 20 (24 h, complete loss of activity) [4]; 23 (immobilized enzyme more stable at 6°C than at 23°C) [14]; 50 (10 min, stable up to) [25, 33]; 60 (10 min, 15% loss of activity) [25, 33]; 70 (5 min, complete loss of activity) [33]; 80 (0.17 M NaCl: 90% loss of activity after 1 min, 0.85 M NaCl: 50% loss of activity after 7 min, 4.27–5.31 M NaCl: no loss of activity after 15 min) [3]

Oxidation

Organic solvent

General stability information

NaCl protects against thermal inactivation, 4.27–5.31 M [3]; Deoxycholate stabilizes [4]; Freezing (slow freezing inactivates [10], rapid freezing in liquid N_2 and thawing at room temperature can be repeated at least ten times without loss of activity [9]) [9, 10]; Immobilized enzyme is more labile than soluble [14]; Sucrose stabilizes [25]; Stable to prolonged dialysis [33]

Storage

4°C, prolonged storage [4, 9]; 5°C, 0.1 M Tris-HCl, 10% sucrose, pH 8.8, 24 h, 10% loss of activity [25]; Frozen, several months [33]

6 CROSSREFERENCES TO STRUCTURE DATABANKS

PIR/MIPS code

Brookhaven code

7 LITERATURE REFERENCES

[1] Van'T Riet, J., Planta, R.J.: FEBS Lett.,5,249–252 (1969)
[2] Rosso, J.-P., Forget, P., Pichinoty, F.: Biochim. Biophys. Acta,321,443–455 (1973)
[3] Marquez, E.D., Brodie, A.F.: Biochim. Biophys. Acta,321,84–89 (1973)
[4] Van 'T Riet, J., Planta, R.J.: Biochim. Biophys. Acta,379,81–94 (1975)
[5] Kemp, M.B., Haddock, B.A., Garland, P.B.: Biochem. J.,148,329–333 (1975)
[6] Van 'T Riet, J., Van Ee, J.H., Wever, R., Van Gelder, B.F., Planta, R.J.: Biochim. Biophys. Acta,405,306–317 (1975)
[7] Clegg, R.A.: Biochem. Soc. Trans.,3,691–694 (1975)
[8] Garland, P.B., Downie, J.A., Haddock, B.A.: Biochem. J.,152,547–559 (1975)
[9] Clegg, R.A.: Biochem. J.,153,533–541 (1976)
[10] MacGregor, C.H., Schnaitman, C.A., Normasell, D.E., Hodgins, M.G.: J. Biol. Chem.,249,5321–5327 (1974)
[11] Seki-Chiba, S., Ishimoto, M.: J. Biochem.,82,1663–1671 (1977)
[12] Graham, A., Boxer, D.H.: FEBS Lett.,113,15–20 (1980)
[13] McPherson, M.J., Baron, A.J., Pappin, D.J.C., Wootton, J.C.: FEBS Lett.,177,260–264 (1984)
[14] Schiller, J.G., Liu, C.C.: Biotechnol. Bioeng.,18,1643–1645 (1976)
[15] Radcliff, B.C., Nicholas, D.J.D.: Biochim. Biophys. Acta,205,273–287 (1970)
[16] Forget, P.: Eur. J. Biochem.,18,442–450 (1971)
[17] Turner, N., Ballard, A.L., Bray, R.C., Ferguson, S.: Biochem. J.,252,925–926 (1988)
[18] Byrne, M.D., Nicholas, D.J.D.: Biochim. Biophys. Acta,915,120–124 (1987)
[19] Ballard, A.L., Ferguson, S.J.: Eur. J. Biochem.,174,207–212 (1988)

Enzyme Handbook © Springer-Verlag Berlin Heidelberg 1994
Duplication, reproduction and storage in data banks are only allowed with the prior permission of the publishers

[20] Craske, A., Ferguson, S.J.: Eur. J. Biochem.,158,429–436 (1986)
[21] Morpeth, F.F., Boxer, D.H.: Biochemistry,24,40–46 (1985)
[22] Chikwem, J.O., Downey, R.J.: Anal. Biochem.,126,74–80 (1982)
[23] Krüger, B., Meyer, O., Nagel, M., Andreesen, J.R., Meincke, M., Bock, E., Blümle, S., Zumft, W.G.: FEMS Microbiol. Lett.,48,225–227 (1987)
[24] Adams, M.W.W., Mortenson, L.E.: J. Biol. Chem.,257,1791–1799 (1982)
[25] Aoki, K., Shinke, R., Nishira, H.: Agric. Biol. Chem.,45,817–822 (1981)
[26] Taniguchi, S., Itagaki, E.: Biochim. Biophys. Acta,44,263–279 (1960)
[27] Van 'T Riet, J., Wientjes, F.B., Van Doorn,J., Planta, R.J.: Biochim. Biophys. Acta,576,347–360 (1979)
[28] Wientjes, F.B., Kolk, A.H.J., Nanninga, N., Van 'T Riet, J.: Eur. J. Biochem.,95,61–67 (1979)
[29] Bell, L.C., Richardson, D.J., Ferguson, S.J.: FEBS Lett.,265,85–87 (1990)
[30] Subramanian, K.N., Sorger, G.J.: J. Bacteriol.,110,538–546 (1972)
[31] Boxer, D., Malcolm, A., Graham, A.: Biochem. Soc. Trans.,10,480–481 (1982) (Review)
[32] Nason, A. in "The Enzymes",2nd. Ed. (Boyer, P.D., Lardy, H., Myrbäck, K., eds.) 7,587–607 (1963) (Review)
[33] Sadana, J.C., McElroy, W.D.: Arch. Biochem. Biophys.,67,16–34 (1957)
[34] Sundermeyer-Klinger, H., Meyer, W., Warninghoff, B., Bock, E.: Arch. Microbiol.,140,153–158 (1984)

1 NOMENCLATURE

EC number
1.9.99.1

Systematic name
Ferrocytochrome-c:Fe^{3+} oxidoreductase

Recommended name
Iron-cytochrome-c reductase

Synonymes
Reductase, iron-cytochrome c
Iron-cytochrome c reductase

CAS Reg. No.
37256-52-3

2 REACTION AND SPECIFICITY

Catalysed reaction
Ferrocytochrome c + Fe^{3+} →
→ ferricytochrome c + Fe^{2+}

Reaction type
Redox reaction

Natural substrates
Ferrocytochrome c + Fe^{3+} (close association of enzyme with respiration chain of Ferrobacillus ferrooxidans) [1]

Substrate spectrum
1 Ferrocytochrome c + Fe^{3+}

Product spectrum
1 Ferricytochrome c + Fe^{2+}

Inhibitor(s)
Reduced glutathione [1]; p-Hydroxymercuribenzoate [1]; Cysteine [1]; Reduced cytochrome c [1]; Fe^{3+} [1]; Cu^{2+} [1]; Sodium sulfide [1]; Ni^{2+} (slight) [1]; Cr^{2+} (slight) [1]; More (not: atebrin, amytal, antimycin A) [1]

Enzyme Handbook © Springer-Verlag Berlin Heidelberg 1994
Duplication, reproduction and storage in data banks are only allowed with the prior permission of the publishers

Cofactor(s)/prostethic group(s)
More (most highly purified fraction of the enzyme contains cytochrome c and apparently nonfunctional cytochrome b) [1]

Metal compounds/salts
Iron (hemoprotein) [1]

Turnover number (min^{-1})

Specific activity (U/mg)
0.006 [1]

K_m**-value** (mM)
0.0275 (cytochrome c) [1]; 1.6 (Fe^{2+}) [1]

pH-optimum
5.7–6.2 [1]

pH-range
5–6.7 (5: about 50% of activity maximum, 6.7: about 25% of activity maximum) [1]

Temperature optimum (°C)

Temperature range (°C)

---

3 ENZYME STRUCTURE

Molecular weight

Subunits

Glycoprotein/Lipoprotein
–

---

4 ISOLATION/PREPARATION

Source organism
Ferrobacillus ferrooxidans [1]

Source tissue
Cell [1]

Localisation in source

Purification
Ferrobacillus ferrooxidans (partial) [1]

Crystallization

–

Cloned

–

Renaturated

–

5 STABILITY

pH

Temperature (°C)

100 (1 min, complete loss of activity) [1]; 90 (1 min, 74% loss of activity) [1]; 60 (5 min, 46% loss of activity) [1]; 40 (20 min, 10% loss of activity) [1]; 22 (48 h, no loss of activity) [1]; 0.5 (several weeks, no loss of activity) [1]

Oxidation

Organic solvent

General stability information

Storage

0–5°C, several weeks [1]

6 CROSSREFERENCES TO STRUCTURE DATABANKS

PIR/MIPS code

Brookhaven code

7 LITERATURE REFERENCES

[1] Yates, M.-G., Nason, A.: J. Biol. Chem.,241,4872–4880 (1966)

Enzyme Handbook © Springer-Verlag Berlin Heidelberg 1994
Duplication, reproduction and storage in data banks are only allowed with the prior permission of the publishers

1 NOMENCLATURE

EC number
1.10.1.1

Systematic name
(+,-)-trans-Acenaphthene-1,2-diol:$NADP^+$ oxidoreductase

Recommended name
trans-Acenaphthene-1,2-diol dehydrogenase

Synonymes
Dehydrogenase, trans-acenaphthylene-1,2-diol
trans-1,2-Acenaphthenediol dehydrogenase

CAS Reg. No.
51901-21-4

2 REACTION AND SPECIFICITY

Catalysed reaction
(+,-)-trans-Acenaphthene-1,2-diol + $NADP^+$ →
→ acenaphthenequinone + NADPH

Reaction type
Redox reaction

Natural substrates

Substrate spectrum
1 (-)trans-Acenaphthene-1,2-diol + $NADP^+$ (inactive with (+)-form [1]) [1]
2 Acenaphthene-1-ol + $NADP^+$ [1]
3 trans-1,2-Dihydronaphthalene-1,2-diol + $NADP^+$ [1]

Product spectrum
1 Acenaphthenequinone + NADPH (i.e. acenaphthylene-1,2-dione)
2 1-Acenaphthenone + NADPH (i.e. 2H-acenaphthylene-1-one)
3 1,2-Naphthoquinone + NADPH

Inhibitor(s)
(+)-trans-Acenaphthene-1,2-diol [1]; Hg^{2+} [1]; p-Chloromercuribenzoate [1]; KCN [1]; NADPH [1]; Catechol [1]

Enzyme Handbook © Springer-Verlag Berlin Heidelberg 1994
Duplication, reproduction and storage in data banks are only allowed with the prior permission of the publishers

Cofactor(s)/prostethic group(s)
NADP$^+$ (inactive with NAD$^+$) [1]; Ethanol (activation) [1]

Metal compounds/salts
Mg^{2+} (activation) [1]

Turnover number (min^{-1})

Specific activity (U/mg)
More [1]

K_m-value (mM)

pH-optimum

pH-range

Temperature optimum (°C)

Temperature range (°C)

---

3 ENZYME STRUCTURE

Molecular weight

Subunits

Glycoprotein/Lipoprotein
–

---

4 ISOLATION/PREPARATION

Source organism
Rat [1, 2]; Rabbit [1]; Mouse [1]; Guinea pig [1]; Hamster [1]; Dog [1]; Cat [1]

Source tissue
Liver [1, 2]

Localisation in source
Cytosol [1]

Purification
Rat [1]

Crystallization
–

Cloned

–

Renaturated

–

5 STABILITY

pH

Temperature (°C)

Oxidation

Organic solvent

General stability information

Storage

6 CROSSREFERENCES TO STRUCTURE DATABANKS

PIR/MIPS code

Brookhaven code

7 LITERATURE REFERENCES

[1] Hopkins, R.P., Drummond, E.C., Callaghan, P.: Biochem. Soc. Trans.,1,989–991 (1973)
[2] Drummond, E.C, Callaghan, P., Hopkins, R.P.: Xenobiotica,2,529–538 (1972)

Enzyme Handbook © Springer-Verlag Berlin Heidelberg 1994
Duplication, reproduction and storage in data banks are only allowed with the prior permission of the publishers

1 NOMENCLATURE

EC number
1.10.2.1

Systematic name
L-Ascorbate:ferricytochrome-b_5 oxidoreductase

Recommended name
L-Ascorbate-cytochrome-b_5 reductase

Synonymes
Reductase, ascorbate-cytochrome b_5
Ascorbate-cytochrome b_5 reductase

CAS Reg. No.
37237-57-3

2 REACTION AND SPECIFICITY

Catalysed reaction
L-Ascorbate + ferricytochrome b_5 →
→ monodehydroascorbate + ferrocytochrome b_5

Reaction type
Redox reaction

Natural substrates
L-(+)-Ascorbate + ferricytochrome b_5 (fatty acid desaturation [1])

Substrate spectrum
1 L-(+)-Ascorbate + ferricytochrome b_5 (r [7])

Product spectrum
1 (+)-Monodehydroascorbate + ferrocytochrome b_5

Inhibitor(s)
Hg^{2+} [7]

Cofactor(s)/prostethic group(s)
Cytochrome b_5

Enzyme Handbook © Springer-Verlag Berlin Heidelberg 1994
Duplication, reproduction and storage in data banks are only allowed with the prior permission of the publishers

Metal compounds/salts

Turnover number (min^{-1})

Specific activity (U/mg)
0.029 [6]; 0.0088–0.013 (depending on organism) [5]

K_m-value (mM)
0.0014–0.009 (cytochrome b_5, depending on organism) [5]; 0.12 (L-(+)-ascorbic acid) [7]; 0.005 (ferro- and ferricytochrome b_5) [7]

pH-optimum
6.4–6.7 [7]; 6.5 (in solubilized microsomes) [5]; 6.8 (substrate cytochrome b_5 bound to Sepharose) [4]; 6.9 [2]

pH-range

Temperature optimum (°C)

Temperature range (°C)

---

3 ENZYME STRUCTURE

Molecular weight

Subunits

Glycoprotein/Lipoprotein
–

---

4 ISOLATION/PREPARATION

Source organism
Rat [1, 2, 5–7]; Pig [3, 5]; Rabbit [5]

Source tissue
Liver

Localisation in source
Microsomes

Purification
Rat (partial) [2]

Crystallization
–

Cloned

–

Renaturated

–

5 STABILITY

pH

Temperature (°C)

45 (denaturation above) [7]

Oxidation

Organic solvent

General stability information

Inactivation by isoelectric focusing [2]

Storage

4°C, 0.02 M phosphate buffer, pH 8.1, 035% w/v lensodel NP 40 [2]

6 CROSSREFERENCES TO STRUCTURE DATABANKS

PIR/MIPS code

Brookhaven code

7 LITERATURE REFERENCES

[1] Scherer, G., Weis, W.: Hoppe-Seyler's Z. Physiol. Chem.,359,1527–1530 (1978)
[2] Scherer, G., Weis, W.: Hoppe-Seyler's Z. Physiol. Chem.,358,1499–1503 (1977)
[3] Weber, H., Weis, W., Wolf, B.: Hoppe-Seyler's Z. Physiol. Chem.,355,595–599 (1974)
[4] Scherer, G., Weber, H., Weis, W.: Hoppe-Seyler's Z. Physiol. Chem.,355,1350–1354 (1974)
[5] Weber, H., Weis, W., Schaeg, W., Staudinger, H.: Hoppe-Seyler's Z. Physiol. Chem.,354,1277–1284 (1973)
[6] Weber, H., Weis, W., Staudinger, H.: Hoppe-Seyler's Z. Physiol. Chem.,353,1415–1419 (1972)
[7] Everling, F.B., Weis, W., Staudinger, H.: Hoppe-Seyler's Z. Physiol. Chem.,350,1485–1492 (1969)

Enzyme Handbook © Springer-Verlag Berlin Heidelberg 1994
Duplication, reproduction and storage in data banks are only allowed with the prior permission of the publishers

1 NOMENCLATURE

EC number
1.10.2.2

Systematic name
Ubiquinol:ferricytochrome-c oxidoreductase

Recommended name
Ubiquinol-cytochrome-c reductase

Synonymes
Reductase, ubiquinol-cytochrome c
Coenzyme Q-cytochrome c reductase
Dihydrocoenzyme Q-cytochrome c reductase
Reduced ubiquinone-cytochrome c reductase, Complex III (mitochondrial electron transport)
Ubiquinone-cytochrome c reductase
Ubiquinol-cytochrome c oxidoreductase
Reduced coenzyme Q-cytochrome c reductase
Ubiquinone-cytochrome c oxidoreductase
Reduced ubiquinone-cytochrome c oxidoreductase
Mitochondrial electron transport complex III
Ubiquinol-cytochrome c-2 oxidoreductase
Ubiquinone-cytochrome b-c_1 oxidoreductase
Ubiquinol-cytochrome c_2 reductase
Ubiquinol-cytochrome c_1 oxidoreductase
$CoQH_2$–cytochrome c oxidoreductase
Ubihydroquinol:cytochrome c oxidoreductase
Coenzyme QH_2–cytochrome c reductase [15]
QH_2:cytochrome c oxidoreductase [38]
More (formerly: b-c_2 complex [15])

CAS Reg. No.
9027-03-6

2 REACTION AND SPECIFICITY

Catalysed reaction
Ubiquinol + 2 ferricytochrome c →
→ Ubiquinone + 2 ferrocytochrome c (discussion of mechanism [1, 2, 20, 29])

Enzyme Handbook © Springer-Verlag Berlin Heidelberg 1994
Duplication, reproduction and storage in data banks are only allowed with the prior permission of the publishers

Reaction type

Redox reaction

Natural substrates

Coenzyme Q_{10} + cytochrome c [28]
Ubiquinol-50 + cytochrome c [25]
More (link of electron transfer from ubiquinol to cytochrome c with proton translocation across inner mitochondrial membrane) [3]

Substrate spectrum

1 Ubiquinol + cytochrome c (ubiquinols with different lengths of side chain [25, 36, 37, 39, 42, 47], duroquinol can only partially substitute: reduction of Fe-S-center and cytochrome b, but not efficient in reoxidizing cytochrome b [2], cytochrome c from different sources [37, 47]) [25, 36, 37, 39, 42, 47]
2 2,3-Dimethoxy-5-methyl-6-(10-bromodecyl)-1,4-benzoquinol ($Q_0C_{10}BrH_2$) + cytochrome c [25, 35]
3 2,3-Dimethoxy-5-methyl-6-decyl-1,4-benzoquinol ($Q_0C_{10}H_2$) + cytochrome c [25]
4 More (no electron donor: dihydroubiquinone-1 [42], ferricyanide as electron acceptor [35], electron transfer activity [31, 39]) [31, 35, 39, 42]

Product spectrum

1 Ubiquinone + reduced cytochrome c
2 2,3-Dimethoxy-5-methyl-6-(10-bromodecyl)-1,4-benzoquinone ($Q_0C_{10}Br$) + reduced cytochrome c
3 2,3-Dimethoxy-5-methyl-6-decyl-1,4-benzoquinone (Q_0C_{10}) + reduced cytochrome c
4 ?

Inhibitor(s)

Antimycin A (not: [35], interaction with ubiquinone site of cytochrome b [10], reversed by ubiquinone, not by azido ubiquinone derivatives [24]) [1, 2, 9–11, 16, 23–25, 31, 36, 37, 39, 41, 42, 47, 49, 51, 52]; Myxothiazol [2, 9, 16, 20, 23, 29, 49]; 5-(n-Undecyl)-6-hydroxy-4,7-oxobenzothiazole (UHDBT, binding to Fe-S-protein [12]) [2, 12, 16, 25, 37, 49]; 2-(n-Heptyl)-4-hydroxyquinoline N-oxide (HQNO) [2, 16, 23, 49]; Triton X-100 [6, 37]; Diuron [10]; Stigmatellin [16, 49]; N, N'-Dicyclohexylcarbodiimide [18]; 2-Iodo-6-isopropyl-3-methyl-2,2'-trinitrophenyl ether [37]; Thenoyl trifluoroacetate [39]; 2-Alkyl-4-hydroxyquinoline N-oxide [41]; 2-Alkyl-3-hydroxyl-1,4-naphthoquinone (SN5949) [42]; More (not inhibitory: p-chloromercuribenzoate) [25]

Cofactor(s)/prostethic group(s)

Cytochrome c_1 heme [1, 2, 6, 9, 15, 36, 40–43, 47, 49]; Cytochrome b heme (b-561 [29], b-565 [5], b-562 [5]) [1, 2, 5, 6, 9, 16, 29, 36, 40–43, 47, 49]; Fe-S protein [1, 2, 6, 9, 16, 49]; Ubiquinone [9, 16, 36, 41–43]; Semiquinone anion [38]; Soybean lecithin (activation) [6]; Dodecyl maltoside (activation) [6, 49]; More (no flavins) [1, 51]

Metal compounds/salts

Non-heme iron [1, 6, 9, 36, 41–43, 47]; Sulfide [41]; More (no Cu) [51]

Turnover number (min^{-1})

450–240000 (cytochrome c, depending on method of solubilization) [2]; 15000–21000 (cytochrome c) [6]; 4000 (2,3-dimethoxy-5-methyl-6-(10-bromodecyl)-1,4-benzoquinol) [9]; More [16, 20, 28, 36, 43]

Specific activity (U/mg)

0–1000 (depending on preparation method) [45]; More (assay method) [42, 45]

K_m-value (mM)

0.006 (ubiquinol-10) [25]; 0.038 (ubiquinol-10) [5]; 0.08 (ubiquinol-1) [11]; 0.1 (ubiquinol-2) [23]; 0. 12–0.175 (ubiquinol-2) [30]; 0.006–0.03 (cytochrome c from different sources) [11]; 0.004 (cytochrome b-c_1 complex) [25]; 0.0005–0.0012 (2,3-dimethoxy-5-methyl-6-decyl-benzoquinol, depending on ionic strength) [28]; 0.175 (cytochrome c) [43, 51]; More (comparison of values from different organisms and purification procedures [30]) [30, 39, 51]

pH-optimum

7.0 [11]; 8.0 [28]; 8.5 [35]; More (below 10.0) [29]

pH-range

7.5–8.7 [35]

Temperature optimum (°C)

Temperature range (°C)

More (temperature dependency of activity of complex III in preparations with phospholipids replaced by dimyristoylglycerophosphocholine) [27]

Enzyme Handbook © Springer-Verlag Berlin Heidelberg 1994
Duplication, reproduction and storage in data banks are only allowed with the prior permission of the publishers

3 ENZYME STRUCTURE

Molecular weight

550000 (Neurospora crassa, hydrodynamic measurements) [31]
440000 (bovine heart, hydrodynamic measurements) [31]
400000 (bovine heart, sedimentation equilibrium centrifugation) [44]
285000–290000 (Neurospora crassa, calculation from content of prosthetic groups, calculation from subunit composition) [31]
200000–250000 (bovine heart, calculation from concentration of prosthetic groups [19, 31, 44], calculation from subunit composition [31, 46] light scattering [42]) [19, 31, 42, 44, 46]
121000 (Paracoccus denitrificans, calculation from cytochrome content) [16]
88400 (Rhodospirillum rubrum, calculation from subunit composition, assuming stoechiometry of 1:1:1) [9]

Subunits

Dimer of undecamer (bovine heart mitochondria, 11 subunits, core protein I: 49000, SDS-PAGE [48], core protein II: 47000, SDS-PAGE [48], cytochrome b: 42500, amino acid sequence [63], cytochrome c_1: 27874, amino acid sequence [54], Rieske Fe-S-protein: 21708, amino acid sequence [56], subunit VI, ubiquinone-binding protein: 13389, amino acid sequence [59], subunit VII: 9507, amino acid sequence [55], subunit VIII: 9175, amino acid sequence [58], subunit IX: 7998, amino acid sequence [60], subunit X: 7189, amino acid sequence [57], subunit XI: 6363, SDS-PAGE [48], ratio of polypeptides [1, 2, 46], subunit arrangement [69, 73], subunit arrangement in membrane crystals [31], values from SDS-PAGE [19, 24, 30, 41, 42, 44–46]) [1, 2, 19, 24, 30, 31, 41, 42, 44–46, 48, 54–60, 63, 69, 73]
Dimer of multimer (Neurospora crassa, 9 subunits, core protein I: 52000, core protein II: 2 × 45000, cytochrome b: 2 × 30000, cytochrome c_1: 31000, Rieske Fe-S-protein: 25000, small subunits: 1 × 14000, 1 × 12000, 1 × 9000 SDS-PAGE [2, 61], similar values deduced from gene sequence: subunit VI: 14500, subunit VII: 14400, subunit VIII: 12300, subunit IX: 7300, overview [53], values from SDS-PAGE [30, 40], possible function of subunits [3], Saccharomyces cerevisiae 8–9 subunits, core protein I: 47000–50000, core protein II: 39000–45000, cytochrome b: 31000, cytochrome c_1: 31000, Rieske Fe-S-protein: 23000–25000, small subunits: 1 × 14000–17000, 1 × 11500–13000, 1 × 11000–11200, SDS-PAGE [2, 32, 33, 62], subunit 9: 97300, deduced from gene sequence [4], gene sequences of subunit II [74], Rieske Fe-S-protein [75], subunit VIII [76]) [2–4, 30, 32, 33, 40, 53, 61, 62, 74–76]

Multimer (4 subunits: Rhodobacter sphaeroides, cytochrome b: 2 × 40000, cytochrome c_1: 34000, Rieske Fe-S-protein 24000, SDS-PAGE [6], subunit IV: 14000, amino acid analysis, SDS-PAGE [7], similar values from SDS-PAGE [36, 37], Rhodopseudomonas sphaeroides, cytochrome b: 48000, cytochrome c_1: 30000, subunit III: 24000, subunit IV: 12000, SDS-PAGE with and without mercaptoethanol [25], amino acid sequence [68]) [6, 7, 25, 36, 37, 68]
Trimer (Rhodospirillum rubrum, cytochrome b: 35000, cytochrome c_1: 31000, Rieske Fe-S-protein: 22400, Western blotting/SDS-PAGE [9], Paracoccus denitrificans, cytochrome b: 39000, cytochrome c_1: 62000, Rieske Fe-S-protein: 20000, SDS-PAGE [16, 49]) [9, 16, 49]

Glycoprotein/Lipoprotein
Phospholipoprotein [6, 16, 21, 24, 25, 36, 39, 41]; Lipoprotein [2, 42]; No phospholipoprotein [1]

4 ISOLATION/PREPARATION

Source organism
Eukaryotes (lower and higher) [53]; Bacteria (using O_2, N_2, sulfur-compounds as terminal electron acceptors [53], oxygenic and anoxygenic photosynthetic [2, 53], gram positive and gram negative respires [2], facultative chemoautotrophs [2]) [2, 53]; Bovine (calf [30]) [2, 5, 18, 19, 21, 22, 24, 27, 28, 30, 31, 38, 39, 41, 42, 44–46, 48, 54–60, 63]; Neurospora crassa [3, 14, 17, 20, 30, 31, 34, 40, 61, 71]; Saccharomyces cerevisiae [4, 8, 10, 15, 32, 43, 50, 51, 62, 72, 77]; Rhodobacter sphaeroides [6, 7, 25, 29]; Rhodospirillum rubrum [9, 67]; Rhodopseudomonas palustris [11]; Rhodopseudomonas capsulata [13]; Paracoccus denitrificans [16, 49, 64]; Helianthus tuberosus (Jerusalem artichoke) [23]; Rhodopseudomonas sphaeroides [25, 29, 35–38, 68]; Saccharomyces carlsbergensis [32]; Bradyrhizobium japonicum [65]; Trypanosoma brucei [66]; Leishmania tartentolae [66]; Rhodopseudomonas viridis [67]

Source tissue
Heart (bovine) [2, 5, 18, 19, 21, 22, 24, 27, 28, 30, 31, 38, 39, 41, 42, 44–46, 48, 54–60, 63]; Liver (calf) [30]

Localisation in source
Mitochondrial membrane [2, 4, 15, 18–24, 26, 28, 30, 38–44, 46–48, 50–56]; Chromatophores [6, 7, 25, 29, 35–37]; Plasma membrane [16, 49, 53]

Enzyme Handbook © Springer-Verlag Berlin Heidelberg 1994
Duplication, reproduction and storage in data banks are only allowed with the prior permission of the publishers

Purification

Bovine (ubiquinone-binding protein [5, 19, 59], Rieske Fe-S-protein [19, 56], core protein II [19], cytochrome c_1 containing protein [19], cytochrome b containing protein [19], 9.5 kDa protein [55], 11 kDa protein [58], 8 kDa subunit [60], smallest subunit [57], 11 subunits [48], several subunits, comparison of methods [45, 46]) [5, 19, 30, 39, 42, 45, 46, 48, 55–60]; Rhodobacter sphaeroides [6, 7]; Rhodospirillum rubrum [9, 67]; Paracoccus denitrificans [16, 49]; Neurospora crassa [20, 30, 40, 61]; Helianthus tuberosus (partial) [23]; Rhodopseudomonas sphaeroides [25, 35, 36]; Saccharomyces cerevisiae [43, 50, 51, 62]; Rhodopseudomonas viridis [67]

Crystallization

(single-layer membrane crystals [17], multi-layer membrane crystals [34], modelling of three-dimensional structure of cytochrome b [70]) [17, 34, 70]

Cloned

[4, 8, 13, 15, 26, 33, 53, 57, 64, 65, 71, 72, 77]

Renaturated

(reconstitution) [1, 2, 20, 22]

---

5 STABILITY

pH

5.0 (denaturation below) [25]; 9.5 (19 h, 40% denaturation) [25]

Temperature (°C)

0 (several h [25], 12 h, 50% inactivation [35]) [25, 35]; 37 (15 min, 50% inactivation) [25]; 20 (1 h, 13% inactivation) [25]

Oxidation

Organic solvent

Stable in chloroform and hexane at 20°C [35]

General stability information

Unstable in aqueous solutions [35]; Stable to repeated freezing/thawing [42]

Storage

–80°C or –70°C, 50 mM phosphate buffer, pH 8.0, 20% glycerol, 1 mM EDTA, several weeks [25, 36]; –80°C, lyophilized in 50% v/v glycerol [7]; –80°C [18]; –70°C, 50 mM Tris-HCl buffer, pH 8.0, 200 mM NaCl, 0.1 mg/ml dodecylmaltoside, 50% glycerol [6]; –70°C [42]; –20°C, 50% w/v glycerol, 2 months: unaltered spectral properties, decrease in electron transfer ratio [30]; –20°C, glycerol, 6 months completely stable [16]; –20°C, precipitated or dissolved in 50% v/v glycerol, several months [39, 49]; –20°C [9, 55, 56]

6 CROSSREFERENCES TO STRUCTURE DATABANKS

PIR/MIPS code

PIR2: B29413 ((cytochrome b) Paracoccus denitrificans); PIR3: A34591 ((cytochrome b) Rhodobacter sphaeroides (fragment)); PIR3: B34591 ((cytochrome c_1) Rhodobacter sphaeroides (fragment)); PIR2: C29413 ((cytochrome c_1) precursor Paracoccus denitrificans); PIR2: JQ0021 (Euglena gracilis mitochondrion); PIR1: CCBO11 (11K protein Bovine); PIR2: S00219 (11K protein precursor Human); PIR1: RDBYUN (14K protein Yeast (Saccharomyces cerevisiae)); PIR1: RDBYUC (17K protein Yeast (Saccharomyces cerevisiae)); PIR2: A27535 (40K chain II precursor Yeast (Saccharomyces cerevisiae)); PIR3: A38325 (7.3K chain 9 Yeast (Saccharomyces cerevisiae)); PIR2: A24011 (8K protein Bovine); PIR2: A24864 (9.5K protein Bovine); PIR3: A35823 (9.5K protein Bovine (fragments)); PIR3: S05860 (chain VIII Yeast (Saccharomyces cerevisiae)); PIR2: S02075 (cytochrome c_1 Euglena gracilis (fragment)); PIR3: S16220 (cytochrome c_1 precursor Bovine); PIR3: S16221 (cytochrome c_2 precursor Bovine); PIR3: S14093 (II Bovine); PIR2: S00003 (iron-sulfur protein Bovine); PIR1: RDNCUF (iron-sulfur protein Neurospora crassa); PIR2: A29413 (iron-sulfur protein Paracoccus denitrificans); PIR2: S22633 (iron-sulfur protein Rhodobacter capsulatus (fragment)); PIR3: C34591 (iron-sulfur protein Rhodobacter sphaeroides (fragment)); PIR1: RDQFBR (iron-sulfur protein Rhodospirillum rubrum); PIR2: A34660 (iron-sulfur protein precursor mitochondrial Bovine)

Brookhaven code

7 LITERATURE REFERENCES

[1] Rieske, J.S.: Biochim. Biophys. Acta,456,195–247 (1976) (Review)
[2] Hauska, G., Hurt, E., Gabellini, N., Lockau, W.: Biochim. Biophys. Acta,726,97–133 (1983) (Review)
[3] Weiss, H., Leonard, K., Neupert, W.: Trends Biochem. Sci.,15,178–180 (1990)
[4] Schmitt, M.E., Phillips, J.D., Trumpower, B.L.: Biochim. Biophys. Acta,1018,119–123 (1990)
[5] Usui, S., Yu, L., Yu, C.-A.: Biochemistry,29,4618–4626 (1990)
[6] Andrews, K.M., Crofts, A.R., Gennis, R.B.: Biochemistry,29,2645–2651 (1990)
[7] Purvis, D.J., Theiler, R, Niederman, R.A.: J. Biol. Chem.,265,1208–1215 (1990)
[8] Wu, M., Tzagoloff, A.: J. Biol. Chem.,254,11122–11130 (1989)
[9] Kriauciunas, A., Yu, L., Yu, C.-A., Wynn, R.M., Knaff, D.B.: Biochim. Biophys. Acta,976,70–76 (1989)
[10] di Rago, J.P., Colson, A.-M.: J. Biol. Chem.,263,12564–12570 (1988)
[11] Takamiya, K.-i.: J. Biochem.,103,755–758 (1988)
[12] Meinhardt, S.W., Yang, X., Trumpower, B.L., Ohnishi, T.: J. Biol. Chem.,262,8702–8706 (1987)
[13] Daldal, F., Davidson, E., Cheng, S.: J. Mol. Biol.,195,1–12 (1987)

Enzyme Handbook © Springer-Verlag Berlin Heidelberg 1994
Duplication, reproduction and storage in data banks are only allowed with the prior permission of the publishers

[14] Weiss, H., Linke, P., Haiker, H., Leonard, K.: Biochem. Soc. Trans.,15,100–103 (1987)
[15] Tzagoloff, A., Wu, M., Crivellone, M.: J. Biol. Chem.,261,17163–17169 (1986)
[16] Yang, X., Trumpower, B.L.: J. Biol. Chem.,261,12282–12289 (1986)
[17] Weiss, H., Hovmöller, S., Leonard, K.: Methods Enzymol.,126,191–201 (1986)
[18] Nalecz, M.J., Casey, R.P., Azzi, A.: Methods Enzymol.,125,86–108 (1986)
[19] Shimomura, Y., Nishikimi, M., Ozawa, T.: Anal. Biochem.,153,126–131 (1986)
[20] Linke, P., Weiss, H.: Methods Enzymol.,126,201–210 (1986)
[21] Gwak, S.-H., Yu, L., Yu, C.-A.: Biochim. Biophys. Acta,809,187–198 (1985)
[22] Moser, C.C., Giangiacomo, K.M., Matsuura, K., de Vries, S., Dutton, P.L.: Methods Enzymol.,126,293–305 (1986)
[23] Eposti, M.D., Flamini, E., Zannoni, D.: Plant Physiol.,77,758–764 (1985)
[24] Yu, L., Yang, F.-D., Yu, L.-A.: J. Biol. Chem.,260,963–973 (1985)
[25] Yu, L., Mei, Q.-C., Yu, C.-A.: J. Biol. Chem.,259,5752–5760 (1984)
[26] van Loon, A.P.G.M., Maarse, A.C., Riezman, H., Grivell, L.A.: Gene,26,261–272 (1983)
[27] Froud, R.J., Ragan, C.I.: Biochem. J.,217,561–571 (1984)
[28] Speck, S.H., Margoliash, E.: J. Biol. Chem.,259,1064–1072 (1984)
[29] Glaser, E.G., Meinhardt, S.W., Crofts, A.R.: FEBS Lett.,178,336–342 (1984)
[30] Engel, W.D., Schägger, H., von Jagow, G.: Hoppe-Seyler's Z. Physiol. Chem.,364,1753–1763 (1983)
[31] Wikström, M., Krab, K., Saraste, M.: Annu. Rev. Biochem.,50,623–655 (1981) (Review)
[32] van Loon, A.P.G.M., Kreike, J., de Ronde, A., van der Horst, G.T., Gasser, S.M., Grivell, L.A.: Eur. J. Biochem.,135,457–463 (1983)
[33] van Loon, A.P.G.M., de Groot, R., van Eyk, E., van der Horst, G.T.J., Grivell, L.A.: Gene,20,323–337 (1982)
[34] Hovmöller, S., Slaughter, M., Berriman, J., Karlson, B., Weiss, H., Leonard, K.: J. Mol. Biol.,165,401–406 (1983)
[35] Yu, L., Yu, C.-A.: Biochem. Biophys. Res. Commun.,112,450–457 (1983)
[36] Yu, L., Yu, C.-A.: Biochem. Biophys. Res. Commun.,108,1285–1292 (1982)
[37] Gabellini, N., Bowyer, J.R., Hurt, E., Melandri, B.A. , Hauska, G.: Eur. J. Biochem.,126,105–110 (1982)
[38] de Vries, S., Berden, J.A., Slater, E.C.: FEBS Lett.,122,143–148 (1980)
[39] Engel. W.D., Schägger, H., von Jagow, G.: Biochim. Biophys. Acta,592,211–222 (1980)
[40] Weiss, H., Juchs, B., Ziganke, B.: Methods Enzymol.,53,98–112 (1978)
[41] Hatefi, Y., Galante, Y.M., Stiggall, D.L., Ragan, C. I.: Methods Enzymol.,56,577–602 (1979) (Review)
[42] Hatefi, Y.: Methods Enzymol.,53,35–40 (1978)
[43] Palmer, G.: Methods Enzymol.,53,113–121 (1978)
[44] von Jagow, G., Schägger, H., Engel, W.D., Riccio, P., Kolb, H.J., Klingenberg, M.: Methods Enzymol.,53,92–98 (1978)
[45] Nelson, B.D., Gellerfors, P.: Methods Enzymol.,53,80–91 (1978)
[46] Marres, C.A.M., Slater, E.C.: Biochim. Biophys. Acta,462,531–548 (1977)
[47] Hatefi, Y. in "The Enzymes of Biological Membranes" (Martonosi, A., ed.) 4,3–42, Plenum N.Y. (1976) (Review)
[48] Schägger, H., Link, T.A., Engel, W.D., von Jagow, G.: Methods Enzymol.,126,224–237 (1986)
[49] Yang, X., Trumpower, B.L.: J. Biol. Chem.,261,12282–12289 (1986)
[50] Katan, M.B., Pool, L., Groot, G.S.P.: Eur. J. Biochem.,65,95–105 (1976)

[51] Siedow, J.N., Power, S., de la Rosa, F.F., Palmer, G. : J. Biol. Chem.,253,2392–2399 (1978)
[52] Slater, E.C.: Biochim. Biophys. Acta,301,129–154 (1973) (Review)
[53] Trumpower, B.L.: Microbiol. Rev.,54,101–129 (1990) (Review)
[54] Wakabayashi, S., Matsubara, H., Kim, C.H., Kawai, K., King, T.E.: Biochem. Biophys. Res. Commun.,97,1548–1554 (1980)
[55] Borchart, U., Machleidt, W., Schägger, H., Link, T.A. , von Jagow, G.: FEBS Lett.,200,81–86 (1986)
[56] Schägger, H., Borchart, U., Link, T.A., von Jagow, G. : FEBS Lett.,219,161–168 (1987)
[57] Schägger, H., von Jagow, G., Borchart, U., Machleidt, G.: Hoppe-Seyler's Z. Physiol. Chem.,364,307–311 (1983)
[58] Wakabayashi, S., Takeda, H., Matsubara, H., Kim. C.H. , King, T.E.: J. Biochem.,91,2077–2085 (1982)
[59] Wakabayashi, S., Takao, T., Shimonishi, Y., Kuramitsu, S., Matsubara, H., Wang, T.-y., Zhang, Z.-p., King, T.E.: J. Biol. Chem.,260,337–343 (1985)
[60] Borchart, U., Machleidt, W., Schägger, H., Link, T.A. , von Jagow, G.: FEBS Lett.,191,125–130 (1985)
[61] Weis, H., Kolb, H.J.: Eur. J. Biochem.,99,139–149 (1979)
[62] Siedow, J.N., Power, S., de la Rosa, F.F., Palmer, G. : J. Biol. Chem.,253,2392–2399 (1978)
[63] Anderson, S., de Brujin, M.H.L., Coulson, A.R., Eperon, I.C., Sanger, F., Young, I.G.: J. Mol. Biol.,156,683–717 (1982)
[64] Kurowski, B., Ludwig, B.: J. Biol. Chem.,262,13805–13811 (1987)
[65] Thöny-Meyer, L., Stax, D., Hennecke, H.: Cell,57,683–697 (1989)
[66] Michelotti, E.F., Hajduk, S.L.: J. Biol. Chem.,262,927–932 (1987)
[67] Wynn, R.M., Gaul, D.F., Choi, W.K., Shaw, R.W, Knaff, D.B: Photosynth. Res.,9,181–195 (1986)
[68] Gabellini, N., Sebald, W.: Eur. J. Biochem.,154,569–579 (1986)
[69] von Jagow, G., Link, A.T., Schägger, H. in "Advances in Membrane Biochemistry and Bioenergetics" (Kim, C.H., Tedeschi, H., Diwan, J.J., Salerno, J.C., eds.) Plenum, N. Y. (1987) (Review)
[70] Brasseur, R.: J. Biol. Chem.,263,12571–12575 (1988)
[71] Schulte, U., Arretz, M., Schneider, H., Tropschug, M. , Wachter, E., Neupert, W., Weiss, H.: Nature,339,147–149 (1989)
[72] de Vries, S., Marres, C.A.M.: Biochim. Biophys. Acta,895,205–239 (1987) (Review)
[73] Gonzales-Halphen, D., Lindorfer, A., Capaldi, M.A.: Biochemistry,27,7021–7031 (1988)
[74] Oudshoorn, P., van Steeg, H., Swinkels, B.W., Shoppink, P., Grivell, L.A.: Eur. J. Biochem.,163,97–103 (1987)
[75] Beckmann, J.D., Ljungdahl, P.O., Lopez, J.L., Trumpower, B.L.: J. Biol. Chem.,262,8901–8909 (1987)
[76] Maarse, A.C., Bowyer, J.R., Ohnishi, T., Dutton, P.L. : Eur. J. Biochem.,165,419–425 (1987)
[77] van Lon, A.P.G.M., de Groot, R.J., de Haan, M., Dekker, A., Grivell, A.L.: EMBO J.,3,1039–1043 (1984)

Enzyme Handbook © Springer-Verlag Berlin Heidelberg 1994
Duplication, reproduction and storage in data banks are only allowed with the prior permission of the publishers

1 NOMENCLATURE

EC number
1.10.3.1

Systematic name
1,2-Benzenediol:oxygen oxidoreductase

Recommended name
Catechol oxidase

Synonymes
Diphenol oxidase
o-Diphenolase
Phenolase
Polyphenol oxidase
Tyrosinase
Pyrocatechol oxidase
Dopa oxidase
Catecholase
o-Diphenol:oxygen oxidoreductase
o-Diphenol oxidoreductase

CAS Reg. No.
9002-10-2 (not distinguished from EC 1.14.18.1)

2 REACTION AND SPECIFICITY

Catalysed reaction
2 Catechol + O_2 →
→ 2 1,2-benzoquinone + 2 H_2O (this reaction, called catecholase activity is also catalyzed by EC 1.14.18.1. As catecholase always is associated with cresolase activity (EC 1.14.18.1) all enzymes except one with only catecholase activity are summerized under EC 1.14.18.1 [4])

Reaction type
Redox reaction

Natural substrates

Enzyme Handbook © Springer-Verlag Berlin Heidelberg 1994
Duplication, reproduction and storage in data banks are only allowed with the prior permission of the publishers

Substrate spectrum

1 Catechol + O_2 [1, 2]
2 d-Catechin + O_2 [1, 2]
3 Chlorogenic acid + O_2 [1, 2]
4 Dopamine + O_2 [1, 2]
5 Dopa + O_2
6 Gallic acid (3,4,5-trihydroxybenzoic acid) + O_2 [1, 2]
7 4-Methylcatechol + O_2 [1, 2]
8 l-Catechin + O_2 [2]
9 Caffeic acid + O_2 [2]
10 Epinephrine + O_2 [2]
11 Quercetin + O_2 [2]
12 Rutin + O_2 [2]
13 2,3-Dihydroxybenzoic acid [2]
14 Ferulic acid + O_2 [2]
15 Pyrogallol + O_2 [2]
16 Tannic acid + O_2 [2]
17 Shikimic acid + O_2 [2]
18 Phloroglucinol + O_2 [2]
19 More (no monophenols as substrates [1], oxidation of o-diphenols to o-benzoquinones are summerized under EC 1.14.18.1)

Product spectrum

1 1,2-Benzoquinone + H_2O
2 ?
3 ?
4 4-(2-Aminoethyl)-1,2-benzoquinone
5 Dopaquinone
6 ?
7 4-Methyl-o-benzoquinone
8 ?
9 ?
10 ?
11 ?
12 ?
13 ?
14 ?
15 ?
16 ?
17 ?
18 ?
19 ?

Inhibitor(s)

$NaHSO_3$ [1, 2]; Ascorbic acid [1, 2]; Glutathione [1]; Diethyldithiocarbamate [1, 2]; CN^- [1, 2]; 8-Hydroxyquinoline [2]; o-Phenanthroline [2]; Hg^{2+} [2]; Ag^+ [2]; Al^{3+} [2]; Sn^{2+} [2]; Xanthogenate [2]; Thiourea [2]; Cysteine [2]; 1-Phenyl-2-thiourea [2]; Cuprizone [2]; Sodium azide [2]; Isoascorbic acid [2]; 2-Mercaptobenzothiazole [2]; 2-Mercaptoethanol [2]

Cofactor(s)/prostethic group(s)

Metal compounds/salts

Cu^{2+} (stimulation) [2]; Cu^+ (stimulation) [2]; Pb^{2+} (stimulation) [2]; Co^{2+} (stimulation) [2]; Zn^{2+} (stimulation) [2]

Turnover number (min^{-1})

Specific activity (U/mg)

More (unit defined as change in absorbance) [1, 3]

K_m-value (mM)

pH-optimum

4.7 [3]

pH-range

4.0–5.8 [3]

Temperature optimum (°C)

30–40 [3]

Temperature range (°C)

10–45 [3]

---

3 ENZYME STRUCTURE

Molecular weight

88000 (Alternaria tenuis, gel filtration) [3]

Subunits

Glycoprotein/Lipoprotein

–

---

4 ISOLATION/PREPARATION

Source organism

Prunus persica (peach) [1]; Alternaria tenuis [2, 3]

Enzyme Handbook © Springer-Verlag Berlin Heidelberg 1994
Duplication, reproduction and storage in data banks are only allowed with the prior permission of the publishers

Source tissue
Fruit [1]; Culture filtrate [2, 3]

Localisation in source

Purification
Prunus persica [1]; Alternaria tenuis [3]

Crystallization
–

Cloned
–

Renaturated
–

5 STABILITY

pH
4.6–6.3 [3]

Temperature (°C)
55 (half life 2.2–14.6 min, depending on isozyme) [1]; 35 (stable up to) [3]

Oxidation

Organic solvent

General stability information

Storage

6 CROSSREFERENCES TO STRUCTURE DATABANKS

PIR/MIPS code
PIR2: JH0665 (A2 Fruit fly (Drosophila melanogaster)); PIR2: S19805 (precursor Fava bean)

Brookhaven code

7 LITERATURE REFERENCES

[1] Wong, T.C., Luh, B.S., Whitaker, J.R.: Plant Physiol.,48,19–23 (1971)
[2] Motoda, S.: J. Ferment. Technol.,57,79–85 (1979)
[3] Motoda, S.: J. Ferment. Technol.,57,71–78 (1979)
[4] Literature for enzymes with o-diphenol oxidoreductase activity (i.e. catecholase) as well as monophenol hydroxylase activity (I.E. cresolase) is summarized under EC 1.14.18.1. Both activities are located on one enzyme molecule, frequently referred to as tyrosinase.

Enzyme Handbook © Springer-Verlag Berlin Heidelberg 1994
Duplication, reproduction and storage in data banks are only allowed with the prior permission of the publishers

1 NOMENCLATURE

EC number
1.10.3.2

Systematic name
Benzenediol:oxygen oxidoreductase

Recommended name
Laccase

Synonymes
Urishiol oxidase
Urushiol oxidase
p-Diphenol oxidase

CAS Reg. No.
80498-15-3

2 REACTION AND SPECIFICITY

Catalysed reaction
4 Benzenediol + O_2 →
→ 4 benzosemiquinone + 2 H_2O (discussion of mechanism [1, 2], non-enzymatic disproportionation of 2 semiquinones to quinone and quinol [2])

Reaction type
Redox reaction

Natural substrates
More (discussion of physiological role) [1]

Enzyme Handbook © Springer-Verlag Berlin Heidelberg 1994
Duplication, reproduction and storage in data banks are only allowed with the prior permission of the publishers

Substrate spectrum

1 p-Cresol + O_2 (or m-cresol [14], or o-cresol [12, 14]) [1, 12, 14, 17, 18]
2 o-Diphenol + O_2 [1]
3 Ferulic acid + O_2 [3, 13]
4 Sinapic acid + O_2 [3, 13]
5 d-Catechin + O_2 [6, 39]
6 l-Epicatechin + O_2 [6]
7 Chlorogenic acid + O_2 [6]
8 Caffeic acid + O_2 [12, 17, 39]
9 Catechol + O_2 [12, 27, 35, 41]
10 4-Methylcatechol + O_2 [12, 17, 18]
11 Quinol + O_2 [12, 17]
12 Gallic acid + O_2 [12, 17, 27, 32]
13 Pyrogallol + O_2 [12, 18, 27, 32, 35, 41]
14 1,2,4-Benzenetriol + O_2 [12]
15 Guaiacol + O_2 (i.e. 1-hydroxy-2-methoxybenzene) [12, 13, 18, 32, 39]
16 3,5-Dimethoxy-hydroxy-benzaldazine + O_2 [12]
17 p-Phenylenediamine + O_2 [12, 19]
18 4,5-Dimethyl-o-phenylenediamine + O_2 [12]
19 4-Amino-N, N'-dimethylaniline + O_2 [12]
20 Ascorbate + O_2 [12, 27, 32, 37, 41]
21 1-Naphthol + O_2 [12, 14, 27]
22 2-Naphthol + O_2 [12]
23 p-Methoxyphenol + O_2 [14]
24 2,6-Dimethoxyphenol + O_2 [14]
25 o-Chlorophenol + O_2 (or p-chlorophenol or m-chlorophenol) [14]
26 2,4-Dichlorophenol + O_2 (or 2,6-dichlorophenol) [14]
27 2,6-Dimethylphenol + O_2 [14]
28 Phenol + O_2 [14]
29 4-Chloro-2-methylphenol + O_2 [14]
30 p-Aminophenol + O_2 [19]
31 Ferrocyanide + O_2 (not: [32]) [17, 41]
32 Hydroquinone + O_2 [35]
33 Dopa + O_2 [37]
34 Brenzcatechin + O_2 [37]
35 More (no diphenols [27], overview [12, 14, 18, 27, 39]) [12, 14, 18, 27, 37, 39, 41]

Product spectrum

1 ?
2 Benzosemiquinone
3 ?
4 ?
5 ?
6 ?
7 ?
8 ?
9 ?
10 ?
11 ?
12 ?
13 ?
14 ?
15 ?
16 ?
17 ?
18 ?
19 ?
20 ?
21 ?
22 ?
23 ?
24 ?
25 ?
26 ?
27 ?
28 ?
29 ?
30 ?
31 ?
32 ?
33 ?
34 ?
35 ?

Enzyme Handbook © Springer-Verlag Berlin Heidelberg 1994
Duplication, reproduction and storage in data banks are only allowed with the prior permission of the publishers

Inhibitor(s)

Potassium xanthogenate [6]; Thiourea [6]; Cysteine [6, 35]; 1-Phenyl-2-thiourea [6, 33, 40]; CN^- [6, 12, 18, 27, 33, 35, 40]; Azide [6, 12, 27, 35, 39]; Ascorbic acid [6]; Isoascorbic acid [6]; 2-Mercaptobenzothiazole [6]; 2-Mercaptoethanol [6]; Sodium metabisulfite [6]; Sodium bisulfite [6]; Hg^{2+} [6]; Ferrous sulfate [6]; Diethyldithiocarbamate [12, 33, 35, 40]; EDTA [18]; Propylgallate [18]; Cetyltriammonium bromide [27]; Cetylpyridinium bromide [27]; p-Nitrophenol [33]; Neocuprein [33]; 4,5-Methyl-o-phenylenediamine [33]; F^- [34], Salicylaldoxime [35]; Glutathione [35]; 8-Hydroxyquinoline [39]; 2,3-Dimercaptopropanol [39]; Cationic detergents [39]

Cofactor(s)/prostethic group(s)

Metal compounds/salts

Copper (4 gatom per mol [1, 18, 24, 30], 1.8 gatom per mol [27], 2 gatom per mol [33], 18.3 gatom per mol [37], X-ray absorption edge study of type I, type II and binuclear type III copper center [5], preparation of type II-depleted enzyme [8], chemical and spectral studies of binuclear copper site [10], X-ray absorption spectra [16], ENDOR (i.e. electron nuclear double resonance) studies [20]) [1, 5, 8, 10, 11, 16, 18, 20, 24, 25, 27, 30, 31, 33, 37, 38]

Turnover number (min^{-1})

Specific activity (U/mg)

1.66–28.67 (micromol O_2/min/mg depending on organism and isozyme) [12]; 77.5 [23]; More [3, 4, 7, 13, 15, 18, 21, 23, 24, 32, 33, 40]

K_m-value (mM)

0.133 (O_2, laccase 1) [12]; 0.533 (O_2, laccase 2) [12]; 0.02 (O_2) [18]; 4.5 (4-methylcatechol) [18]; 0.25 (O_2, extracellular enzyme) [23]; 0.37 (O_2, intracellular enzyme) [23]; 1.25 (quinol) [24]; 0.1 (p-phenylenediamine) [27]; 0.05 (catechol) [35]; 0.029 (hydroquinone) [35]; 3.3 (dopa) [37]; 1.03 (ferrocyanide) [37]; 0.192 (ascorbic acid) [37]; 2.27 (brenzcatechin) [37]; 2 (4-methylcatechol) [40]; 12 (quinol) [40]; More (comparison of values for free and immobilized laccase [3], kinetic studies of electron transfer [28]) [3, 28]

pH-optimum

2.5–3.0 (laccase 1) [12]; 3.5 (substrate p-phenylenediamine has two optima: 3.5 and 5.6) [27]; 4.0 (laccase 2 [12], laccase 3 [13]) [12, 13]; 3.8–4.8 (depending on substrate and organism) [3]; 4.5 [22]; 5.0 (substrate hydroquinone [35]) [7, 35]; 5.4 (substrate catechol) [35]; 5.6 (laccase 1 [13], substrate p-phenylendiamine has two optima: 3.5 and 5.6, N,N'-dimethyl-p-phenylenediamine, guaiacol [27]) [13, 27]; 5.8–5.9 [41]; 6.0 [40]; 6.2 [24]; 6.5 [39]; More (overview [1], dependency on isozyme and organism [3]) [1, 3]

pH-range

4.5–5.5 [7]; 5.0 (decline of activity above) [12]

Temperature optimum (°C)

38 (immobilized) [3, 4]; 52 (free enzyme) [3, 4]; 55 (laccase 1 [12]) [7, 12]; 60 (laccase 2) [12]

Temperature range (°C)

45–60 [7]; More (temperature dependency of reduction potential) [22]

3 ENZYME STRUCTURE

Molecular weight

96000–110000 (Chaetomium thermophile, gel filtration [7], Botrytis cinerea, gel filtration [12], Acer pseudoplatanus, sedimentation and diffusion constants [18], Aspergillus nidulans, gel filtration [21], Schinus molle, sedimentation and diffusion constants [24], Agaricus bisporus, sucrose density centrifugation, gel filtration [27], Rhus vernicifera, sedimentation velocity, calculation from copper content [38]) [7, 12, 18, 21, 24, 27, 38]
383000–390000 (Podospora anserina, laccase I, disc gel electrophoresis [26], sedimentation and diffusion constants, sedimentation equilibrium centrifugation [37]) [26, 37]
80000 (Podospora anserina, laccase IV, disc gel electrophoresis) [26]
73500 (Prunus persica) [33]
70000 (Rigidoporus lignosus, laccase 3, gel filtration) [10]
64800 (Neurospora crassa, sedimentation equilibrium centrifugation) [32]
55000 (Rigidoporus lignosus, laccase 2, gel filtration) [10]
52000 (Rigidoporus lignosus, laccase 1, gel filtration) [10]
35500–38000 (Botrytis cinerea, laccase I and II, ultracentrifugation [17], sedimentation and diffusion data [23]) [17, 23]
More (overview) [1]

Enzyme Handbook © Springer-Verlag Berlin Heidelberg 1994
Duplication, reproduction and storage in data banks are only allowed with the prior permission of the publishers

Subunits

Monomer (1 × 54000–55000, Rigidoporus lignosus, laccase 1 and 2, SDS-PAGE) [13]
Oligomer (x × 21400–56000, Agaricus bisporus, SDS-PAGE [27], x × 72400, Botrytis cinerea, SDS-PAGE, probably anomalous migration on SDS-gels [12]) [12, 27]
? (x × 36000, Chaetomium thermophile, SDS-PAGE [6], x × 59000, Pleurotus ostreatus, SDS-PAGE [15]) [6, 15]

Glycoprotein/Lipoprotein

Glycoprotein (sugar content laccase 1: 86%, laccase 2: 91% [12], laccase I: 80%, laccase II: 70% [17], laccase: 45% [18], 12% [21], 11% [32], 25% [33], depending on method of analysis [23, 24]) [1, 6, 12, 15, 17, 18, 21, 23, 24, 32, 33, 37]

---

4 ISOLATION/PREPARATION

Source organism

Fungi (overview) [1]; Plants (overview) [1, 2]; Trametes versicolor [3, 4]; Fomes fomentarius [3]; Rhus vernicifera [5, 10, 16, 20, 25, 28, 34, 38, 42]; Chaetomium thermophile [6, 7]; Polyporus versicolor [8, 22, 34]; Neurospora crassa [9, 30, 32]; Coriolus versicolor [11]; Botrytis cinerea [12, 17, 23]; Rigidoporus lignosus [13]; Phellinus noxius [13]; Rhizoctonia praticola [14]; Pleurotus ostreatus [15]; Acer pseudoplatanus (sycamore) [18]; Aspergillus nidulans [19, 21]; Schinus molle [24, 29]; Podospora anserina [26, 31, 37]; Agaricus bisporus [27]; Magnifera indica [29]; Pistacia palaestina [29]; Pleiogynium timoriense [29]; Prunus persica (peach) [33, 40]; Ganoderma lucidum [35]; Glomerella cingulata [39]; Lactarius piperatus [41]; Russula delica [43]; Aesculus sp. (horse-chestnut tree) [44]

Source tissue

Culture filtrate [3, 7, 13, 15, 27, 30, 32, 35]; Secretions [29]; Fruit [33, 40]; Lacquer [38]; Latex [42]; More (tissue localization) [19]

Localisation in source

Extracellular [1, 3, 4, 7, 13, 15, 17, 18, 23, 27, 30, 32, 35]; Intracellular [1, 32]; Cytoplasm (of plants) [1]

Purification

Trametes versicolor [3, 4]; Fomes fomentarius [3]; Chaetomium thermophile [7]; Rigidoporus lignosus [13]; Pleurotus ostreatus [15]; Botrytis cinerea [17, 23]; Acer pseudoplatanus [18]; Aspergillus nidulans [21]; Schinus molle [24]; Agaricus bisporus [27]; Neurospora crassa [30, 32]; Podospora anserina (laccase I-II [31], laccase I [37]) [31, 37]; Prunus persica [33, 40]; Ganoderma lucidum (partial) [35]; Rhus vernicifera [38, 42]; Lactarius piperatus [41]

Crystallization

–

Cloned

[9]

Renaturated

(reconstitution of Cu-depleted enzyme) [36]

---

5 STABILITY

pH

4–10 [18]; 4.0–11.0 [7]; More (unstable at alkaline pH) [1]

Temperature (°C)

55 (stable below) [7]; 50 (half-life 3 h) [27]; 60 (half-life 40 min) [27]; 70 (half-life 10 min) [27]; More (laccase 2 shows higher thermostability than laccase 1) [12]

Oxidation

Organic solvent

General stability information

Freezing/thawing causes inactivation [4]

Storage

–30°C [38]; –20°C, 0.01 M phosphate buffer, pH 7.0, at least 1 year [27]; –20°C, 0.1 M phosphate buffer, pH 5 [37]; –20°C, several months stable [12]; 2°C, several weeks [24]; 4°C, pH 7, several days, or frozen several months stable [18]; 4°C, immobilized, 2 years, no loss of activity [4]

Enzyme Handbook © Springer-Verlag Berlin Heidelberg 1994
Duplication, reproduction and storage in data banks are only allowed with the prior permission of the publishers

6 CROSSREFERENCES TO STRUCTURE DATABANKS

PIR/MIPS code

PIR2: A29762 (Neurospora crassa (fragment)); PIR1: KSASL1 (I Emericella nidulans); PIR1: KSNCLO (precursor Neurospora crassa (strain OR)); PIR1: KSNCLT (precursor Neurospora crassa (strain TS))

Brookhaven code

7 LITERATURE REFERENCES

[1] Mayer, A.M.: Phytochemistry,26,11–20 (1987) (Review)
[2] Mayer, A.M., Harel, E.: Phytochemistry,18,193–215 (1979) (Review)
[3] Rogalski, J., Wojtas-Wasilewska, M., Apalovic, R., Leonowicz, A.: Biotechnol. Bioeng.,37,770–777 (1991)
[4] Rogalski, J., Dawidowicz, A.L., Leonowicz, A.: Acta Biotechnol.,10,261–269 (1990)
[5] Cole, J.L., Tan, G.O., Yang, E.K., Hodgson, K.O., Solomon, E.I.: J. Am. Chem. Soc.,112,2243–2249 (1990)
[6] Ishigami, T., Hirose, Y., Yamada, Y.: J. Gen. Appl. Microbiol.,34,401–407 (1988)
[7] Ishigami, T., Yamada, Y.: J. Gen. Appl. Microbiol.,32 ,293–301 (1986)
[8] Hanna, P.M., McMillin, D.R., Pasenkiewicz-Gierula, M., Antholine, W.E., Reinhammar, B.: Biochem. J.,253,561–568 (1988)
[9] Germann, U.A., Müller, G., Hunziker, P.E., Lerch, K.: J. Biol. Chem.,263,885–896 (1988)
[10] Spira-Solomon, D.J., Solomon, E.I.: J. Am. Chem. Soc.,109,6421–6432 (1987)
[11] Wrigley, S.K., Gibson, J.F.: Biochim. Biophys. Acta,916,259–264 (1987)
[12] Zouari, N., Romette, J.-L., Thomas, D.: Appl. Biochem. Biotechnol.,15,213–225 (1987)
[13] Geiger, J.P., Rio, B., Nandris, D., Nicole, M.: Appl. Biochem. Biotechnol.,12,121–133 (1986)
[14] Shuttleworth, K.L., Bollag, J.-M.: Enzyme Microb. Technol.,8,171–177 (1986)
[15] Sannia, G., Giardina, P., Luna, M., Rossi, M., Buonocore, V.: Biotechnol. Lett.,8,797–800 (1986)
[16] Woolery, G.L., Powers, L., Peisach, J., Spiro, T.G.: Biochemistry,23,3428–3434 (1984)
[17] Marbach, I., Harel, E., Mayer, A.M.: Phytochemistry,23,2713–2717 (1984)
[18] Bligny, R., Douce, R.: Biochem. J.,209,489–496 (1983)
[19] Herman, R.E., Berman Kurtz, M., Champe, S.P.: J. Bacteriol.,154,955–964 (1983)
[20] Cline, J., Reinhammer, B., Jensen, P., Venters, R., Hoffman, B.M.: J. Biol. Chem.,258,5124–5128 (1983)
[21] Berman Kurtz, M., Champe, S.P.: J. Bacteriol.,151,1338–1345 (1982)
[22] Taniguchi, V.T., Malmström, B.G., Anson, F.C., Gray, H.B.: Proc. Natl. Acad. Sci. USA,79,3387–3389 (1982)
[23] Gigi, O., Marbach, I., Mayer, A.M.: Phytochemistry,20,1211–1213 (1981)
[24] Bar-Nun, N., Mayer, A.M., Sharon, N.: Phytochemistry,20,407–408 (1981)
[25] Morpurgo, L., Calabrese, L., Desideri, A., Rotilio, G.: Biochem. J.,193,639–642 (1981)
[26] Durrens, P.: Arch. Microbiol.,130,121–124 (1981)
[27] Wood, D.A.: J. Gen. Microbiol.,117,327–338 (1980)
[28] Clemmer, J.D., Gilliland, B.L., Bartsch, R.A.: Biochim. Biophys. Acta,568,307–320 (1979)

[29] Joel, D.M., Marbach, I., Mayer, A.M.: Phytochemistry,17,796–797 (1978)
[30] Lerch, K., Deinum, J., Reinhammer, B.: Biochim. Biophys. Acta,534,7–14 (1978)
[31] Molitoris, H.P., Reinhammer, B.: Biochim. Biophys. Acta,386,493–502 (1975)
[32] Froehner, S.C., Eriksson, K.-E.: J. Bacteriol.,120,458–465 (1974)
[33] Lehman, E., Harel, E., Mayer, A.M.: Phytochemistry,13,1713–1717 (1974)
[34] Bränden, R., Malmström, B.G., Vänngard, T.: Eur. J. Biochem.,36,195–200 (1973)
[35] Lalitha Kumari, H., Sirsi, M.: Arch. Mikrobiol.,84,350–357 (1972)
[36] Ando, K.: J. Biochem.,68,501–508 (1970)
[37] Molitoris, H.P., Esser, K.: Arch. Mikrobiol.,72,267–296 (1970)
[38] Reinhammer, B.: Biochim. Biophys. Acta,205,35–47 (1970)
[39] Walker, J.R.L.: Phytochemistry,7,1231–1240 (1968)
[40] Mayer, A.M., Harel, E.: Phytochemistry,7,1253–1256 (1968)
[41] Iwasaki, H., Matsubara, T., Mori, T.: J. Biochem.,61 ,814–816 (1967)
[42] Nakamura, T.: Biochim. Biophys. Acta,30,44–52 (1958)
[43] Matsubara, T., Iwasaki, H.: Bot. Mag. Tokyo,85,71–83 (1972)
[44] Wosilait, W.D., Nason, A., Terrel, A.J.: J. Biol. Chem.,206,271–282 (1954)

Enzyme Handbook © Springer-Verlag Berlin Heidelberg 1994
Duplication, reproduction and storage in data banks are only allowed with the prior permission of the publishers

1 NOMENCLATURE

EC number
1.10.3.3

Systematic name
L-Ascorbate:oxygen oxidoreductase

Recommended name
L-Ascorbate oxidase

Synonymes
Oxidase, ascorbate
Ascorbic acid oxidase
Ascorbate oxidase
Ascorbase
Ascorbic oxidase
Ascorbate dehydrogenase
L-Ascorbic acid oxidase
AAO [1]
L-Ascorbate:O_2 oxidoreductase [4]
AA oxidase [40]

CAS Reg. No.
9029-44-1

2 REACTION AND SPECIFICITY

Catalysed reaction
2 L-Ascorbate + O_2 →
→ 2-dehydroascorbate + H_2O

Reaction type
Redox reaction

Natural substrates
L-Ascorbate + O_2 (possibly a kind of pathogenesis-related protein [21])
More [1]

Enzyme Handbook © Springer-Verlag Berlin Heidelberg 1994
Duplication, reproduction and storage in data banks are only allowed with the prior permission of the publishers

Substrate spectrum

1 L-Ascorbate + O_2
2 D-Glucoascorbic acid + O_2 [1]
3 D-Isoascorbic acid + O_2 [1]
4 More (substrate specificity: overview [38], highly specific for ascorbic acid (and a few of its analogs and O_2) [40], anionic form of the substrate is an important requirement of the enzyme specificity [38], mechanism: Cu^{2+} is reduced to Cu^+, which is then reoxidized by oxygen [1], double displacement mechanism (enzymatic memory) [28], low reaction rate with bilirubin [30], at pH 5.7: oxidation of leuco 2,6-dichloroindophenol to the blue quinoid dye [38], oxidation of 2,6- and 2,5-dichlorohydroquinone and hydroxyquinone at a rate about 1/12 of ascorbic acid [38], no oxidation of hydroquinone [38], not appreciably oxidized: p-phenylenediamine [1], $Na_2S_2O_3$ [1], glutathione [1], cysteine [1], ascorbate oxidase activity of caeruloplasmin [39]) [1, 28, 30, 38–40]

Product spectrum

1 L-Dehydroascorbate + H_2O
2 Glucodehydroascorbate + H_2O
3 2-Dehydroisoascorbate + H_2O
4 ?

Inhibitor(s)

Hg^{2+} [1]; Ni^{2+} (some authors report inhibition, others not) [1]; Zn^{2+} (some authors report inhibition, others not) [1]; Cu^{2+} [1]; Cyanide (very slight [40]) [1, 12, 38]; Diethyldithiocarbamate (not [40]) [1, 5, 38]; KNO_3 [40]; Nitrofurantoin (slight) [40]; H_2S [1]; Ethylxanthate [1, 40]; 8-Hydroxyquinoline [1, 5]; Pyridine-KNCS [1]; Thiourea [1, 32]; Salicylaldoxime [1]; Azide (mixed-type inhibition above pH 6, competitive inhibition at pH 5.6 [33], weak [32, 33]) [1, 2, 5–7, 32, 33]; Leucocyanidol [1]; 3,4-Dichlorophenylserine [1]; Tetraethylthiuramidisulfide [1]; Cupferron [1]; Carotenes [1]; SO_2 [1]; Deoxycorticosterone [1]; Anthocyanin pigments [1]; H_2O_2 (Inhibition at 5.6 mM, stimulation at 0.56 mM [40]) [1, 40]; alpha-Tocopherol [1]; Propylgallate [1]; Nordihydroguaiaretic acid [1]; Thiamine [1]; Auxin analogs [1]; Organic mercurials [1]; Iodoacetate [1]; p-Substituted mercuribenzoate (some authors report inhibition, others not) [1]; F^- [2]; Citrate (univalent anion) [9]; Piperazine N, N'-bis(2-ethanesulfonic acid) (anions) [9]; $NaNO_2$ [40]; Lauryl sulfate [40]; Fenton's reagent (Fe^{2+} + H_2O_2 + 2 H^+) [40]; Urea (effect on various molecular forms) [36]; More (natural inhibitors: cabbage extract, tomato extract, strawberry juice, extract of lemons, oranges, parsley, hips, leeks [1], Myrothecium verrucaria, reaction inactivation: progressive loss of activity during oxidation of ascorbic acid) [40]

Cofactor(s)/prostethic group(s)

H_2O_2 (stimulation at low concentration, 0.056 mM) [40]; o-Phenanthroline (stimulation) [40]; Bipyridyl (stimulation) [40]; Ethanol (stimulation) [40]; p-Chloromercuribenzoate (stimulation) [40]; Iodoacetate (stimulation) [40]

Metal compounds/salts

Copper (a multicopper protein, 6 atoms of copper per enzyme molecule [1], contains 8 atoms of copper per enzyme molecule (of MW 132000 [6], 140000 [7, 16, 34]), contains 10–12 atoms of copper per enzyme molecule of MW 140000 [8], mononuclear blue copper in domain 3 and trinuclear copper between domain 1 and 3 [11], measurement of intramolecular electron transfer between type I and type III copper centers in the multi-copper enzyme [12], contains two type 1, two type 2 and four type 3 copper ions [16], electronic structure of blue copper sites [26], coordination environment of type 2 copper [27], principal active site comprised of one type I, one type II and a pair of type III coppers [29]) [1, 6–8, 11, 12, 16, 24, 26, 27, 29, 34]; More (enzyme from Myrothecium verrucaria: no support of a metal in the enzyme) [40]

Turnover number (min^{-1})

Specific activity (U/mg)

More [2, 7, 8, 40]

K_m-value (mM)

0.100–0.350 (L-ascorbate, free enzyme) [31]; 0.200 (L-ascorbate, spectrophotometric method, Cucurbita pepo condensa) [36]; 0.98 (L-ascorbate, Warburg method, Cucurbita pepo condensa) [36]; 0.36 (L-ascorbate, spectrophotometric method, Cucurbita pepo medullosa) [36]; 2.21 (L-ascorbate, Warburg method) [36]; 0.181 (L-ascorbate, spectrophotometric method, Cucumis sativus) [36]; 1.125 (L-ascorbate, Warburg method, Cucumis sativus) [36]; More (K_m of the native enzyme and various deglycosylated forms [22], of K_m ascorbate and O_2 is insensitive to pH in the range 5–8.5 [35]) [1, 5, 35]

pH-optimum

5.5–7.0 [6, 32]; 5.6 (free and immobilized enzyme [31]) [1, 30, 31]; 6.0 [5, 23]

pH-range

4.5–8.3 (at pH 4.5 and 8.3: about 50% of activity maximum) [6]; 5–9.0 (5.0: about 60% of activity maximum (free enzyme), about 80% of activity maximum (immobilized enzyme), 9.0: less than 10% of activity maximum (free enzyme), about 45% of activity maximum (immobilized enzyme)) [23]

Enzyme Handbook © Springer-Verlag Berlin Heidelberg 1994
Duplication, reproduction and storage in data banks are only allowed with the prior permission of the publishers

Temperature optimum (°C)

Temperature range (°C)

3 ENZYME STRUCTURE

Molecular weight

132000 (Cucumis sativus, sedimentation equilibrium) [6]
140000 (Cucurbita pepo medullosa, sedimentation and difussion studies) [2, 8]
150000 (Cucurbita maxima, gel filtration) [32]

Subunits

Monomer (1 × 30000, Cucumis sativus, SDS-PAGE, enzyme also exists as dimer and tetramer [36], 1 × 35000, Cucurbita pepo, SDS-PAGE, enzyme also exists as tetramer, octamer, dodecamer and polymer [36]) [36]
Dimer (2 × 70000, MW 70000 subunit consists of 2 polypeptide chains: MW 30000 and 40000, Cucurbita pepo medullosa [2], 2 × 65000, MW 65000 subunit consists of 2 chains: A chain (38000) and B chain (28000), Cucurbita pepo medullosa [37], 2 × 30000, Cucumis sativus, SDS-PAGE, enzyme exists as monomer, dimer and tetramer [36]) [2, 36, 37]
Tetramer (4 × 30000, Cucumis sativus, SDS-PAGE, enzyme exists as monomer, dimer and tetramer, 4 × 35000, Cucurbita pepo, SDS-PAGE, enzyme exists as monomer, tetramer, octamer, dodecamer and polymer) [36]
Octamer (8 × 35000, Cucurbita pepo, SDS-PAGE, enzyme exists as monomer, tetramer, octamer, dodecamer and polymer) [36]
Dodecamer (12 × 35000, Cucurbita pepo, SDS-PAGE, enzyme exists as monomer, tetramer, octamer, dodecamer and polymer) [36]
Polymer (x × 35000, Cucurbita pepo, MW between 670000 and 2000000, SDS-PAGE, enzyme exists as monomer, tetramer, octamer, dodecamer and polymer) [36]
More (quarternary structure) [37]

Glycoprotein/Lipoprotein

Glycoprotein (enzyme may be a protein-copper carbohydrate complex [1], 2.4% carbohydrate [7], deglycosylation [22]) [1, 7, 22]

4 ISOLATION/PREPARATION

Source organism

Cucurbita pepo condensa (yellow summer crookneck, yellow summer squash, 5 isoenzymes [4], 5 molecular forms: monomer, tetramer, octamer, dodecamer, polymer [36]) [1, 2, 4, 9, 35, 36]; Cucumis sativus (cucumber, 3 molecular forms: monomer, dimer, tetramer [36]) [2, 6, 18, 34, 36]; Cucurbita pepo medullosa [2, 7, 8, 10, 12, 13, 20, 22, 24, 33, 37]; Myrothecium verrucaria [3, 40]; Brassica oleracea (cabbage) [5]; Sinapis alba (mustard) [17]; Cucurbita spp. (Ebisu Nankin) [19, 21, 25]; Cucurbita moschata [23]; Cucurbita maxima [32]; More (ascorbate oxidase activity of human caeruloplasmin) [39]

Source tissue

Medium (of cultured cells of Cucumis sativus [18], Cucurbita spp. [19, 21, 25]) [18, 19, 21, 25]; Seeds [17]; Leaves [5]; Peel [4, 6, 34, 36]; Fruit [2, 8, 32]; Cotyledons [17]; Cultured cells (of Cucumis sativus) [18]; Flesh [36]

Localisation in source

Soluble [5, 26]; Cell-wall [5, 20]; Cytoplasm [20]; Extracellular (cultured cells) [10, 19, 21, 25]

Purification

Cucumis sativus [6]; Cucurbita pepo medullosa [2, 7, 8]; Myrothecium verrucaria [40]

Crystallization

(Cucurbita pepo medullosa [10, 13], X-ray crystal structure [14]) [10, 13, 14]

Cloned

[15]

Renaturated

–

5 STABILITY

pH

4 (irreversible loss of activity below) [1]

Enzyme Handbook © Springer-Verlag Berlin Heidelberg 1994
Duplication, reproduction and storage in data banks are only allowed with the prior permission of the publishers

Temperature (°C)

0–40 (30 min, stable) [32]; 12 (stable for at least 30 days) [31]; 15 (immobilized enzyme retains full activity for 3 months at pH 5–7, free enzyme: pH 5–7, 40–70% loss of activity within one day) [23]; 40–50 (conversion of octamer and heavier forms to a dimer) [36]; 60 (20 min, 90% loss of activity (free enzyme), 10% loss of activity (immobilized enzyme)) [23]; 80 (10 min, complete loss of activity) [8]; 100 (1 min, all forms of enzyme inactivated) [36]; More (role of copper in heat stability [24], different molecular forms vary in resistance to heat inactivation: tetramer of squashes and dimer of cucumber being most resistant [36]) [24, 36]

Oxidation

Organic solvent

General stability information

Role of copper in stability [24]; Gelatin protects against inactivation [1]; Catalase protects against inactivation [1]; Peroxidase protects against inactivation [1]; Methemoglobin protects against inactivation [1]; Immobilization within 6% Ca-alginate gel beads improves stability [23]; 20–25% retention of activity after immobilization, at 12°C, stable for at least 30 days [31]; Stable to dialysis against EDTA or cyanide [40]

Storage

4°C, concentrated solution, 3 months [8]

6 CROSSREFERENCES TO STRUCTURE DATABANKS

PIR/MIPS code

PIR2: A30066 (Zucchini (fragments)); PIR1: KSKVAO (precursor Cucumber)

Brookhaven code

7 LITERATURE REFERENCES

[1] Stark, G.R., Dawson, C.R. in "The Enzymes",2nd Ed. (Boyer, P.D., Lardy, H., Myrbäck, K., eds.) 8,297–311 (1963) (Review)
[2] Lee, M.H., Dawson, C.R.: Methods Enzymol.,62,30–39 (1979) (Review)
[3] Lillehoj, E.B., Smith, F.G.: Plant Physiol.,41,1553–1560 (1966)
[4] Amon, A., Markakis, P.: Phytochemistry,8,997–998 (1969)
[5] Hallaway, M., Phethean P.D., Taggart, J.: Phytochemistry,9,935–944 (1970)
[6] Nakamura, T., Makino, N., Ogura, Y.: J. Biochem.,64,189–195 (1968)
[7] Marchesini, A., Kroneck, P.M.H.: Eur. J. Biochem.,101 ,65–76 (1979)
[8] Lee, M.H., Dawson, C.R.: J. Biol. Chem.,248,6596–6602 (1973)
[9] Gerwin, B., Burstein, S.R., Westley, J.: J. Biol. Chem.,249,2005–2008 (1974)
[10] Bolognesi, M., Gatti, G., Coda, A., Avigliano, L., Marcozzi, G., Finazzi-Agro, A.: J. Mol. Biol.,169,351–352 (1983)
[11] Messerschmidt, A., Huber, R.: Eur. J. Biochem.,187,341–352 (1990)
[12] Meyer, T.E., Marchesini, A., Cusanovich, M.A., Tollin, G.: Biochemistry,30,4619–4623 (1991)
[13] Ladenstein, R., Marchesini, A., Palmieri, S.: FEBS Lett.,107,407–408 (1979)
[14] Messerschmidt, A., Rossi, A., Ladenstein, R., Huber, R., Bolognesi, M., Gatti, G., Marchesini, A., Petruzzelli, R., Finazzi-Agro, A.: J. Mol. Biol.,206,513–529 (1989)
[15] Esaka, M., Hattori, T., Fujisawa, K., Sakajo, S., Asahi, T.: Eur. J. Biochem.,191,537–541 (1990)
[16] Morpurgo, L., Savini, I., Gatti, G., Bolognesi, M., Avigliano, L.: Biochem. Biophys. Res. Commun.,152,623–628 (1988)
[17] Leaper, L., Newbury, H.J.: Plant Sci.,64,79–90 (1989)
[18] Cho, H.-J., Aimi, T., Paik, S.-Y., Murooka, Y.: J. Ferment. Bioeng.,68,193–199 (1989)
[19] Esaka, M., Nishitani, I., Fukui, H., Suzuki, K., Kubota, K.: Phytochemistry,28,2655–2658 (1989)
[20] Chichiricco, G., Ceru, M.P., D'Alessandro, A., Oratore, A., Avigliano, L.: Plant Sci.,64,61–66 (1989)
[21] Esaka, M., Suzuki, K., Kubota, K.: Phytochemistry,29 ,1547–1549 (1990)
[22] D'Andrea, G., Maccarrone, M., Oratore, A., Avigliano, L., Messerschmidt, A.: Biochem. J.,264,601–604 (1989)
[23] Esaka, M., Suzuki, K., Kubota, K.: Agric. Biol. Chem.,49,2955–2960 (1985)
[24] Savini, I., D'Alessio, S., Giartosio, A., Morpurgo, L., Avigliano, L.: Eur. J. Biochem.,190,491–495 (1990)
[25] Esaka, M., Fukui, M., Suzuki, K., Kubota, K.: Phytochemistry,28,117–119 (1989)
[26] Dooley, D.M., Dawson, J.H., Stephens, P.J., Gray, H. B.: Biochemistry,20,2024–2028 (1981)
[27] Dawson, J.H., Dooley, D.M., Gray, H.B.: Proc. Natl. Acad. Sci. USA,77,5028–5031 (1980)
[28] Katz, M., Westley, J.: J. Biol. Chem.,254,9142–9147 (1979)
[29] Sakurai, T., Sawada, S., Suzuki, S., Nakahara, A.: Biochem. Biophys. Res. Commun.,131,647–652 (1985)
[30] Tanaka, N., Murao, S.: Agric. Biol. Chem.,47,1627–1628 (1983)
[31] Bradberry, C.W., Borchardt, R.T., Decedue, C.J.: FEBS Lett.,146,348–352 (1982)
[32] Bezerra Carvalho, L., Lima, C.J., Medeiros, P.H.: Phytochemistry,20,2423–2424 (1981)

Enzyme Handbook © Springer-Verlag Berlin Heidelberg 1994
Duplication, reproduction and storage in data banks are only allowed with the prior permission of the publishers

[33] Sheline, R.R., Strothkamp, K.G.: Biochem. Biophys. Res. Commun.,96,1343–1348 (1980)
[34] Aikazyan, V.T., Nalbandyan, R.M.: FEBS Lett.,104,127–130 (1979)
[35] Strothkamp, R.E., Dawson, C.R.: Biochem. Biophys. Res. Commun.,85,655–661 (1978)
[36] Amon, A., Markakis, P.: Phytochemistry,12,2127–2132 (1973)
[37] Strothkamp, K.G., Dawson, C.R.: Biochemistry,13,434–440 (1974)
[38] Dayan, J., Dawson, C.R.: Biochem. Biophys. Res. Commun.,73,451–458 (1976)
[39] Curzon, G., Young, S.N.: Biochim. Biophys. Acta,268,41–48 (1972)
[40] White, G.A., Krupka, R.M.: Arch. Biochem. Biophys.,110,448–461 (1965)

1 NOMENCLATURE

EC number
1.10.3.4

Systematic name
2-Aminophenol:oxygen oxidoreductase

Recommended name
o-Aminophenol oxidase

Synonymes
Oxidase, o-aminophenol
Isophenoxazine synthase (o-aminophenol:O_2 oxidoreductase)

CAS Reg. No.
9013-85-8

2 REACTION AND SPECIFICITY

Catalysed reaction
2 2-Aminophenol + 3 O_2 →
→ 2 isophenoxazine + 6 H_2O (discussion of mechanism [3], isophenoxazine may be formed by a secondary condensation from the initial oxidation product)

Reaction type
Redox reaction

Natural substrates

Substrate spectrum
1 2-Aminophenol + O_2 (oxidation of o-aminophenol to o-quinoneimine and subsequent condensation of o-aminophenol to 2-amino-3H-isophenoxazin-3-one in absence of Mn^{2+} requires about 1 mol of riboflavin 5'-phosphate per mol of o-aminophenol oxidized, in presence of Mn^{2+} riboflavin 5'-phosphate acts as catalyst [2], highly specific for o-aminophenol, not: related compounds (e.g. 3-hydroxyanthranilic acid, 3-hydroxy-kynurenine, p-aminophenol, catechol [3])) [2, 3]

Product spectrum
1 Isophenoxazine + H_2O

Enzyme Handbook © Springer-Verlag Berlin Heidelberg 1994
Duplication, reproduction and storage in data banks are only allowed with the prior permission of the publishers

Inhibitor(s)

Ascorbic acid [2, 3]; $FeSO_4$ [2]; $NaBH_4$ [2]; Cysteine [2, 3]; Glutathione [2, 3]; N-Ethylmaleimide (no effect [3]) [2]; Na_3AsO_3 [2]; $CuSO_4$ [2]; Anthranilic acid (competitive) [3]; 3-Hydroxyanthranilic acid (competitive) [3]; 2,3-Dimercaptopropanol [3]; p-Hydroxymercuribenzoate (reversed by glutathione or cysteine) [3]; Hg^{2+} [1, 3]; Co^{2+} [1]; Mg^{2+} [1]; Fe^{3+} [1]; Atebrin (slight [3]) [1, 3]; Cu^{2+} [3]; Ag^{+} [3]; Fe^{2+} [3]; p-Chloromercuribenzoate (inhibition reversed by glutathione) [1, 2]; Cyanide [1, 3]; Azide [3]; o-Aminophenol (substrate inhibition above 0.6 mM) [2]

Cofactor(s)/prostethic group(s)

FAD (flavoprotein [1], requires FAD for maximal activity [1], no action as cofactor [2]) [1]; Riboflavin 5'-monophosphate (holoenzyme reconstituted by addition of riboflavin 5'-phosphate and Mn^{2+}) [2]; More (no cofactor requirement) [3]

Metal compounds/salts

Mn^{2+} (activates [1], holoenzyme reconstituted by addition of riboflavin and Mn^{2+} [2], required for maximal activity, ineffective in absence of FMN) [2]; More (no metal requirement) [3]

Turnover number (min^{-1})

Specific activity (U/mg)

56.1 [3]; More [2]

K_m-value (mM)

0.264 (FMN) [2]; 0.75 (o-aminophenol) [3]

pH-optimum

6.2 [1, 3]

pH-range

4.6–7.6 (4.6: about 50% of activity maximum, 7.6: about 65% of activity maximum) [1]

Temperature optimum (°C)

40 [3]; 45 [1]

Temperature range (°C)

3 ENZYME STRUCTURE

Molecular weight

Subunits

Glycoprotein/Lipoprotein

–

---

4 ISOLATION/PREPARATION

Source organism

Tecoma stans [1]; Pycnoporus coccineus (wood-rotting fungus) [2]; Bauhenia monandra (leguminous plant) [3]

Source tissue

Leaves [1]

Localisation in source

Purification

Pycnoporus coccineus (partial) [2]; Bauhenia monandra [3]

Crystallization

(apoenzyme) [2]

Cloned

–

Renaturated

–

---

5 STABILITY

pH

Temperature (°C)

Oxidation

Organic solvent

General stability information

Stable to extensive dialysis against cyanide or EDTA [3]

Storage

–20°C, for at least 1 month [3]

Enzyme Handbook © Springer-Verlag Berlin Heidelberg 1994
Duplication, reproduction and storage in data banks are only allowed with the prior permission of the publishers

6 CROSSREFERENCES TO STRUCTURE DATABANKS

PIR/MIPS code

Brookhaven code

7 LITERATURE REFERENCES

[1] Nair, P.M., Vaidyanathan, C.S.: Biochim. Biophys. Acta,81,507–516 (1964)
[2] Nair, P.M., Vining, I.C.: Biochim. Biophys. Acta,96,318–327 (1965)
[3] Rao, P.V.S., Vaidyanathan, C.S.: Arch. Biochem. Biophys.,118,388–394 (1967)

1 NOMENCLATURE

EC number
1.10.3.5

Systematic name
3-Hydroxyanthranilate:oxygen oxidoreductase

Recommended name
3-Hydroxyanthranilate oxidase

Synonymes
Oxidase, 3-hydroxyanthranilate
3-Hydroxyanthranilic acid oxidase

CAS Reg. No.
37256-53-4

2 REACTION AND SPECIFICITY

Catalysed reaction
3-Hydroxyanthranilate + O_2 →
→ 6-imino-5-oxocyclohexa-1,3-dienecarboxylate + H_2O_2

Reaction type
Redox reaction

Natural substrates
3-Hydroxyanthranilate + O_2

Substrate spectrum
1 3-Hydroxyanthranilate + O_2 [1, 2]

Product spectrum
1 6-Imino-5-oxocyclohexa-1,3-dienecarboxylate + H_2O_2

Inhibitor(s)
4-Chloro-3-hydroxyanthranilate [1, 2]

Cofactor(s)/prostethic group(s)

Metal compounds/salts

Enzyme Handbook © Springer-Verlag Berlin Heidelberg 1994
Duplication, reproduction and storage in data banks are only allowed with the prior permission of the publishers

Turnover number (min^{-1})

Specific activity (U/mg)

K_m-value (mM)

pH-optimum

pH-range

Temperature optimum (°C)

Temperature range (°C)

---

3 ENZYME STRUCTURE

Molecular weight

Subunits

Glycoprotein/Lipoprotein
–

---

4 ISOLATION/PREPARATION

Source organism
Rat [1, 2]; Guinea pig [2]

Source tissue
Liver [1]

Localisation in source

Purification

Crystallization
–

Cloned
–

Renaturated
–

5 STABILITY

pH

Temperature (°C)

Oxidation

Organic solvent

General stability information

Storage

6 CROSSREFERENCES TO STRUCTURE DATABANKS

PIR/MIPS code

Brookhaven code

7 LITERATURE REFERENCES

[1] Cook, J.S., Pogson, C.I.: Biochem. J.,214,511–516 (1983)
[2] Parli, C.J., Krieter, P., Schmidt, B.: Arch. Biochem. Biophys.,203,161–166 (1980)

Enzyme Handbook © Springer-Verlag Berlin Heidelberg 1994
Duplication, reproduction and storage in data banks are only allowed with the prior permission of the publishers

1 NOMENCLATURE

EC number
1.10.3.6

Systematic name
Rifamycin-B:oxygen oxidoreductase

Recommended name
Rifamycin-B oxidase

Synonymes
Oxidase, rifamycin B
Rifamycin B oxidase

CAS Reg. No.
84932-52-5

2 REACTION AND SPECIFICITY

Catalysed reaction
Rifamycin B + O_2 →
→ rifamycin O + H_2O_2

Reaction type
Redox reaction
Oxidative cyclization [1]

Natural substrates

Substrate spectrum
1 Rifamycin B + O_2 [1]
2 Rifamycin SV + O_2 [1]
3 p-Hydroquinone + O_2 [1]
4 Pyrogallol + O_2 [1]
5 Catechol + O_2 [1]
6 p-Hydroxyphenoxy acetic acid + O_2 [3]

Enzyme Handbook © Springer-Verlag Berlin Heidelberg 1994
Duplication, reproduction and storage in data banks are only allowed with the prior permission of the publishers

Product spectrum

1 Rifamycin O + H_2O_2 (rifamycin O spontaneously hydrolyzed to rifamycin S in neutral aqueous milieu) [1]
2 ?
3 p-Benzoquinone + H_2O_2
4 ?
5 ?
6 ?

Inhibitor(s)

Fe^{2+} [3]; Hg^{2+} [3]; Rifamycin B (substrate inhibition above 2 mM [3]) [3, 4]

Cofactor(s)/prostethic group(s)

Metal compounds/salts

Turnover number (min^{-1})

Specific activity (U/mg)

3.2 [1]

K_m-value (mM)

0.05 (rifamycin B) [3]; 0.3 (rifamycin B, enzyme in acetone-defatted cells) [4]; 0.6 (rifamycin B, enzyme in immobilized acetone-defatted cells) [4]

pH-optimum

6.5 [2]; 7.5 (enzyme in acetone-defatted cells) [4]; 7.8 (enzyme in immobilized acetone-defatted cells) [4]; 7.8–8.0 [3]

pH-range

5.5–7.0 (about 50% of activity maximum at pH 5.5 and 7.0) [2]; 6.2–8.5 (about 50% of activity maximum at pH 6.2. and 8.5) [4]

Temperature optimum (°C)

45 [3]; 50 (enzyme in acetone-defatted cells and immobilized acetone-defatted cells [4]) [2, 4]

Temperature range (°C)

30–58 (about 50% of activity maximum at 30°C and 58°C) [2]; 30–60 (30°C: about 50% of activity maximum (immobilized acetone-defatted cells), about 65% of activity maximum (acetone defatted cells), 60°C: about 65% of maximal activity (immobilized acetone-defatted cells)) [4]; 20–55 (20°C: about 50% of activity maximum, 55°C: about 60% of activity maximum) [2]

3 ENZYME STRUCTURE

Molecular weight

Subunits

Glycoprotein/Lipoprotein

–

4 ISOLATION/PREPARATION

Source organism

Monocillium sp. ATCC 20621 [1]; Curvularia lunata var. aeri [2]; Humicola sp. ATCC 20620 [3, 4]

Source tissue

Localisation in source

Intracellular (Monocillium sp., Humicola sp. [3]) [2, 3]; Extracellular (Curvularia lunata) [2]

Purification

Monocillium sp. ATCC 20621 [1]; Humicola sp. ATCC 20620 [2]

Crystallization

–

Cloned

–

Renaturated

–

5 STABILITY

pH

8 (enzyme in immobilized acetone-defatted cells, highest stability) [4]

Temperature (°C)

40 (pH 8, half-life: 8 days, enzyme in immobilized acetone-defatted cells) [4]

Oxidation

Organic solvent

General stability information

Enzyme Handbook © Springer-Verlag Berlin Heidelberg 1994
Duplication, reproduction and storage in data banks are only allowed with the prior permission of the publishers

Storage

4°C, pH 7.8, enzyme in free and immobilized whole cells stable [4]

6 CROSSREFERENCES TO STRUCTURE DATABANKS

PIR/MIPS code

Brookhaven code

7 LITERATURE REFERENCES

[1] Han, M.H., Seong, B.-L., Son, H.-J., Mheen, T.-I.: FEBS Lett.,151,36–40 (1983)
[2] Vohra, R.M., Banerjee, U.C., Das, S., Dube, S.: Biotechnol. Lett.,11,851–854 (1989)
[3] Seong, B.L., Son, H.J., Mheen, T.I., Park, Y.H., Han, M.H.: J. Ferment. Technol.,63,515–522 (1985)
[4] Lee, G.M., Choi, C.Y.: Biotechnol. Lett.,6,143–148 (1984)

1 NOMENCLATURE

EC number
1.10.3.7

Systematic name
Sulochrin:oxygen oxidoreductase (cyclizing, (+)-specific)

Recommended name
Sulochrin oxidase ((+)-bisdechlorogeodin-forming)

Synonymes
Oxidase, sulochrin

CAS Reg. No.
82469-87-2

2 REACTION AND SPECIFICITY

Catalysed reaction
2 Sulochrin + O_2 →
→ 2 (+)-bisdechlorogeodin + 2 H_2O

Reaction type
Redox reaction

Natural substrates
Sulochrin + O_2 (main function: synthesis of bisdechlorogeodin, involved in biosynthesis of mold metabolites related to the antibiotic griseofulvin) [1]

Substrate spectrum
1 Sulochrin + O_2 [1]
2 Hydroquinone + O_2 [1]
3 Catechol + O_2 [1]
4 p-Phenylenediamine + O_2 [1]
5 o-Phenylenediamine + O_2 [1]
6 N,N-Dimethyl-p-phenylenediamine + O_2 [1]
7 N,N,N',N',-tetramethyl-p-phenylenediamine + O_2 [1]

Enzyme Handbook © Springer-Verlag Berlin Heidelberg 1994
Duplication, reproduction and storage in data banks are only allowed with the prior permission of the publishers

Product spectrum

1 (+)-Bisdechlorogeodin + H_2O [1]
2 ?
3 ?
4 ?
5 ?
6 ?
7 ?

Inhibitor(s)

Sodium diethyldithiocarbamate [1]

Cofactor(s)/prostethic group(s)

Copper (contains 6 copper atoms) [1]

Metal compounds/salts

Turnover number (min^{-1})

Specific activity (U/mg)

1.1 [1]

K_m-value (mM)

0.063 (O_2) [1]; 0.0065 (sulochrin) [1]; 60 (hydroquinone) [1]; 40 (catechol) [1]; 2 (p-phenylenediamine) [1]; 5 (o-phenylenediamine) [1]; 0.4 (N,N-dimethyl-p-phenylenediamine) [1]; 1 (N,N,N',N'-tetramethyl-p-phenylenediamine) [1]; 150 ((+)-ascorbic acid) [1]

pH-optimum

5.8 [1]

pH-range

3.5–7 [1]

Temperature optimum (°C)

25 (assay at) [1]

Temperature range (°C)

3 ENZYME STRUCTURE

Molecular weight

157000 (Penicillium frequentans, gel chromatography) [1]

Subunits

Dimer (1 × 85000 + 1 × 27800, Penicillium frequentans, SDS-PAGE, possible structure: 2 noncovalently bound identical subunits, each of which consists of 3 smaller subunits) [1]

Glycoprotein/Lipoprotein

Glycoprotein (19.5% carbohydrate) [1]

4 ISOLATION/PREPARATION

Source organism

Penicillium frequentans [1]

Source tissue

Mycelium [1]

Localisation in source

Intracellular (no extracellular activity) [1]

Purification

Penicillium frequentans [1]

Crystallization

–

Cloned

–

Renaturated

–

5 STABILITY

pH

Temperature (°C)

Oxidation

Organic solvent

General stability information

Storage

–20°C, pH 5.8, citrate-phosphate buffer, several months [1]

Enzyme Handbook © Springer-Verlag Berlin Heidelberg 1994
Duplication, reproduction and storage in data banks are only allowed with the prior permission of the publishers

6 CROSSREFERENCES TO STRUCTURE DATABANKS

PIR/MIPS code

Brookhaven code

7 LITERATURE REFERENCES

[1] Nordlöv, H., Gatenbeck, S.: Arch. Microbiol., 131, 208–211 (1982)

1 NOMENCLATURE

EC number
1.10.3.8

Systematic name
Sulochrin:oxygen oxidoreductase (cyclizing, (-)-specific)

Recommended name
Sulochrin oxidase ((-)-bisdechlorogeodin-forming)

Synonymes
Oxidase, sulochrin

CAS Reg. No.
82469-87-2

2 REACTION AND SPECIFICITY

Catalysed reaction
2-Sulochrin + O_2 →
→ 2 (-)-bisdechlorogeodin + 2 H_2O

Reaction type
Redox reaction

Natural substrates
Sulochrin + O_2 (main function: synthesis of bisdechlorogeodin [1], involved in biosynthesis of mold metabolites related to antibiotic griseofulvin)

Substrate spectrum
1 Sulochrin + O_2 [1]
2 N,N,N',N'-Tetramethyl-p-phenylenediamine + O_2 [1]
3 Hydroquinone + O_2 [1]

Product spectrum
1 (-)-Bisdechlorogeodin + H_2O
2 ?
3 ?

Inhibitor(s)
Sodium diethyldithiocarbamate [1]

Enzyme Handbook © Springer-Verlag Berlin Heidelberg 1994
Duplication, reproduction and storage in data banks are only allowed with the prior permission of the publishers

Cofactor(s)/prostethic group(s)

Metal compounds/salts
Copper (6 copper atoms per molecule) [1]

Turnover number (min^{-1})

Specific activity (U/mg)
1.2 [1]

K_m-value (mM)
1.5 (N, N, N', N'-tetramethyl-p-phenylenediamine) [1]; 0.0085 (sulochrin) [1]; 53 (hydroquinone) [1]

pH-optimum
5.8 [1]

pH-range
3.5–7 [1]

Temperature optimum (°C)
25 (assay at) [1]

Temperature range (°C)

3 ENZYME STRUCTURE

Molecular weight
128000 (Oospora sulphurea-ochracea, gel chromatography) [1]

Subunits

Glycoprotein/Lipoprotein
Glycoprotein [1]

4 ISOLATION/PREPARATION

Source organism
Oospora sulphurea-ochracea [1]

Source tissue
Mycelium [1]

Localisation in source
Intracellular (no extracellular activity) [1]

Purification

Oospora sulphurea-ochracea [1]

Crystallization

–

Cloned

–

Renaturated

–

5 STABILITY

pH

Temperature (°C)

Oxidation

Organic solvent

General stability information

Storage

–20°C, pH 5.8, several months, citrate-phosphate buffer [1]

6 CROSSREFERENCES TO STRUCTURE DATABANKS

PIR/MIPS code

Brookhaven code

7 LITERATURE REFERENCES

[1] Nordlöv, H., Gatenbeck, S.: Arch. Microbiol.,131,208–211 (1982)

Enzyme Handbook © Springer-Verlag Berlin Heidelberg 1994
Duplication, reproduction and storage in data banks are only allowed with the prior permission of the publishers

1 NOMENCLATURE

EC number
1.10.99.1

Systematic name
Plastoquinol:oxidized-plastocyanin oxidoreductase

Recommended name
Plastoquinol-plastocyanin reductase

Synonymes
Reductase, plastoquinol-plastocyanin
Plastoquinol/plastocyanin oxidoreductase
Cytochrome b_6/f complex
Cytochrome b_6–f complex

CAS Reg. No.
79079-13-3

2 REACTION AND SPECIFICITY

Catalysed reaction
Plastoquinol-1 + 2 oxidized plastocyanin →
→ plastoquinone + 2 reduced plastocyanin (mechanism [12])

Reaction type
Redox reaction

Natural substrates
Plastoquinol + plastocyanin

Substrate spectrum
1 Plastoquinol + plastocyanin (best electron donor: plastoquinol-9 [2, 27], decylplastoquinol [29], best electron acceptors: plastocyanin from Anabaena variabilis, cytochrome c-553 from Anabaena variabilis, cytochrome c from horse heart [2, 27], ferricyanide [29])
2 More (proton translocation with b_6-f complex incorporated into liposomes) [9, 28]

Product spectrum
1 Plastoquinone + reduced plastocyanin
2 ?

Enzyme Handbook © Springer-Verlag Berlin Heidelberg 1994
Duplication, reproduction and storage in data banks are only allowed with the prior permission of the publishers

Inhibitor(s)

2,5-Dibromomethylisopropylbenzoquinone (DBMIB, reactivation by 3-chloro-5-hydroxyl-2-methyl-6-decyl-1,4-benzoquinone [1]) [1, 2, 5, 12–14, 17, 18, 27]; Stigmatellin [2]; 2-Iodo-6-isopropyl-3-methyl-2', 4,4'-trinitrodiphenyl ether (DNP-INT) [2, 6, 10–12, 14, 18]; 5-Undecyl-6-hydroxy-4,7-dioxobenzothiazole (UHDBT) [2, 10, 12, 17]; 3-Azido-2-methyl-5-methoxy-6-(3,7-dimethyloctyl)-1,4-benzoquinone [17]; Aurachin C and D [15]

Cofactor(s)/prostethic group(s)

3-Chloro-5-hydroxyl-2-methyl-6-decyl-1,4-benzoquinone (activation with substrate 2,3-dimethyl-6-geranyl-1,4-benzoquinone) [1]

Metal compounds/salts

Fe (2.1 mol nonheme iron per mol of cytochrome f [5], subunit IV: 3.1 mol nonheme iron per mol [13]) [5, 13]

Turnover number (min^{-1})

840–2040 (cytochrome f) [5, 14]; More [2, 5, 12, 29]

Specific activity (U/mg)

More (14–20 micromol cytochrome c-552 reduced per nmol cytochrome f per h [14], assay method [2]) [2, 14, 18, 25, 27]

K_m-value (mM)

0.009 (plastoquinol) [10]; 0.02 (plastoquinol-1) [27]; 0.04 (plastoquinol-9) [27]; 0.0032–0.0039 (2,3-dimethyl-6-geranyl-1,4-benzoquinol, depending on presence of activator) [1]

pH-optimum

7.5 [27]; 8.0 [14]

pH-range

Temperature optimum (°C)

Temperature range (°C)

---

3 ENZYME STRUCTURE

Molecular weight

185000 (Spinacia oleracea, gel filtration) [12]
160000 (Spinacia oleracea, calculation from ultraviolet spectrum assuming subunit stoichiometry of 1:2:1:2) [12]
142000 (Spinacia oleracea, calculation from cytochrome f content) [12, 14]
95000 (Spinacia oleracea, calculation from subunit size assuming stoichiometry of 1:1:1:1) [5]

Subunits

More (4 or 5 different subunits, 33000–37300, cytochrome f [5, 8, 10, 21], 19500–23500, cytochrome b_{563} [5, 8, 10, 21], 19000–21000, Rieske Fe-S protein [5, 10, 21], 12500–17500, subunit IV (plastoquinone binding side [17]) [5, 10, 17], 19500, subunit V [21], ratio cytochrome b_{563}: cytochrome f: Rieske Fe-S protein 2: 1: 1 [2, 5, 12, 18], structure-function relationship of subunits [7]) [2, 5, 7, 8, 10, 12, 14, 17, 18, 21, 25, 27]

Glycoprotein/Lipoprotein

Glycolipoprotein [17]

4 ISOLATION/PREPARATION

Source organism

Spinacia oleracea (spinach) [1, 4–7, 9–14, 17, 19, 20, 23–26, 28]; Cyanophora paradoxa [3]; Anabaena variabilis [2, 9, 19, 27]; Pisum sativum (pea) [8]; Nostoc sp. PCC 7906 [16]; Chlamydomonas reinhardtii [18, 21, 22]; Dunaliella saline [18]; Scenedesmus obliquus [18]

Source tissue

Leaves [1, 4–14, 17, 19, 20, 23–26, 28]; Cell [3, 18, 27]

Localisation in source

Thylakoid membranes (of chloroplasts, regional localization [22–24, 26]) [1–28]

Purification

Spinacia oleracea (90% pure [5], Rieske Fe-S protein [13], five polypeptides [14], plastoquinol and phospholipid deficient complex [25]) [5, 10, 13, 14, 19, 25]; Anabaena variabilis [2, 27]; Pisum sativum [8]; Cyanobacteria [18]

Crystallization

–

Cloned

(subunit V [3]) [3, 16]

Renaturated

(reconstitition of Rieske Fe-S protein into cytochrome b_6–f complex [4], reconstitution of lipid- and plastoquinone depleted complex [25]) [4, 25]

Enzyme Handbook © Springer-Verlag Berlin Heidelberg 1994
Duplication, reproduction and storage in data banks are only allowed with the prior permission of the publishers

5 STABILITY

pH

Temperature (°C)

Oxidation

Organic solvent

General stability information

PMSF prevents proteolysis during purification [10]; Most stable in cholate buffer [10]

Storage

–156°C, best stability, or –80°C or –20°C [2]; –70°C, 30 mM Tris-succinate buffer, pH 6.5, 1% sodium cholate, 10% glycerol [25]; –20°C [18]; Concentrated, partially reduced form [10]

6 CROSSREFERENCES TO STRUCTURE DATABANKS

PIR/MIPS code

Brookhaven code

7 LITERATURE REFERENCES

[1] Gu, L.-Q., Yu, L., Yu, C.-A.: J. Biol. Chem.,264,4506–4512 (1989)
[2] Malkin, R.: Methods Enzymol.,167,341–349 (1988)
[3] Stirewalt, V.L., Bryant, D.A.: Nucleic Acids Res.,17,10095 (1989)
[4] Adam, Z., Malkin, R.: FEBS Lett.,225,67–71 (1987)
[5] Black, M.T., Widger, W.R., Cramer, W.A.: Arch. Biochem. Biophys.,252,655–661 (1987)
[6] Malkin, R.: FEBS Lett.,208,317–320 (1986)
[7] Lam, E.: Biochim. Biophys. Acta,848,324–332 (1986)
[8] Phillis, A.L., Gray, J.C.: Eur. J. Biochem.,137,553–560 (1983)
[9] Hurt, E.C., Gabellini, N., Shahak, Y., Lochau, W., Hauska, G.: Arch. Biochem. Biophys.,225,879–885 (1983)
[10] Clark, R.D., Hind, G.: J. Biol. Chem.,258,10348–10354 (1983)
[11] Oettmeier, W., Masson, K., Olschewski, E.: FEBS Lett.,155,241–244 (1983)
[12] Hauska, G., Hurt, E., Gabellini, N., Lockau, W.: Biochim. Biophys. Acta,726,97–133 (1983) (Review)
[13] Hurt, E., Hauska, G., Malkin, R.: FEBS Lett.,134,1–5 (1981)
[14] Hurt, E., Hauska, G.: Eur. J. Biochem.,117,591–599 (1981)
[15] Oettmeier, W., Dostatni, R., Majewski, C., Höfle, G., Fecker, T., Kunze, B., Reichenbach, H.: Z. Naturforsch.,45c,322–328 (1990)
[16] Kallas, T., Malkin, R.: Methods Enzymol.,167,779–794 (1988)

[17] Doyle, M.P., Li, L.-B., Yu, L., Yu, C.-A.: J. Biol. Chem.,264,1387–1392 (1989)
[18] Wynn, R.M., Bertsch, J., Bruce, B.D., Malkin, R.: Biochim. Biophys. Acta,935,115–122 (1988)
[19] Hauska, G.: Methods Enzymol.,126,271–285 (1986)
[20] Morrissey, P.J., McCauley, S.W., Melis, A.: Eur. J. Biochem.,160,389–393 (1986)
[21] Lemaire, C., Girard-Bascou, J., Wollman, F.-A., Bennoun, P.: Biochim. Biophys. Acta,851,229–238 (1986)
[22] Olive, J., Vallon, O., Wollman, F.-A., Recouvreur, M. , Bennoun, P.: Biochim. Biophys. Acta,851,239–248 (1986)
[23] Melis, A., Svensson, P., Albertsson, P.-A.: Biochim. Biophys. Acta,850,402–412 (1986)
[24] Allred, D.R., Staehelin, L.A.: Biochim. Biophys. Acta,849,94–103 (1986)
[25] Doyle, M.F., Yu, C.-A.: Biochem. Biophys. Res. Commun.,131,700–706 (1985)
[26] Ortiz, W., Malkin, R.: Biochim. Biophys. Acta,808,164–170 (1985)
[27] Krinner, M., Hauska, G., Hurt, E., Lockau, W.: Biochim. Biophys. Acta,681,110–117 (1982)
[28] Hurt, E.C., Hauska, G., Shahak, Y.: FEBS Lett.,149,211–216 (1982)
[29] Rich, P., Heathcote, P., Moss, D.A.: Biochim. Biophys. Acta,892,138–151 (1987)

Enzyme Handbook © Springer-Verlag Berlin Heidelberg 1994
Duplication, reproduction and storage in data banks are only allowed with the prior permission of the publishers

1 NOMENCLATURE

EC number
1.11.1.1

Systematic name
NADH:hydrogen-peroxide oxidoreductase

Recommended name
NADH peroxidase

Synonymes
DPNH peroxidase
NAD peroxidase
Peroxidase, nicotinamide adenine dinucleotide
Diphosphopyridine nucleotide peroxidase
NADH-peroxidase
Nicotinamide adenine dinucleotide peroxidase

CAS Reg. No.
9032-24-0

2 REACTION AND SPECIFICITY

Catalysed reaction
NADH + H_2O_2 →
→ NAD^+ + 2 H_2O

Reaction type
Redox reaction

Natural substrates
NADH + H_2O_2 (elimination of toxic H_2O_2, regeneration of oxidized pyridine nucleotide which is essential to the strictly fermentative metabolism [8]) [3, 8]

Enzyme Handbook © Springer-Verlag Berlin Heidelberg 1994
Duplication, reproduction and storage in data banks are only allowed with the prior permission of the publishers

Substrate spectrum

1 NADH + H_2O_2 (ir [1], H_2O_2 can be replaced by 1,4-naphthoquinone which is one third as effective as H_2O_2 [3]) [1, 3]
2 Ferricyanide + H_2O_2 [3, 6]
3 More (NADH cannot be replaced by reduced cytochrome c, H_2O_2 cannot be replaced by Cu^{2+}, Fe^{3+}, SeO_3^{2+}, NO_3^-, NO_2^-, oxidized glutathione, cystine, lipoic acid disulfide, dehydroascorbic acid, cytochrome c, FAD, flavin mononucleotide, riboflavin, 2,6-dichlorophenolindophenol, methylene blue, brilliant cresyl blue) [3]

Product spectrum

1 NAD^+ + H_2O [1, 3]
2 Ferrocyanide + H_2O
3 ?

Inhibitor(s)

NADPH [2]; NADH [5]; Ag^+ [3]; Pb^{2+} [3]; Cu^{2+} [3]; Co^{2+} [3]; Hg^{2+} [3]; p-Chloromercuribenzoate [3]; H_2O_2 [3]; NaN_3 [3]; p-Hydroxymercuribenzoate [6]; N-Ethylmaleimide [6]; Ethyl hydroperoxide [8]

Cofactor(s)/prostethic group(s)

FAD (2 mol flavin per mol) [2, 3, 5]

Metal compounds/salts

More (no Mo, Cu, Mn, Zn) [3]

Turnover number (min^{-1})

4000 (H_2O_2, calculated per flavin) [5]

Specific activity (U/mg)

More (8500–9000 U/ml, calculated as change of 0.01 of optical density) [3]

K_m-value (mM)

0.002 (NADH) [1]; 0.0068 (NADH) [5]; 0.02 (H_2O_2) [5]; 0.08 (H_2O_2) [1]; 0.1 (ferricyanide) [3]

pH-optimum

4.5 (ferricyanide) [3]; 5.5 [1]; 5.2–5.4 (substrate H_2O_2 in acetate or phosphate buffer) [3]; 5.4 [6]

pH-range

5.0–8.5 (90% of maximal activity at pH 5.0, 32% of maximal activity at pH 8.5) [1]

Temperature optimum (°C)
30 [1]; 26 [3]

Temperature range (°C)

3 ENZYME STRUCTURE

Molecular weight
120000 (Streptococcus faecalis, calculation from FAD content) [3]

Subunits
Dimer (2 × 60000, Streptococcus faecalis, calculation from FAD content [5])
Tetramer (4 × 46000, identical, Streptococcus faecalis [13, 14] structure of active site [10, 11, 13]) [10, 11, 13, 14]

Glycoprotein/Lipoprotein
–

4 ISOLATION/PREPARATION

Source organism
Streptococcus faecalis [1–3, 5, 7, 8, 10–15]; Streptococcus lactis (effect of growth conditions on enzyme formation) [4]; Lactobacillus casei [6]; Streptococcus mutans [9]

Source tissue
Cell [4]

Localisation in source

Purification
Streptococcus faecalis [3, 7, 8]; Lactobacillus casei [6]

Crystallization
[2, 10]

Cloned
[15]

Renaturated
–

Enzyme Handbook © Springer-Verlag Berlin Heidelberg 1994
Duplication, reproduction and storage in data banks are only allowed with the prior permission of the publishers

5 STABILITY

pH

5.4 (45 min stable) [8]

Temperature (°C)

5 (1 month) [6]; 50 (5 min) [6]; 60 (5 min, 20% loss of activity) [6]; 70 (complete loss of activity) [6]

Oxidation

Organic solvent

General stability information

At 25°C without H_2O_2 in assay system loses 85% of activity in 30 min [5]

Storage

5°C, pH 7.0, 1 month [6]; –20°C, 50 mM phosphate buffer pH 7.0 [8]

6 CROSSREFERENCES TO STRUCTURE DATABANKS

PIR/MIPS code

PIR3: A34320 (Enterococcus faecalis (fragment)); PIR3: A36440 (Enterococcus hirae (fragments))

Brookhaven code

1NPX (Streptococcus faecalis)

7 LITERATURE REFERENCES

[1] Cox, C., Camus, P., Buret, J., Duvivier, J.: Anal. Biochem.,119,185–193 (1982)
[2] Dolin, M.I.: J. Biol. Chem.,250,310–317 (1975)
[3] Dolin, M.I.: J. Biol. Chem.,225,557–573 (1957)
[4] Hansson, L., Häggström, M.H.: Curr. Microbiol.,10,345–352 (1984)
[5] Dolin, M.I.: Biochem. Biophys. Res. Commun.,78,393–400 (1977)
[6] Walker, G.A., Kilgour, G.L.: Arch. Biochem. Biophys.,111,534–539 (1965)
[7] Dolin, M.I.: Arch. Biochem. Biophys.,55,415–435 (1955)
[8] Miller, H., Poole, L.B., Claiborne, A.: J. Biol. Chem.,265,9857–9863 (1990)
[9] Thomas, E.L., Pera, K.A.: J. Bacteriol.,154,1236–1244 (1983)
[10] Schiering, N., Stoll, V.S., Blanchart, J.S., Pai, E.F.: J. Biol. Chem.,264,21144–21145 (1989)
[11] Claiborne, A., Ahmed, S.A., Ross, P., Miller, H. in "Flavins and Flavoproteins", Proc. Int. Symp.,10th., Meeting Date1990 (Curti, B., Ronchi S., Zanetti, G., eds.) 667–670, de Gruyter, Berlin, New York (1991) (Review)
[12] Stehle, T., Ahmed, S.A., Claiborne, A., Schulz, G.E.: J. Mol. Biol.,221,1325–1344 (1991)

[13] Stehle, T., Schulz, G.E., Ahmed, S.A., Claiborne, A. in "Flavins and Flavoproteins", Proc. Int. Symp.,10th., Meeting Date1990 (Curti, B., Ronchi, S., Zanetti, eds.) 651–654, de Gruyter, Berlin, New York (1991)
[14] Poole, L.B., Claiborne, A.: J. Biol. Chem.,264,12330–12338 (1989)
[15] Ahmed, A.S., Ross, P., Miller, H., Claiborne, A. in "Flavins and Flavoproteins", Proc. Int. Symp.,10th., Meeting Date1990 (Curti, B., Ronchi, S., Zanetti, G., eds) 647–650, de Gruyter, Berlin, New York (1991)

Enzyme Handbook © Springer-Verlag Berlin Heidelberg 1994
Duplication, reproduction and storage in data banks are only allowed with the prior permission of the publishers

1 NOMENCLATURE

EC number
1.11.1.2

Systematic name
NADPH:hydrogen-peroxide oxidoreductase

Recommended name
NADPH peroxidase

Synonymes
TPNH peroxidase [1]
Peroxidase, nicotinamide adenine dinucleotide phosphate
NADP peroxidase
Nicotinamide adenine dinucleotide phosphate peroxidase
TPN peroxidase
Triphosphopyridine nucleotide peroxidase

CAS Reg. No.
9029-51-0

---

2 REACTION AND SPECIFICITY

Catalysed reaction
NADPH + H_2O_2 →
→ $NADP^+$ + 2 H_2O

Reaction type
Redox reaction

Natural substrates

Substrate spectrum
1 NADPH + H_2O_2 [1]

Product spectrum
1 $NADP^+$ + H_2O

Inhibitor(s)
Catalase [1]; Ascorbic acid [1]

Cofactor(s)/prostethic group(s)
Heme [1]; $NADP^+$ [1]; More (no cytochrome c) [1]

Enzyme Handbook © Springer-Verlag Berlin Heidelberg 1994
Duplication, reproduction and storage in data banks are only allowed with the prior permission of the publishers

Metal compounds/salts
Fe (hemoprotein) [1]

Turnover number (min^{-1})

Specific activity (U/mg)

K_m-value (mM)

pH-optimum

pH-range

Temperature optimum (°C)

Temperature range (°C)

3 ENZYME STRUCTURE

Molecular weight

Subunits
More (composed of two protein fragments of which the slower moving component shows slight NADPH oxidase and peroxidase activity and is enhanced by the faster moving component) [1]

Glycoprotein/Lipoprotein
–

4 ISOLATION/PREPARATION

Source organism
Wheat [1]

Source tissue
Germ [1]

Localisation in source
Cytoplasm [1]

Purification

Crystallization
–

Cloned
–

Renatured

–

5 STABILITY

pH

Temperature (°C)

100 (1 min, complete inactivation) [1]

Oxidation

Organic solvent

General stability information

Storage

–15°C, several months or dialyzed at neutral pH several days [1]

6 CROSSREFERENCES TO STRUCTURE DATABANKS

PIR/MIPS code

Brookhaven code

7 LITERATURE REFERENCES

[1] Conn, E.E., Kraemer, L.M., Liu, P.-N., Vennesland, B.: J. Biol. Chem.,194,143–151 (1952)

Enzyme Handbook © Springer-Verlag Berlin Heidelberg 1994
Duplication, reproduction and storage in data banks are only allowed with the prior permission of the publishers

1 NOMENCLATURE

EC number
1.11.1.3

Systematic name
Hexadecanoate:hydrogen-peroxide oxidoreductase

Recommended name
Fatty-acid peroxidase

Synonymes
Long chain fatty acid peroxidase [2]
Peroxidase, fatty acid
Fatty acid peroxidase

CAS Reg. No.
9029-52-1

2 REACTION AND SPECIFICITY

Catalysed reaction
Palmitate + 2 H_2O_2 →
→ pentadecanal + CO_2 + 3 H_2O

Reaction type
Redox reaction

Natural substrates
Fatty acids + H_2O_2 (alpha oxidation of fatty acids) [2]

Substrate spectrum
1 Palmitate + H_2O_2 [2]

Product spectrum
1 Pentadecanal + CO_2 + H_2O [2]

Inhibitor(s)
Imidazole [2]; CO [2]; Cyanide [2]; 1,2,4-Triazole [2]; 1-Methylimidazole [2]; 4-Hydroxymethylimidazole hydrochloride [2]; Imidazole aldehyde [2]; 2-Benzylimidazole [2]; 4,5-Imidazole dicarboxylic acid [2];
1-Hydroxyethyl-2-methylimidazole [2]; Imidazole lactic acid [2]; 2-Mercapto-4,5-imidazole dicarboxylic acid [2]

Enzyme Handbook © Springer-Verlag Berlin Heidelberg 1994
Duplication, reproduction and storage in data banks are only allowed with the prior permission of the publishers

Cofactor(s)/prostethic group(s)

Metal compounds/salts

Turnover number (min^{-1})

Specific activity (U/mg)
More (assay method) [3]

K_m-value (mM)
0.091 (palmitate) [2]

pH-optimum
7.5 [2]

pH-range

Temperature optimum (°C)

Temperature range (°C)

---

3 ENZYME STRUCTURE

Molecular weight

Subunits

Glycoprotein/Lipoprotein
–

---

4 ISOLATION/PREPARATION

Source organism
Peanut [2]; Carthamus tinctorius (safflower) [2]; Pea [2]; Rat (not detected [2]) [1]; More (not in mammalia) [2]

Source tissue
Liver [1]; Cotyledons [2]

Localisation in source
Mitochondria [1, 2]; Microsomes [2]; Cytoplasm [2]

Purification

Crystallization
–

Cloned

–

Renaturated

–

5 STABILITY

pH

Temperature (°C)

55 (destroyed after 5 min) [2]

Oxidation

Organic solvent

General stability information

Storage

5°C, 3 months [2]

6 CROSSREFERENCES TO STRUCTURE DATABANKS

PIR/MIPS code

Brookhaven code

7 LITERATURE REFERENCES

[1] Haeffner, E.W., Privett, O.S.: Lipids,10,75–81 (1975)
[2] Martin, R.O., Stumpf, P.K.: J. Biol. Chem.,234,2548–2554 (1959)
[3] Placer, Z.A., Cushman, L.L., Johnson, B.C.: Anal. Biochem.,16,359–364 (1966)

Enzyme Handbook © Springer-Verlag Berlin Heidelberg 1994
Duplication, reproduction and storage in data banks are only allowed with the prior permission of the publishers

1 NOMENCLATURE

EC number
1.11.1.5

Systematic name
Ferrocytochrome-c:hydrogen-peroxide oxidoreductase

Recommended name
Cytochrome-c peroxidase

Synonymes
Cytochrome peroxidase [5]
Cytochrome c-551 peroxidase [3]
Apocytochrome c peroxidase [7]
Mesocytochrome c peroxidase azide
Mesocytochrome c peroxidase cyanide
Mesocytochrome c peroxidase cyanate
Cytochrome c-H_2O oxidoreductase [5]
Peroxidase, cytochrome c
Cytochrome c peroxidase

CAS Reg. No.
9029-53-2

---

2 REACTION AND SPECIFICITY

Catalysed reaction
2 Ferrocytochrome c + H_2O_2 →
→ 2 ferricytochrome c + 2 OH^- (reaction scheme [2])

Reaction type
Redox reaction

Natural substrates

Enzyme Handbook © Springer-Verlag Berlin Heidelberg 1994
Duplication, reproduction and storage in data banks are only allowed with the prior permission of the publishers

Substrate spectrum

1 Ferrocytochrome c-551 + H_2O_2 [3, 5]
2 Ferrocytochrome c + H_2O_2 (ir [7], horse heart [2, 6, 7], yeast [7], H_2O_2 can be substituted by ethyl peroxide [7]) [2, 6, 7]
3 Ferrocytochrome c_4 + H_2O_2 [5]
4 NADH + H_2O_2 [6]
5 Azurin + H_2O_2 (blue copper protein) [3, 5]
6 NADPH + H_2O_2 [6]
7 Ferrocytochrome c + menadione (menadione can be substituted by 1,4-naphthoquinone) [6]
8 Ascorbate + H_2O_2 [7]
9 Pyrogallol + H_2O_2 [7]
10 Guaiacol + H_2O_2 [7]
11 Hydroquinone + H_2O_2 [7]
12 Ferrocyanide + H_2O_2 [7]
13 More (no oxidation of ferrocytochrome c of bacteria, no mammalian ferrocytochrome b, b_5, c_1) [7]

Product spectrum

1 Ferricytochrome c-551 + OH^- [3, 5]
2 Ferricytochrome c + OH^- [2, 6]
3 Ferricytochrome c_4 + OH^- [5]
4 NAD^+ + H_2O [6]
5 Oxidized azurin + ? [5]
6 $NADP^+$ + H_2O
7 Ferricytochrome + oxidized menadione
8 Dehydroascorbate + H_2O
9 ?
10 2-Methoxy-cyclohexa-2,5-dienone + H_2O
11 Benzoquinone + H_2O
12 Ferricyanide + OH^-
13 ?

Inhibitor(s)

Azide [6, 9]; Cyanide [6, 9]; CO (40% inhibition) [6]; Hg^{2+} (inhibition of NADH oxidizing activity) [6]; Pb^{2+} (inhibition of NADH oxidizing activity) [6]; Cu^{2+} (inhibition of NADH oxidizing activity) [6]; Ag^+ (inhibition of NADH oxidizing activity) [6]; F^- [9]; Cytochrome c-551 (above 0.05 mM, substrate inhibition) [5]; NO [16]

Cofactor(s)/prostethic group(s)

Heme (2.41 mol heme per mol [1], 2 mol heme c per mol [5], protoheme [6], protoporphyrin IX, 1 mol per mol, bound noncovalently [7]) [1, 5–7]

Metal compounds/salts
Iron (heme prosthetic group) [1, 5–7]

Turnover number (min^{-1})
180 (ascorbate) [9]; 264 (guaiacol) [9]; 480 (pyrogallol) [9]; 90000 (Saccharomyces cerevisiae ferrocytochrome c) [9]; 120000 (horse heart ferrocytochrome c) [9]; More [7]

Specific activity (U/mg)
135 [3]; 18.16 [6]

K_m-value (mM)
0.0041 (horse heart ferrocytochrome c with electron acceptor ethyl peroxide) [7]; 0.0045 (horse heart ferrocytochrome c with electron acceptor H_2O_2) [7]; 0.005 (horse heart ferrocytochrome) [9]; 0.01 (yeast ferrocytochrome c) [9] 0.023 (yeast ferrocytochrome c with electron acceptor ethyl peroxide) [7]; 0.025 (yeast ferrocytochrome with electron acceptor H_2O_2) [7]; 5 (pyrogallol) [9]; 10 (guaiacol) [9]; 11 (ascorbate) [9]

pH-optimum

pH-range
5–7 [3]

Temperature optimum (°C)

Temperature range (°C)

3 ENZYME STRUCTURE

Molecular weight
32000 (Thiobacillus thiooxidans, gel filtration) [6]
34100–35235 (Saccharomyces cerevisiae, sedimentation and diffusion constants, partial specific volume [7, 11], amino acid sequence of apoprotein: 33419, of holoenzyme: 34036 [14], amino acid composition of tryptic peptides [4]) [4, 7, 11, 14]
44000 (Pseudomonas aeruginosa, amino acid analysis) [13]
63000 (Pseudomonas denitrificans, gel filtration) [5]

Subunits
Monomer (1 × 32500, Saccharomyces cerevisiae, SDS-PAGE) [4]

Glycoprotein/Lipoprotein
No glycoprotein [13]

Enzyme Handbook © Springer-Verlag Berlin Heidelberg 1994
Duplication, reproduction and storage in data banks are only allowed with the prior permission of the publishers

4 ISOLATION/PREPARATION

Source organism

Pseudomonas stutzeri [1]; Pseudomonas aeruginosa [2, 3, 13]; Pseudomonas denitrificans [5]; Thiobacillus thiooxidans [6]; Saccharomyces cerevisiae (aerobically grown) [4, 7–9, 11, 12, 14–18]; Saccharomyces carlsbergensis (aerobically grown) [7, 10]

Source tissue

Cells (acetone dried cells [2, 5]) [1–3, 5, 6]

Localisation in source

Mitochondria [7, 10]

Purification

Pseudomonas stutzeri [1]; Pseudomonas aeruginosa [3]; Saccharomyces cerevisiae (affinity chromatography) [7, 9, 15] Pseudomonas denitrificans [5]; Thiobacillus thiooxidans [6]

Crystallization

(apo- and holoenzyme [7, 8, 12], crystal structure [18], structure of NO-inhibited enzyme [16], structure of fluoride-inhibited enzyme [17]) [7, 8, 12, 16–18]

Cloned

–

Renaturated

(reconstitution of holoenzyme) [12]

---

5 STABILITY

pH

2.0–10.0 [6]

Temperature (°C)

70 (15 min, 80% loss of peroxidizing activity, 50% loss of NADH oxidizing activity) [6]; 90 (inactivation) [6]

Oxidation

Organic solvent

General stability information

No dimerization after 7 years [4]; Unstable during degassing under vacuum except in presence of detergent [3]; Crystals stable in water-saturated atmosphere for more than 5 h at 23°C [8]

Storage

−20°C, 0.1 M phosphate buffer, pH 7.0, two months [6]; −20°C, 0.5 M phosphate buffer, pH 6 [9]

6 CROSSREFERENCES TO STRUCTURE DATABANKS

PIR/MIPS code

PIR2: S05001 (Pseudomonas aeruginosa); PIR1: OPBYC (precursor Yeast (Saccharomyces cerevisiae))

Brookhaven code

0CCI (Baker, S yeast (Saccharomyces cerevisiae)); 1CCP (Yeast (Saccharomyces cerevisiae) expressed in (Escherichia)

7 LITERATURE REFERENCES

[1] Villalain, J., Moura, I., Liu, M.C., Payne, W.J., LeGall, J., Xavier, A.V., Moura, J.J.G.: Eur. J. Biochem.,141,305–312 (1984)
[2] Ellfolk, N., Rönnberg, M., Aasa, R., Andreasson, L.-E. , Vänngard, T.: Biochim. Biophys. Acta,743,23–30 (1983)
[3] Foote, N., Thompson, A.C., Barber, D., Greenwood, C.: Biochem. J.,209,701–707 (1983)
[4] Takio, K., Yonetani, T.: Arch. Biochem. Biophys.,203,605–614 (1980)
[5] Coulson, A.F.W., Oliver, R.I.C.: Biochem. J.,181,159–169 (1979)
[6] Tano, T., Sakai, K., Sugio, T., Imai, K.: Agric. Biol. Chem.,41,323–330 (1977)
[7] Yonetani, T. in "The Enzymes",3rd. Ed. (Boyer, P.D., ed.) 13,345–361 (1976)
[8] Yonetani, T., Chance, B., Kajiwara, S.: J. Biol. Chem.,241,2981–2982 (1966)
[9] Yonetani, T., Ray, G.S.: J. Biol. Chem.,240,4503–4508 (1965)
[10] Yonetani, T., Ohnishi, T.: J. Biol. Chem.,241,2983–2984 (1966)
[11] Ellfork, N.: Acta Chem. Scand.,21,1921–1924 (1967)
[12] Yonetani, T.: J. Biol. Chem.,242,5008–5013 (1967)
[13] Soininen, R., Ellfolk, N.: Acta Chem. Scand.,27,2193–2198 (1973)
[14] Takio, K., Titani, K., Ericsson, L.H., Yonetani, T.: Arch. Biochem. Biophys.,203,615–629 (1980)
[15] Azzi, A., Bill, K., Broger, C.: Proc. Natl. Acad. Sci. USA,79,2447–2450 (1982)
[16] Edward, S.L., Kraut, J., Poulos, T.L.: Biochemistry,27,8074–8081 (1988)
[17] Edwards, S.L., Poulos, T.L., Kraut, J.: J. Biol. Chem.,259,12984–12988 (1984)
[18] Finzel, B.C., Poulos, T.L., Kraut, J.: J. Biol. Chem.,259,13027–13036 (1984)

Enzyme Handbook © Springer-Verlag Berlin Heidelberg 1994
Duplication, reproduction and storage in data banks are only allowed with the prior permission of the publishers

1 NOMENCLATURE

EC number
1.11.1.6

Systematic name
Hydrogen-peroxide:hydrogen-peroxide oxidoreductase

Recommended name
Catalase

Synonymes
Equilase
Caperase
Optidase
Catalase-Peroxidase
CAT [12]

CAS Reg. No.
9001-05-2

2 REACTION AND SPECIFICITY

Catalysed reaction
$2\ H_2O_2 \rightarrow$
$\rightarrow 2\ H_2O + O_2$ (catalatic activity)
$ROOH + AH_2 \rightarrow$
$\rightarrow H_2O + ROH + A$ (peroxidatic activity) [4, 9]

Reaction type
Redox reaction

Natural substrates
H_2O_2 (regulation of H_2O_2 levels, protective function for hemoglobin and other SH-containing proteins) [6]

Enzyme Handbook © Springer-Verlag Berlin Heidelberg 1994
Duplication, reproduction and storage in data banks are only allowed with the prior permission of the publishers

Substrate spectrum

1 H_2O_2
2 Methanol + H_2O_2 (in presence of glucose and glucose oxidase) [3]
3 Methylhydrogen peroxide + H_2O_2 [7]
4 Guaiacol + H_2O_2 (not [1, 11, 13, 14]) [8]
5 Pyrogallol + H_2O_2 [1, 8]
6 o-Dianisidine + H_2O_2 [1]
7 Diaminobenzidine + H_2O_2 [1]
8 Catechol + H_2O_2 [8]
9 beta-(3,4-Dihydroxyphenyl)-L-alanine + H_2O_2 [3, 8]
10 More (not: ascorbate [1], L-dopa [11]) [1, 9, 11]

Product spectrum

1 $H_2O + O_2$
2 Formaldehyde + ? [3]
3 ?
4 ?
5 ?
6 ?
7 ?
8 ?
9 ?
10 ?

Inhibitor(s)

3-Amino-1-H-1,2,4-triazole (not inhibited: E. coli, Rhodopseudomonas capsulata [2]) [2, 10, 12]; Cyanide [1, 6, 7, 12, 14]; Azide [1, 6, 7, 12, 14]; Hydroxylamine [1]; Cyanogenbromide [6]; Nitrite [14]; Fluoride [7]; beta-Mercaptoethanol [12]; Dithiothreitol [12]; Dianisidine (inhibition of catalatic activity) [16]; Nitrate [14]; H_2O_2 (2 mM, E. coli, Rhodopseudomonas capsulata) [2]

Cofactor(s)/prostethic group(s)

Ferric protoporphyrin IX (heme) [1, 2, 4, 7–9]

Metal compounds/salts

Fe (1 ferriheme per subunit) [13]

Turnover number (min^{-1})

980000 (H_2O_2) [16]

Specific activity (U/mg)

7800 [1]; 282000 [3]; 121000 [5]; More (overview assay methods [4, 6]) [4, 6, 8, 12–14, 17, 18]

K_m-value (mM)
1.93 (H_2O_2) [8]; 25 (H_2O_2) [3]; 83 (methanol, in presence of glucose and glucose oxidase) [3]; 3.7 (H_2O_2, hydroxyperoxidase I) [10]; 18.2 (H_2O_2, hydroxyperoxidase II, pH 6.8) [10]; 10 (H_2O_2, hydroxyperoxidase II, pH 10.5) [10]; 19.2 (H_2O_2) [13]

pH-optimum
5.5–10.0 (Rhodopseudomonas sphaeroides independent of pH) [2]; 6–6.5 (Chromatium vinosum, Rhodopseudomonas capsulata, E. coli) [2]; 6.0–9.0 [13]; 6.5 (sharp) [8]; 6.8 (hydroxyperoxidase I has one optimum, hydroxyperoxidase II has 2 optima, pH 6.8 and 10.5) [10]; 7.2 [3]; 7.5 [18]; 7–9 [12]; 10.5 (hydroxyperoxidase II has 2 optima, pH 6.8 and 10.5) [10]

pH-range
5–9 [3]; 5–10 [2]; 5–12 (isozymes CAT 1 and 3) [12]; 4–13 (isozyme CAT 2) [12]

Temperature optimum (°C)

Temperature range (°C)
40–67 (Micrococcus luteus, increase of activity up to 155% over range) [2]; 40–55 (Rhodospirillum rubrum, increase of activity up to 188% over range) [2]; 70 (isozyme hydroxyperoxidase II activated by elevated temperatures) [9]; More (E. coli, Rhodopseudomonas capsulata: no activation by elevated temperatures) [2]

---

3 ENZYME STRUCTURE

Molecular weight
385000 (Aspergillus niger, sedimentation equilibrium) [7]
337000 (E. coli, ultracentrifugation) [16]
288000–314000 (Neurospora crassa, sedimentation equilibrium, gel filtration) [14]
210000–240000 (Kloeckera sp., ultracentrifugation [3], Cucurbita sp., sucrose density centrifugation [5], Zea mays, calculation from subunit composition [12], Candida tropicalis, ultracentrifugal analysis [13], goat, gel filtration [17], Lens culinaris, zonal centrifugation [18]) [3, 5, 12, 13, 17, 18]
150000 (Comamonas compransoris, gel filtration, sucrose gradient, sedimentation equilibrium centrifugation) [8]

Enzyme Handbook © Springer-Verlag Berlin Heidelberg 1994
Duplication, reproduction and storage in data banks are only allowed with the prior permission of the publishers

Subunits

Tetramer (4 × 54000–62000, Kloeckera sp., SDS-PAGE [3], Cucurbita sp., SDS-PAGE [5], Zea mays, SDS-PAGE [12], Candida tropicalis, SDS-PAGE [13], goat, SDS-PAGE [17], Lens culinaris, SDS-PAGE [18], 4 × 80000–85000, Neurospora crassa, SDS-PAGE [14], 4 × 97000, Aspergillus niger, SDS-PAGE [7]) [3, 5, 7, 12–14, 17, 18]
Dimer (2 × 75000, Comamonas compransoris, SDS-PAGE) [8]

Glycoprotein/Lipoprotein

Glycoprotein (Aspergillus niger, 8.3% neutral sugar, 1.9% glucosamine) [7]

4 ISOLATION/PREPARATION

Source organism

Rhodopseudomonas capsulata [1, 2]; Comamonas compransoris [8]; E. coli [2, 10, 16, 22]; Rhodospirillum rubrum [2]; Micrococcus luteus [2]; Kloeckera sp. (strain 2201, a strain of Candida boidinii) [3]; Cucurbita sp. (pumpkin) [5]; Aspergillus niger [7]; Bovine [11, 23]; Candida tropicalis [13]; Neurospora crassa [14, 15]; Goat [17]; Lens culinaris [18]; Zea mays (corn) [12]; Cucumis sativus (cucumber) [19]; Sweet potato [20]; Saccharomyces cerevisiae [21]; More (widely distributed in nature, all aerobic micro-organism plant and animal cells) [6]

Source tissue

Cell [1]; Cotyledons [5, 19]; Liver [6, 11, 17, 23]; Kidney [6]; Leaves [18]; Roots [20]; Seeds [12]; Scrutella [12]; Epicotyl tissue [12]; Erythrocytes [6]; More (high activity in liver and kidney, low activity in connective tissues) [6]

Localisation in source

Glyoxysomes [4]; Peroxisomes [13, 18]; More (soluble in erythrocytes, particle-bound in connective tissues) [6]

Purification

Rhodopseudomonas capsulata [1]; Comamonas compransoris [8]; Aspergillus niger [7]; Kloeckera sp. [3]; Cucurbita sp. [5]; Candida tropicalis [13]; Neurospora crassa [14]; E. coli [16, 22]; Goat [17]; Lens culinaris [18]; Zea mays (3 isozymes) [12]; Bovine [23]; Cucumis sativus [19]; Sweet potato [20]

Crystallization

[3]

Cloned

–

Renaturated

–

5 STABILITY

pH

7–11 (isozyme CAT 2) [12]; 7–10 (isozyme CAT 1) [12]; 6–9 (isozyme CAT 3) [12]; 9 (optimal value) [18]

Temperature (°C)

45 (half-lives for isozymes: CAT 1 (12.5 min), CAT 3 (18 min), CAT 2 (stable)) [12]; 50 (half-life 10 min) [1]; 70 (hydroxperoxidase II, complete inactivation) [10]

Oxidation

Photooxidation inactivates (catalase-peroxidase) [1, 2]; H_2O_2 (above 0.1 M, inactivation) [4, 8]

Organic solvent

Ethanol (catalase stable, catalase-peroxidase not) [1, 2]; Chloroform (catalase stable, catalase-peroxidase not) [1, 2]

General stability information

Stabilization by ethanol, glycerol, serum albumin [18]; Slight stabilization by EDTA [18]; Decrease of stability in presence of NaCl, beta-mercaptoethanol, dithiothreitol [18]

Storage

4°C, pH 7.2, more than 6 months [3]; 4°C, 1 month or –196°C, indefinitely [14]

---

6 CROSSREFERENCES TO STRUCTURE DATABANKS

PIR/MIPS code

PIR3: S18972 (Arabidopsis thaliana); PIR2: JH0532 (Bacillus subtilis); PIR1: CSBO (Bovine); PIR3: S16231 (Castor bean (fragment)); PIR1: CSFF (Fruit fly (Drosophila melanogaster)); PIR2: S18346 (Garden pea); PIR2: A23646 (Human); PIR1: CSHU (Human (fragment)); PIR3: A40367 (Listeria seeligeri); PIR3: S18819 (Maize); PIR3: A36695 (Mouse); PIR2: A25001 (Penicillium sp.); PIR1: CSRT (Rat); PIR1: CSRZ (Rice); PIR1: CSSY (Soybean); PIR3: S07124 (Sweet potato); PIR3: S10770 (Upland cotton); PIR2: S23422 (Yeast (Hansenula polymorpha)); PIR2: JA0090 (chain 1 Maize); PIR2: S10395 (chain 1 Upland cotton); PIR2: JA0091 (chain 3 Maize); PIR1: CSECHP (HPI Escherichia coli); PIR1: CSEBHT (HPI Salmonella typhimurium); PIR3: A39129 (HPII Escherichia coli); PIR2: JS0520 (I Bacillus stearothermophilus); PIR1: CSBYT (T Yeast (Saccharomyces cerevisiae)); PIR1: CSCKPT (peroxisomal Imperfect fungus (Candida tropicalis)); PIR1: CSBYP (peroxisomal Yeast (Saccharomyces cerevisiae))

Enzyme Handbook © Springer-Verlag Berlin Heidelberg 1994
Duplication, reproduction and storage in data banks are only allowed with the prior permission of the publishers

Brookhaven code

4CAT (Penicillium vitale); 7CAT (Beef (Bos taurus) liver); 8CAT (Beef (Bos taurus) liver)

7 LITERATURE REFERENCES

[1] Hochman, A., Shemesh, A.: J. Biol. Chem.,262,6871–6876 (1987)
[2] Nadler, V., Goldberg, I., Hochman, A.: Biochim. Biophys. Acta,882,234–241 (1986)
[3] Mozaffar, S., Ueda, M., Kitatsuji, K., Shimizu, S., Osumi, M., Tanaka, A.: Eur. J. Biochem.,155,527–531 (1986)
[4] Aebi, H.: Methods Enzymol.,105,121–126 (1984)
[5] Yamaguchi, J., Nishimura, M.: Plant Physiol.,74,261–267 (1984)
[6] Aebi, H.E. in "Methods Enzym. Anal.",3rd Ed. (Bergmeyer, H.U., ed.) 3,273–286 (1983)
[7] Kikuchi-Torii, K., Hayashi, S., Nakamoto, H., Nakamura, S.: J. Biochem.,92,1449–1456 (1982)
[8] Nies, D., Schlegel, H.G.: J. Gen. Appl. Microbiol.,28 ,311–319 (1982)
[9] Schonbaum, G.R., Chance, B. in "The Enzymes",3rd Ed. (Boyer, PD., ed.) 13,363–408 (1976) (Review)
[10] Meir, E., Yagil, E.: Curr. Microbiol.,12,315–320 (1985)
[11] Sichak, S.P., Dounce, A.L.: Arch. Biochem. Biophys.,249,286–295 (1986)
[12] Chandlee, J.M., Tsaftaris, A.S., Scandalios, J.G.: Plant Sci. Lett.,29,117–131 (1983)
[13] Yamada, T., Tanaka, A., Fukui, S.: Eur. J. Biochem.,125,517–521 (1982)
[14] Jacob, G.S., Orme-Johnson, W.H.: Biochemistry,18,2967–2975 (1979)
[15] Jacob, G.S., Orme-Johson, W.H.: Biochemistry,18,2975–2980 (1979)
[16] Claiborne, A., Fridovich, I.: J. Biol. Chem.,254,4245–4252 (1979)
[17] Miyahara, T., Takeda, A., Hachimori, A., Samejima, T.: J. Biochem.,84,1267–1276 (1978)
[18] Schiefer, S., Teifel, W., Kindl, H.: Hoppe-Seyler's Z. Physiol. Chem.,357,163–175 (1976)
[19] Lamb, J.E., Riezman, H., Becker, W.M.: Plant Physiol.,62,754–760 (1978)
[20] Esaka, M., Asahi, T.: Plant Cell Physiol.,23,315–322 (1982)
[21] Seah, T.C.M., Kaplan, J.G.: J. Biol. Chem.,248,2889–2893 (1973)
[22] Claiborne, A., Malinowski, D.P., Fridovich, I.: J. Biol. Chem.,254,11664–11668 (1978)
[23] Shirakawa, M.: J. Agric. Chem. Soc. Jpn.,23,361–362 (1950)

1 NOMENCLATURE

EC number
1.11.1.7

Systematic name
Donor:hydrogen-peroxide oxidoreductase

Recommended name
Peroxidase

Synonymes
Myeloperoxidase
Lactoperoxidase
Verdoperoxidase
Guaiacol peroxidase
Thiocyanate peroxidase
Eosinophil peroxidase
Japanese radish peroxidase [8]
Horseradish peroxidase (HRP) [6]
Extensin peroxidase
Heme peroxidase
MPO
Oxyperoxidase
Protoheme peroxidase
Pyrocatechol peroxidase
Scopoletin peroxidase

CAS Reg. No.
9003-99-0

2 REACTION AND SPECIFICITY

Catalysed reaction
Donor + H_2O_2 →
→ oxidized donor + 2 H_2O (reaction intermediates [3, 6])

Reaction type
Redox reaction

Natural substrates
More (role in biosynthesis of thyroxine [3])

Enzyme Handbook © Springer-Verlag Berlin Heidelberg 1994
Duplication, reproduction and storage in data banks are only allowed with the prior permission of the publishers

Substrate spectrum

1 H_2O_2 + phenol [1, 2]
2 H_2O_2 + 4-chlorophenol [1]
3 H_2O_2 + 2,4-dichlorophenol [1]
4 H_2O_2 + pyrocatechol [1]
5 H_2O_2 + 2,4-dibromophenol [1]
6 H_2O_2 + 2,4,6-trichlorophenol [1]
7 H_2O_2 + 4-hydroxybenzoate [1]
8 H_2O_2 + 2-cresol [2]
9 H_2O_2 + di-o-anisidine [9, 15, 17]
10 H_2O_2 + guaiacol [2, 4, 9, 17]
11 H_2O_2 + 2,2'-azino-di[3-ethylbenzthiazolin-(6)-sulfonic acid] [2, 9, 17]
12 H_2O_2 + 4-aminophenazone [17]
13 H_2O_2 + pyrogallol [2, 17]
14 H_2O_2 + 2-toluidine [2]
15 H_2O_2 + homovanillinic acid [2]
16 H_2O_2 + hydroquinone [2]
17 H_2O_2 + 1,2-phenylenediamine [2]
18 H_2O_2 + 1,4-phenylenediamine [2]
19 H_2O_2 + leukomalachite green [2]
20 H_2O_2 + reduced dichlorophenolindophenol [2]
21 H_2O_2 + 4,4'-diaminodiphenylamine [2]
22 H_2O_2 + propionylpromazine [2]
23 H_2O_2 + benzidine [2]
24 H_2O_2 + o-toluidine [2]
25 H_2O_2 + hydroxycinnamic acid [15]
26 H_2O_2 + sphagnum acid (i.e. p-hydroxy-(beta-carboxymethyl)-cinnamic acid) [15]
27 More (halogenation [3], oxidation of iodide by lactoperoxidase [3], relatively specific for H_2O_2, only acetyl-, methyl- and ethyl-hydroperoxide can substitute [2], N- and O-demethylations [21]) [2, 3, 21]

Product spectrum

1 ?
2 ?
3 ?
4 ?
5 ?
6 ?
7 ?
8 ?
9 ?
10 ? (more than 1 product formed [2, 9])
11 ? (formation of radical cation or di-cation) [2]

12 ?
13 ?
14 ?
15 ?
16 ?
17 ?
18 ?
19 ?
20 ?
21 ?
22 ?
23 ?
24 ?
25 ? (degradation to non-phenolic compounds) [15]
26 ?
27 ?

Inhibitor(s)

Azide [15, 17]; Cyanide [15, 17]; Benzhydroxamic acid [7]; Dithionite [17]; Hydroxylamine [17]; Diethyldithiocarbamate [17]; Borohydride [17]; Iodoacetamide [17]; 1,10-Phenanthroline [17]

Cofactor(s)/prostethic group(s)

Heme (non-covalently bound to smaller subunit [19], 1 mol protoheme IX per mol [1, 3], myeloperoxidase: 2 mol heme per mol [3], no protoporphyrin IX [11]) [1, 3, 11, 19]

Metal compounds/salts

Fe (hemoprotein)

Turnover number (min^{-1})

Specific activity (U/mg)

290 [1]; 262.1 [4]; More (assay method [2, 9]) [2, 7, 9, 11, 13]

K_m-value (mM)

0.052 (H_2O_2, + guaiacol) [4]; 44 (guaiacol) [4]; 0.004 (H_2O_2, + KI) [4]; 181.7 (KI) [4]; 0.21 (2,2'-azino-di-[3-ethylbenzothiazoline-(6)-sulfonic acid]) [17]; 0.48 (o-dianisidine) [17]; 83.33 (4-aminophenazone) [17]; 10.50 (guaiacol) [17]; 23.80 (pyrogallol) [17]; More (comparison of values for various isozymes and phenolic compounds) [16]

pH-optimum

4.5 [17]; 5.0 [15]; 5.5–7.5 (chlorinating activity, pH-optimum is shifted to higher values in presence of ascorbic acid) [12]; 6.0 [1]; More (comparison of values for various isozymes and substrates) [16]

Enzyme Handbook © Springer-Verlag Berlin Heidelberg 1994
Duplication, reproduction and storage in data banks are only allowed with the prior permission of the publishers

pH-range

Temperature optimum (°C)
40 [1]; 52 (increase of activity up to) [15]

Temperature range (°C)

3 ENZYME STRUCTURE

Molecular weight
120000–150000 (human myeloperoxidase, SDS-PAGE) [19]
85000 (bovine non-heme lactoperoxidase, sucrose density centrifugation) [11]
104000 (mammalian myeloperoxidase) [3]
53000 (Nicotiana tabacum, isozyme Af, gel filtration) [16]
49000–51000 (pig, gel filtration, SDS-PAGE) [7]
43000–45000 (rat, gel filtration, SDS-PAGE) [7]
41000 (Arthromyces ramosus, sedimentation equilibrium) [1]
40000 (horseradish) [3]
28000 (Nicotiana tabacum, isozyme Cn, gel filtration) [16]

Subunits
Monomer (1 × 41000, Arthromyces ramosus, SDS-PAGE [1], 1 × 43000–51000, mammalia, SDS-PAGE [7], 1 × 85000, bovine, non-heme lactoperoxidae [11], 1 × 30000, Nicotiana tabacum, isozyme Cn, 1 × 54000, Nicotiana tabacum, isozyme Af, SDS-PAGE [16]) [1, 7, 11, 16]
Tetramer (2 × 57000, carbohydrate-rich subunit + 2 × 15000, heme-containing subunit, human myeloperoxidase, SDS-PAGE) [19]
? (x × 57000 + x × 15000, rat myeloperoxidase, SDS-PAGE) [4]
More (overview myeloperoxidase [3, 19], lactoperoxidase [3]) [3, 19]

Glycoprotein/Lipoprotein
Glycoprotein (5% carbohydrate [1], horseradish peroxidase: 18% carbohydrate [3], myeloperoxidase: no glycoprotein [3]) [1, 3, 11, 16, 17]

4 ISOLATION/PREPARATION

Source organism
Arthromyces ramosus [1]; Horseradish [3, 6, 14, 18, 22]; Bovine (cow) [3, 11]; Rat (myeloperoxidase) [4, 7]; Human (myeloperoxidase [5, 12, 17], eosinophil peroxidase [5]) [5, 12, 17, 19]; Arachis hypogaea (peanut) [13, 20] Raphanus sativus (japanese radish) [8]; Pig [7]; [13]; Sphagnum magellani [15]; Nicotiana tabacum [16]; More (overview animals, higher plants [2, 10], overview mammals, plants, yeast, fungi, bacteria [9], overview myeloperoxidase [19]) [2, 9, 10, 19]

Source tissue

Leukocytes [2, 19]; Milk [2, 3, 11]; Liver [2]; Spleen [2]; Salivary glands [2]; Gutwall [2]; Lung [2]; Bone marrow [4]; Blood [5]; Intestine [7]; Neutrophils [12]; Cell culture [13, 16]; Cervical mucus [17]; Roots [18]; Commercial preparation [14]; More (overview) [2, 9]

Localisation in source

Extracellular [1]; Particle-bound [7]; Lysosomes [19]

Purification

Arthromyces ramosus [1]; Human (eosinophil peroxidase, myeloperoxidase [5]) [5, 17]; Rat [4, 7]; Bovine (non-heme lactoperoxidase) [11]; Arachis hypogaea [13]; Horseradish (5 isozymes from commercial preparation) [14]; Nicotiana tabacum (2 isozymes) [16]; Sphagnum magellani (partial) [15]

Crystallization

[1, 3, 10]

Cloned

–

Renaturated

(regeneration after phenol inactivation) [8]

---

5 STABILITY

pH

5.0–9.0 [1, 8]

Temperature (°C)

50 (30 min stable) [1]; 52 (inactivation above) [15]; 60 (inactivation above) [8]; 65 (up to) [5]; 75 (5 min, 10% loss of activity) [5]

Oxidation

Organic solvent

General stability information

Guanidium chloride, above 3.0 M, inactivation [5]; Urea, up to 6.5 M, stable [5]; No unfolding of protein in 10 mM urea or 5 mM guanidium chloride [19]

Storage

–20°C, 0.1 M sodium acetate buffer, pH 5.6, 0.5 M $CaCl_2$, 0.05% w/v cetyltriammonium bromide, 75% w/v glycerol, or 0.1 M phosphate buffer, pH 7.5, several months [5]; In liquid N_2 [15]

Enzyme Handbook © Springer-Verlag Berlin Heidelberg 1994
Duplication, reproduction and storage in data banks are only allowed with the prior permission of the publishers

6 CROSSREFERENCES TO STRUCTURE DATABANKS

PIR/MIPS code

PIR3: JN0092 (precursor Basidiomycete (Phanerochaete chrysosporium)); PIR3: A32457 (heavy chain Bovine (fragment)); PIR3: B32457 (light chain Bovine (fragment)); PIR1: OPHUM (precursor Human); PIR2: S18064 (Barley (fragment)); PIR3: A39889 (Common tobacco); PIR3: S16332 (Cucumber); PIR2: S11870 (Cucumber (fragment)); PIR3: S25522 (Fruit fly (Drosophila melanogaster)); PIR1: OPRHC (Horseradish); PIR3: A34650 (Human (fragment)); PIR3: S26066 (Human (fragment)); PIR3: S21746 (Inky cap fungus (Coprinus cinereus)); PIR2: S26672 (Large-leaved lupine (fragment)); PIR1: OPNB7 (Turnip); PIR2: S04862 (Wheat (fragment)); PIR2: S04763 (1 precursor Tomato); PIR2: PH0218 (2 Rice (fragments)); PIR3: PH0222 (2 and/or 4 Rice (fragment)); PIR2: S04764 (2 precursor Tomato); PIR2: S08221 (2 thyroid Human); PIR2: PH0219 (4 Rice (fragments)); PIR2: PH0220 (6 Rice (fragment)); PIR2: PH0221 (7 Rice (fragment)); PIR3: JC1249 (BP-2A precursor Barley); PIR2: S22505 (BP1 precursor Barley); PIR2: JU0457 (C Arabidopsis thaliana); PIR2: S00625 (C1 precursor Horseradish); PIR2: JH0149 (C2 precursor Horseradish); PIR2: JH0150 (C3 precursor Horseradish); PIR2: S00627 (C3 precursor Horseradish (fragment)); PIR2: JU0458 (E Arabidopsis thaliana); PIR2: A32630 (H4 precursor manganese-dependent Basidiomycete (Phanerochaete chrysosporium)); PIR2: S08200 (lignin 1 Basidiomycete (Phanerochaete chrysosporium) (fragment)); PIR2: S08201 (lignin 2 Basidiomycete (Phanerochaete chrysosporium) (fragment)); PIR2: S08202 (lignin 3 Basidiomycete (Phanerochaete chrysosporium) (fragment)); PIR2: S08203 (lignin 4 Basidiomycete (Phanerochaete chrysosporium) (fragment)); PIR2: S08204 (lignin 5 Basidiomycete (Phanerochaete chrysosporium) (fragment)); PIR3: A33271 (MnP-1 Basidiomycete (Phanerochaete chrysosporium)); PIR3: A35828 (precursor Bovine); PIR2: A29467 (precursor Human); PIR2: S06068 (precursor Mouse); PIR2: S07407 (precursor Potato (fragment)); PIR3: S22087 (precursor Rice); PIR2: S13325 (precursor Wheat); PIR3: A38265 (precursor cationic (clone PNC1) Peanut); PIR3: B38265 (precursor cationic (clone PNC2) Peanut); PIR2: S04746 (precursor eosinophil Human (fragment)); PIR2: S13375 (precursor pathogen-induced Wheat); PIR3: A34408 (eosinophil Human); PIR3: S14268 (neutral Horseradish); PIR3: A35415 (thyroid (Grave's disease) Human (fragment)); PIR2: S07047 (thyroid Rat)

Brookhaven code

7 LITERATURE REFERENCES

[1] Shinmen, Y., Asami, J., Amachi, T., Shimizu, S., Yamada, H.: Agric. Biol. Chem.,50,247–249 (1986)
[2] Pütter, J., Becker, R. in "Methods Enzym. Anal.",3,286–293 (1983)
[3] Hewson, W.D., Hager, L.P. in "The Porphyrins" (Dolphin, D., ed.) 7,295–332, Academic Press, N.Y. (1979) (Review)
[4] Kariya, K., Lee, E., Hirouchi, M., Hosokawa, M., Sayo, H.: Biochim. Biophys. Acta,911,95–101 (1987)
[5] Olsen, R.L., Little, C.: Biochem. J.,209,781–787 (1983)
[6] Shahangian, S., Hager, L.P.: J. Biol. Chem.,257,11529–11533 (1982)
[7] Kimura, S., Jellinck, P.H.: Biochem. J.,205,271–279 (1982)
[8] Tamura, Y., Morita, Y.: J. Biochem.,78,561–571 (1975)
[9] Pütter, J. in "Methods Enzym. Anal.",1,725–731 (1974)
[10] Paul, K.G. in "The Enzymes",2nd. Ed. (Boyer, P.D., Lardy, H., Myrbäck, K., eds.) 8,227–274 (1963) (Review)
[11] Dumentet, C., Rousset, B.: J. Biol. Chem.,258,14166–14172 (1983)
[12] Bolscher, B.G.J.M., Zoutberg, G.R., Cuperus, R.A., Wever, R.: Biochim. Biophys. Acta,784,189–191 (1984)
[13] Chibbar, R.N., van Huystee, R.B.: Plant Physiol.,75,956–958 (1984)
[14] Aibara, S., Yamashita, H., Mori, E., Kato, M., Morita, Y.: J. Biochem.,92,531–539 (1982)
[15] Tutschek, R.: Phytochemistry,18,1437–1439 (1979)
[16] Kim, S.S., Wender, S.H., Smith, E.C.: Phytochemistry,19,165–168 (1980)
[17] Shindler, J.S., Childs, R.E., Bardsley, W.G.: Eur. J. Biochem.,65,325–331 (1976)
[18] Clarke, J., Shannon, L.M.: Biochim. Biophys. Acta,427,428–442 (1976)
[19] Olsen, R.L., Little, C.: Biochem. J.,222,701–709 (1984)
[20] Van Huystee, R.B. in "Peroxidases Chem. Biol." (Everse, J., Everse, K.E., Grisham, M.B., eds.) 2,155–170, CRC, Boca Raton, Fla. (1991)
[21] Meunier, B. in "Peroxidases Chem. Biol." (Everse, J., Everse, K.E., Grisham, M.B., eds.) 2,201–217, CRC, Boca Raton Fla. (1991)
[22] Dunford, H.B. in "Peroxidases Chem. Biol." (Everse, J., Everse, K.E., Grisham, M.B., eds.) 2,1–24, CRC, Boca Raton Fla. (1991)

Enzyme Handbook © Springer-Verlag Berlin Heidelberg 1994
Duplication, reproduction and storage in data banks are only allowed with the prior permission of the publishers

1 NOMENCLATURE

EC number
1.11.1.8

Systematic name
Iodide:hydrogen-peroxide oxidoreductase

Recommended name
Iodide peroxidase

Synonymes
Iodoperoxidase (heme type)
Iodotyrosine deiodase
Iodinase
Thyroid peroxidase [1]
Iodide peroxidase-tyrosine iodinase [4]
Peroxidase, iodide
Iodotyrosine deiodinase
Monoiodotyrosine deiodinase
Thyroperoxidase
Tyrosine iodinase

CAS Reg. No.
9031-28-1

2 REACTION AND SPECIFICITY

Catalysed reaction
Iodide + H_2O_2 →
→ iodine + 2 H_2O (reaction intermediates [3])

Reaction type
Redox reaction
Iodination
Peroxidation

Natural substrates
Thyroglobulin + I^- + H_2O_2 [5]

Enzyme Handbook © Springer-Verlag Berlin Heidelberg 1994
Duplication, reproduction and storage in data banks are only allowed with the prior permission of the publishers

Substrate spectrum

1 Iodide + H_2O_2
2 Guaiacol + H_2O_2 [3]
3 Tyrosine + H_2O_2 + KI [4]
4 Thyroglobulin + H_2O_2 + KI [5]
5 More (iodination of tyrosyl moieties in proteins [3], insulin [5], bovine serum albumin [4]) [3–5]

Product spectrum

1 Iodine + H_2O
2 Tetraguaiacol [3]
3 Monoiodotyrosine + diiodotyrosine (ratio: 4:1) [4]
4 Thyroglobulin with 3-monoiodotyrosine, 3,5-diiodotyrosine + thyroxine [5]
5 ?

Inhibitor(s)

Cyanide [1, 3–5]; Aminotriazole [1]; Azide [1, 3–5]; Diiodotyrosine (inhibition or stimulation depending on concentration) [3]; Bisulfite [3]; Borohydrite [3]; p-Chloromercuribenzoate [3, 5]; H_2O_2 (above 100 micromol) [4]; F^- [4]; Thiourea [4, 5]; Thiouracil [4, 5]; Ascorbic acid [4]; Glutathione [4, 5]; Resorcinol [5]; p-Aminobenzoic acid [5]; 1-Methyl-2-mercaptoimidazole [5]; Phloroglucinol [5]; Sulfathiazole [5]

Cofactor(s)/prostethic group(s)

Heme [3, 5]; Protoheme IX [1]

Metal compounds/salts

Fe^{3+} (hemoprotein) [1, 3, 5]

Turnover number (min^{-1})

50000 (guaiacol oxidation) [3]; 32000 (guaiacol 3-idodide formation) [3]

Specific activity (U/mg)

381 (guaiacol) [2]; 138 (guaiacol) [3]; More [4]

K_m-value (mM)

0.1 (protein iodination); 0.003 (Glu-Tyr-Glu iodination); 6 (iodide oxidation) [3]

pH-optimum

7.0 (iodide oxidation) [7]; 7.4 (guaiacol oxidation) [7]; 4.0 [4]; 5–7 (no defined optimum) [5]

pH-range

3.5–4.8 (iodinase, less than half-maximal activity above and below) [4]

Temperature optimum (°C)

Temperature range (°C)

3 ENZYME STRUCTURE

Molecular weight

71000 (porcine thyroid gland, HPLC) [1]
92000–135000 (porcine thyroid gland, value depending on solubilization method) [3]
64000 (porcine thyroid gland, gel filtration) [5]

Subunits

Monomer or dimer (60000 + 24000, porcine thyroid gland, SDS-PAGE, monomer [7], dimer [6]) [3, 6, 7]
? (x × 10000, human, SDS-PAGE) [2]

Glycoprotein/Lipoprotein

Glycoprotein (10% carbohydrate) [2]

4 ISOLATION/PREPARATION

Source organism

Pig [1, 3, 5, 6, 8, 9]; Human [2]; Goat [4]

Source tissue

Thyroid gland [1–3, 5–9]; Submaxillary gland [4]

Localisation in source

Microsomes [2]; Membrane-bound (orientated towards luminal side of microsomal vesicles) [9]; Soluble part of cell [4]

Purification

Human [2]; Goat [4]; Pig (proteolytic and detergent solubilization methods [3]) [3, 5, 8]

Crystallization

–

Cloned

–

Renaturated

–

Enzyme Handbook © Springer-Verlag Berlin Heidelberg 1994
Duplication, reproduction and storage in data banks are only allowed with the prior permission of the publishers

5 STABILITY

pH

Temperature (°C)

Oxidation

Organic solvent

General stability information

KI: at 37°C, protection against inactivation only in partially purified preparations [7]

Storage

–20°C, 4 years [6]; Inactivation in 3–5 d [7]; –70°C, presence of trypsin: several months stable, presence of detergent: 1 month stable [3]; Liquid N_2, 10 mM Tris-HCl buffer, pH 7.5, 1 M KCl [1]

6 CROSSREFERENCES TO STRUCTURE DATABANKS

PIR/MIPS code

PIR1: OPHUIT (precursor thyroid Human); PIR1: OPPGIT (precursor thyroid Pig); PIR2: A41408 (thyroid Rat (fragment))

Brookhaven code

7 LITERATURE REFERENCES

[1] Ohtaki, S., Nakagawa, H., Nakamura, S., Nakamura, M., Yamazaki, I.: J. Biol. Chem.,260,441–448 (1985)
[2] Czarnocka, B., Ruf, J., Ferrand, M., Carayon, P., Lissitzky, S.: FEBS Lett.,190,147–152 (1985)
[3] Neary, J.T., Soodak, M., Maloof, F.: Methods Enzymol.,107,445–475 (1984)
[4] Mahajani, M., Haldar, I., Datta, A.G.: Eur. J. Biochem.,37,541–552 (1973)
[5] Coval, M.L., Taurog, A.: J. Biol. Chem.,242,5510–5523 (1967)
[6] Rawitch, A.B., Taurog, A., Chernoff, S.B., Dorris, M.L.: Arch. Biochem. Biophys.,194,244–257 (1979)
[7] Alexander, N.M.: Endocrinology,100,1610ff. (1977)
[8] Ohtaki, S., Nakagawa, H., Nakamura, M., Yamazaki, I.: J. Biol. Chem.,257,13398–13403 (1982)
[9] Nakagawa, H., Ohtaki, S.: J. Biochem.,94,155–162 (1983)

1 NOMENCLATURE

EC number
1.11.1.9

Systematic name
Glutathione:hydrogen-peroxide oxidoreductase

Recommended name
Glutathione peroxidase

Synonymes
GSH peroxidase [1]
Selenium-glutathione peroxidase [13]
Peroxidase, glutathione
Reduced glutathione peroxidase

CAS Reg. No.
9013-66-5

2 REACTION AND SPECIFICITY

Catalysed reaction
2 Glutathione + H_2O_2 →
→ oxidized glutathione + 2 H_2O (2 GSH + ROOH →
→ ROH + H_2O + GSSG [2], mechanism [25])

Reaction type
Redox reaction

Natural substrates
Glutathione + peroxide (detoxification of organic hydroperoxides [2, 3], discussion of biological role [1, 25]) [1–3, 25]

Enzyme Handbook © Springer-Verlag Berlin Heidelberg 1994
Duplication, reproduction and storage in data banks are only allowed with the prior permission of the publishers

Substrate spectrum

1 Glutathione + ROOH (highly specific for glutathione [9, 15, 17], dithiothreitol has 10% of glutathione activity [17], gamma-glutamyl-L-cysteine-methylester has 26% of glutathione activity [22], ROOH can be: aliphatic or aromatic peroxide or H_2O_2 [2], hydroperoxyarachidonate [1], cumene hydroperoxide [5, 6, 9, 17], tert.-butylhydroperoxide [5, 6], menthane hydroperoxide [6], diisopropyl hydroperoxide [6], ethyl hydroperoxide [6, 9], linoleic acid hydroperoxide [6, 9], pregnelone 17alpha-hydroperoxide [9], cholesterol 7beta-hydroperoxide [9], thymine hydroperoxide [9], peroxidized DNA [9], lauryl hydroperoxide [17], reduction of hydroperoxides increases with hydrophobicity of substrates [6], specificity for donor-substrate: free amino group near the SH group inhibits the reaction, free alpha-carboxylic group of glutamic acid residue in glutathione analogs increases reaction rate [22], broad specificity for hydroperoxides [23, 25]) [1–25]

Product spectrum

1 Oxidized glutathione + ROH + H_2O [2]

Inhibitor(s)

Mercaptosuccinate [3]; alpha-Mercaptopropionylglycine [3]; Penicillamine [3]; Gold(I)-thioglucose [3]; Ag^+ [9]; Iodoacetate (inhibitory only after preincubation with glutathione, not after preincubation with H_2O_2 [15]) [9, 15]; Chloroacetate [9]; CN^- (no inhibition if simultaneously incubated with 2-mercaptoethanol, glutathione or dithiothreitol) [14]; Iodoacetamide (inhibitory only after preincubation with glutathione, not after preincubation with H_2O_2) [15]; Adenosine (inhibition probably due to inhibition of glutathione reductase in coupled assay) [15]; NADPH (inhibition probably due to inhibition of glutathione reductase in coupled assay) [15]; Polyvalent anions (e.g. phosphate, sulfate, maleate) [25]; O_2^- [25]; More (not: ophthalmic acid) [25]

Cofactor(s)/prostethic group(s)

Selenocysteine [1–4, 11, 19, 20, 25]

Metal compounds/salts

Selenium (selenocysteine) [1–4, 11, 19, 20, 25]

Turnover number (min^{-1})

Specific activity (U/mg)

1407 [4]; 488 [6]; 1852 [8]; More (assay methods [9, 18, 25]) [9, 10, 12, 14–18, 25]

K_m-value (mM)
0.006 (cumene hydroperoxide) [5]; 0.024 (tert.-butylhydroperoxide) [5]; 1.33 (glutathione) [5]; 0.144 (cumene hydroperoxide) [10]; More [13]

pH-optimum
8.0 [6]; 8.5 [10]; 8.7 [24]; 8.8 [9]; 8.8–9.0 [17]

pH-range
7.5–8.5 [6]; 7.0 (less than 20% of maximal activity below) [17]

Temperature optimum (°C)
42 [9, 10]

Temperature range (°C)

3 ENZYME STRUCTURE

Molecular weight
100000 (rainbow trout, gel filtration) [7]
75000–85500 (rat, gel filtration [8, 12, 16, 17], sedimentation and diffusion constants, partial specific volume, sedimentation equilibrium centrifugation [12], bovine, gel filtration [9], sedimentation equilibrium centrifugation [21], human, gel filtration [14]) [8, 9, 12, 14, 16, 17, 21]

Subunits
Tetramer (4 × 21900, identical, bovine, nucleotide sequence [19, 25], SDS-PAGE [9, 21], 4 × 17000–20000, rat, SDS-PAGE [8, 12, 16, 17], amino acid composition [12], 4 × 22000, human, SDS-PAGE [14]) [8, 9, 12, 14, 16, 17, 19, 21, 25]
? (x × 23000, hamster, SDS-PAGE [6], x × 44000, rainbow trout, SDS-PAGE [7]) [6, 7]

Glycoprotein/Lipoprotein
–

4 ISOLATION/PREPARATION

Source organism
Mammals (not in guinea pig [2]) [1–3, 9, 23–25]; Sheep [4]; Human [5, 14]; Hamster [6]; Rainbow trout [7]; Rat [8, 10, 12, 13, 15–17]; Bovine [9, 11, 19–22]; Birds [25]; Reptiles [25]; Fish [25]; More (overview [25], not in plants [9]) [9, 25]

Enzyme Handbook © Springer-Verlag Berlin Heidelberg 1994
Duplication, reproduction and storage in data banks are only allowed with the prior permission of the publishers

Source tissue

All tissues so far examined [1–3, 5, 7, 25]; Blood [4, 9, 11, 20–22]; Liver [5–8, 10, 12, 13, 15, 16]; Placenta [14]; Lung [17]

Localisation in source

Cytosol (75% of activity [25]) [2, 8, 10, 16, 17, 25]; Mitochondrial matrix (25% of activity [25]) [2, 15, 25]; More (not in peroxisomes) [25]

Purification

Sheep (oxidized and reduced form) [4]; Hamster [6]; Rat [8, 15–17]; Bovine [9, 21]; Human [14]

Crystallization

(crystal structure [11, 20]) [9, 11, 20]

Cloned

–

Renaturated

–

5 STABILITY

pH

2 (complete inactivation but reactivation by neutralization) [3]; 5.0 (extremely unstable below) [14]; 7–10 [9]

Temperature (°C)

100 (complete inactivation) [3]

Oxidation

Autooxidation (reactivation by glutathione) [9]

Organic solvent

General stability information

Stable to H_2O_2, linoleic hydroperoxide, peroxidizing linolate, methylethylketone peroxide [3]; Stabilization by glutathione or dithiothreitol [12, 14]; Stabilization by 2-mercaptoethanol [14]

Storage

–20°C, 10 mM phosphate buffer, pH 7.0, 1 mM EDTA, 0.1 M NaCl, 50% glycerol [8]; 4°C, 50 mM potassium phosphate buffer, pH 7.2, 10% ethanol [4]; 4°C, precipitated in 2.5 M potassium phosphate [21]; 4°C, 1 month, less than 50% loss of activity [17]; 4°C, more than 4 months [14]; Frozen, 1 mM dithiothreitol, 3 months stable, or 2 mM glutathione, 1 month, 15–22% loss of activity [12]

6 CROSSREFERENCES TO STRUCTURE DATABANKS

PIR/MIPS code

PIR1: OPBOE (Bovine); PIR1: OPHUE (Human); PIR1: OPMSE (Mouse); PIR2: S03723 (Rabbit); PIR3: B40464 (homolog Rat); PIR1: OPRTE (I Rat); PIR2: JX0176 (precursor Rat); PIR3: S18912 (precursor epididymal Crab-eating macaque); PIR3: S18913 (precursor epididymal Rat); PIR3: JQ0476 (plasma Human); PIR2: S15877 (plasma Human (fragments))

Brookhaven code

1GP1 (Bovine (Bos taurus) erythrocyte)

7 LITERATURE REFERENCES

[1] Flohe, L.: Curr. Top. Cell. Regul.,27,473–478 (1985) (Review)
[2] Mannervik, B.: Methods Enzymol.,113,490–495 (1985) (Review)
[3] Tappel, A.L.: Curr. Top. Cell. Regul.,24,87–96 (1984) (Review)
[4] Ganther, H.E., Kraus, R.J.: Methods Enzymol.,107,593–602 (1984)
[5] Charmagnol, F., Sinet, P.M., Jerome, H.: Biochim. Biophys. Acta,759,49–57 (1983)
[6] Chaudiere, J., Tappel, A.L.: Arch. Biochem. Biophys.,226,448–457 (1983)
[7] Bell, J.G., Cowey, C.B.: Biochem. Soc. Trans.,10,451 (1982)
[8] Yoshida, M., Iwami, K., Yasumoto, K.: Agric. Biol. Chem.,46,41–46 (1982)
[9] Wendel, A.: Methods Enzymol.,77,325–333 (1981) (Review)
[10] Yoshimura, S., Komatsu, N., Watanabe, K.: Biochim. Biophys. Acta,621,130–137 (1980)
[11] Ladenstein, R., Epp, O., Bartels, K., Jones, A., Huber, R.: J. Mol. Biol.,134,199–218 (1979)
[12] Nakamura, W., Hosoda, S., Hayashi, K.: Biochim. Biophys. Acta,358,251–261 (1974)
[13] Splittgerber, A.G., Tappel, A.L.: J. Biol. Chem.,254 ,9807–9813 (1979)
[14] Awasthi, Y.C., Dao, D.D., Lal, A.K., Srivastava, S.K. : Biochem. J.,177,471–476 (1979)
[15] Zakowski, J.J., Tappel, A.L.: Biochim. Biophys. Acta,526,65–76 (1978)
[16] Stults, F.H., Forstrom, J.W., Chiu, D.T.Y., Tappel, A.L.: Arch. Biochem. Biophys.,183,490–497 (1977)
[17] Chiu, D.T.Y., Stults, F.H., Tappel, A.L: Biochim. Biophys. Acta,445,558–566 (1976)
[18] Flohe, L., Günzler, W.A.: Methods Enzymol.,105,114–121 (1984)
[19] Günzler, W.A., Steffens, G.J., Grossmann, A., Kim, S.-M.A., Ötting, F., Wendel, A., Flohe, L.: Hoppe-Seyler's Z. Physiol. Chem.,365,195–212 (1984)
[20] Epp, O., Ladenstein, R., Wendel, A.: Eur. J. Biochem.,133,51–69 (1983)
[21] Flohe, L., Eisele, B., Wendel, A.: Hoppe-Seyler's Z. Physiol. Chem.,352,151–158 (1971)
[22] Flohe, L., Günzler, W., Jung, G., Schaich, E., Schneider, F.: Hoppe-Seyler's Z. Physiol. Chem.,352,159–169 (1971)
[23] Flohe, L. in "Free Radicals in Biology" (Pryor, W.A., ed.) 5,223–254, Academic Press, N.Y. (1982) (Review)
[24] Flohe, L., Loschen, G., Günzler, W.A. Eichele, E.: Hoppe-Seyler's Z. Physiol. Chem.,353,987–999 (1972)
[25] Flohe, L. in "Glutathione, Chemical, Biochemical, and Medical Aspects", Part A (Dolphin, D., Poulson, R., Avramovic, O., eds.) pp643–731, John Wiley & Sons (1989) (Review)

Enzyme Handbook © Springer-Verlag Berlin Heidelberg 1994
Duplication, reproduction and storage in data banks are only allowed with the prior permission of the publishers

1 NOMENCLATURE

EC number
1.11.1.10

Systematic name
Chloride:hydrogen-peroxide oxidoreductase

Recommended name
Chloride peroxidase

Synonymes
Chloroperoxidase
Peroxidase, chloride

CAS Reg. No.
9055-20-3

2 REACTION AND SPECIFICITY

Catalysed reaction
$RH + X^- + H_2O_2 + H^+ \rightarrow$
$\rightarrow RX + 2\,H_2O$ (halogenation, R: nucleophile, e.g. beta-diketone, beta-ketoacid, X^-: chloride, bromide, iodide, but not fluoride) [8]
$BH_2 + H_2O_2 \rightarrow$
$\rightarrow B + 2\,H_2O$ (peroxidation, B: o-dianisidine, pyrogallol, guaiacol) [8];
$2\,H_2O_2 \rightarrow$
$\rightarrow 2\,H_2O + O_2$ (catalase) [8]

Reaction type
Redox reaction
Peroxidation
Halogenation

Natural substrates
More (involvement in biosynthesis of caldariomycin) [9]

Enzyme Handbook © Springer-Verlag Berlin Heidelberg 1994
Duplication, reproduction and storage in data banks are only allowed with the prior permission of the publishers

Substrate spectrum

1 trans-Cinnamic acid + H_2O_2 + Br^- + H^+ [18]
2 trans-4-Methoxy-cinnamic acid + Br^- + H^+ H_2O_2 [18]
3 trans-4-Hydroxycinnamic acid + Br^- + H^+ + H_2O_2 (Br^- can be substituted by Cl^-) [18]
4 trans-3,4-Dimethoxycinnamic acid + Br^- + H_2O_2 + H^+ (Br^- can be substituted by Cl^-) [18]
5 Cytosine + H_2O_2 + Cl^- + H^+ (Cl^- can be substituted by Br^-) [1]
6 Uracil + H_2O_2 + Cl^- + H^+ (Cl^- can be substituted by Br^-, I^-) [1]
7 Thymine + H_2O_2 + Cl^- + H^+ (Cl^- can be substituted by Br^-) [1]
8 Cytidine + H_2O_2 + Cl^- + H^+ [1]
9 2'-Deoxyuridine + H_2O_2 + Br^- + H^+ [1]
10 Guanosine + H_2O_2 + Br^- + H^+ [1]
11 Adenine + H_2O_2 + Cl^- + H^+ [1]
12 Adenosine + H_2O_2 + Cl^- + H^+ [1]
13 Monochlorodimedon + H_2O_2 + Cl^- [10]
14 Tyrosine + H_2O_2 + I^- (or Br^-) [10]
15 2-exo-Methylbicyclo[2.2.1]hept-2-endo-carboxylic acid + H_2O_2 + Br^- [19]
16 2-Methyl-4-propyl-1,3-cyclopentanedione + H_2O_2 + Cl^- [19]
17 Bicyclo[3.2.0]hept-2-en-6-one + H_2O_2 + Br^- (racemic) [19]
18 More (no reaction with nonsubstituted pyrimidine, purine and pyrazole [1], overview and stereoselectivity [19], phenols and substituted phenols [12], chlorite and chlorine dioxide [6]) [1, 6, 12, 19]

Product spectrum

1 trans-1-Bromo-2-phenylethylene + erythro-2-bromo-3-hydroxy-3-phenylpropionic acid + H_2O [18]
2 2,3-Dihydroxy-3-(4-methoxyphenyl)ethane + DL-1,1-dibromo-2-hydroxy-2-(4-hydroxyphenyl)ethane + H_2O [18]
3 trans-1-Bromo-2-(4-hydroxyphenyl)ethylene + H_2O (or corresponding chlorinated product) [18]
4 DL-1,1-Dibromo-2-hydroxy-2-(3,4-dimethoxy-5-bromophenyl)ethane + DL-1,1-dibromo-2-hydroxy-2-(3,4-dimethoxyphenyl)ethane + 2-bromo-3-hydroxy-3-(3,4-dimethoxyphenyl)propionic acid + H_2O (with Cl^-: trans-1-chloro-2-(3,4-dimethoxy-5-chlorophenyl)ethylene + trans-1-chloro-2-(3,4-dimethoxyphenyl)ethylene + DL-1,1-dichloro-2-hydroxy-2-(3,4-dimethoxyphenyl)ethane) [18]
5 5-Chlorocytosine + H_2O [1]
6 5-Chlorouracil + H_2O [1]
7 5-Bromo-6-hydroxy-5,6-dihydrothymine + H_2O [1]
8 5-Bromocytidine + H_2O [1]
9 5-Bromo-2'-deoxyuridine + H_2O [1]

10 8-Bromoguanosine + H_2O [1]
11 ? (unstable product which gradually reverts to the substrate in a few h) [1]
12 ? (unknown product which gradually reverts to the substrate in a few h) [1]
13 1,1-Dimethyl-4,4-dichloro-3,5-cyclohexanedione + H_2O [10]
14 Monoiodotyrosine + H_2O (or monobromotyrosine + dibromotyrosine) [10]
15 ? (a delta-lactone) [19]
16 2-Chloro-2-methyl-4-propyl-1,3-cyclopentanedione (racemic diastereomers) [19]
17 2-exo-Bromo-3-endo-hydroxybromohydrin (racemic) [19]
18 ?

Inhibitor(s)
F^- [8, 10]; CN^- [8]; N_3^- [8]; Thiourea [11]; 1,1,3-Tricyano-1-amino-1-propene [11]; Thiocyanate [11]; CO [14]

Cofactor(s)/prostethic group(s)
Ferri-protoporphyrin IX (His close to heme iron [2], spectral study [4], EXAFS study [17]) [2, 4, 8, 9, 17]

Metal compounds/salts
Fe^{3+} (ferri-protoporphyrin IX) [4, 8]

Turnover number (min^{-1})
67000 (formation of dichlorodimedon from monochlorodimedon + Cl^- + H_2O_2) [10]; 7500 (chlorination of tyrosine) [10]

Specific activity (U/mg)
26000 [7]; More [1, 8, 20]

K_m-value (mM)
10 (chlorite) [6]; More (kinetic analysis) [13, 14]

pH-optimum
2.8 (decomposition of chlorite and chlorine dioxide [6], halide-dependent decomposition of H_2O_2 [8]) [6, 8]; 3.0 (halogenation [8]) [3, 8]; 4–7 (peroxidase) [8]; 4.8 (decomposition of H_2O_2) [8]

pH-range
2.6–3.6 [3]; 2.6–7 [8]

Temperature optimum (°C)

Temperature range (°C)

Enzyme Handbook © Springer-Verlag Berlin Heidelberg 1994
Duplication, reproduction and storage in data banks are only allowed with the prior permission of the publishers

3 ENZYME STRUCTURE

Molecular weight

32974 (Caldariomyces fumago, sequence of cDNA) [16]
42000 (Caldariomyces fumago, sedimentation equilibrium centrifugation, calculation from heme content [8], sedimentation equilibrium centrifugation [9]) [8, 9]
46000 (Caldariomyces fumago, gel filtration) [7]

Subunits

Glycoprotein/Lipoprotein

Glycoprotein (7–10% carbohydrate [8], 25–30% carbohydrate [9]) [3, 8, 9]

4 ISOLATION/PREPARATION

Source organism

Caldariomyces fumago [1–20]

Source tissue

Mycelium [1, 18]; Culture medium [7, 9]

Localisation in source

Extracellular [3, 20]

Purification

Caldariomyces fumago (2 isozymes [20]) [1, 7–9, 20]

Crystallization

(2 isozymes [7]) [7, 9]

Cloned

–

Renaturated

–

5 STABILITY

pH

3–6.5 [8]; 1.5 (presence of 0.1 M F^-, several h) [8]; 7 (4°C, 2 h, complete inactivation) [20]

Temperature (°C)

Oxidation

Organic solvent

General stability information

Storage

–20°C, 20 mM potassium phosphate buffer, pH 4.0 [18]

6 CROSSREFERENCES TO STRUCTURE DATABANKS

PIR/MIPS code

PIR2: A31956 (Imperfect fungus (Caldariomyces fumago) (fragment)); PIR2: A28557 (precursor Imperfect fungus (Caldariomyces fumago))

Brookhaven code

7 LITERATURE REFERENCES

[1] Itoh, N., Izumi, Y., Yamada, H.: Biochemistry,26,282–289 (1987)
[2] Sono, M., Smith Eble, K., Dawson, J.H., Hager, L.P.: J. Biol. Chem.,260,15530–15535 (1985)
[3] Hashimoto, A., Pickard, M.A.: J. Gen. Microbiol.,130,2051–2058 (1984)
[4] Lambeir, A.-M., Dunford, H.B.: Arch. Biochem. Biophys.,220,549–556 (1983)
[5] Libby, R.D., Thomas, J.A., Kaiser, L.W., Hager, L.P.: J. Biol. Chem.,257,5030–5037 (1982)
[6] Shahangian, S., Hager, L.P.: J. Biol. Chem.,256,6034–6040 (1981)
[7] Sae, A.S.W., Cunningham, B.A.: Phytochemistry,18,1785–1787 (1979)
[8] Hallenberg, P.F., Hager, L.P.: Methods Enzymol.,52,521–529 (1978)
[9] Morris, D.R., Hager, L.P.: J. Biol. Chem.,241,1763–1768 (1966)
[10] Hager, L.P., Morris, D.R., Brown, F.S., Eberwein, H.: J. Biol. Chem.,241,1769–1777 (1966)
[11] Morris, D.R., Hager, L.P.: J. Biol. Chem.,241,3582–3589 (1966)
[12] Carmichael, R., Fedorak, P.M., Pickard, M.A.: Biotechnol. Lett.,7,289–294 (1985)
[13] Lambeir, A.-M., Dunford, H.B.: J. Biol. Chem.,258,13558–13563 (1983)
[14] Campbell, B.N., Araiso, T.A., Reinisch, A.L., Yue, K.T., Hager, L.P.: Biochemistry,21,4343–4349 (1982)
[15] Dunford, H.B., Lambeir, A.-M., Kashem, M.A., Pickard, M.: Arch. Biochem. Biophys.,252,292–302 (1987)
[16] Fang, G.-H., Kenigsberg, P., Axley, M.J., Nuell, M., Hager, L.P.: Nucleic Acids Res.,14,8061–8071 (1986)
[17] Dawson, J.H., Kau, L.-S., Penner-Hahn, J.E., Sono, M., Smith Eble, K., Bruce, G.S., Hager, L.P., Hodgson, K.O.: J. Am. Chem. Soc.,108,8114–8116 (1986)
[18] Yamada, H., Itoh, N., Izumi, Y.: J. Biol. Chem.,260,11962–11969 (1985)
[19] Ramakrishnan, K., Oppenhiuzen, M.E., Saunders, S., Fisher, J.: Biochemistry,22,3271–3277 (1983)
[20] Gonzalez-Vergara, E., Ales, D.C., Goff, H.M.: Prep. Biochem.,15,335–348 (1985)

Enzyme Handbook © Springer-Verlag Berlin Heidelberg 1994
Duplication, reproduction and storage in data banks are only allowed with the prior permission of the publishers

1 NOMENCLATURE

EC number
1.11.1.11

Systematic name
L-Ascorbate:hydrogen-peroxide oxidoreductase

Recommended name
L-Ascorbate peroxidase

Synonymes
L-Ascorbic acid peroxidase
L-Ascorbic acid-specific peroxidase [4, 6]
Peroxidase, ascorbate
Ascorbate peroxidase
Ascorbic acid peroxidase

CAS Reg. No.

2 REACTION AND SPECIFICITY

Catalysed reaction
L-Ascorbate + H_2O_2 →
→ dehydroascorbate + 2 H_2O

Reaction type
Redox reaction
Peroxidation

Natural substrates
L-Ascorbate + H_2O_2 [4]

Substrate spectrum
1 L-Ascorbate + H_2O_2 [1, 4]
2 Pyrogallol + H_2O_2 [1, 4]
3 Guaiacol + H_2O_2 [1, 4]
4 o-Dianisidine + H_2O_2 [1]
5 D-Araboascorbic acid + H_2O_2 [4]
6 2,2'-Azino-di-[3-ethylbenzothiazoline-(6)-sulfonic acid] + H_2O_2 [1]
7 Iodide + H_2O_2 [4]
8 Reductic acid + H_2O_2 (i.e. 2,3-dihydroxy-2-cyclopenten-1-one) [4]
9 L-Ascorbic acid + tert.-butylhydroperoxide [4]
10 L-Ascorbic acid + cumene hydroperoxide [4]

Enzyme Handbook © Springer-Verlag Berlin Heidelberg 1994
Duplication, reproduction and storage in data banks are only allowed with the prior permission of the publishers

Product spectrum

1 Dehydroascorbate + 2 H_2O
2 ?
3 ?
4 ?
5 ?
6 ?
7 ?
8 ?
9 ?
10 ?

Inhibitor(s)

CN^- [1, 2, 4]; N_3^- [1, 2, 4]; CO [1]; C_2H_2 [1]; 2-Mercaptoethanol [2]; Dithiothreitol [2]; Reduced glutathione [2]; Hg^{2+} [4]; Mn^{2+} [4]; Br^- [4]; Zn^{2+} [4]; Al^{3+} [4]; Mg^{2+} [4]; Ca^{2+} [4]; Ni^{2+} [4]; Li^+ [4]; I^- [4]; F^- [4]; EDTA [4]

Cofactor(s)/prostethic group(s)

Heme [1, 2, 4, 6]

Metal compounds/salts

Fe (hemoprotein) [1, 2, 4, 6]

Turnover number (min^{-1})

Specific activity (U/mg)

254 [4]; 34.2 [1]; More [3]

K_m-value (mM)

2.9 (ascorbate, enzyme form C) [2]; 6.5 (ascorbate, enzyme form B) [2]; 0.41 (L-ascorbate) [4]; 7.4 (guaiacol, enzyme form C) [2]; 0.056 (H_2O_2) [4]; 9.6 (pyrogallol) [4]; More (Hill plot) [1]

pH-optimum

5.2 (enzyme form B) [2]; 6.2 (enzyme form C [2]) [2, 3]

pH-range

5–8 [2]

Temperature optimum (°C)

37 [4]

Temperature range (°C)

3 ENZYME STRUCTURE

Molecular weight

47000 (Glycine max, gel filtration) [1]
57000 (Pisum sativum, gel filtration) [2]
76000 (Euglena gracilis, gel filtration) [4]
30000 (Spinacia oleracea) [6]

Subunits

Monomer (1 × 30000, Glycine max, SDS-PAGE, discrepancy to value from gel filtration probably due to enzyme conformation) [1]

Glycoprotein/Lipoprotein

–

4 ISOLATION/PREPARATION

Source organism

Glycine max [1]; Vigna unguiculata [1]; Vicia faba [1]; Vicia sativa [1]; Arachis hypogaea [1]; Lupinus alba (low activity) [1]; Medicago sativa [1]; Trifoleum subterraneum [1]; Pisum sativum (pea) [1, 2, 5]; Alnus rubra [1]; Euglena gracilis [3, 4]; Wheat [5]; Spinacia oleracea (spinach) [6]

Source tissue

Root (nodules) [1]; Leaves [5, 6]; Shoots [2]

Localisation in source

Chloroplast [6]; Cytosol [1, 3]; More (not in mitochondria, chloroplasts, microsomes) [3]

Purification

Glycine max [1]; Pisum sativum [2]; Euglena gracilis [4]; Spinacia oleracea [6]

Crystallization

–

Cloned

–

Renaturated

–

Enzyme Handbook © Springer-Verlag Berlin Heidelberg 1994
Duplication, reproduction and storage in data banks are only allowed with the prior permission of the publishers

5 STABILITY

pH

Temperature (°C)

4 (half-life 10 h) [1]; 40 (inactivated after 5 min) [4]; 52 (complete inactivation) [4]; 70 (5 min, complete inactivation) [2]

Oxidation

Not stable under aerobic conditions [6]; Stabilization by ascorbate and sorbitol [6]

Organic solvent

General stability information

Storage

–20°C or –80°C, crude extracts of root nodules, 50 mM potassium phosphate buffer, pH 7.0, several months [1]

6 CROSSREFERENCES TO STRUCTURE DATABANKS

PIR/MIPS code

PIR2: $S_1$5878 (isozyme II Spinach (fragment)); PIR3: S20866 (precursor Arabidopsis thaliana (fragment))

Brookhaven code

7 LITERATURE REFERENCES

[1] Dalton, D.A., Hanus, F.J., Russell, St.A., Evans, H.J. : Plant Physiol.,83,789–794 (1987)

[2] Gerbling, K.-P., Kelly, G.J., Fischer, K.-H., Latzko, E.: J. Plant Physiol.,115,59–67 (1984)

[3] Shigeoka, S., Nakano, Y., Kitaoka, S.: Biochem. J.,186,377–380 (1980)

[4] Shigeoka, S., Nahana, Y., Kitaoka, S.: Arch. Biochem. Biophys.,201,121–127 (1980)

[5] Kelly, G.J., Latzko, E.: Naturwissenschaften,66,67–68 (1979)

[6] Nakano, Y., Asada, K.: Plant Cell Physiol.,28,131–140 (1987)

1 NOMENCLATURE

EC number
1.11.1.12

Systematic name
Glutathione:lipid-hydroperoxide oxidoreductase

Recommended name
Phospholipid-hydroperoxide glutathione peroxidase

Synonymes
PHGPX [3, 8]
Peroxidation-inhibiting protein: Peroxidase, glutathione (phosphol hydroperoxide-reducing)
Phospholipid hydroperoxide glutathione peroxidase
Hydroperoxide glutathione peroxidase [1]

CAS Reg. No.
97089-70-8

2 REACTION AND SPECIFICITY

Catalysed reaction
2 Glutathione + a lipid hydroperoxide →
→ oxidized glutathione + lipid + 2 H_2O (ping-pong mechanism wit formation of ternary complexes [4])

Reaction type
Redox reaction

Natural substrates
Glutathione + lipid hydroperoxide (enzyme reduces amphiphilic pe possibly at the water-lipid interface [2], protection of biomembranes against oxidative damage [5], role in cellular detoxification of a wide of lipid hydroperoxides in membranes and internalized lipoproteins [2, 5, 7]

Enzyme Handbook © Springer-Verlag Berlin Heidelberg 1994
Duplication, reproduction and storage in data banks are only allowed with the prior permission of the publishers

Substrate spectrum

1 Glutathione + lipid hydroperoxide (wide specificity for lipid hydroperoxides [4], e.g. linoleic acid hydroperoxide [1, 2], all phospholipid hydroperoxides [4], fatty acid hydroperoxides [4], tert.-butyl hydroperoxide [4, 5], H_2O_2 [4], cholesterol hydroperoxides [4, 7], phosphatidyl choline [5], cumene [5], the oxidized active site of the enzyme can also be reduced by mercaptoethanol [4]) [1, 2, 4, 5, 7]
2 More (the products of EC 1.13.11.12 on phospholipids can act as acceptor, H_2O_2 can also act, but much more slowly)

Product spectrum

1 Oxidized glutathione + lipid + H_2O
2 ?

Inhibitor(s)

Iodoacetate [1]; Deoxycholate [2]; Unsaturated fatty acids [2]

Cofactor(s)/prostethic group(s)

Metal compounds/salts

Selenium (enzyme contains a selenocysteine residue [4], contains 1 selenium atom at the active site [4], 1 gatom Se per mol protein [1], Se in selenol form [1]) [1, 4]

Turnover number (min^{-1})

Specific activity (U/mg)

310.15 [4]

K_m-value (mM)

pH-optimum

pH-range

Temperature optimum (°C)

37 (assay at) [4]

Temperature range (°C)

---

3 ENZYME STRUCTURE

Molecular weight

20000 (pig, gel filtration) [5]

Subunits

Monomer (1 × 23000 [1], 1 × 20000 [4, 5], pig, SDS-PAGE [1, 4, 5]) [1, 4, 5]

Glycoprotein/Lipoprotein

–

4 ISOLATION/PREPARATION

Source organism

Pig [1, 4, 5–7]; Rat [3, 4, 8]; Dog [4]; Bovine [4]; Fish [4]

Source tissue

Muscle [8]; Spermatozoa [4]; Heart [1, 4–7, 8]; Brain [3, 4, 8]; Liver [4, 5, 8]; Kidney [4, 8]; Testis [4]; Lung [8]

Localisation in source

Mitochondria (inner membrane) [3]; Cytosol [4]

Purification

Pig [1, 4, 6]

Crystallization

–

Cloned

–

Renaturated

–

5 STABILITY

pH

Temperature (°C)

Oxidation

Organic solvent

General stability information

Storage

–20°C, 10% glycerol, stable for several months [4]

6 CROSSREFERENCES TO STRUCTURE DATABANKS

PIR/MIPS code

Brookhaven code

Enzyme Handbook © Springer-Verlag Berlin Heidelberg 1994
Duplication, reproduction and storage in data banks are only allowed with the prior permission of the publishers

7 LITERATURE REFERENCES

[1] Ursini, F., Maiorini, M., Gregolin, C.: Biochim. Biophys. Acta,839,62–70 (1985)
[2] Maiorino, M., Roveri, A., Gregolin, C., Ursini, F.: Arch. Biochem. Biophys.,251,600–605 (1986)
[3] Panfili, E., Sandri, G., Ernster, L.: FEBS Lett.,290,35–37 (1991)
[4] Maiorino, M., Gregolin, C., Ursini, F.: Methods Enzymol.,186,448–457 (1990)
[5] Ursini, F., Maiorino, M., Valente, M., Ferri, L., Gregolin, C.: Biochim. Biophys. Acta,710,197–211 (1982)
[6] Maiorino, M., Ursini, F., Leonelli, M., Finato, N., Gregolin, C.: Biochem. Int.,5,575 (1982)
[7] Thomas, J.P., Geiger, P.G., Maiorino, M., Ursini, F., Girotti, A.W.: Biochim. Biophys. Acta,1045,252–260 (1990)
[8] Zhang, L., Maiorini, M., Roveri, A., Ursini, F.: Biochim. Biophys. Acta,1006,140–143 (1989)

1 NOMENCLATURE

EC number
1.11.1.13

Systematic name
Mn(II):hydrogen-peroxide oxidoreductase

Recommended name
Manganese peroxidase

Synonymes
Peroxidase, manganese
Peroxidase-M2 [1]
Mn-dependent (NADH-oxidizing) peroxidase [15]

CAS Reg. No.
114995-15-2

2 REACTION AND SPECIFICITY

Catalysed reaction
$2\,Mn(II) + 2\,H^+ + H_2O_2 \rightarrow$
$\rightarrow 2\,Mn(III) + 2\,H_2O$ (shows properties of a peroxidase and an oxidase [1], mechanism [2–4, 10, 12, 15, 17], ping-pong mechanism [9], structural properties [8])

Reaction type
Redox reaction

Natural substrates
$Mn(II) + H^+ + H_2O_2$ (acts together with ligninase in lignin-degradation of white-rot fungi [13, 20], the product, Mn(III), is involved in the oxidative degradation of lignin in white-rot basidiomycetes, induced by veratryl alcohol [1], by Mn(II) [6, 11, 16], the mechanism enables the fungus to oxidize structures within woods which are inaccessible to enzymes [2], active site inaccessible to large substrates [9], Mn(III) is stabilized by chelating agents, malonate is the most effective physiological chelator excreted by the fungus [9], produced under lignolytic conditions [19], in absence of H_2O_2 it may play a role in fungal peroxide production [20]) [1–6, 9, 13, 16, 19, 20]

Enzyme Handbook © Springer-Verlag Berlin Heidelberg 1994
Duplication, reproduction and storage in data banks are only allowed with the prior permission of the publishers

Substrate spectrum

1 Mn(II) + H^+ + H_2O_2 (ir [9], specifically oxidizes Mn(II) [2, 20] in presence of H_2O_2 to a higher oxidation state [1], free divalent Mn is the substrate, not Mn(II)-complexes [9], no organic substrates are oxidized in absence of Mn(II) [2], Mn(III)-chelate-complexes catalyze decarboxylation and demeth(ox)ylation of aromatic substrates [17]) [1–22]
2 Co(II) + H^+ + H_2O_2 (oxidation at 2% the rate of Mn(II) oxidation) [2]
3 More (primary reaction product of peroxidation with H_2O_2: enzyme compound I, formation of compound II from I follows second-order kinetic [4], substrates are oxidized via delta-meso heme edge of the enzyme [12], in presence of H_2O_2 and Mn the enzyme shows peroxidase activity against a variety of vinyl- and syringyl-side chain substituted substrates [1, 10, 21], various organic dyes (e.g. phenol red [2, 4, 11], after or during NADH-oxidation without exogenously added H_2O_2 [15], poly B-411 dye [2]) [2, 4, 11, 15, 20], large substrates have no ready access to the catalytic center [3], in absence of H_2O_2 the enzyme shows Mn-dependent oxidase activity against GSH, DTT, dihydroxymaleic acid [1, 2, 21], NADPH [1, 2, 15, 20, 21], NADH [15, 20, 21], producing H_2O_2 [1, 2, 15], poor substrates: benzoate, benzaldehyde or benzyl alcohol [15]) [1–4, 10–12, 15, 20, 21]

Product spectrum

1 Mn(III) + H_2O (the diffusible product Mn(III) [1, 2] complexed to lactate or other alpha-hydroxy acids [2] oxidizes phenols (requires O_2 and H_2O_2, phenolic hydroxyl group is essential [15]) [1, 3, 4, 15, 20], not non-phenolic compounds [15], amines [4, 21] and phenolic lignin model-compounds [1, 10, 18, 20], phenol rings are oxidized to phenoxy-radicals [6, 19], methoxy benzenes, i.e. 1,2,4-tri-, 1,2,3,5-tetra-, 1,2,4,5-tetra-, penta methoxybenzene [13], guaiacol [2, 3, 11, 21], pinacyanol [2], vanillyl acetone [11, 18, 21], curcumin [21] and o-dianisidine [3, 21] are oxidized, Mn(III) oxidizes thiols to thiyl radicals which undergo radical coupling to form disulfides [19], veratryl alcohol [18, 19] Is oxidized by thiyl-radicals derived from Mn(III)-oxidation of GSH, not directly by Mn(III) [13]) [1–22]
2 Co(III) + H_2O [2]
3 ?

Inhibitor(s)

Fe^{3+} (0.1 mM) [1]; Fe^{2+} (0.1 mM [1]) [1, 21]; Co^{2+} (0.1 mM [1]) [1, 21]; Cu^{2+} (0.1 mM [1], inhibits Mn(III)-tartrate formation [14], not: [18]) [1, 14]; Ascorbic acid [1]; Superoxide dismutase (inhibits phenol red oxidation to 50%, vanillyl acetone oxidation to 90%, NADH oxidation at high concentrations partially [15]) [1, 15]; H_2O_2 (above 0.2 mM [1]) [1, 17]; Nitrilotriacetate (competitive to Mn(III)-malonate formation) [9]; Cellobionate (competitive to Mn(III)-malonate formation) [9]; 1,10-Phenanthroline (competitive to Mn(III)-malonate formation) [9]; GSH [19]; NaN_3 (complete inactivation within 2 min in presence of H_2O_2, accompanied by azidyl radical formation, prosthetic heme is converted to meso-azido adduct which is more rapidly oxidized by H_2O_2 than prosthetic heme, 2 equivalents of azide are oxidized before the enzyme molecule is inactivated, inactivation is slightly stimulated by Co^{2+}) [12]; Phenylhydrazine (rapid inactivation, concentration-dependent, no heme adducts detectable, inactivation is slightly stimulated by Co^{2+}) [12]; Methylhydrazine (slow inactivation in presence of H_2O_2, concentration- and time-dependent, slightly stimulated by Co^{2+}) [12]; Ethylhydrazine (inactivation via delta-meso-ethylheme adduct at pH 7.0, slightly stimulated by Co^{2+}) [12]; Co^{2+} (in the presence of H_2O_2 competitive inhibitor to Mn^{2+}, slightly stimulates NaN_3 or alkylhydrazine inactivation [12], not: [18]) [12]; Formate [14]; Acetate (100 mM) [14]; Catalase (inhibits oxidation of vanillyl acetone completely, oxidation of NADH at high concentrations partially) [15]; Sodium diphosphate (inhibits phenol red oxidation, NADH oxidation at high concentrations partially) [15]; More (no inhibition: 0.1 mM salicylic acid [1, 18], ethanol (5% v/v) [1], succinate [9], sodium diphosphate, mannitol, Mn(III)-chelators [15]) [1, 9, 15, 18]

Cofactor(s)/prostethic group(s)

H_2O_2 (requirement) [1–22]; GSH (activation) [18]; DTT (activation) [18]

Metal compounds/salts

Iron (heme-iron protein, 0.7 heme per enzyme molecule [1], high-spin ferri heme complex [1], the enzyme has a single Mn(II)-binding site near the heme [9], substrates are oxidized via delta-meso heme edge of the enzyme [12], protoporphyrin IX [15], 1 heme group per enzyme molecule as iron protoporphyrin IX) [1, 4, 9, 12, 15, 18, 22]; alpha-Hydroxy acids (activation by complexing and stabilizing Mn(III) rather than activating the enzyme) [2]; Lactate (activation, stabilizes Mn(III) in aqueous solution with a relatively high redox potential [9, 19]) [2, 4, 9, 19, 20]; Citrate (activation, stabilizes Mn(III) in aqueous solution with a relatively high redox potential [9], no chelator of Mn(III) [18]) [4, 9, 20]; Oxaloacetate (activation, chelator of Mn(III)) [18]; Methylmalonate (activation, chelator of Mn(III)) [18]; Polyglutamate (activation, stabilizes Mn(III) in aqueous solution with a relatively high redox potential) [9]; L-Tartrate (activation, stabilizes Mn(III) in aqueous solution with a relatively high redox potential [9]) [9, 14, 18, 20]; Succinate (activation, stabilizes Mn(III) less effective as citrate or lactate [4], no activation by succinate [9, 18]) [4]; L-Malate (activation, chelator of Mn(III)) [14, 18, 20]; Malonate (activation, stabilizes Mn(III) in aqueous solution with a relatively high redox potential, most effective physiological chelating agent, excreted by the fungus [9]) [9, 18]; Oxalate (activation,

Enzyme Handbook © Springer-Verlag Berlin Heidelberg 1994
Duplication, reproduction and storage in data banks are only allowed with the prior permission of the publishers

chelator of Mn(III)) [18]; Bulk protein (activation) [20]; More (chelating organic acids, C2- and C3-dicarboxylic or alpha-hydroxyl acids, facilitate the dissociation of Mn(III) from manganese-enzyme complex, greater activation with weakly binding chelators with a low binding constant, e.g. lactate or tartrate [9], these complexes oxidize a variety of substrates, formate and acetate do not stabilize Mn(III) [14], no chelating of Mn(III): acetate, propionate, succinate, citrate, D-malate, ethylene glycol [18]) [9, 14, 18]

Turnover number (min^{-1})

5900 (Mn(III)-tartrate, preparation M1) [14]; 7100 (Mn(III)-tartrate, preparation M2) [14]

Specific activity (U/mg)

More (216–467, 1 U: initial increase in absorbance of 1.0 per min at 465 nm [14], 284–422 (5 isozymes), expressed as spectral change in deltaA per min and mg of heme-containing protein [22]) [14, 22]; 38.5 [16]; 100 [21]; 180 (Mn(III)-lactate formation) [2]; 660 (o-toluidine as substrate for Mn(III)-oxidation) [18]

K_m-value (mM)

0.036 (Mn(II)) [14]; 0.057–0.059 (H_2O_2) [14]; 0.08 (Mn(II)) [2, 20]; 0.14 (H_2O_2) [2]

pH-optimum

More (5 isozymes, pI: 2.93–3.17 [22], pI: 3.2 [18], pI: 3.8 [15], pI: 4.2 (isozyme H5), pI: 4.5 (isozyme H4) [11], pI: 4.9 (H3 [11]) [11, 12, 16]) [11, 12, 16, 18, 22]; 3.5–5.5 [18] 4.5 (2,2-azino-bis-3-ethyl-6-benzothiazoline-sulfonate-oxidation) [2]; 4.8 [14]; 5.0 (Mn(III)-lactate-formation) [2]

pH-range

3.1–8.3 (in presence of H_2O_2, enzyme component I reacts independently from changes in pH) [4]; 4.0–5.5 (about half-maximal activity at pH 4.0 and 5.5, 2,2-azino-bis-3-ethyl-6-benzothlazolinesulfonate-oxidation) [2]; 4.5–5.7 (about 65% maximal activity at pH 4.5 and about 50% at pH 5.7, Mn(III)-lactate formation) [2]

Temperature optimum (°C)

22 (assay at) [21]; 23 (assay at) [17]; 25 (assay at) [9, 10]; 30 (assay at) [1]; 40 (assay at) [18]

Temperature range (°C)

20–65 (progressive enhancement of activity from 20°C to 65°C) [18]

3 ENZYME STRUCTURE

Molecular weight

Subunits

? (x × 44000, Trametes versicolor, isozyme TvMP4, SDS-PAGE [22], x × 44500, Trametes versicolor, isozyme TvMP2 and 5, SDS-PAGE [22], x × 44600, Lentinula edodes, SDS-PAGE [18], x × 45000, Trametes versicolor, isozyme TvMP1 and 3, SDS-PAGE [22], x × 45000–47000, Phanerochaete chrysosporium, SDS-PAGE [1], x × 46000, Phanerochaete chrysosporium, SDS-PAGE [20], x × 49000, Phlebia radiata, SDS-PAGE [15]) [1, 15, 18, 20, 22]

Glycoprotein/Lipoprotein

Glycoprotein (17% [1], 3–6% [7] (w/w) carbohydrates, glucosamine and mannose [7]) [1, 6, 7, 16]

4 ISOLATION/PREPARATION

Source organism

Phanerochaete chrysosporium (strains BKM-F-1767 [1, 5, 11, 21], PSBL-1, overproducing strain, isozyme profile depends on type of nutrient limitation and growth phase: no H5 in C-limited cultures, manganese increases H3-production [8, 11], OGC101 [9, 10, 12, 19], VKM F-1767 [14, 17]) [1–5, 8–14, 16, 17, 19–21]; Phlebia radiata (strain 79) [15]; Phlebia brevispora (strain HHB-7030-sp) [6]; Ceriporiopsis subvermispora (strains L-14807, L-15225, FP-104027, FP-90031-sp, FP-105752) [6]; Trametes versicolor (syn. Coriolus versicolor, strain PRL572) [7, 22]; Lentinula edodes (syn. Lentinus, heterodicaryon strain 861) [18]

Source tissue

Culture supernatant [1–22]

Localisation in source

Extracellular (appears to be closely related to mycelium [1]) [1–22]

Purification

Phanerochaete chrysosporium (partial [16], 3 isozymes, affinity chromatography [20] on blue agarose: isozyme H4 95% pure, followed by preparative isoelectric focusing: isozymes H3 and H5 pure [11]) [1, 11, 16, 20]; Trametes versicolor (5 isozymes [7, 22], amino acid analysis and partial sequencing reveals more hetergeneous relationship: 70% degree of identity of 3 studied isozymes [7]) [7, 22]; Phlebia brevispora (partial) [6]; Lentinula edodes (hydrophobic interaction HPLC) [18]

Enzyme Handbook © Springer-Verlag Berlin Heidelberg 1994
Duplication, reproduction and storage in data banks are only allowed with the prior permission of the publishers

Crystallization

–

Cloned

–

Renaturated

–

5 STABILITY

pH

3.0–6.0 (stable) [15]

Temperature (°C)

40 (stable below) [17]; 100 (complete inactivation after 5 min [1, 18], 80% of original activity recovered after storage of heat-inactivated protein for 1 h at 0°C) [1, 18]

Oxidation

Organic solvent

General stability information

Heat-inactivated protein, 100°C for 5 min, regains 80% of original activity by storage at 0°C for 1 h [1]; Prolonged dialysis inactivates [16]; CHAPS, stabilizes [16]; Bovine serum albumin stabilizes [17]; Stable to lyophilization [21]

Storage

–20°C, freeze-dried preparation up to a year stable [21]; –20°C, concentrated preparation, 1 mg/ml, stable for at least 3 months [20]; Frozen, stable for at least 6 months in tartrate buffer, pH 4.5 [1]; 5°C, unstable [1]; 23°C, crude, stable [17]

6 CROSSREFERENCES TO STRUCTURE DATABANKS

PIR/MIPS code

Brookhaven code

7 LITERATURE REFERENCES

[1] Paszczynski, A., Huynh, V.-B., Crawford, R.L.: Arch. Biochem. Biophys.,244,750–765 (1986)

[2] Glenn, J.K., Akileswaran, L., Gold, M.H.: Arch. Biochem. Biophys.,251,688–696 (1986)

[3] Wariishi, H., Akileswaran, L., Gold, M.H.: Biochemistry,27,5365–5370 (1988)

[4] Wariishi, H., Dunford, H.B., MacDonald, I.D., Gold, M.H.: J. Biol. Chem.,264,3335–3340 (1989)

[5] Dass, S.B., Reddy, C.A.: FEMS Microbiol. Lett.,69,221–224 (1990)

[6] Rüttimann, C., Schwember, E., Salas, L., Cullen, D., Vicũa, R.: Biotechnol. Appl. Biochem.,16,64–76 (1992)

[7] Johansson, T., Welinder, K.G., Nyman, P.O.: Arch. Biochem. Biophys.,300,57–62 (1993)

[8] Banci, L., Bertini, I., Pease, E.A., Tien, M., Turano, P.: Biochemistry,31,10009–10017 (1992)

[9] Wariishi, H., Valli, K., Gold, M.H.: J. Biol. Chem.,267,23688–23695 (1992)

[10] Tuor, U., Wariishi, H., Schoemaker, H.E., Gold, M.H.: Biochemistry,31,4986–4995 (1992)

[11] Pease, E.A., Tien, M.: J. Bacteriol.,174,3532–3540 (1992)

[12] Harris, R.Z., Wariishi, H., Gold, M.H., Ortiz de Montellano, P.R.: J. Biol. Chem.,266,8751–8758 (1991)

[13] Popp, J.L., Kirk, T.K.: Arch. Biochem. Biophys.,288,145–148 (1991)

[14] Aitken, M., Irvine, R.L.: Arch. Biochem. Biophys.,276,405–414 (1990)

[15] Karhunen, E., Kantelinen, A., Niku-Paavola, M.-L.: Arch. Biochem. Biophys.,279,25–31 (1990)

[16] Datta, A., Bettermann, A., Kirk, T.K.: Appl. Environ. Microbiol.,57,1453–1460 (1991)

[17] Aitken, M., Irvine, R.L.: Biotechnol. Bioeng.,34,1251–1260 (1989)

[18] Forrester, I.T., Grabski, A.C., Mishra, C., Kelley, B.D., Strickland, W.N., Leatham, G.F., Burgess, R.R.: Appl. Microbiol. Biotechnol.,33,359–365 (1990)

[19] Wariishi, H., Valli, K., Renganathan, V., Gold, M.H.: J. Biol. Chem.,264,14185–14191 (1989)

[20] Gold, M.H., Glenn, J.K.: Methods Enzymol.,161,258–264 (1988) (Review)

[21] Paszczynski, A., Crawford, R.L., Huynh, V.-B.: Methods Enzymol.,161,264–270 (1988) (Review)

[22] Johansson, T., Nyman, P.O.: Arch. Biochem. Biophys.,300,49–56 (1993)

Enzyme Handbook © Springer-Verlag Berlin Heidelberg 1994
Duplication, reproduction and storage in data banks are only allowed with the prior permission of the publishers

1 NOMENCLATURE

EC number
1.11.1.14

Systematic name
Diarylpropane:oxygen,hydrogen-peroxide oxidoreductase (C-C-bond-cleaving)

Recommended name
Diarylpropane peroxidase

Synonymes
Oxygenase, diarylpropane
Diarylpropane oxygenase
Ligninase I

CAS Reg. No.
93792-13-3

2 REACTION AND SPECIFICITY

Catalysed reaction
1,2-Bis(3,4-dimethoxyphenyl)propane-1,3-diol + H_2O_2 →
→ veratraldehyde + 1-(3,4-dimethylphenyl)ethane-1,2-diol + 4 H_2O
(mechanism [1, 3, 4, 8])

Reaction type
Redox reaction

Natural substrates
1-(3', 4'-Diethoxyphenyl)-1,3-dihydroxy-2-(4''-methoxyphenyl)-propane + O_2 + H_2O_2 (i.e. diarylpropane, involved in the oxidative breakdown of lignin in white rot basidiomycetes, induced by veratryl alcohol [1]) [1–10]

Enzyme Handbook © Springer-Verlag Berlin Heidelberg 1994
Duplication, reproduction and storage in data banks are only allowed with the prior permission of the publishers

Substrate spectrum

1 1,2-Bis(3,4-dimethoxyphenyl)propane-1,3-diol + H_2O_2 [1]
2 3,4-Dimethoxybenzyl alcohol + H_2O_2 (i.e. veratryl alcohol) [1, 4, 7, 10]
3 1-(3', 4'-Diethoxyphenyl)-1,3-dihydroxy-2-(4''-methoxy-phenyl)propane + O_2 + H_2O_2 (i.e. diarylpropane, lignin-model compound, alpha,beta-cleavage with insertion of a single atom of oxygen from O_2 into the alpha-position of the product 1-(4'-methoxyphenyl)-1,2-dihydroxyethane [4]) [3, 4]
4 2-Keto-4-thiomethylbutyric acid + H_2O_2 (only in the presence of veratryl alcohol, it possibly reacts with a veratryl alcohol radical to produce ethylene) [3]
5 1-(4'-Ethoxy-3'-methoxyphenyl)propane + O_2 + H_2O_2 [3]
6 1-(4'-Ethoxy-3'-methoxyphenyl)-1,2-propene + O_2 + H_2O_2 (olefinic hydroxylation) [3, 4]
7 Non-phenolic substrates + H_2O_2 (e.g. 1,2,4-trimethoxybenzene, 4,4'-dimethoxy-biphenyl, isoeugenol methylether, 1-(3,4-dimethoxyphenyl)-2-(2, 4-dichlorophenoxyl)-ethanol, guaiacyl glycerolether) [1]
8 More (oxidation of various phenolic and non-phenolic lignin model-compounds [1–7] with concomitant insertion of 1 atom of molecular oxygen [2, 4], catalyzes non-specifically several oxidations in the alkyl-side-chains of lignin-related compounds, C_{alpha}-C_{beta} cleavage in lignin model-compounds [4] of the aryl-C_{alpha}HOH-C_{beta}HR-$C_{gamma}H_2$OH-type (R being aryl or O-aryl) [2], oxidation of benzyl alcohols to aldehydes or ketones [2, 4], intradiol cleavage in phenylglycol structures, hydroxylation of benzylic methylene groups, oxidative coupling of phenols, all reactions require H_2O_2, C_{alpha}-C_{beta} cleavage and methylene hydroxylation involve substrate oxygenation, the oxygen atom originates from O_2 not H_2O_2: thus the enzyme acts as oxygenase which requires H_2O_2 [2, 4]) [1–7]

Product spectrum

1 3,4-Dimethoxybenzaldehyde + 1-(3,4-dimethyl-phenyl)ethane-1,2-diol + H_2O
2 3,4-Dimethoxybenzaldehyde + H_2O [4, 7, 10]
3 1-(4'-Methoxyphenyl)-1,2-dihydroxyethane + 3,4-diethoxybenzaldehyde [3, 4]
4 ?
5 1-(4'-Ethoxy-3'-methoxyphenyl)1-hydroxypropane [3]
6 1-(4'-Ethoxy-3'-methoxyphenyl)-1,2-dihydroxypropane [3]
7 ?
8 ?

Inhibitor(s)

H_2O_2 (above 5 mM [2], second-order process [8]) [2, 8]

Cofactor(s)/prostethic group(s)

H_2O_2 (requirement, 0.15 mM or above [1]) [1–7]

Metal compounds/salts

Iron (hemoprotein containing protoporphyrin IX, high-spin ferri-heme-complex, 0.7 heme per enzyme molecule [1], 1 mol protoheme IX per mol enzyme [2, 4], 1.08 atom Fe per enzyme molecule [2]) [1–7]

Turnover number (min^{-1})

Specific activity (U/mg)

2.2 (isozyme TvLP2) [9]; 8.4 (veratryl alcohol) [2]; 8.48 (isozyme III) [3]; 10.6 (TvLP12) [9]; 11.4 (1,2-bis(3,4-di-methoxyphenyl)propane-1,3-diol) [2]; 11.8 (isozyme II) [3]; 12.5 [1]; 18.1 (isozyme TvLP9) [9]; 18.75 (isozyme I) [3]; 27 (isozyme LI) [8]; 28 (isozyme L2) [8]

K_m-value (mM)

0.03 (H_2O_2) [2]; 0.032–0.049 (H_2O_2) [3]; 0.055–0.095 (veratryl-alcohol) [3]

pH-optimum

More (pI: 3.5 [2], pI: 3.09–3.74 (15 isozymes) [9], pI: 3.2 (isozyme LI), 3.8 (LII), 3.9 (LIII) [7]) [2, 7, 9]; 3.0 (diarylpropane [5]) [2, 3, 5]; 4.5 (1-(4'-ethoxy-3'-methoxy-phenyl)glycerol-beta-guaiacylether) [5]

pH-range

Temperature optimum (°C)

23 (assay at) [8]; 37 (assay at) [2, 5]

Temperature range (°C)

---

3 ENZYME STRUCTURE

Molecular weight

Enzyme Handbook © Springer-Verlag Berlin Heidelberg 1994
Duplication, reproduction and storage in data banks are only allowed with the prior permission of the publishers

Subunits

? (x × 39000 Phanerochaete chrysosporium, isozyme I, SDS-PAGE [3], x × 40900–45700, Trametes versicolor, 15 isozymes, SDS-PAGE [9], x × 41000, Phanerochaete chrysosporium (isozyme II [3]), SDS-PAGE [3–5], x × 42000, Phanerochaete chrysosporium, SDS-PAGE [2], x × 42000–43000, Phanerochaete chrysosporium, SDS-PAGE [1], x × 43000 Phanerochaete chrysosporium, isozyme III, SDS-PAGE [3], x × 42000, Phanerochaete chrysosporium, SDS-PAGE [2]) [1–5, 9]

Glycoprotein/Lipoprotein

Glycoprotein (21% [1], 3–6% [6] (w/w) carbohydrates, glucosamine and mannose [6], isozyme II: 6% (w/w) neutral carbohydrates [3]) [1, 3, 6]

---

4 ISOLATION/PREPARATION

Source organism

Phanerochaete chrysosporium (strains BKM-F-1767 [1, 2, 10], VKM F-1767 [8], different isozyme pattern depending on growth conditions [10]) [1–5, 8, 10]; Trametes versicolor (syn. Coriolus versicolor, strain PRL572) [6, 9]; Phlebia radiata (strain 79) [7]

Source tissue

Culture supernatant [1–9]

Localisation in source

Extracellular (or losely associated to cell membrane [7]) [1–7]

Purification

Phanerochaete chrysosporium (ultrafiltration, chromatofocusing column chromatography [1], affinity chromatography, 3 isozymes by ion-exchange chromatography [3]) [1–5]; Trametes versicolor (amino acid analysis and partial sequencing [6], 16 isozymes: TvLP1–16, differing in primary structure) [6, 9]

Crystallization

–

Cloned

(partially) [10]

Renaturated

–

5 STABILITY

pH

3.0 (rapid inactivation) [8]

Temperature (°C)

23 (retains nearly 30% of initial activity after 18 h, in presence of veratryl alcohol) [8]; 35 (stable below) [8]

Oxidation

Organic solvent

General stability information

Lyophilization, stable to [2]; Veratryl alcohol stabilizes at pH 3.5 and 23°C, lag-phases are not observed [8]; High protein concentrations stabilize [8]

Storage

–196°C, indefinitely stable [5]; –20°C, stable as lyophilized powder [2]; Frozen, in crude concentrates of growth medium complete loss of activity within a month [1]

6 CROSSREFERENCES TO STRUCTURE DATABANKS

PIR/MIPS code

Brookhaven code

7 LITERATURE REFERENCES

[1] Paszczynski, A., Huynh, V.-B., Crawford, R.L.: Arch. Biochem. Biophys.,244,750–765 (1986)
[2] Tien, M., Kirk, T.K.: Proc. Natl. Acad. Sci. USA,81,2280–2284 (1984)
[3] Renganathan, V., Miki, K., Gold, M.H.: Arch. Biochem. Biophys.,241,304–314 (1985)
[4] Andersson, L.A., Renganathan, V., Chiu, A.A., Loehr, T.M., Gold, M.H.: J. Biol. Chem.,260,6080–6087 (1985)
[5] Gold, M.H., Kuwahara, M., Chiu, A.A., Glenn, J.K.: Arch. Biochem. Biophys.,234,353–362 (1984)
[6] Johansson, T., Welinder, K.G., Nyman, P.O.: Arch. Biochem. Biophys.,300,57–62 (1993)
[7] Karhunen, E., Kantelinen, A., Niku-Paavola, M.-L.: Arch. Biochem. Biophys.,279,25–31 (1990)
[8] Aitken, M., Irvine, R.L.: Biotechnol. Bioeng.,34,1251–1260 (1989)
[9] Johansson, T., Nyman, P.O.: Arch. Biochem. Biophys.,300,49–56 (1993)
[10] Dass, S.B., Reddy, C.A.: FEMS Microbiol. Lett.,69,221–224 (1990)

1 NOMENCLATURE

EC number
1.12.1.2

Systematic name
Hydrogen:NAD^+ oxidoreductase

Recommended name
Hydrogen dehydrogenase

Synonymes
Hydrogenase
Dehydrogenase, hydrogen
H_2:NAD^+ oxidoreductase [3]

CAS Reg. No.
37256-54-5

2 REACTION AND SPECIFICITY

Catalysed reaction
$H_2 + NAD^+ \rightarrow$
$\rightarrow H^+ + NADH$ (ping pong mechanism [7])

Reaction type
Redox reaction

Natural substrates
$H_2 + NAD^+$ (utilization of H_2 as energy source during autotrophic growth on hydrogen and CO_2) [1, 3]

Substrate spectrum

1 $H_2 + NAD^+$ (electron acceptors: NAD^+ [8, 10, 11], FMN (only in presence of catalytic amounts of NAD^+ [2], not [10]) [8], FAD (only in presence of catalytic amounts of NAD^+ [2]) [8, 11], janus green [8], 2,6-dichlorophenolindophenol [8], ferricyanide [8, 11], O_2 (not [2]) [8], benzyl viologen [8, 10, 11], phenazine methosulfate [8], menaquinone [8], ubiquinone [8], cytochrome c [8], methylene blue (only in presence of catalytic amounts of NAD^+ [2, 10]) [8, 10, 11], methyl viologen [8, 10, 11]) [1–19]

2 More (Alcaligenes eutrophus: enzyme also has diaphorase and NAD(P)H oxidase activity [8], hydrogenase produces superoxide free radical anions (which are responsible for enzyme inactivation) [17]) [8, 17]

Enzyme Handbook © Springer-Verlag Berlin Heidelberg 1994
Duplication, reproduction and storage in data banks are only allowed with the prior permission of the publishers

Product spectrum

1 H^+ + NADH [1–19]
2 ?

Inhibitor(s)

Mercaptoethanol [2]; NO [4]; Tris-buffer (at high concentration) [10]; Triethanolamine buffer (at high concentration) [10]; Glutathione (above 0.1 mM: inactivation, up to 0.1 mM: activation) [10]; Na^+ [10]; Iodoacetate [10]; p-Chloromercuribenzoate [10, 18]; Sulfide (slight [10]) [2, 10]; Cyanide [10]; O_2 (slight inhibition of NAD^+ reduction and H_2 evolution from NADH [17], no inhibition [18]) [17, 18]; High ionic strength [8]

Cofactor(s)/prostethic group(s)

Flavin (flavoprotein [8, 12, 19], 1 molecule of FMN per enzyme molecule [12], 2 molecules of FMN per enzyme molecule [19], only when FMN or riboflavin are added in combination with Mg^{2+} or Ni^{2+}, NAD-reducing system is activated [10]) [8, 10, 12, 19]; FMN (1 FMN molecule per enzyme molecule [12], 2 FMN molecules per enzyme molecule [19], stimulates [8]) [8, 12, 19]; ATP (stimulates) [8]; Glutathione (up to 0.1 mM: activation, above 0.1 mM: inactivation) [10]

Metal compounds/salts

Iron (non-heme iron protein [8, 12, 15], 13.6 iron atoms per enzyme molecule [12], 11.5 iron atoms per enzyme molecule [18], two identical centres each of the [4Fe-4S] and of the [2Fe-2S] type [18], no increase of activity by addition of Co^{2+}, Mn^{2+}, Ni^{2+} or Fe^{2+} [8]) [8, 12, 15, 18]; $Na_2S_2O_4$ ((+ 101 kPa H_2) activates) [5]; K^+ (activates) [10]; NH_4^+ (activates) [10]; SO_4^{2-} (activates) [10]; Co^{2+} (NAD^+ reduction with H_2 is completely dependent on the presence of divalent metal ions (Ni^{2+}, Co^{2+}, Mg^{2+} or Mn^{2+}) or of high salt concentrations (0.5–1.5 M) [11], no increase of activity by addition of Co^{2+}, Mn^{2+}, Ni^{2+} or Fe^{2+} [8]) [11]; Mn^{2+} (NAD^+ reduction with H_2 is completely dependent on the presence of divalent metal ions (Ni^{2+}, Co^{2+}, Mg^{2+} or Mn^{2+}) or of high salt concentrations (0.5–1.5 M) [11], no increase of activity by addition of Co^{2+}, Mn^{2+}, Ni^{2+} or Fe^{2+} [8]) [11]; Ni^{2+} (nickel protein [12, 15, 16], 3.8 nickel atoms per enzyme molecule [12], 2 nickel atoms per enzyme molecule [16], NAD-reduction: no linear kinetics in absence of metals, 0.5 mM Ni^{2+} and 5 mM Mg^{2+} required [10], NAD^+ reduction with H_2 is completely dependent on the presence of divalent metal ions (Ni^{2+}, Co^{2+}, Mg^{2+} or Mn^{2+}) or of high salt concentrations (0.5–1.5 M) [11], most specific effect by $NiCl_2$, optimal concentration: 1 mM [11], no increase of activity by addition of Co^{2+}, Mn^{2+}, Ni^{2+} or Fe^{2+} [8]) [10–12, 15, 16]; Mg^{2+} (NAD-reduction: no linear kinetics in absence of metals, 0.5 mM Ni^{2+} and 5 mM Mg^{2+} required [10], NAD^+ reduction with H_2 is completely dependent on the presence of divalent metal ions (Ni^{2+}, Co^{2+}, Mg^{2+} or

Mn^{2+}) or of high salt concentrations (0.5–1.5 M) [11]) [10, 11]; Phosphate (increases NAD-reduction rate) [10]; More (no increase of activity by addition of Co^{2+}, Mn^{2+}, Ni^{2+} or Fe^{2+} [8], stimulation of activity by salts is the greater the less chaotrophic the anion [11]) [8, 11]

Turnover number (min^{-1})

Specific activity (U/mg)
36.5 [2]; 54.46 [8]; 103 [10]; 57 [18]

K_m-value (mM)
0.19 (H_2) [2]; 0.037 (H_2) [8]; 0.56 (NAD^+) [8]; 0.123 (NAD^+) [8]; 0.063 (H_2) [10]; 0.15 (NAD^+) [11]; 0.38 (FAD) [11]; 0.90 (methyl viologen, pH 8.0) [11]; 0.05 (methyl viologen, pH 9.4) [11]; 0.44 (methylene blue) [11]; 4.00 (ferricyanide) [11]

pH-optimum
6.0 (NADH + H^+) [11]; 6.1 (NAD^+ + H_2, no addition of Ni^{2+}, Tris/MES buffer) [11]; 7.0 (H_2 + methylene blue (+ catalytic amounts of NAD^+)) [10]; 7.8–8.0 (NAD^+ + H_2, in triethanolamine/HCl buffer containing $NiCl_2$) [11]; 7.9 (potassium phosphate buffer) [2]; 8.0 (H_2 + NAD^+ [10]) [8, 10]; 8.5 (Tris-HCl buffer) [2]

pH-range
6.9–9.1 (6.9: 20% of activity maximum (Tris-HCl buffer and potassium phosphate buffer), 9.1: 40% of activity maximum (Tris-HCl buffer), 65% of activity maximum (potassium phosphate buffer)) [2]; 7.0–9.5 (7.0: 40–45% of activity maximum (triethanolamine buffer, Tris buffer), 9.5: 60–75% of activity maximum (triethanolamine buffer, Tris buffer)) [10]; More [11]

Temperature optimum (°C)
33 [8]; 36 [2]

Temperature range (°C)

---

3 ENZYME STRUCTURE

Molecular weight
178000–200000 (Nocardia opaca, gel filtration, sucrose density gradient centrifugation) [11]
205000 (Alcaligenes eutrophus H16, sucrose density gradient centrifugation, gel filtration) [8]

Enzyme Handbook © Springer-Verlag Berlin Heidelberg 1994
Duplication, reproduction and storage in data banks are only allowed with the prior permission of the publishers

Subunits

Tetramer (1 × 64000 + 1 × 56000 + 1 × 31000 + 1 × 27000, Nocardia opaca 1b, native hydrogenase dissociates into two subunit dimers: 1. 64000 MW sununit + 31000 MW subunit, 2. 56000 MW subunit + 27000 MW subunit, SDS-PAGE) [11]

Glycoprotein/Lipoprotein

–

4 ISOLATION/PREPARATION

Source organism

Hydrogenomonas sp. H16 [2]; Alcaligenes eutrophus (H16 [4, 8, 9, 15, 17], N9A [9], Z1 [6, 7, 13, 14], B19 [9], G27 [9]) [3–9, 13–19]; Nocardia opaca 1b [10–12]

Source tissue

Cells [2]

Localisation in source

Soluble [2–4, 8, 9, 11, 15–19]; Cytoplasm (located as inclusion bodies in the DNA region of the cell) [3]

Purification

Hydrogenomonas sp. H16 [2]; Alcaligenes eutrophus (H16 [8]) [8, 18]; Nocardia opaca 1b [10]

Crystallization

–

Cloned

[1]

Renaturated

–

5 STABILITY

pH

6.0 (1 day: 9% loss of activity in presence of 0.5 mM $NiCl_2$ + 5 mM $MgCl_2$, 56% loss of activity in absence of $NiCl_2$ and $MgCl_2$, 3 days: 44% loss of activity in presence of 0.5 mM $NiCl_2$ + 5 mM $MgCl_2$, 80% loss of activity in absence of $NiCl_2$ and $MgCl_2$) [11]; 6.5 (highest stability, 4°C, H_2–atmosphere, 3 days) [10]; 7.0 (1 day: no loss of activity in presence of 0.5 mM $NiCl_2$ + 5 mM $MgCl_2$, 62% loss of activity in absence of $NiCl_2$ and $MgCl_2$, 3 days: 21%

loss of activity in presence of 0.5 mM $NiCl_2$ + 5 mM $MgCl_2$, 87% loss of activity in absence of $NiCl_2$ and $MgCl_2$) [11]; 8.0 (1 day: no loss of activity in presence of 0.5 mM $NiCl_2$ + 5 mM $MgCl_2$, 86% loss of activity in absence of $NiCl_2$ and $MgCl_2$, 3 days: 27% loss of activity in presence of 0.5 mM $NiCl_2$ + 5 mM $MgCl_2$, complete loss of activity in absence of $NiCl_2$ and $MgCl_2$) [11]

Temperature (°C)

45 (10 min, 90% loss activity of enzyme prepared in Tris or triethanolamine buffer, 30% loss of activity of enzyme prepared in K-phosphate buffer) [10]; 50 (thermoinactivation by dissociation of the native enzyme tetramer into constituent heterodimers) [6]

Oxidation

Oxidized hydrogenase as purified under aerobic conditions is of high stability, but is not reactive, reductive activation of the enzyme by H_2 in presence of catalytic amounts of NADH or by reducing agents destabilizes [8]; Redox-dependent inactivation and activation [13, 14]; Stabilized by oxidation with O_2 and ferricyanide in absence of reducing components, simultaneous presence of H_2, NADH and O_2 causes irreversible inactivation, dependent on O_2 concentration [17]

Organic solvent

General stability information

Addition of 5 mM Mg^{2+} + 0.5 mM Ni^{2+} stabilizes [10]; Ni^{2+}: stabilizes [11]; 5 mM Mg^{2+} and 0.5 mM Ni^{2+} stabilizes [10]

Storage

–20°C, pH 7.0 [8]; –20°C [10]; –20°C, 4 mg/ml protein concentration, pH 7.0 [18]

6 CROSSREFERENCES TO STRUCTURE DATABANKS

PIR/MIPS code

PIR3: A35385 (alpha chain Alcaligenes eutrophus); PIR2: S03941 (alpha chain cytosolic "Nocardia opaca" (fragment)); PIR2: S03942 (alpha chain cytosolic Alcaligenes eutrophus (fragment)); PIR3: D35385 (beta chain Alcaligenes eutrophus); PIR2: S03943 (beta chain cytosolic "Nocardia opaca" (fragment)); PIR2: S03944 (beta chain cytosolic Alcaligenes eutrophus (fragment)); PIR3: C35385 (delta chain Alcaligenes eutrophus); PIR2: S03947 (delta chain cytosolic "Nocardia opaca" (fragment)); PIR2: S03948 (delta chain cytosolic Alcaligenes eutrophus (fragment)); PIR3: B35385 (gamma chain Alcaligenes eutrophus); PIR2: S03945 (gamma chain cytosolic "Nocardia opaca" (fragment)); PIR2: S03946 (gamma chain cytosolic Alcaligenes eutrophus (fragment))

Enzyme Handbook © Springer-Verlag Berlin Heidelberg 1994
Duplication, reproduction and storage in data banks are only allowed with the prior permission of the publishers

Brookhaven code

7 LITERATURE REFERENCES

[1] Tran-Betcke, A., Warnecke, U., Böcker, C., Zaborosch, C., Friedrich, B.: J. Bacteriol.,172,2920–2929 (1990)
[2] Pfitzner, J., Linke, H.A.B., Schlegel, H.G.: Arch. Mikrobiol.,71,67–78 (1970)
[3] Rohde, M., Johannssen, W., Mayer, F.: FEMS Microbiol. Lett.,36,83–86 (1986)
[4] Hyman, M.R., Arp, D.J.: Biochem. J.,254,469–475 (1988)
[5] Hyman, M.R., Fox, C.A., Arp, D.J.: Biochem. J.,254,463–468 (1988)
[6] Popov, V.O., Ovchinnikov, A.N., Utkin, I.B., Gazaryan, I.G., Egorov, A.M., Berezin, I.V.: Biochim. Biophys. Acta,831,297–301 (1985)
[7] Popov, V.O., Gazaryan, I.G., Egorov, A.M., Berezin, I. V.: Biochim. Biophys. Acta,827,466–471 (1985)
[8] Schneider, K., Schlegel, H.G.: Biochim. Biophys. Acta,452,66–80 (1976)
[9] Schneider, K., Schlegel, H.G.: Arch. Microbiol.,112,229–238 (1977)
[10] Aggag, M., Schlegel, H.G.: Arch. Microbiol.,100,25–39 (1974)
[11] Schneider, K., Schlegel, H.G., Jochim, K.: Eur. J. Biochem.,138,553–541 (1984)
[12] Schneider, K., Cammack, R., Schlegel, H.G.: Eur. J. Biochem.,142,75–84 (1984)
[13] Petrov, R.R., Utkin, I.B., Popov, V.O.: Arch. Biochem. Biophys.,268,298–305 (1989)
[14] Petrov, R.R., Utkin, I.B., Popov, V.O.: Arch. Biochem. Biophys.,268,287–297 (1989)
[15] Friedrich, C.G., Suetin, S., Lohmeyer, M.: Arch. Microbiol.,140,206–211 (1984)
[16] Friedrich, C.G., Schneider, K., Friedrich, B.: J. Bacteriol.,152,42–48 (1982)
[17] Schneider, K., Schlegel, H.G.: Biochem. J.,193,99–107 (1981)
[18] Schneider, K., Cammack, R., Schlegel, H.G., Hall, D. O.: Biochim. Biophys. Acta,578,445–461 (1979)
[19] Schneider, K., Schlegel, H.G.: Biochem. Biophys. Res. Commun.,84,564–571 (1978)

1 NOMENCLATURE

EC number

1.12.2.1

Systematic name

Hydrogen:ferricytochrome-c_3 oxidoreductase

Recommended name

Cytochrome-c_3 hydrogenase

Synonymes

Hydrogenase
Hydrogenase, cytochrome
Cytochrome c_3 hydrogenase
Cytochrome c_3 reductase
Cytochrome hydrogenase
H_2:ferricytochrome c_3 oxidoreductase
More (may be identical to EC 1.18.3.1, hydrogenase which can use a cytochrome c (not precisely characterized) as electron acceptor: Paracoccus denitrificans [20], Alcaligenes latus [21], Alcaligenes eutrophus [22]) [20–22]

CAS Reg. No.

9079-91-8

2 REACTION AND SPECIFICITY

Catalysed reaction

H_2 + ferricytochrome c_3 →
→ 4 H^+ + ferrocytochrome c_3

Reaction type

Redox reaction

Natural substrates

H_2 + ferricytochrome c_3 (enzyme participates in production or consumption of H_2 in bacterial metabolism) [6]
H^+ + ferrocytochrome c_3 (enzyme participates in production or consumption of H_2 in bacterial metabolism) [6]

Enzyme Handbook © Springer-Verlag Berlin Heidelberg 1994
Duplication, reproduction and storage in data banks are only allowed with the prior permission of the publishers

Substrate spectrum

1 H_2 + ferricytochrome c_3 (r [4, 5, 15, 18], H_2–evolution: Desulfovibrio multispirans enzyme is more active in hydrogen evolution rather than in hydrogen uptake assay [15], Desulfovibrio vulgaris [1, 6, 7], Desulfovibrio gigas [1, 14], Desulfovibrio desulfuricans [1, 4, 5], Desulfovibrio baculatus [12], H_2–consumption: Desulfovibrio vulgaris [1, 6, 7], Desulfovibrio desulfuricans [1, 4, 5], Desulfovibrio multispirans (more active in hydrogen evolution rather than in hydrogen uptake assay) [15], other electron carriers: methyl viologen [4, 5, 18], benzyl viologen [4, 5], methylene blue [4, 5], ferricyanide [18], 2,6-dichlorophenolindophenol [18], no activity as electron carriers: $NADP^+$ [18], FAD [18], FMN [18], O_2 [18], Desulfovibrio vulgaris: reduces cytochrome c_3 but not cytochrome c_{553} [1], Desulfovibrio gigas: cytochrome c_3 required for reduction of ferredoxin, rubredoxin and cytochrome c_3 with hydrogen and for the evolution of hydrogen from dithionite [1]) [1, 4–7, 12, 14, 15, 18]

2 More (hydrogenases which can use a cytochrome c (not precisely characterized) as electron acceptor: Paracoccus denitrificans [20], Alcaligenes latus [21], Alcaligenes eutrophus [22]) [20–22]

Product spectrum

1 H^+ + ferrocytochrome c_3 (r [4, 5, 15, 18])

2 ?

Inhibitor(s)

CO ((NiFe)hydrogenase from Desulfovibrio gigas has lower sensitivity to CO inhibition than (Fe)hydrogenase from Desulfovibrio vulgaris [1], (Fe) and (NiFeSe)hydrogenase sensitive, (NiFe)enzyme slightly inhibited [16]) [1, 4–6, 16]; O_2 (Desulfovibrio desulfuricans (Norway), reversible inactivation) [1]; Bromosuccinimide [4]; $HgCl_2$ [4]; SDS [6]; p-Chloromercuribenzoate [6]; Dimethylsulfoxide [7]; NO ((NiFe)hydrogenase, (Fe) hydrogenase and (NiFeSe)hydrogenase sensitive) [16]; NO_2^- ((Fe)hydrogenase and (NiFeSe)hydrogenase sensitive, (NiFe) hydrogenase uneffected) [16]; High ionic strength [4, 7]; More (resistant to inhibition by thiol-blocking and metal-complexing reagents) [4]

Cofactor(s)/prostethic group(s)

Cytochrome c_3 (rate enhancement in H_2–evolution assay with methyl viologen) [4]

Metal compounds/salts

Iron (Desulfovibrio gigas: (NiFe)protein, nonheme iron protein [1], 11–12 gatom of nonheme iron and sulfide per mol [1, 14], 3 4Fe-4S clusters [14], Desulfovibrio vulgaris: (Fe)protein, 12.5 gatom of nonheme iron and 11.5 gatom of acid-labile sulfide per mol, at least 2 (4Fe-4S) clusters of the ferredoxin type [13], has a (NiFe) hydrogenase and a (NiFeSe)hydrogenase [13], Desulfovibrio vulgaris (Miyazaki): 7–9 gatom of iron per mol [6], Desulfovibrio desulfuricans (Norway): 6 iron atoms and 6 acid-labile sulphur groups per molecule [4], 5–10 iron atoms and 6 acid-labile sulfur groups per molecule [11], Desulfovibrio baculatus: nickel-[iron-sulfur]-selenium protein [12], Desulfovibrio multispirans: (NiFe)protein, 0.9 atoms of nickel, 11 atoms of iron and 10 atoms of labile sulfide per molecule [15], Thiocapsa roseopersicina: 1 mol Ni^{2+} per mol of enzyme, 4 mol Fe^{2+} per mol of enzyme [19]) [1, 4, 6, 11–15, 19]; Nickel (Desulfovibrio gigas: (NiFe)protein, 1 mol of nickel per mol of enzyme [1], Thiocapsa roseopersicina: 1 mol Ni^{2+} per mol of enzyme, 4 mol Fe^{2+} per mol of enzyme [19], Desulfovibrio vulgaris (Hildenborough): (Fe) protein does not contain nickel as essential constituent [1], has a (NiFe)hydrogenase and a (NiFeSe)hydrogenase [13], Desulfovibrio baculatus: nickel-[iron-sulfur]-selenium containing enzyme [12], Desulfovibrio multispirans: (NiFe)protein, 0.9 atoms of nickel per molecule [15]) [1, 12, 13, 15, 19]; Selenium (Desulfovibrio desulfuricans (Norway): 0.5–0.8 atoms of selenium per molecule [11], Desulfovibrio baculatus: nickel-[iron-sulfur]-selenium-containing enzyme [12], Desulfovibrio vulgaris: has a (NiFe)hydrogenase and a (NiFeSe)hydrogenase [13]) [11–13]

Turnover number (min^{-1})

25500 (methyl viologen) [11]

Specific activity (U/mg)

440 (Desulfovibrio gigas, H_2–evolution assay) [1]; 4600 (Desulfovibrio vulgaris, H_2–evolution assay) [1]; 70 (Desulfovibrio desulfuricans, H_2–evolution assay) [4]; 12.7 (H_2–evolution assay) [12]; 90.8 (H_2–evolution assay) [14]; More [5, 11, 18]

K_m-value (mM)

0.002 (cytochrome c_3 (H_2–evolution)) [4]; 0.003 (methyl viologen (H_2–evolution)) [4]; 0.080 (benzyl viologen (H_2–evolution)) [4]; 0.075 (ferrocytochrome c_3 (H_2–evolution)) [5]; 0.0014 (H_2) [18]

Enzyme Handbook © Springer-Verlag Berlin Heidelberg 1994
Duplication, reproduction and storage in data banks are only allowed with the prior permission of the publishers

pH-optimum

4.0 (Desulfovibrio baculatus, periplasm, H_2–production) [12]; 4.5 (Desulfovibrio baculatus, cytoplasm, H_2–production) [12]; 5.5 (soybean root nodules infected with Rhizobium japonicum) [18]; 5–6 (Desulfovibrio desulfuricans Norway, cytochrome c_3-dependent H_2–evolution) [5]; 6 [7]; 6.5 (Desulfovibrio desulfuricans Norway, H_2–evolution) [4]; 8–9 (Desulfovibrio desulfuricans Norway, cytochrome c_3 reduction by H_2) [5]; More (enzyme encapsulated in sodium di-2-ethylhexylsulfosuccinate micelles: pH-optimum shift of 1 pH unit in the basic direction) [8]

pH-range

4–7 (cytochrome c_3-dependent H_2–evolution, 4: about 65% of activity maximum, 7: about 40% of activity maximum) [5]; 5–9 (H_2–evolution, 5: about 30% of activity maximum, 9: about 40% of activity maximum) [14]; 6–10 (cytochrome c_3 reduction by H_2, 6: about 25% of activity maximum, 10: about 30% of activity maximum) [5]

Temperature optimum (°C)

65–70 [18]

Temperature range (°C)

---

3 ENZYME STRUCTURE

Molecular weight

50000 (Desulfovibrio vulgaris (Hildenborough)) [1]
58000 (Desulfovibrio desulfuricans (Norway), gel filtration) [1, 4]
60000 (Desulfovibrio desulfuricans, gel filtration) [5]
65000 (soybean root nodules infected with Rhizobium japonicum, gel filtration, SDS-PAGE) [18]
67000 (Desulfovibrio desulfuricans, gel filtration) [11]
68000 (Thiocapsa roseopersicina) [19]
89000 (Desulfovibrio vulgaris (Miyazaki), low-speed sedimentation equilibrium) [6]
89500 (Desulfovibrio gigas, analytical ultracentrifugation and equilibrium sedimentation) [1, 14]
100000 (Desulfovibrio baculatus, periplasmic, cytoplasmic and membrane-bound enzyme, non-dissociating HPLC gel filtration) [12]

Subunits

Monomer (1 × 50000, Desulfovibrio vulgaris (Hildenborough), (Fe)hydrogenase [1], 1 × 58000, Desulfovibrio desulfuricans Norway [1, 4]) [1, 4]
Dimer (1 × 62000 + 1 × 26000, Desulfovibrio gigas, (nonheme iron protein), SDS-PAGE [1, 14], 1 × 59000 + 1 × 28000, Desulfovibrio vulgaris (Miyazaki), SDS-PAGE [6], 1 × 56000 + 1 × 29000, Desulfovibrio desulfuricans (Norway), SDS-PAGE [11], 1 × 47000 + 1 × 25000, Thiocapsa roseopersicina [19], 1 × 54000 + 1 × 27000, Desulfovibrio baculatus, cytoplasmic, SDS-PAGE [12], 1 × 49000 + 1 × 26000, Desulfovibrio baculatus, periplasmic, SDS-PAGE [12], 1 × 62000 + 1 × 27000, Desulfovibrio baculatus, membrane-bound, SDS-PAGE [12]) [1, 6, 11, 12, 14, 19]

Glycoprotein/Lipoprotein

–

4 ISOLATION/PREPARATION

Source organism

Desulfovibrio salexigens [16]; Desulfovibrio baculatus (DSM 1743) [12]; Desulfovibrio multispirans [15, 16]; Desulfovibrio vulgaris (Hildenborough [1, 3, 7, 16], Miyazaki [6, 10]) [1–3, 6, 7, 10, 13, 16]; Desulfovibrio gigas [1, 2, 8, 9, 14, 16, 17]; Rhizobium japonicum (soybean root nodules infected with Rhizobium japonicum 110) [18]; Thiocapsa roseopersicina BBS [19]; E. coli [23]; More (hydrogenases which can use a cytochrome c (not precisely characterized) as electron acceptor: Paracoccus denitrificans [20], Alcaligenes latus [21], Alcaligenes eutrophus [22]) [20–22]

Source tissue

Cells [6]; Root nodules (soybean root nodules infected with Rhizobium japonicum) [18]

Localisation in source

Membrane (bound: E. coli [23], Desulfovibrio desulfuricans Norway [1, 4], (NiFe)hydrogenase: located in the near vicinity of the cytoplasmic membrane facing the periplasmic space [13], (NiFeSe)hydrogenase: associated with the cytoplasmic side of the cytoplasmic membrane [13]) [1, 4, 12, 13, 19, 23]; Periplasm (Desulfovibrio gigas, Desulfovibrio vulgaris) [1, 12]; Cytoplasm (Desulfovibrio baculatus [12], Desulfovibrio multispirans [15]) [12, 15]; Soluble (Desulfovibrio desulfuricans Norway [11]) [11, 19]

Purification

Rhizobium japonicum (soybean root nodules infected with Rhizobium japonicum 110) [18]; Thiocapsa roseopersicina BBS [19]

Enzyme Handbook © Springer-Verlag Berlin Heidelberg 1994
Duplication, reproduction and storage in data banks are only allowed with the prior permission of the publishers

Crystallization

(Desulfovibrio vulgaris (Miyazaki)) [10]

Cloned

(Desulfovibrio (Hildenborough), expression in E. coli) [3]

Renaturated

–

5 STABILITY

pH

Temperature (°C)

30 (120 min, stable) [4]; 40 (120 min, 7% loss of activity) [4]; 50 (10 min, stable) [5]; 60 (120 min, 7% loss of activity [4], 10 min, 30% loss of activity [5]) [4, 5]; 65–70 (unstable above) [18]; 70 (10 min, 37% loss of activity) [5]; 77 (120 min, 25% loss of activity) [4]; 80 (10 min, 69% loss of activity) [5]; 85 (120 min, 94% loss of activity) [4]; 90 (10 min, complete loss of activity) [5]

Oxidation

Half-life in O_2 of enzyme from soybean root nodules infected with Rhizobium japonicum 110: 70 min [18]; Stable in O_2 [4]; Desulfovibrio vulgaris, strain Hildenborough: long exposure of oxidized enzyme to O_2 does not irreversibly inactivate the hydrogenase, the reduced form is irreversibly inactivated by O_2 [1]; More (Desulfovibrio gigas: enzyme can exist in 3 different states: 1. active state (active in all assays), 2. unready state (totally inactive), 3. ready state (does not react with hydrogen, but is rapidly activated by strong reductants), the active form of hydrogenase is rapidly inactivated by O_2, producing mostly the unready state, which can be reactivated only slowly [17], Desulfovibrio desulfuricans (Norway): enzyme reversibly inactivated by O_2 [1]) [1, 17]

Organic solvent

Hexane, half-life of enzyme encapsulated in sodium di-2-ethylhexylsulfosuccinate micelles is 13.1 h [9]; Isooctane, half-life of enzyme encapsulated in sodium di-2-ethylhexylsulfosuccinate micelles is 9.5 h [9]; Dodecane, half-life of enzyme encapsulated in sodium di-2-ethylhexylsulfosuccinate micelles: 2.9 h [9]

General stability information

Stable to urea [6]; Operational stability of enzyme encapsulated in sodium di-2-ethylhexyl-sulfosuccinate micelles [9]; Repeated freezing and thawing causes loss of activity [18]

Storage

4°C, half-life: several days [18]; Stored indefinitely in liquid N_2 [18]

6 CROSSREFERENCES TO STRUCTURE DATABANKS

PIR/MIPS code

Brookhaven code

7 LITERATURE REFERENCES

[1] Odom, J.M., Peck, H.D.: Annu. Rev. Microbiol.,38,551–592 (1984) (Review)
[2] DerVartanian, D.V., LeGall, J.: Biochim. Biophys. Acta,346,79–99 (1974)
[3] Voordouw, G., Hagen, W.R., Krüse-Wolters, K.M., van Berkel-Arts, A., Veeger, C.: Eur. J. Biochem.,162,31–36 (1987)
[4] Lalla-Maharajh, W.V., Hall, D.O., Cammack, R., Rao, K. K.: Biochem. J.,209,445–454 (1983)
[5] Yagi, T., Honya, M., Tamiya, N.: Biochim. Biophys. Acta,153,699–705 (1968)
[6] Yagi, T., Kimura, K., Daidoji, H., Sakai, F., Tamura, S., Inokuchi, H.: J. Biochem.,79,661–671 (1976)
[7] Grande, H.J., van Berkel-Arts, A., Bregh, J., van Dijk, K., Veeger, C.: Eur. J. Biochem.,131,81–88 (1983)
[8] Castro, M.J.M., Cabral, J.M.S.: Enzyme Microb. Technol.,11,6–11 (1989)
[9] Castro, M.J.M., Cabral, J.M.S.: Enzyme Microb. Technol.,11,668–672 (1989)
[10] Higuchi, Y., Yasuoka, N., Kakudo, M., Katsube, Y., Yagi, T., Inokuchi, H.: J. Biol. Chem.,262,2823–2825 (1987)
[11] Rieder, R., Cammack, R., Hall, D.O.: Eur. J. Biochem.,145,637–643 (1984)
[12] Teixeira, M., Fauque, G., Moura, I., Lespinat, P.A., Berlier, Y., Prickril, B., Peck, H.D., Xavier, A.V., Le Gall, J., Moura, J.J.G.: Eur. J. Biochem.,167,47–58 (1987)
[13] Rohde, M., Fürstenau, U., Mayer, F., Przybyla, A.E., Peck, H.D., Le Gall, J., Choi, E.S., Menon, N.K.: Eur. J. Biochem.,191,389–396 (1990)
[14] Hatchikian, E.C., Bruschi, M., Le Gall, J.: Biochem. Biophys. Res. Commun.,82,451–461 (1978)
[15] Czechowski, M.H., He, S.H., Nacro, M., DerVartanian, D.V., Peck, H.D., LeGall, J.: Biochem. Biophys. Res. Commun.,125,1025–1032 (1984)
[16] Berlier, Y., Fauque, G.D., LeGall, J., Choi, E.S., Peck, H.D., Lespinat, P.A.: Biochem. Biophys. Res. Commun.,146,147–153 (1987)
[17] Fernandez, V.M., Hatchikian, E.C., Cammack, R.: Biochim. Biophys. Acta,832,69–79 (1985)
[18] Arp, D.J., Burris, R.H.: Biochim. Biophys. Acta,570,221–230 (1979)
[19] Gogotov, I.N.: Arch. Microbiol.,140,86–90 (1984)
[20] Knüttel, K., Schneider, K., Schlegel, H.G., Müller, A.: Eur. J. Biochem.,179,101–108 (1989)
[21] Pinkwart, M., Schneider, K., Schlegel, H.G.: Biochim. Biophys. Acta,745,267–278 (1983)
[22] Podzuweit, H.G., Schneider, K., Knüttel, H.: Biochim. Biophys. Acta,905,435–446 (1987)
[23] Adams, M.W.W., Hall, D.O.: Biochem. Soc. Trans.,6,1339–1341 (1978)

Enzyme Handbook © Springer-Verlag Berlin Heidelberg 1994
Duplication, reproduction and storage in data banks are only allowed with the prior permission of the publishers

1 NOMENCLATURE

EC number
1.12.99.1

Systematic name
Hydrogen:(acceptor) oxidoreductase

Recommended name
Coenzyme F_{420} hydrogenase

Synonymes
Hydrogenase, coenzyme F_{420}
Coenzyme F_{420}-dependent hydrogenase
8-Hydroxy-5-deazaflavin-reducing hydrogenase [3]
F_{420}-reducing hydrogenase [5]
More (a similar enzyme from Streptomyces aureofaciens and Streptomyces rimosus reduces 8-hydroxy-5-deazaisoalloxazine derivatives using NADPH) [16]

CAS Reg. No.
65099-08-3

2 REACTION AND SPECIFICITY

Catalysed reaction
H_2 + coenzyme F_{420} →
→ reduced coenzyme F_{420} (two-site hybrid ping-pong mechanism [3], the natural hydrogen acceptor, coenzyme F_{420} is a deazaflavin derivative (7,8-didemethyl-8-hydroxy-5-deazariboflavin 5'-phosphoryllactylglutamyl-glutamate) [7])

Reaction type
Redox reaction

Natural substrates
H_2 + coenzyme F_{420} (one of the key enzymes in hydrogen-dependent light-electron reduction of CO_2 to methane [8]) [3, 8]

Enzyme Handbook © Springer-Verlag Berlin Heidelberg 1994
Duplication, reproduction and storage in data banks are only allowed with the prior permission of the publishers

Substrate spectrum

1 H_2 + coenzyme F_{420} (i.e. 7,8-didemethyl-8-hydroxy-5-deazariboflavin 5'-phosphoryllactylglutamylglutamate [7], other electron acceptors: methyl viologen [3, 4, 11, 13], flavins [11], 7,8-didemethyl-8-hydroxy-5-deazariboflavin [11, 13]) [1–16]
2 More (F_{420} hydrogenase and methyl viologen hydrogenase in Methanobacterium formicium are distinct enzymes) [6]

Product spectrum

1 Reduced coenzyme F_{420} [1–16]
2 ?

Inhibitor(s)

Cofactor(s)/prostethic group(s)

Coenzyme F_{420} [1–16]; Flavin (flavoprotein, FMN not detected) [4]; FAD (2.3 mol of FAD per mol of 170000 MW protein [13], 1 mol per mol of 109000 MW protein [11], 1 mol per mol of 105000 MW protein [4], 0.8–0.9 mol FAD per mol of 115000 MW unit [7], FAD required [9]) [4, 7, 9, 11, 13]

Metal compounds/salts

Iron (4.5 mol per mol of 105000 MW protein [4], 13–14 iron atoms in iron-sulfur centers per molecule 115000 MW protein [7], 12–14 iron atoms per molecule of 109000 MW protein [11], 33 atoms of Fe per molecule of 170000 MW protein [11]) [4, 7, 11]; Nickel (nickel protein [4, 5, 7, 8], 0.63 mol of nickel per mol of 105000 MW protein [4], 6–7 gatom of Ni^{2+} per mol of enzyme [5], 0.6–0.7 nickel atoms per molecule of 115000 MW protein [7], 1 mol of nickel per mol of 109000 MW protein [11]) [4, 5, 7, 8, 11]; Selenium (Methanococcus vannielii: 3.8 atoms of selenium per molecule of enzyme, selenium is present exclusively in a 420000 subunit, chemical form is selenocysteine [15], 0.66 mol of selenium per mol of 105000 MW protein [4], enzyme contains no selenium [11]) [4, 15]; Mg^{2+} (stimulates) [5]; Ca^{2+} (stimulates) [5]

Turnover number (min^{-1})

43500 (F_{420}) [13]; 13500 (F_{420}) [3]; 1050 (8-hydroxy-5-deazariboflavin) [4]

Specific activity (U/mg)

10 [4]; 49 (F_{420}) [7]; 16 (methyl viologen) [12]; 5.2 (F_{420}) [12]; 0.847 (F_{420}) [5]; 11.1 (benzyl viologen) [5]; More [9]

K_m-value (mM)

0.012 (H_2 (+ 8-hydroxy-5-deazariboflavin)) [3]; 0.007 (H_2 (+ methyl viologen)) [3]; 0.087 (FAD) [11]; 0.019 (FAD) [13]; 0.016 (F_{420}) [4]; 1.2 (methyl viologen) [4]; 1.56 (methyl viologen) [11]; 0.42 (methyl viologen) [13]; 0.36 (coenzyme F_{420}) [3]; 0.037 (F_{420}) [11]; 0.034 (7,8-didemethyl-8-hydroxy-5-deazaflavin) [13]; 0.010 (H_2 (+ didemethyl-8-hydroxy-5-deazaflavin)) [13]; More [13]

pH-optimum

6.5–7.5 (8-hydroxy-5-deazariboflavin) [7]; 7.0–7.5 (F_{420}) [11]; 10 (methyl viologen reduction rate increases up to pH 11 beyond which the enzyme is not stable) [7]; 11 (methyl viologen reduction rate increases up to pH 11 beyond which the enzyme is not stable) [12]

pH-range

5.5–9.3 (5.5: about 40% of activity maximum, 9.3: about 10% of activity maximum) [7]; 5.5–9.4 (5.5: about 20% of activity maximum, 9.4: about 15% of activity maximum) [11]; 5.8–9.0 (5.8: about 55% of activity maximum, 9.0: about 20% of activity maximum) [12]

Temperature optimum (°C)

55 [11]; 80 (H_2 + coenzyme F_{420}) [12]; 90 (H_2 + methyl viologen) [12]

Temperature range (°C)

25–92 (at 25°C and 92°C: about 20% of activity maximum, H_2 + methyl viologen) [12]; 40–103 (40°C: about 5% of activity maximum, 103°C: about 10% of activity maximum, H_2 + coenzyme F_{420}) [12]

Enzyme Handbook © Springer-Verlag Berlin Heidelberg 1994
Duplication, reproduction and storage in data banks are only allowed with the prior permission of the publishers

3 ENZYME STRUCTURE

Molecular weight

105000 (Methanococcus voltae, sucrose density gradient centrifugation, minimal active unit) [4]
109000 (Methanobacterium formicium, gel filtration, minimal active unit) [11]
115000 (Methanobacterium thermoautotrophicum, gel filtration, native gradient PAGE, minimal active unit [7], Methanococcus jannaschii, native gradient electrophoresis, 2 major bands: MW 115000 and 990000 [12]) [7, 12]
170000 (Methanobacterium thermoautotrophicum, native gradient PAGE, minimal active subunit) [13]
340000 (Methanococcus vannielii, gel filtration) [15]
380000 (Methanobacterium formicium, gel filtration, 2 forms: MW 600000 and 380000) [14]
600000 (Methanobacterium formicium, gel filtration, 2 forms: MW 600000 and 380000) [14]
720000 (Methanospirillum hungatei, estimated from subunits) [5]
800000 (Methanobacterium thermoautotrophicum, high molecular aggregate) [7, 8]
990000 (Methanococcus jannaschii, native gradient electrophoresis, 2 major bands: MW 115000 and 990000) [12]
More (enzyme tends to aggregate) [4, 8, 15]

Subunits

Trimer (1 × alpha (43600) + 1 × beta (36700) + 1 × gamma (28800), Methanobacterium formicium, minimal active unit of MW 109000, SDS-PAGE [11], 1 × alpha (47000) + 1 × beta (31000) + 1 × gamma (26000), Methanobacterium thermoautotrophicum, minimal active unit of MW 115000, SDS-gel electrophoresis [7]) [7, 11]
Oligomer (x × 420000 + x × 34000 + x × 20500, Methanobacterium formicium, SDS-PAGE [14], x × 40000 + x × 31000 + x × 26000, Methanobacterium thermoautotrophicum [13], x × 55000 + x × 45000 + x × 37000 + x × 27000, Methanococcus voltae, SDS-PAGE [4], 5 × alpha (50700) + 15 × 30700, Methanospirillum hungatei, SDS-PAGE (mercaptoethanol) [5]) [4, 5, 13, 14]
Polymer (8 × alpha (47000) + 8 × beta (31000) + 8 × gamma (26000), Methanobacterium thermoautotrophicum, high molecular aggregate, electron microscopy) [8]
More (enzyme tends to aggregate) [4, 8, 15]

Glycoprotein/Lipoprotein

–

4 ISOLATION/PREPARATION

Source organism

Methanococcus voltae [2, 4]; Methanospirillum hungatei [5]; Methanobacterium formicium [1, 6, 9, 11, 14]; Methanobacterium thermoautotrophicum [7, 8, 10, 13]; Methanococcus jannaschii [12]; Methanococcus vannielii [15]

Source tissue

Spheroblast lysate [5]

Localisation in source

Cell membrane [1, 2, 6]; Soluble [8]; More (enzyme is not a soluble periplasmic enzyme [1], F_{420}-hydrogenase associated with membranes, methyl viologen hydrogenase in soluble fraction [6]) [1, 6]

Purification

Methanococcus voltae [4]; Methanospirillum hungatei [5]; Methanobacterium thermoautotrophicum [7, 13]; Methanobacterium formicium [11]; Methanococcus jannaschii (partial) [12]

Crystallization

–

Cloned

(Methanobacterium thermoautotrophicum gene expression in E. coli) [10]

Renaturated

–

5 STABILITY

pH

10 (unstable above) [7]; 11 (unstable above) [12]

Temperature (°C)

65 (30 min, no loss of methyl viologen reducing activity, 76% loss of coenzyme F_{420} reducing activity) [14]; 70 (half-life: 70 min (H_2 + coenzyme F_{420}), 22 min (H_2 + methyl viologen)) [12]

Oxidation

Reductively activated enzyme is rapidly and irreversibly inactivated when exposed to air [7]; Purified F_{420}-hydrogenase requires reductive reactivation before assay [11, 14]

Organic solvent

General stability information

Bovine serum albumin stabilizes [12]

Enzyme Handbook © Springer-Verlag Berlin Heidelberg 1994
Duplication, reproduction and storage in data banks are only allowed with the prior permission of the publishers

Storage

−70°C, more than 1 year, 30% loss of activity of enzyme isolated in aerobic, inactivated form [7]; 4°C, 24 h [12]; −70°C, a few days, loss of high-temperature activity, above 80°C [12]

6 CROSSREFERENCES TO STRUCTURE DATABANKS

PIR/MIPS code

Brookhaven code

7 LITERATURE REFERENCES

[1] Baron, S.F., Williams, D.S., May, H.D., Patel, P.S., Aldrich, H.C., Ferry, J.G.: Arch. Microbiol.,151,307–313 (1989)
[2] Muth, E.: Arch. Microbiol.,150,205–207 (1988)
[3] Livingston, D.J., Fox, J.A., Orme-Johnson, W.H., Walsh, C.T.: Biochemistry,26,4228–4237 (1987)
[4] Muth, E., Mörschel, E., Klein, A.: Eur. J. Biochem.,169,571–577 (1987)
[5] Sprott, G.D., Shaw, K.M., Beveridge, T.J.: Can. J. Microbiol.,33,896–904 (1987)
[6] Baron, S.F., Brown, D.P., Ferry, J.G.: J. Bacteriol.,169,3823–3825 (1987)
[7] Fox, J.A., Livingston, D.J., Orme-Johnson, W.H., Walsh, C.T.: Biochemistry,26,4219–4227 (1987)
[8] Wackett, L.P., Hartwieg, E.A., King, J.A., Orme-Johnson, W.H., Walsh, C.T.: J. Bacteriol.,169,718–727 (1987)
[9] Nelson, M.J.K., Brown, D.P., Ferry, J.G.: Biochem. Biophys. Res. Commun.,120,775–781 (1984)
[10] Alex, L.A., Reeve, J.N., Orme-Johnson, W.H., Walsh, C.T.: Biochemistry,29,7237–7244 (1990)
[11] Baron, S.F., Ferry, J.G.: J. Bacteriol.,171,3846–3853 (1989)
[12] Shah, N.N., Clark, D.S.: Appl. Environ. Microbiol.,56,858–863 (1990)
[13] Jacobson, F.S., Daniels, L., Fox, J.A., Walsh, C.T., Orme-Johnson, W.H.: J. Biol. Chem.,257,3385–3388 (1982)
[14] Jin, S.-L. C., Blanchard, D.K., Chen, J.-S.: Biochim. Biophys. Acta,748,8–20 (1983)
[15] Yamazaki, S.: J. Biol. Chem.,257,7926–7929 (1982)
[16] Novotna, J., Neuzil, J., Hostalek, Z.: FEMS Microbiol. Lett.,59 ,241–246 (1989)

1 NOMENCLATURE

EC number
1.12.99.2

Systematic name
Hydrogen:coenzyme-M-7-mercaptoheptanoylthreonine-phosphate-heterodisulfide oxidoreductase

Recommended name
Coenzyme-M-7-mercaptoheptanoylthreonine-phosphate-heterodisulfide hydrogenase

Synonymes
Reductase, coenzyme M-7-mercaptoheptanoylthreonine phosphate heterodisulfide
Heterodisulfide reductase [3]
More (may be a system of 2 enzymes, a hydrogenase and a disulfide reductase)

CAS Reg. No.
116515-35-6

---

2 REACTION AND SPECIFICITY

Catalysed reaction
Coenzyme M 7-mercaptoheptanoylthreonine-phosphate heterodisulfide + H_2 →
→ coenzyme M + N-(7-mercaptoheptanoyl)threonine O^3-phosphate

Reaction type
Redox reaction

Natural substrates
Coenzyme M 7-mercaptoheptanoylthreonine-phosphate heterodisulfide + H_2 (key reaction in metabolism of methanogenic bacteria [2, 3], physiological electron donor/acceptor not identified, may be a system of 2 enzymes, a hydrogenase and a disulfide reductase) [2, 3]

Enzyme Handbook © Springer-Verlag Berlin Heidelberg 1994
Duplication, reproduction and storage in data banks are only allowed with the prior permission of the publishers

Substrate spectrum

1 Coenzyme M 7-mercaptoheptanoylthreonine-phosphate heterodisulfide + H_2 (r [2], highly specific, L-enantiomer, the D-enantiomer is converted at 35% the rate of the L-enantiomer [2]) [1–4]

2 More (no substrates: heterodisulfide of coenzyme M and 6-mercaptohexanoylthreonine-phosphate [2, 3], homodisulfides of coenzyme M or N-(7-mercaptoheptanoyl)threonine O^3-phosphate, glutathione, cysteine [1–3], thioglycolate [2], coenzyme M alone, N-(7-mercaptoheptanoyl)threonine O^3-phosphate alone, 6-mercaptohexanoylthreonine-phosphate alone [3]) [1–3]

Product spectrum

1 Coenzyme M + N-(7-mercaptoheptanoyl)threonine O^3-phosphate [1–4]

2 ?

Inhibitor(s)

O_2 (prolonged incubation, $t_{1/2}$: 2 h) [3]; 6-Mercaptohexanoyl-L-threonine phosphate (weak inhibition) [3]; 3-(1-Sulfopropyl)-7-mercaptoheptanoyl-L-threonine (weak inhibition) [3]; More (no inhibition: bromoethanesulfonate, 7-bromoheptanoyl-L-threonine phosphate) [3]

Cofactor(s)/prostethic group(s)

FAD (4 mol/mol enzyme) [3]; Aquacobalamin (activation) [1]; Reduced benzyl viologen (activation, electron donor [3], 2 mol are oxidized per mol heterodisulfide [2]) [2, 3]; Methylene blue (activation, acceptor for heterodisulfide formation) [2]; O_2 (activation, enhancement upon 30 min incubation in presence of air at 20°C, longer incubation decreases activity with $t_{1/2}$: 2 h) [2]; Oxidized benzyl viologen (activation) [2]; Ti(III)citrate (activation) [4]; Hydroxocobalamin (i.e. vitamin B_{12}, activation) [4]; More (NADH, NADPH or reduced coenzyme F_{420} do not act as electron donors, physiological electron-donor/acceptor not identified) [3]

Metal compounds/salts

Iron (non-heme iron protein, 72 mol FeS/mol enzyme) [3]

Turnover number (min^{-1})

Specific activity (U/mg)

0.2 [1]; 0.60 (Methanosarcina barkeri, formation of heterodisulfide, crude) [2]; 0.62 (Methanosarcina barkeri, reduction of heterodisulfide, crude) [2]; 5.0 (heterodisulfide formation, Methanobacterium thermoautotrophicum) [2]; 10.0 (reduction, Methanobacterium thermoautotrophicum) [2]; 15.0 (oxidation with methylene blue) [3]; 29.0 (reduction with benzyl viologen) [3]

K_m-value (mM)

0.05 (below, benzyl viologen, Methanobacterium thermoautotrophicum [2], N-(7-mercaptoheptanoyl)threonine [3]) [2, 3]; 0.1 (well below [1], coenzyme M 7-mercaptoheptanoylthreonine-phosphate heterodisulfide, reduction) [1, 3]; 0.13 (coenzyme M 7-mercaptoheptanoyl-L-threonine-phosphate heterodisulfide, Methanobacterium thermoautotrophicum) [2]; 0.2 (coenzyme M) [3]; 0.25 (coenzyme M, MR2, with Ti(III)citrate or hydroxocobalamine as reductive activator) [4]; 3.0 (coenzyme M 7-mercaptoheptanoyl-D-threonine-phosphate heterodisulfide, Methanobacterium thermoautotrophicum) [2]; 5.0 (coenzyme M, MR2) [4]

pH-optimum

More (highest activity in Tricine/KOH-buffer) [3]; 6.3 (piperazine-N, N'-bis(2-ethane sulfonic acid)/KOH-buffer) [1]; 7.6 (Tris/HCl-buffer [1], Tricine/KOH-buffer [3]) [1, 3]

pH-range

Temperature optimum (°C)

40 (assay at, Methanosarcina barkeri) [2]; 60 (assay at, Methanobacterium thermoautotrophicum) [2]; 75–80 [3]

Temperature range (°C)

3 ENZYME STRUCTURE

Molecular weight

550000 (Methanobacterium thermoautotrophicum, gel filtration) [3]

Subunits

Oligomer (x × 66000 + x × 51000 + x × 33000, Methanobacterium thermoautotrophicum MR2, SDS-PAGE, x × 69000 + ? + 1 × 49000 + x × 38000, Methanobacterium thermoautotrophicum MR1 SDS-PAGE) [4]
More (probably a trimer of tetramers, alpha$_4$-beta$_4$-gamma$_4$) [3]

Glycoprotein/Lipoprotein

–

4 ISOLATION/PREPARATION

Source organism

Methanobacterium thermoautotrophicum (strain Marburg [1–3], strain deltaH [4]) [1–4]; Methanosarcina barkeri [2]

Enzyme Handbook © Springer-Verlag Berlin Heidelberg 1994
Duplication, reproduction and storage in data banks are only allowed with the prior permission of the publishers

Source tissue
Cell [1–4]

Localisation in source
Cytoplasm (soluble or losely membrane-associated [3]) [1, 3]

Purification
Methanobacterium thermoautotrophicum (anaerobic chamber [1, 3], 2 genetically distinct isozymes MR1 and 2, distinct in size exclusion chromatographic behaviour, pronounced differences in low-temperature MCD spectra, identical resonance Raman spectra, relative ratio depends on cell growth phase, MR1 dominates in cells grown under limiting gas-supply, MR2 dominates in cells of exponential growth phase [4]) [1, 3, 4]

Crystallization
–

Cloned
–

Renaturated
–

---

5 STABILITY

pH

Temperature (°C)
45–65 (Q_{10}: 1.5) [3]; 70 ($t_{1/2}$: 30 min) [3]; 80 (rapid inactivation above) [1]

Oxidation
Aerobically stored preparations, 80% loss of activity within 12 h at 4°C, under N_2 2 weeks stable at 4°C [3]; More (partially purified, insensitive to O_2, Methanobacterium thermoautotrophicum) [2]

Organic solvent

General stability information
DTT stabilizes during purification [3]

Storage
–80°C, under argon, with 20% oxygen-free ethylene glycol stable [4]; 4°C, Tris/HCl-buffer, pH 7.6, 0.5 M NaCl, under N_2, 20% activity is lost within 24 h [1]; 4°C, under N_2, solution with 1 mg protein/ml 2 weeks [3]; 4°C, aerobically stored preparation, 80% loss of activity within 12 h [3]

6 CROSSREFERENCES TO STRUCTURE DATABANKS

PIR/MIPS code

Brookhaven code

7 LITERATURE REFERENCES

[1] Hedderich, R., Thauer, R.K.: FEBS Lett.,234,223–227 (1988)
[2] Hedderich, R., Berkessel, A., Thauer, R.K.: FEBS Lett.,255,67–71 (1989)
[3] Hedderich, R., Berkessel, A., Thauer, R.K.: Eur. J. Biochem.,193,255–261 (1990)
[4] Brenner, M.C., Ma, L., Johnson, M.K., Scott, R.A.: Biochim. Biophys. Acta,1120,160–166 (1992)

Enzyme Handbook © Springer-Verlag Berlin Heidelberg 1994
Duplication, reproduction and storage in data banks are only allowed with the prior permission of the publishers

SPRINGER NATURE

GPSR Compliance

The European Union's (EU) General Product Safety Regulation (GPSR) is a set of rules that requires consumer products to be safe and our obligations to ensure this.

If you have any concerns about our products, you can contact us on ProductSafety@springernature.com

In case Publisher is established outside the EU, the EU authorized representative is:

Springer Nature Customer Service Center GmbH
Europaplatz 3
69115 Heidelberg, Germany

The manufacturer's authorised representative in the EU is Springer Nature Customer Service Centre GmbH, Europaplatz 3, 69115 Heidelberg, Germany. If you have any concerns regarding our products, please contact ProductSafety@springernature.com

Printed and bound by CPI Group (UK) Ltd, Croydon, CR0 4YY

15/07/2026

02167621-0001